FOUNDATION

GCSE Mathematics for Edexcel

SOPHIE GOLDIE

ALAN SMITH

SERIES CONSULTANT: JEAN LINSKY

Hodder Murray

www.hoddereducation.co.uk

Acknowledgements
The Publishers would like to thank the following for permission to reproduce copyright material:

p.181 © akg-images/ullstein bild

Every effort has been made to trace all copyright holders, but if any have been inadvertently overlooked the Publishers will be pleased to make the necessary arrangements at the first opportunity.

Orders: please contact Bookpoint Ltd, 130 Milton Park, Abingdon, Oxon OX14 4SB. Telephone: (44) 01235 827720. Fax: (44) 01235 400454. Lines are open 9.00–6.00, Monday to Saturday, with a 24-hour message answering service. Visit our website at www.hoddereducation.co.uk

First published in 2006 by
Hodder Murray, an imprint of Hodder Education,
a member of the Hodder Headline Group
338 Euston Road
London NW1 3BH

Impression number 10 9 8 7 6 5 4 3
Year 2010 2009 2008 2007 2006

Cover illustration © David Angel @ Début Art
Typeset in Times Ten 11 on 14 point by Tech-Set Ltd, Gateshead, Tyne and Wear
Printed in Great Britain by CPI Bath

A catalogue record for this title is available from the British Library

ISBN–10: 0340 913 592
ISBN–13: 978 0340 913 598

CONTENTS

Introduction

This book provides complete coverage of the new two-tier Edexcel GCSE Mathematics specification (Foundation) for first teaching from September 2006.

The book is aimed at the linear specification, although it can also be used effectively for students following the modular course.

The written examinations account for 80% of the final grade, with the remaining 20% based on coursework. Although the core skills required for good coursework are covered here, extra advice on writing coursework tasks and projects may be found in the accompanying Assessment Pack.

Students studying the Foundation course may have differing subject knowledge at the start, and this is reflected in the steady progression of the content. The material is laid out in the following order:

Number Chapters 1 to 6
Algebra Chapters 7 to 10
Shape, space and measure Chapters 11 to 18
Data handling Chapters 19 to 22

Some of the early work covered in each of these four sections may be omitted, or used as revision material, as appropriate.

Examination candidates will require a good quality scientific calculator for one of the papers; a calculator is not permitted in the other.

All the exercises and examples in this book use an icon to show you whether calculators are permitted or not, and you should follow this advice carefully in order to build up the necessary mixture of calculator and non-calculator skills before the examination. There are many differences between calculators, so make sure that you are fully familiar with your own particular model – do not buy a new calculator (or borrow one) just before the examination is due!

Each chapter begins with a Starter. This is an exercise, activity or puzzle designed to stimulate thinking and discussion about some of the ideas that underpin the content of the chapter. The main chapter contains explanations of each topic, with numerous worked examples, each followed by a corresponding Exercise of questions. At the end of each

chapter there is a Review Exercise, containing questions on the content for the whole chapter. Many of the Review questions are from past Edexcel GCSE examination papers; this is indicated in the margin.

Additional practice exam-style question papers, with full mark schemes, are available in Hodder Murray's accompanying Assessment Pack.

There then follows a set of Key Points you should know and understand.

Most chapters end with an Internet Challenge, intended to be done (either at school or at home) together with an internet search engine such as *Google*. The Internet Challenge material frequently goes beyond the strict boundaries of the GCSE specification, providing enrichment and leading to a deeper understanding of mainstream topics. The Challenges may look at the history of mathematics and mathematicians, or the role of mathematics in the real world. When doing these, it is hoped that you will not just answer the written questions, but also take the time to explore the subject a little deeper: the internet contains a vast reservoir of very well-written information about mathematics. The reliability of internet information can be variable, however, so it is best to check your answers by referring to more than one site if possible.

The Foundation Tier of the GCSE allows all candidates to access a grade C, and the book is written with this goal in mind. In the earlier part of the book some questions are set at a lower level than this, and to help monitor progression the questions are colour coded to indicate their approximate level of difficulty.

The colour coding is as follows:

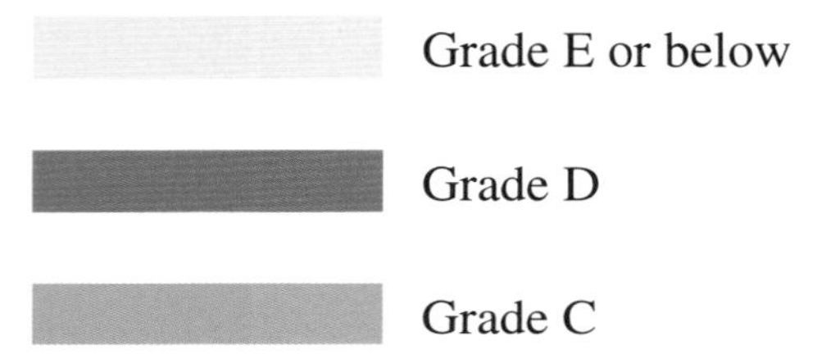

All of the content has been checked very carefully against the new GCSE specification, to ensure that all examination topics are suitably covered. If you have mastered all the relevant topics covered in this book then you will be able to approach the examinations confident in the knowledge that you are fully prepared.

Good luck on the day!

Sophie Goldie
Alan Smith

February 2006

CHAPTER 1

Number review

In this chapter you will **be reminded how to:**

- use place value
- add, subtract, multiply and divide integers (whole numbers)
- multiply and divide by 10, 100 and 1000
- round to the nearest 10, 100 and 1000
- add, subtract, multiply and divide negative numbers
- round to one or more significant figures (s.f.)
- use rounding to find an approximate answer to a problem
- use different mental strategies to solve problems.

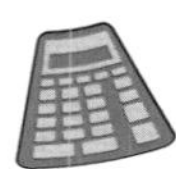

Starter: Little and large

Look at these number cards.

They show the number 45 972

1 Arrange all five number cards so that they make the largest possible number.

2 Arrange all five number cards so that they make the smallest possible number.

3 Use the number cards to make these sums have:
(i) the largest possible answer; **(ii)** the smallest possible answer.

a)

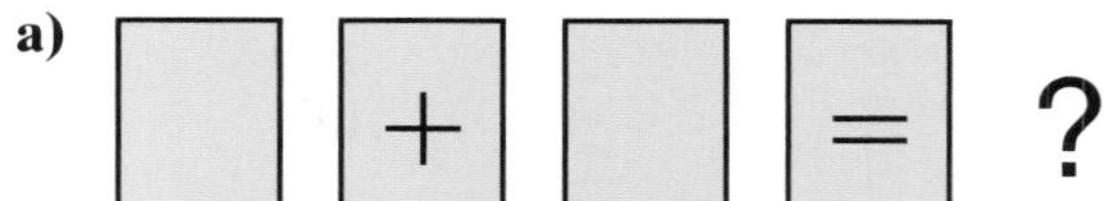

b)

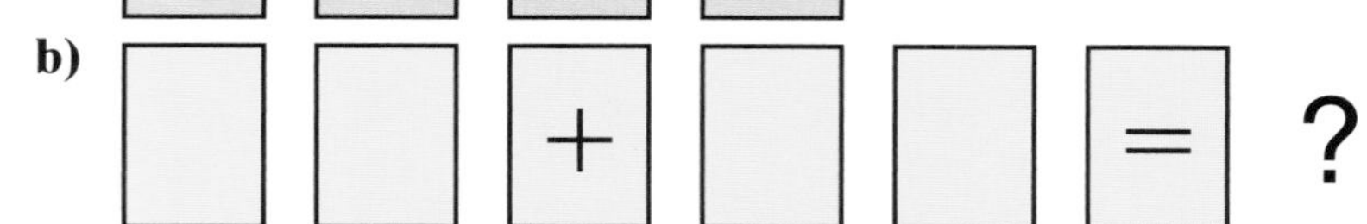

c)

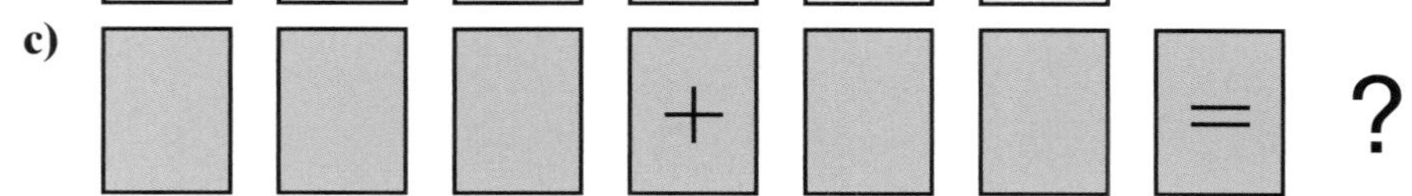

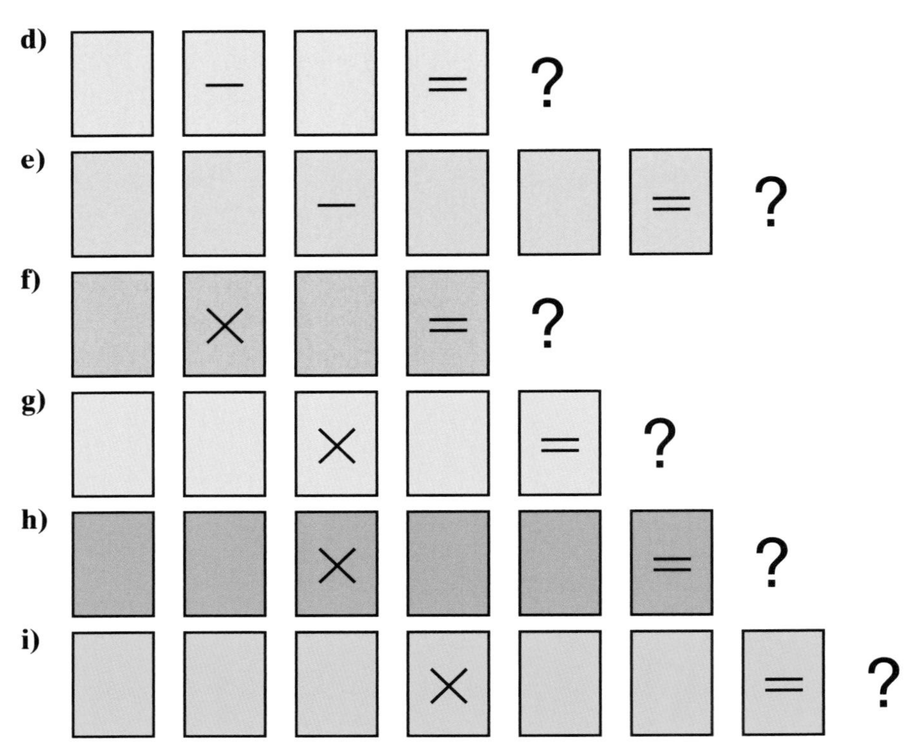

4 What strategies did you use?

1.1 Place value

Look at these numbers:

4	40	400	4000	40 000	400 000

The digit '4' has a different **place value** in each number.

Hundreds of thousands	Tens of thousands	Thousands	Hundreds	Tens	Units	
					4	Four
				4	0	Forty
			4	0	0	Four hundred
		4	0	0	0	Four thousand
	4	0	0	0	0	Forty thousand
4	0	0	0	0	0	Four hundred thousand

EXAMPLE

Write 306 951 in words.

SOLUTION

Hundreds of thousands	Tens of thousands	Thousands	Hundreds	Tens	Units
3	0	6	9	5	1

It helps to write the number on a **place value** diagram.

Three hundred and six thousand, nine hundred and fifty-one

EXAMPLE

Write six point five million in figures.

SOLUTION

Millions	Hundreds of thousands	Tens of thousands	Thousands	Hundreds	Tens	Units
6	5	0	0	0	0	0

0.5 million is 500 000 (five hundred thousand)

6 500 000

EXAMPLE

Write down the value of the digit '4' in the number 3 564 720

SOLUTION

Millions	Hundreds of thousands	Tens of thousands	Thousands	Hundreds	Tens	Units
3	5	6	4	7	2	0

Four thousand or 4000

EXAMPLE

Write these numbers in descending order:

Write the largest number first.

31 760, 3794, 42 780, 5921, 30 999

SOLUTION

Hundreds of thousands	Tens of thousands	Thousands	Hundreds	Tens	Units
	3	1	7	6	0
	4	2	7	8	0
	3	0	9	9	9
		3	7	9	4
		5	9	2	1

The numbers with the most digits before the decimal point are the largest.
So write these on a place value diagram first.

So, in descending order: 42 780, 31 760, 30 999, 5921, 3794

EXERCISE 1.1

1 Write these numbers in words:
a) 472 **b)** 3740 **c)** 2 345 000 **d)** 40 507

2 Write these numbers in figures:
a) six hundred and eight
b) nine thousand and twenty-six
c) four and a half million
d) one thousand, two hundred and ten
e) six and a half thousand
f) two hundred thousand nine hundred and forty.

3 Write these numbers in order of size – start with the smallest number:
a) 106, 60, 6, 160, 600, 100
b) 432, 234, 342, 324, 423, 243
c) 4300, 4000, 3400, 4040, 4030, 3000
d) 58 200, 52 800, 50 028, 58 002, 50 820
e) 591 000, 519 000, 590 100, 510 900, 500 910
f) 73 492, 74 932, 72 943, 39 742, 37 924

1.2 Addition

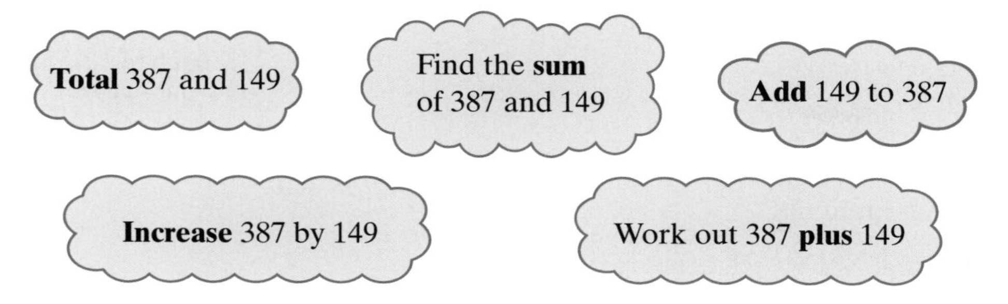

Total 387 and 149
Find the **sum** of 387 and 149
Add 149 to 387
Increase 387 by 149
Work out 387 **plus** 149

These are all ways of asking, 'What is 387 + 149?'

There are a number of different ways you can work this out.
Here are two methods you can use.

Method 1

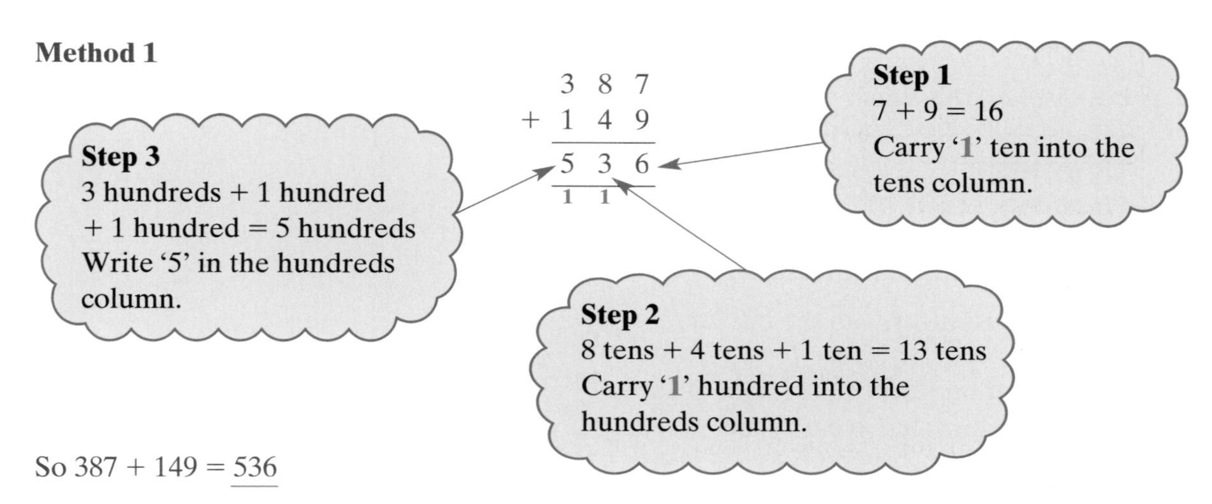

3 8 7
+ 1 4 9
5 3 6
1 1
Step 1
7 + 9 = 16
Carry '1' ten into the tens column.
Step 2
8 tens + 4 tens + 1 ten = 13 tens
Carry '1' hundred into the hundreds column.
Step 3
3 hundreds + 1 hundred + 1 hundred = 5 hundreds
Write '5' in the hundreds column.

So 387 + 149 = 536

Method 2
In this method you break up 149 into 100 + 40 + 9

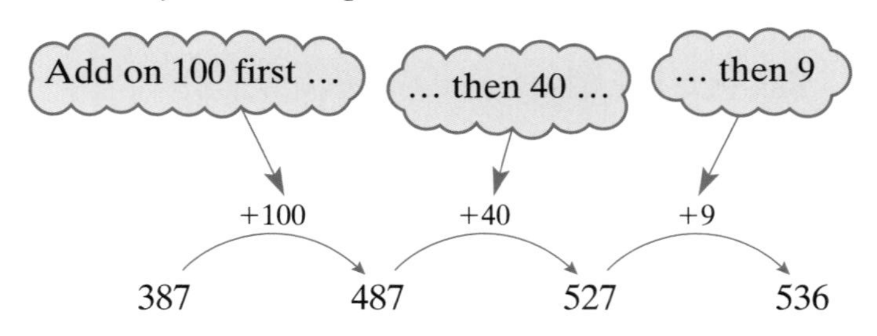

So 387 + 149 = 536

Remember: It doesn't matter in which order you add numbers.

So $100 + 40 + 9 = 140 + 9 = 149$

and $40 + 9 + 100 = 49 + 100 = 149$

and $9 + 100 + 40 = 109 + 40 = 149$

EXERCISE 1.2

1 Work out:

a) 24 + 35 **b)** 123 + 42 **c)** 59 + 63

d) 259 + 74 **e)** 345 + 129 + 46 **f)** 789 + 231 + 67

2 Mary treated her friend to lunch.
Here is the restaurant bill.

Find the total that Mary's bill comes to.

Simply Pasta	
Spaghetti bolognese	£7
Lasagne	£8
Ice cream	£2
Chocolate gateau	£3
Lemonade	£1
Mineral water	£1

3 Find the sum of the numbers 1 to 10.

4 Add 235 to 421.

5 Increase 637 by 184.

6 A bus displays this sign:

Downstairs:	28 seats
Upstairs:	36 seats
Standing room:	17 people

How many passengers can the bus take?

7 These pairs of numbers have a sum of 100
Find the missing value in each pair.

a) 27 and ? **b)** 42 and ? **c)** 78 and ?

d) 54 and ? **e)** 19 and ? **f)** 63 and ?

1.3 Subtraction

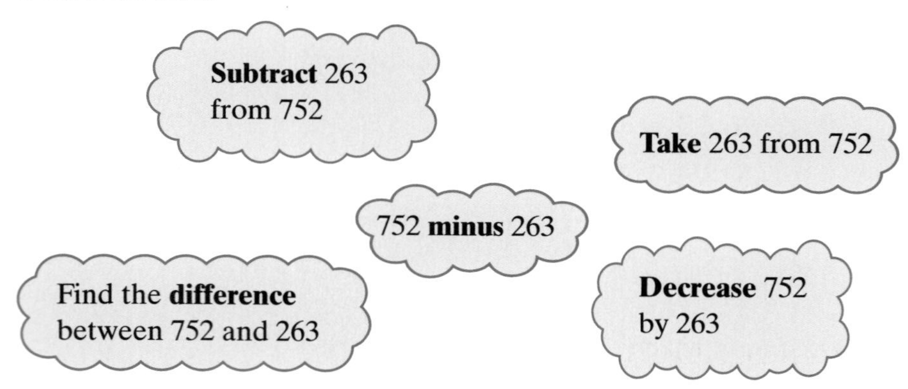

These are all ways of asking, 'What is 752 − 263?'

There are a number of different ways you can work this out.

Here are two methods you can use.

Method 1

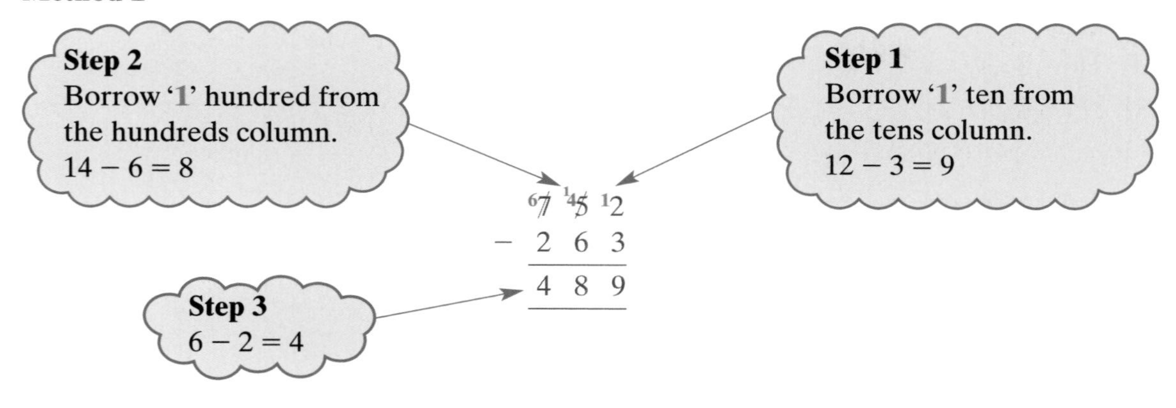

Step 2
Borrow '1' hundred from the hundreds column.
14 − 6 = 8
Step 1
Borrow '1' ten from the tens column.
12 − 3 = 9
Step 3
6 − 2 = 4

So 752 − 263 = 489

Method 2
In this method you work out what you would need to add on to 263 to get 752.

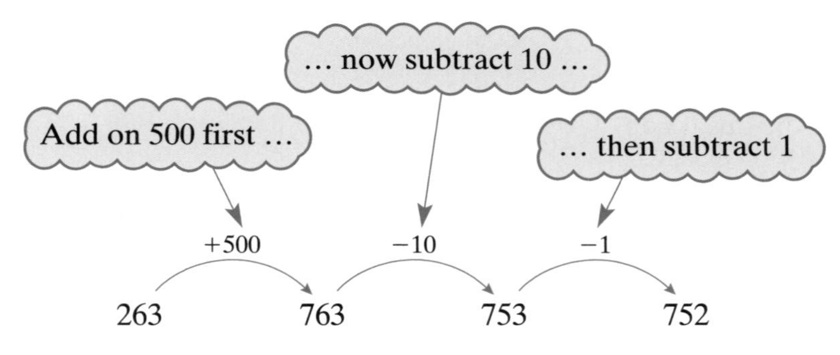

500 – 10 – 1 = 489

So 752 − 263 = 489

EXERCISE 1.3

1 Work out:

a) 53 − 21 **b)** 328 − 117 **c)** 72 − 46
d) 435 − 76 **e)** 745 − 689 **f)** 1356 − 969

2 Decrease 456 by 212

3 What number needs to be added to 67 to make 100?

4 Subtract 158 from 726

5 In a Pop Idol contest, the contestants got the scores shown.
   a) What is the difference between Kofi's and Jon's scores?
   b) How many more votes did Nadine get than Jon?
   c) What is the difference between Nadine's and Sam's scores?
   d) How many votes did Kofi win by?
   e) How many votes were cast altogether?

Nadine	6346 votes
Kofi	7547 votes
Jon	5982 votes
Sam	971 votes

1.4 Multiplication

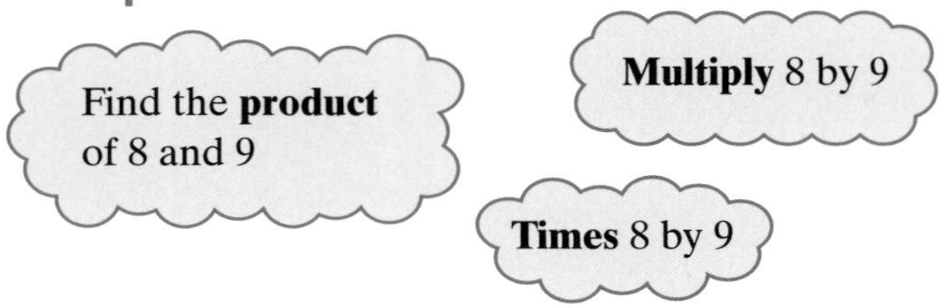
Find the **product** of 8 and 9
Multiply 8 by 9
Times 8 by 9

These are all ways of asking, 'What is 8×9?'

Remember, it doesn't matter in which order you multiply numbers.

So $8 \times 9 = 72$

and $9 \times 8 = 72$

Similarly $3 \times 4 \times 5 = 12 \times 5 = 60$

and $3 \times 5 \times 4 = 15 \times 4 = 60$

and $5 \times 4 \times 3 = 20 \times 3 = 60$

You will need to know your times tables.
They will help you work out problems involving multiplication and division.

EXERCISE 1.4

1 Copy and complete this times table grid:

×	1	2	3	4	5	6	7	8	9	10
1	1	2	3	4	5	6				
2	2	4	6	8	10					
3	3	6	9	12						
4	4	8	12							
5	5	10								
6	6	12								
7	7	14								
8	8								72	
9	9							72		
10	10									

Use the following questions to test yourself on your times tables.
Time yourself answering these questions.
See if you can improve on your time next week.

2 Work out:

a) 5×3 **b)** 7×4 **c)** 8×7 **d)** 6×9
e) 7×7 **f)** 8×8 **g)** 9×9 **h)** 6×8
i) $2 \times 3 \times 7$ **j)** $4 \times 2 \times 9$ **k)** $2 \times 3 \times 10$ **l)** $3 \times 3 \times 8$

3 Find the missing values:

a) $5 \times ? = 40$ **b)** $7 \times ? = 42$ **c)** $? \times 10 = 90$
d) $? \times 9 = 63$ **e)** $3 \times ? = 21$ **f)** $? \times 5 = 45$
g) $7 \times ? = 35$ **h)** $? \times 3 \times 5 = 30$ **i)** $2 \times 4 \times ? = 48$

1.5 Multiplying by 10, 100 and 1000

When you **multiply by 10** you move the decimal point **one place** to the right.

When you **multiply by 100** you move the decimal point **two places** to the right.

When you **multiply by 1000** you move the decimal point **three places** to the right.

EXAMPLE

Work out:

a) 32×30 **b)** 53×400 **c)** 1.6×1000

SOLUTION

a) Multiplying by 30 is the same as multiplying by 3 and then by 10

$32 \times 3 = 96$

96. is the same as 96

Write a decimal point after the 96

$96. \times 10 = 960. = 960$

So $32 \times 30 = \underline{960}$

b) Multiplying by 400 is the same as multiplying by 4 and then by 100

$53 \times 4 = 212$

$50 \times 4 = 200$
$3 \times 4 = 12$
$200 + 12 = 212$

$212. \times 100 = 21\,200. = 21\,200$

So $53 \times 400 = \underline{21\,200}$

c) $1.6 \times 1000 = 1600. = 1600$

So $1.6 \times 1000 = \underline{1600}$

EXERCISE 1.5

1 Multiply the following numbers by 10:

a) 9 **b)** 23 **c)** 51 **d)** 324
e) 0.7 **f)** 1.89 **g)** 6.3 **h)** 0.61

2 Multiply the following numbers by 100:

a) 7 **b)** 36 **c)** 47 **d)** 512
e) 4.31 **f)** 2.6 **g)** 0.1 **h)** 0.01

3 Multiply the following numbers by 1000:

a) 4 **b)** 72 **c)** 123 **d)** 500
e) 15.1 **f)** 0.02 **g)** 4.321 **h)** 0.005

4 Work out:

a) 5×20 **b)** 7×40 **c)** 7×30 **d)** 8×200
e) 3×400 **f)** 80×40 **g)** 200×30 **h)** 9×800
i) 20×30 **j)** 50×60 **k)** 400×30 **l)** 50×20

1.6 Dividing by 10, 100 and 1000

When you **divide by 10** you move the decimal point **one place** to the left.

When you **divide by 100** you move the decimal point **two places** to the left.

When you **divide by 1000** you move the decimal point **three places** to the left.

EXAMPLE

Work out:

a) $720 \div 80$ **b)** $12\,000 \div 400$ **c)** $43.2 \div 1000$

SOLUTION

a) Dividing by 80 is the same as dividing by 10 and then by 8

$720. \div 10 = 72.\overset{\frown}{0} = 72$

$72 \div 8 = 9$

So $720 \div 80 = \underline{9}$

You don't need to write the decimal point.

b) Dividing by 400 is the same as dividing by 100 and then by 4

$12\,000. \div 100 = 120.\overset{\frown}{0}\overset{\frown}{0} = 120$

$120 \div 4 = 30$

So $12\,000 \div 400 = \underline{30}$

c) $43.2 \div 1000 = .\overset{\frown}{0}\overset{\frown}{4}\overset{\frown}{3}2 = 0.0432$

So $43.2 \div 1000 = \underline{0.0432}$

You need to write a '0' before the decimal point.

EXERCISE 1.6

1 Divide the following numbers by 10:

a) 30	**b)** 240	**c)** 3200	**d)** 500 000
e) 24	**f)** 59	**g)** 2.7	**h)** 0.36

2 Divide the following numbers by 100:

a) 4000	**b)** 500	**c)** 66 000	**d)** 1 000 000
e) 44	**f)** 21	**g)** 3.29	**h)** 0.345

3 Divide the following numbers by 1000:

a) 5000	**b)** 12 000	**c)** 340 000	**d)** 2 000 000
e) 600	**f)** 50	**g)** 34	**h)** 6.7

3 Work out:

a) $1000 \div 20$	**b)** $320 \div 80$	**c)** $4200 \div 600$
d) $12\,000 \div 400$	**e)** $3500 \div 50$	**f)** $48\,000 \div 200$
g) $600 \div 300$	**h)** $54\,000 \div 600$	**i)** $720 \div 80$

1.7 Long multiplication

EXAMPLE

Work out 134×26

SOLUTION

Method 1

	1 3 4
	× 2 6
Multiply by 20	2 6 8 0
Multiply by 6	8 0 4
Add	3 4 8 4 (carries: 1, 2, 2)

So $134 \times 26 = \underline{3484}$

$20 = 2 \times 10$
So don't forget the zero.

$6 \times 4 = 24$
Carry '**2**' tens into the tens column.

$6 \times 3 = 18$
18 plus the '**2**' tens carried over = 20
Carry '**2**' hundreds into the hundreds column.

Method 2

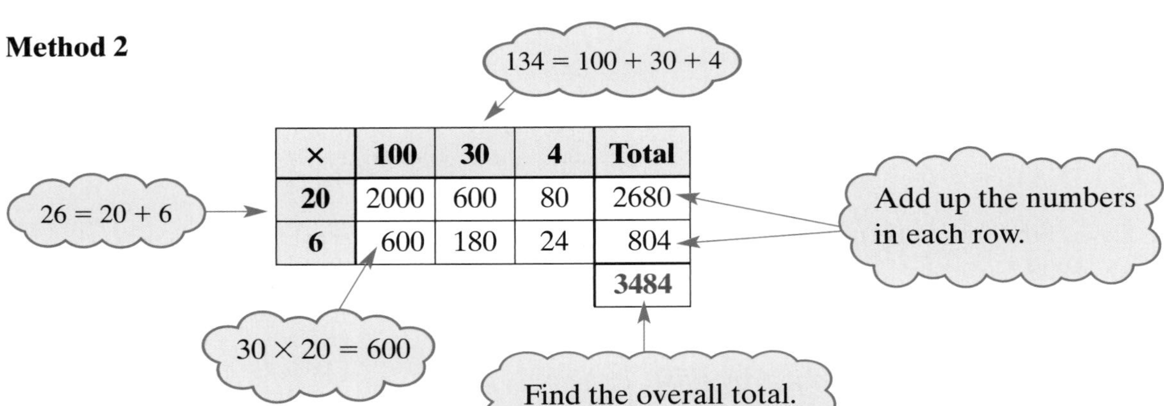

×	100	30	4	Total
20	2000	600	80	2680
6	600	180	24	804
				3484

134 = 100 + 30 + 4
26 = 20 + 6
Add up the numbers in each row.
30 × 20 = 600
Find the overall total.

So 134 × 26 = 3484

Method 3

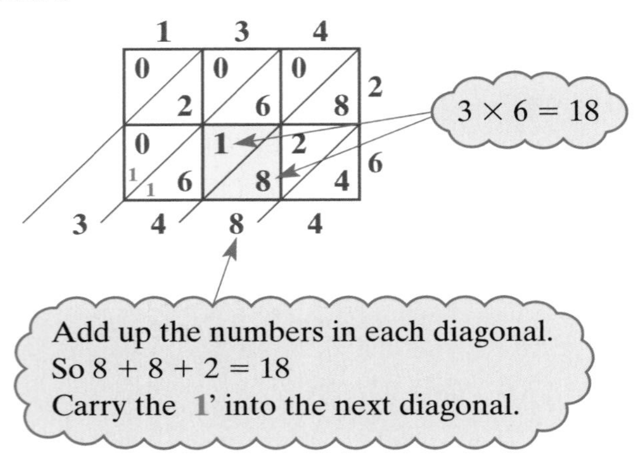
1 3 4
2
6
3 × 6 = 18
3 4 8 4
Add up the numbers in each diagonal.
So 8 + 8 + 2 = 18
Carry the '1' into the next diagonal.

So 134 × 26 = 3484

EXERCISE 1.7

1 Work out:

a) 45 × 23 **b)** 17 × 34 **c)** 45 × 16
d) 146 × 43 **e)** 152 × 13 **f)** 237 × 24
g) 123 × 12 **h)** 276 × 53 **i)** 192 × 36

2 Appleton High School buys 85 new computers.
Each computer costs £765.
How much does Appleton High school spend on the computers?

3 Rovers United football team hire 12 coaches to take fans to an away match.
Each coach can take 56 people.
a) How many fans can go to the match?
b) The football team sell all the tickets for £18 each.
How much money should they get from ticket sales?

1.8 Division

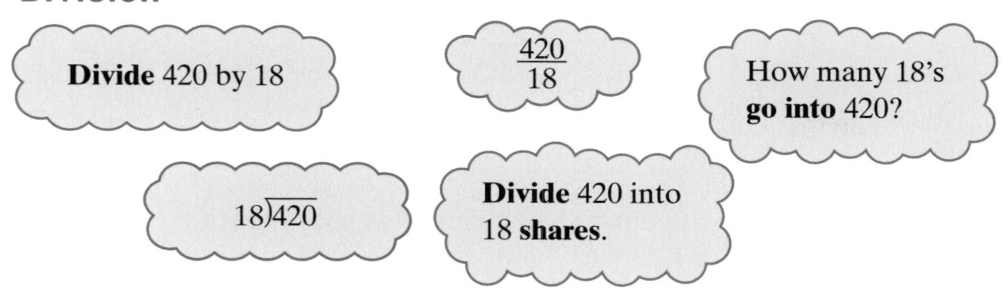

Divide 420 by 18
$\frac{420}{18}$
How many 18's **go into** 420?
$18\overline{)420}$
Divide 420 into 18 **shares**.

These are all ways of asking 'What is $420 \div 18$?'

There are a number of different ways you can work this out.
Here are three methods you can use.

Method 1: Long division

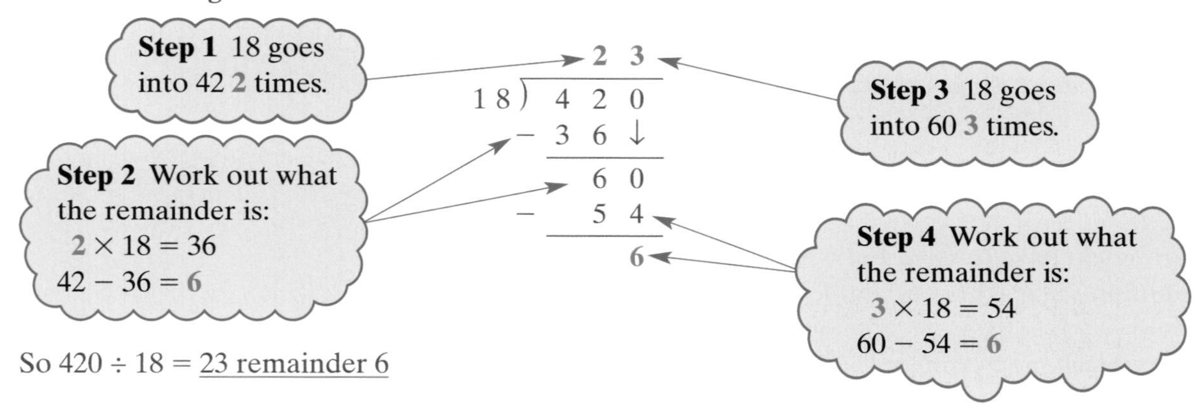

Step 1 18 goes into 42 2 times.
Step 2 Work out what the remainder is: 2 × 18 = 36 42 − 36 = 6
2 3
1 8) 4 2 0
− 3 6 ↓
6 0
− 5 4
6
Step 3 18 goes into 60 3 times.
Step 4 Work out what the remainder is: 3 × 18 = 54 60 − 54 = 6

So $420 \div 18 =$ 23 remainder 6

Method 2: Short division

This is similar to long division but you don't show all your working.

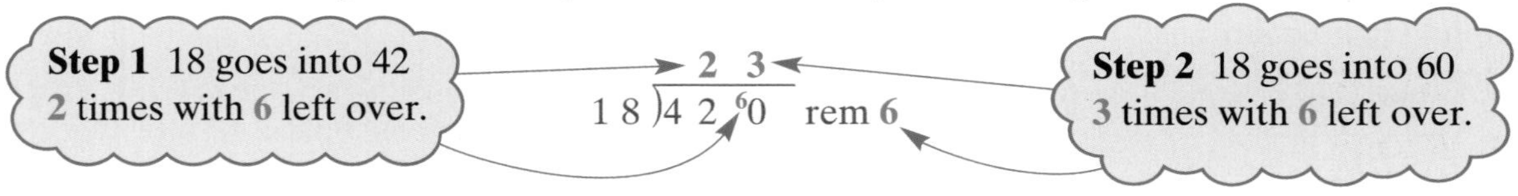

Step 1 18 goes into 42 2 times with 6 left over.
2 3
1 8)4 2 6 0 rem 6
Step 2 18 goes into 60 3 times with 6 left over.

So $420 \div 18 =$ 23 remainder 6

Method 3: Multiplying up

You know that $10 \times 18 = 180$

So $\mathbf{20 \times 18 = 360}$

You still need 60 because $420 - 360 = 60$

You know that $\mathbf{3 \times 18 = 54}$

So $\mathbf{23 \times 18 = 414}$

20 lots of 18 plus 3 lots of 18 $= 360 + 54 = 414$

So $420 \div 18 =$ 23 remainder 6

EXERCISE 1.8

1 Work out:

a) 288 ÷ 4 **b)** 595 ÷ 7 **c)** 136 ÷ 8
d) 288 ÷ 12 **e)** 224 ÷ 14 **f)** 555 ÷ 15
g) 364 ÷ 26 **h)** 344 ÷ 43 **i)** 952 ÷ 56

2 Work out the following and write down the remainder in each question.

a) 75 ÷ 8 **b)** 53 ÷ 6 **c)** 310 ÷ 14
d) 395 ÷ 23 **e)** 579 ÷ 12 **f)** 245 ÷ 11

3 370 students sign up to go on a school trip.
A minibus can take 16 students.
How many minibuses does the school need to hire?

4 Fern has £660 to spend on new chairs for her youth club.
A chair costs £35.
How many chairs can she buy?

1.9 Solving problems

Often you have to work out whether a problem needs you to use addition, subtraction, multiplication or division in order to solve it.

Here are some examples to help you learn how to to decide.

EXAMPLE

The table shows how many students there are in each year group at Alpha High School.

Year 7	Year 8	Year 9	Year 10	Year 11
?	187	217	224	198

a) What is the difference between the number of students in Year 10 and the number of students in Year 11?

b) There are seven tutor groups in Year 7.
Each tutor group has 29 students.
How many students are there in Year 7?

c) What is the total number of students at Alpha High School?

d) There are 28 students in each tutor group in Year 10.
How many tutor groups are there in Year 10?

SOLUTION

a) **Difference** means subtract. So we need to work out 224 − 198

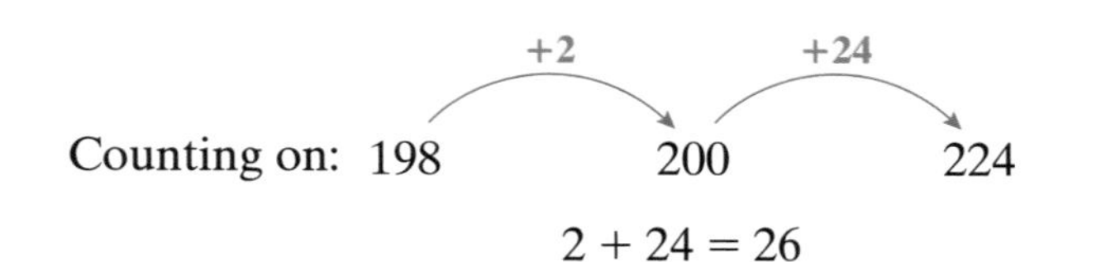

You can use any valid method to work out the answers.

2 + 24 = 26

So there are 26 more students in Year 10 than in Year 11

b) We need to work out 29 × 7

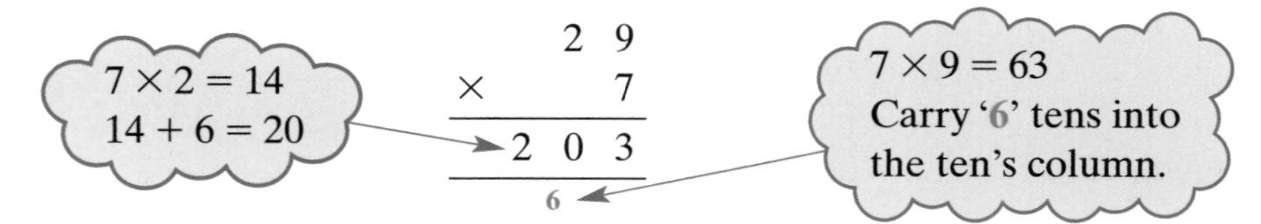

7 × 2 = 14
14 + 6 = 20
2 9
× 7
2 0 3
6
7 × 9 = 63
Carry '6' tens into the ten's column.

There are 203 students in Year 7

c) **Total** means add.
So we need to work out 203 + 187 + 217 + 224 + 198

Add the hundreds first 200 + 100 + 200 + 200 + 100 = 800

Now the tens 80 + 10 + 20 + 90 = 200

Now the units 3 + 7 + 7 + 4 + 8 = 29

800 + 200 + 29 = 1029

So there are 1029 students

d) We need to work out 224 ÷ 28

We know that 10 × 28 = 280

So **5 × 28 = 140**

We still need 84 since 224 − 140 = 84

And **3 × 28 = 84**

So there are 8 tutor groups in Year 10

5 lots of 28 plus
3 lots of 28
= 140 + 84 = 224

EXAMPLE

Some students at Alpha High arrange a school trip to a theme park for the 224 students in Year 10.

a) How many minibuses will the school need to hire?

b) How much will it cost to hire the minibuses?

Minibus Hire	
Capacity:	17 people
Hire charge:	£145 per day

SOLUTION

a) We need to work out 224 ÷ 17

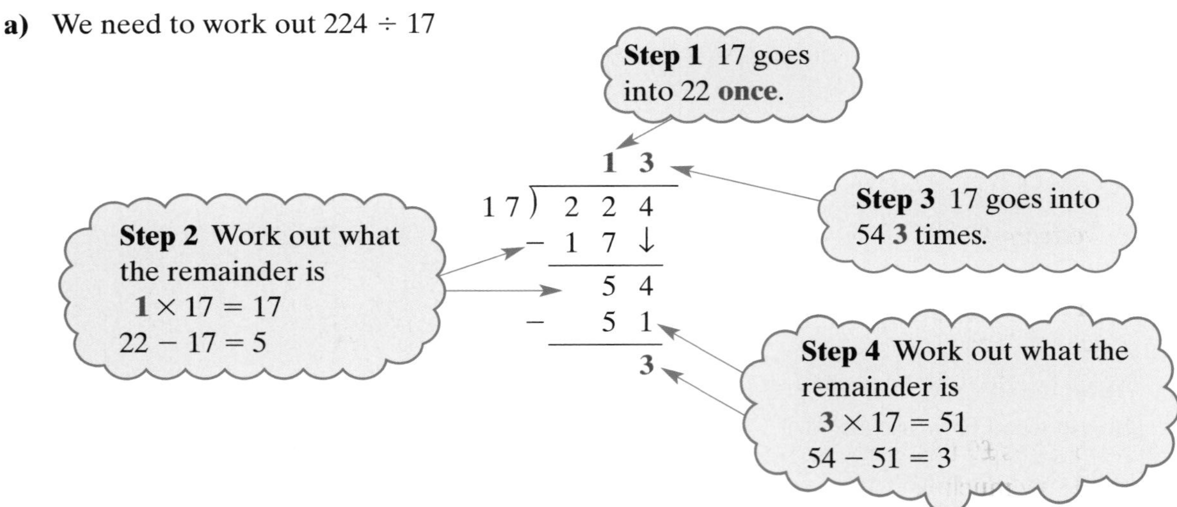

Step 1 17 goes into 22 **once**.
1 3
17) 2 2 4
− 1 7 ↓
5 4
− 5 1
3
Step 2 Work out what the remainder is
1 × 17 = 17
22 − 17 = 5
Step 3 17 goes into 54 **3** times.
Step 4 Work out what the remainder is
3 × 17 = 51
54 − 51 = 3

So 224 ÷ 17 = 13 remainder 3

Now, 13 coaches would mean that three students could not go.
So we need to round our answer up.
<u>14 coaches are needed</u>

b) We need to work out £145 × 14

×	100	40	5	Total
10	1000	400	50	1450
4	400	160	20	580
				2030

<u>The cost will be £2030</u>

EXERCISE 1.9

1

Here is £147.
I want you to share
it equally.

How much money should each person receive?

2 Catherine has been shopping.
She had £200 to spend.
How much change does she
have from £200?

New Fashion	
Jeans	£65
T-shirt	£7
shirt	£22
Skirt	£39
Jumper	£44
Total	

3 Ben buys 65 cans of cola for a party.
Each can costs 27p
How much does he spend?

4 Isobel has £9 to buy chocolate bars for her youth group.
a) How much has she got in pence?
b) How many chocolate bars can she buy?

ONLY **26p**
for ***ANY***
chocolate bar!!

5 Some new tables are ordered for a school canteen.
Each table can seat 12 students.
280 students need to use the canteen at lunch time.
How many tables does the school need?

6 Mrs Smith buys 35 maths textbooks.
Each book costs £12
How much does she spend?

1.10 Adding and subtracting negative numbers

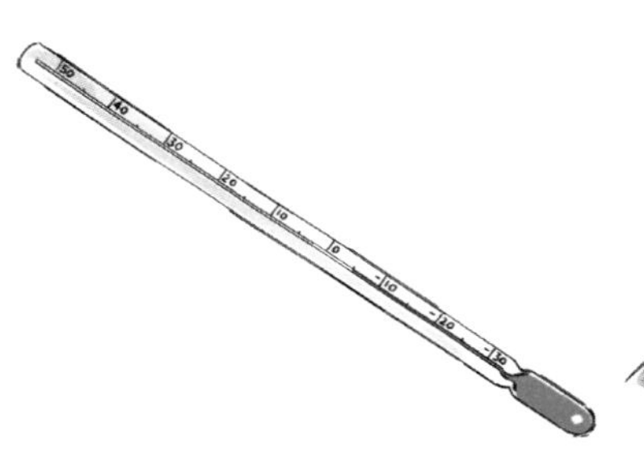

I owe the mobile phone company £27, and my mum £22. Sam owes me £35. How much have I got?

These are examples of using negative numbers.

He owes £27 and £22

Jamie has −£27 + −£22 + £35

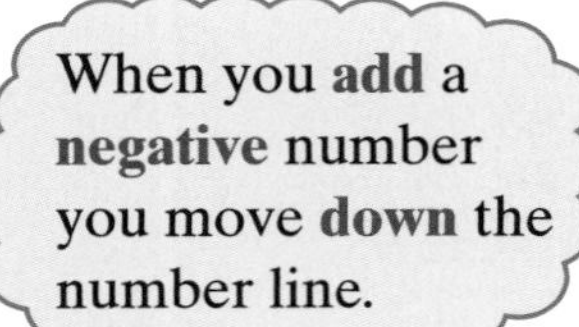

When you **add** a **negative** number you move **down** the number line.

When you **add** a **positive** number you move **up** the number line.

6, 5, 4, 3, 2, 1, 0, −1, −2, −3, −4, −5, −6, −7

When you **subtract** a **negative** number you move **up** the number line.

When you **subtract** a **positive** number you move **down** the number line.

EXAMPLE

Work out:

a) $-3 + 4$ **b)** $-7 + 2$ **c)** $3 + -4$ **d)** $-2 + -5$

> When a number doesn't have a sign written in front it is **positive**.

SOLUTION

a) Start at -3 and move up 4

$\underline{-3 + 4 = 1}$

b) Start at -7 and move up 2

$\underline{-7 + 2 = -5}$

> A '+' and a '−' in the middle make a '−'. So **different signs** mean you **subtract**.

c) Start at 3 and move down 4

$\underline{3 + -4 = -1}$

> … so this is the same as $3 - 4$ …

d) Start at -2 and move down 5

$\underline{-2 + -5 = -7}$

> … and this is the same as $-2 - 5$

EXAMPLE

Work out:

a) $2 - 7$ **b)** $-4 - 3$ **c)** $3 - -4$ **d)** $-2 - -6$

SOLUTION

a) Start at 2 and move down 7

$2 - 7 = \underline{-5}$

> A '−' and a '−', in the middle make a '+'. So **same signs** mean you **add**.

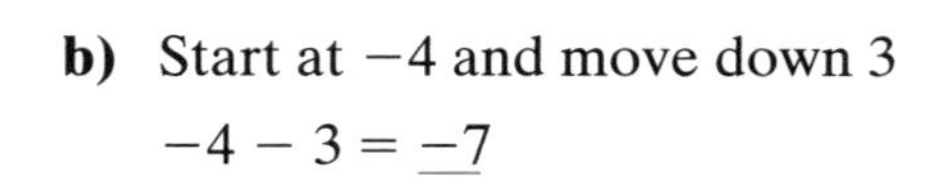

b) Start at -4 and move down 3

$-4 - 3 = \underline{-7}$

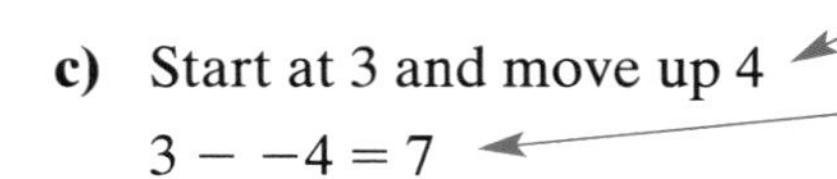

c) Start at 3 and move up 4

$3 - -4 = \underline{7}$

> … so this is the same as $3 + 4$ …

d) Start at -2 and move up 6

$-2 - -6 = \underline{4}$

> … and this is the same as $-2 + 6$

EXERCISE 1.10

Do not use your calculator for Questions 1–5.

1 Work out:

a) $2 + 5$ **b)** $-3 + 6$ **c)** $-2 + 7$
d) $-5 + 3$ **e)** $6 + -4$ **f)** $-3 + -4$

2 Work out:

a) $7 - 3$ **b)** $-3 - 2$ **c)** $-4 - 5$
d) $5 - -2$ **e)** $5 - -4$ **f)** $-2 - -4$

3 Work out:

a) $4 + -5$ **b)** $-5 - -3$ **c)** $-7 + -2$
d) $-4 - 6$ **e)** $-7 - -7$ **f)** $-4 + -6$

4 Write these numbers in order of size.
Write the smallest number first.

a) $4, 0, -3, 2, -4, -2$ **b)** $3, -1, 2, -2, 4, -5$
c) $2, -1, -7, -4, -6, 3$ **d)** $-2, -1, -8, -3, -4, -6$

5 Sally wrote down the temperature at different times on 1st January.

Time	Temperature (°C)
Midnight	−6
4 am	−10
8 am	−4
Noon	7
3 pm	6
7 pm	−2

a) Write down:
(i) the **highest** temperature
(ii) the **lowest** temperature.

b) Work out the **difference** in the temperature between:
(i) 4 am and 8 am **(ii)** 3 pm and 7 pm

At 11 pm that day the temperature had fallen by 5 °C from its value at 7 pm.

c) Work out the temperature at 11 pm.

Edexcel

You can use your calculator for Questions 6 and 7.

6 How much money does Jamie (see page 18) have?

7 An architect has drawn up plans for a new hotel.

a) The lift goes from the ground floor to the 3rd floor. How far has it moved?
b) The lift goes from the ground floor to car park 1. How far has it moved?
c) The lift goes from car park 1 to the 2nd floor. How far has it moved?
d) The lift goes from the attic to car park 2. How far has it moved?

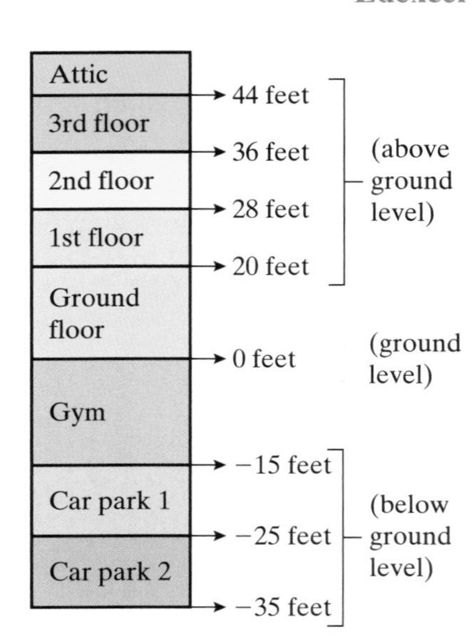

1.11 Multiplying and dividing directed numbers

When you work out 3×-4 you are really finding:

$$-4 + -4 + -4 = -12$$

Notice that the answer is negative:

positive number × negative number = negative number

Also $-4 \times 3 = -12$ as it doesn't matter which order you multiply:

negative number × positive number = negative number

You already know that $3 \times 4 = 12$:

positive number × positive number = positive number

Now by symmetry $-3 \times -4 = 12$:

negative number × negative number = positive number

The same rules work when you are dividing.

You will find this table helpful.

Signs of numbers	Sign of answer
+ and +	+
− and −	+
+ and −	−
− and +	−

same signs … **+ve answer** *different* signs … **−ve answer**

EXAMPLE

Work out:

a) -5×4 **b)** $36 \div -9$

c) $-72 \div -12$ **d)** -15×-10

SOLUTION

a) $-5 \times 4 = \underline{-20}$ **b)** $36 \div -9 = \underline{-4}$

c) $-72 \div -12 = \underline{6}$ **d)** $-15 \times -10 = \underline{150}$

EXERCISE 1.11

Work out the following:

1 7×-6	**2** -4×-11	**3** 12×-5
4 $-81 \div -9$	**5** $-100 \div 5$	**6** -20×3
7 $48 \div -8$	**8** -6×-6	**9** $-52 \div 4$
10 -9×-6	**11** -7×8	**12** -11×12
13 $-49 \div -7$	**14** $42 \div -6$	**15** -7×-9

Be careful not to mix up all the different rules you have learnt.

Work out:

16 $-56 \div 7$	**17** $-2 + -5$	**18** $-3 - -4$
19 $8 \div -2$	**20** -3×-3	**21** $-16 \div -4$
22 $-6 + -5$	**23** $-3 - -2$	**24** $1 + -1$
25 2×-3	**26** $4 \div -2$	**27** $-5 + -3$

1.12 Rounding

We often need to **round** answers to calculations.

EXAMPLE

Round 8275 to the nearest:

a) 10 **b)** 100 **c)** 1000

SOLUTION

a)

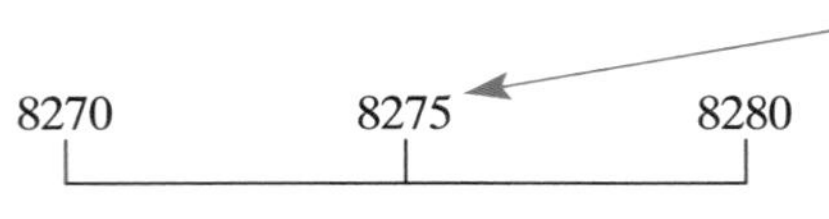

8275 is half way between 8270 and 8280. **Half way or more** means you round **up.**

8275 rounded to the nearest 10 is <u>8280</u>

b) 8200 8275 8300

8275 rounded to the nearest 100 is <u>8300</u>

c)

8000 8275 9000

Less than half way means you round **down.**

8275 rounded to the nearest 1000 is <u>8000</u>

EXAMPLE

Round 6.4 to the nearest integer.

An **integer** is a whole number.

SOLUTION

6	6.4	7

6.4 is nearer to 6 than 7

So 6.4 = 6 to the nearest integer

Less than 0.5 means **round down.**
0.5 or more means **round up.**

The 1st significant figure (s.f.) of a number is the first non-zero digit in that number (moving from left to right).

The 2nd significant figure is the digit (including zero) immediately to the right of the 1st significant figure, and so on.

EXAMPLE

Round these numbers to 1 significant figure:

a) 7243 **b)** 86

SOLUTION

> Underline the 1st significant figure. Look at the 2nd significant figure.
> If it is **4 or less, leave** the 1st significant figure **as it is.**
> If it is **5 or more**, then **round** the 1st significant figure up.
> **Replace** any other digits before the decimal point with a **0**.

a) 7243 = 7000 to 1 s.f. **b)** 86 = 90 to 1 s.f.

EXAMPLE

Round these numbers to 2 significant figures:

a) 48 253 **b)** 506

SOLUTION

> Underline the first 2 significant figures. Look at the 3rd significant figure.
> If it is **4 or less, leave** the 2nd significant figure **as it is.**
> If it is **5 or more**, then **round** the 2nd significant figure **up**.
> **Replace** any digits after the 2nd significant figure before the decimal point with a **0**.

a) 48 253 = 48 000 to 2 s.f. **b)** 506 = 510 to 2 s.f.

EXERCISE 1.12

1 Round these numbers to the nearest 10:

a) 17 **b)** 188 **c)** 395 **d)** 221

2 Round these numbers to the nearest 100:

a) 117 **b)** 150 **c)** 271 **d)** 1959

3 Round these numbers to the nearest 1000:

a) 980 **b)** 500 **c)** 3457 **d)** 9199

4 Round these numbers to the nearest integer:

a) 3.2 **b)** 16.7 **c)** 8.5 **d)** 9.931

5 Round these numbers to 1 significant figure:

a) 42 **b)** 17 **c)** 159 **d)** 364
e) 1234 **f)** 750 **g)** 1956 **h)** 5739

6 Round these numbers to 2 significant figures:

a) 652 **b)** 728 **c)** 479 **d)** 371
e) 4567 **f)** 6783 **g)** 2788 **h)** 8346

1.13 Using rounding

What is $\frac{54(7.6 + 2.3)}{63 - 37}$?
373.4
No, it is 20.6

You can use rounding to 1 significant figure to get an **estimate** without having to find an exact answer.

$$\frac{54 \times (7.6 + 2.3)}{63 - 37} \approx \frac{50 \times (8 + 2)}{60 - 40}$$

'≈' means 'is approximately (or roughly) equal to'

$$\frac{50 \times (8 + 2)}{60 - 40} = \frac{50 \times 10}{20}$$

$$= \frac{500}{20}$$

$$= 25$$

So $\frac{54 \times (7.6 + 2.3)}{63 - 37} \approx \underline{25}$

EXERCISE 1.13

1 By rounding to 1 significant figure, work out an approximate answer to:

a) 129×11 **b)** $129 + 873$ **c)** $\frac{802}{38}$

d) $\frac{32 \times 17}{95}$ **e)** $\frac{39 \times 107}{816 - 298}$ **f)** $\frac{28 \times 176}{873 - 588}$

g) $\frac{1.8 \times 6.1}{3.4}$ **h)** $\frac{3.8(2.93 + 7.1)}{8.2}$ **i)** $\frac{7.5(9.81 - 7.1)}{6.3 - 2.1}$

2 A cinema has 78 rows of 29 seats.

a) Estimate how many people can be seated at the cinema.

a) A cinema ticket costs £6
Estimate how much money the cinema makes when it is fully booked.

REVIEW EXERCISE 1

Do not use your calculator for Questions 1–14.

1 Find the sum of:

a) 326 and 89 **b)** 126, 319 and 742
c) 117 and 325 **d)** 2453 and 1219

2 Find the difference between:

a) 628 and 340 **b)** 221 and 179
c) 5312 and 2348 **d)** 2015 and 127

3 Find the product of:

a) 34 and 17 **b)** 159 and 26 **c)** 125 and 18
d) 57 and 43 **e)** 578 and 35 **f)** 37 and 64

4 Work out:

a) $208 \div 16$ **b)** $243 \div 9$ **c)** $486 \div 18$

d) $9\overline{)288}$ **e)** $345 \div 23$ **f)** $\frac{672}{56}$

5 a) Write the number **three hundred thousand, five hundred and ninety-four** in figures.
b) Round the number 7863 to the nearest 10
c) Work out 3452×4
d) Write down the value of the 6 in 32 657 **[Edexcel]**

6 Write these numbers in order of size – start with the smallest number:

a) 91, 109, 17, 140, 83
b) $-4, 4, 1, -8, -2$ **[Edexcel]**

7 **a)** Write the number **seventeen thousand, two hundred and fifty-two** in figures.
b) Write the number 5367 correct to the nearest hundred.
c) Write down the value of the '4' in the number 274 863 **[Edexcel]**

8 David buys 65 DVDs for £8 each.
What is the total cost of the DVDs?

9 Molly has £460 to spend on books for the school library.
How many books can she buy?

> **Special Offer**
> *Any* hardback book
> only **£8**

10 Amrita has £300 to spend.
She buys:

2 pairs of jeans for £47 each
3 T-shirts for £8 each
3 skirts for £22 each
1 pair of shoes for £55

How much change does she have from £300?

11 Look how Amy multiplies a two digit number by 11:

Add the digits together.

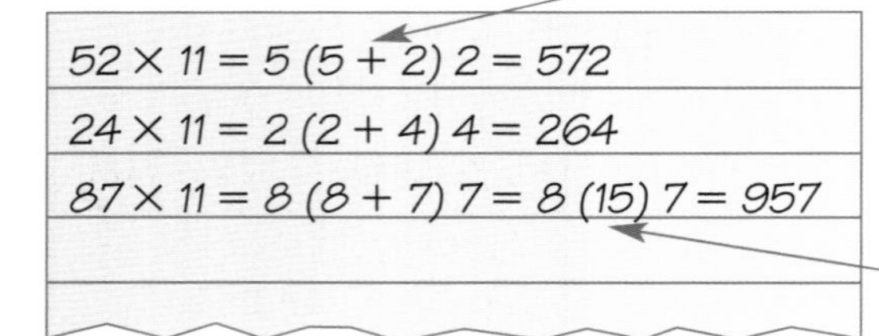

Carry the 1 over.

Use Amy's method to multiply the following numbers by 11:
a) 36 **b)** 17 **c)** 81 **d)** 68

12 Use rounding to 1 significant figure to estimate the answers to:
a) $\dfrac{27(176 + 15)}{18}$ **b)** $\dfrac{247 \times 196}{366 - 189}$ **c)** $\dfrac{5.1(198 + 231)}{99 + 312}$

13 Work out an estimate for the value of $\dfrac{637}{3.2 \times 9.8}$ **[Edexcel]**

14 Estimate the value of $\dfrac{67 \times 403}{197}$

You can use your calculator for Questions 15–18.

15 Fiona has four cards.
Each card has a number written on it.

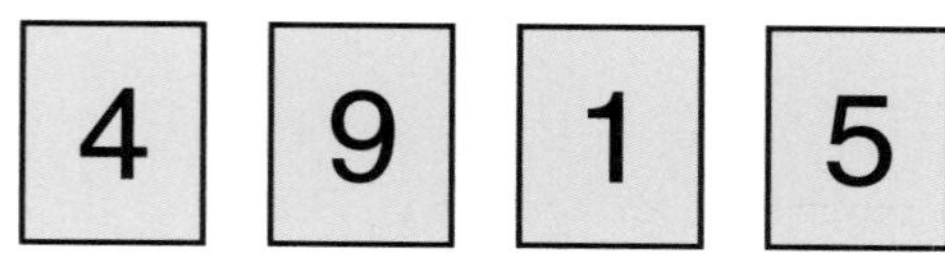

Fiona puts all four cards on the table to make a number.

a) **(i)** What is the smallest number Fiona can make with all four cards?
(ii) What is the largest number Fiona can make with all four cards?

Fiona uses the cards to make a true statement.

b) Write the number on the cards to make the statement true.
Use each of Fiona's cards **once**.

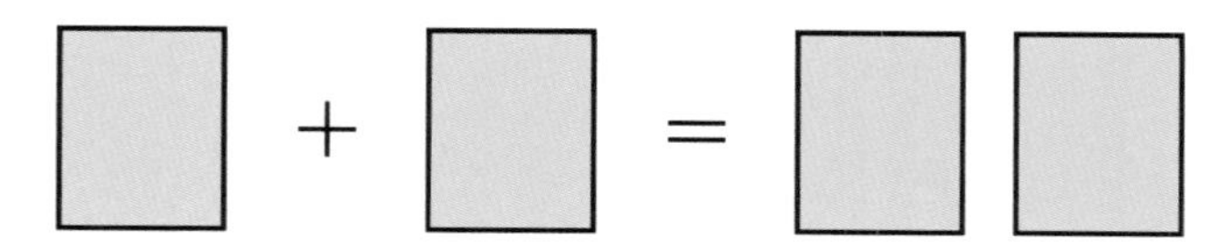

A fifth card is needed to show the result of the multiplication 4915×10

c) Write down the number that should be on the fifth card. **[Edexcel]**

16 54 327 people watched a concert.

a) Write 54 327 to the nearest thousand.

b) Write down the value of the 5 in the number 54 327 **[Edexcel]**

17 The table shows the lowest temperatures during five months in 2004 in a town in Scotland.

Month	Temperature (°C)
January	−16
March	−6
May	−1
July	4
September	7

a) Work out the difference in lowest temperature between January and March.

b) Work out the difference in lowest temperature between March and July.

c) In which month was the lowest temperature 5 °C higher than the lowest temperature in May?

The lowest temperature in November was 10 °C lower than the lowest temperature in May.

d) Work out the lowest temperature in November. **[Edexcel]**

18 The table shows the temperature on the surface of each of five planets.

Planet	Temperature (°C)
Venus	480
Mars	−60
Jupiter	−150
Saturn	−180
Uranus	−210

a) Work out the difference in temperature between Mars and Jupiter.
b) Work out the difference in temperature between Venus and Mars.
c) Which planet has a temperature 30 °C higher than the temperature on Saturn?

The temperature on Pluto is 20 °C lower than the temperature on Uranus.
d) Work out the temperature on Pluto. **[Edexcel]**

KEY POINTS

1. When you have to write down a number in words or figures use a place value diagram to help you.
2. When you order numbers in descending order start with the largest.
3. When you order numbers in ascending order start with the smallest.
4. Add, sum, plus, increase and total all mean addition.
5. Subtract, take, difference, decrease and minus all mean subtraction.
6. Multiply, times and product all mean multiplication.
7. Divide, share and 'go into' all mean division.
8. When you multiply by 100 move the decimal point two places to the right.
9. When you divide by 100 move the decimal point two places to the left.
10. When you are multiplying or dividing directed numbers:
    - same signs .... answer is positive
    - different signs .... answer is negative.
11. When there is a remainder in a division question you need to decide whether to:
    - round up, or
    - round down, or
    - give an exact answer.
12. You can use rounding to 1 significant figure to find an approximate answer to a problem.

CHAPTER 2

Ratio and proportion

In this chapter you will **learn how to:**

- simplify ratios
- divide a quantity in a given ratio
- solve problems involving ratios and proportion
- use unitary methods to solve problems
- use maps and scales.

Starter: Squares

Shade in the squares on a copy of these shapes using the following rules.

1 For every 1 square you shade red, shade 2 in blue.

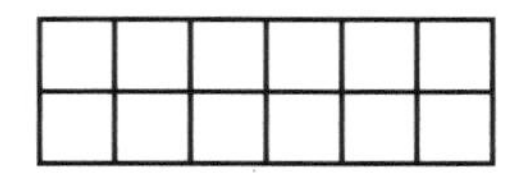

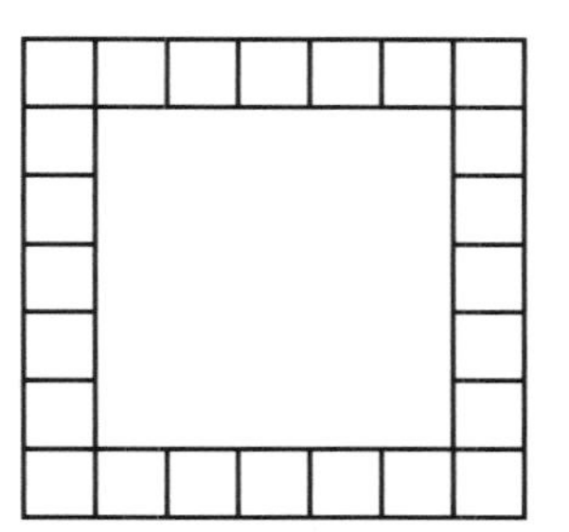

2 For every 1 square you shade red, shade 3 in blue.

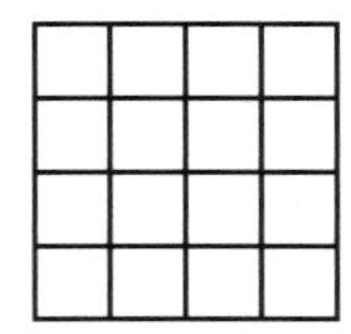

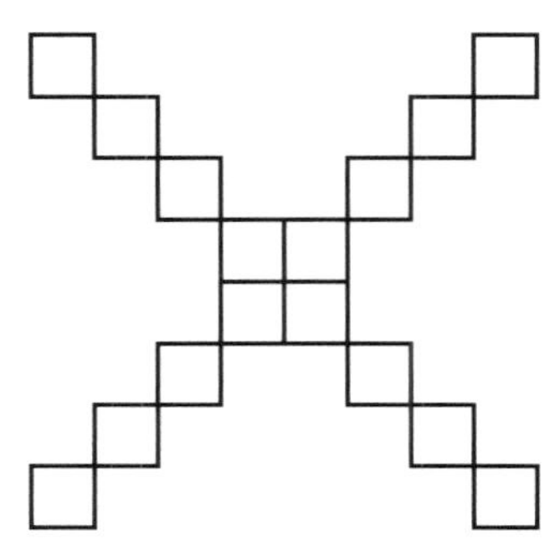

3 For every 1 square you shade red, shade 2 in blue and 3 in green.

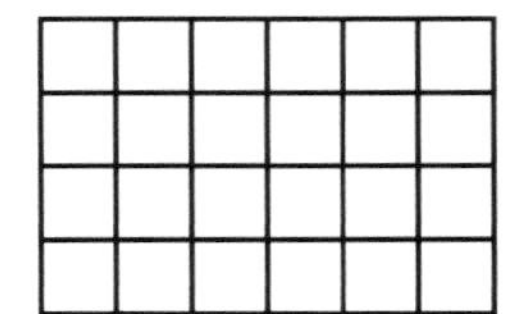

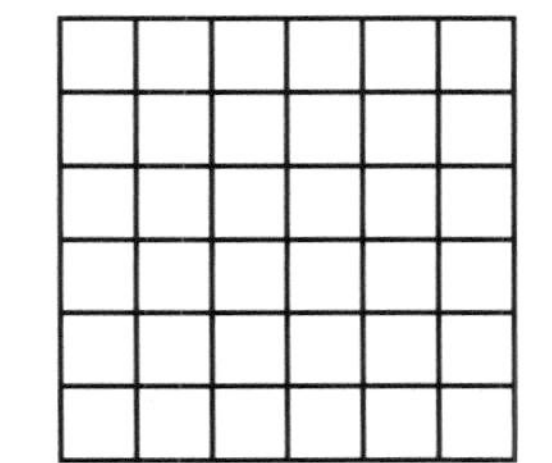

2.1 Simplifying ratios

A bag contains **red** counters and **green** counters in the **ratio 4 : 8**
You could also say that the ratio of **green** to **red** counters is **8 : 4**
All of these sets of counters have **red** to **green** counters in the ratio **4 : 8**

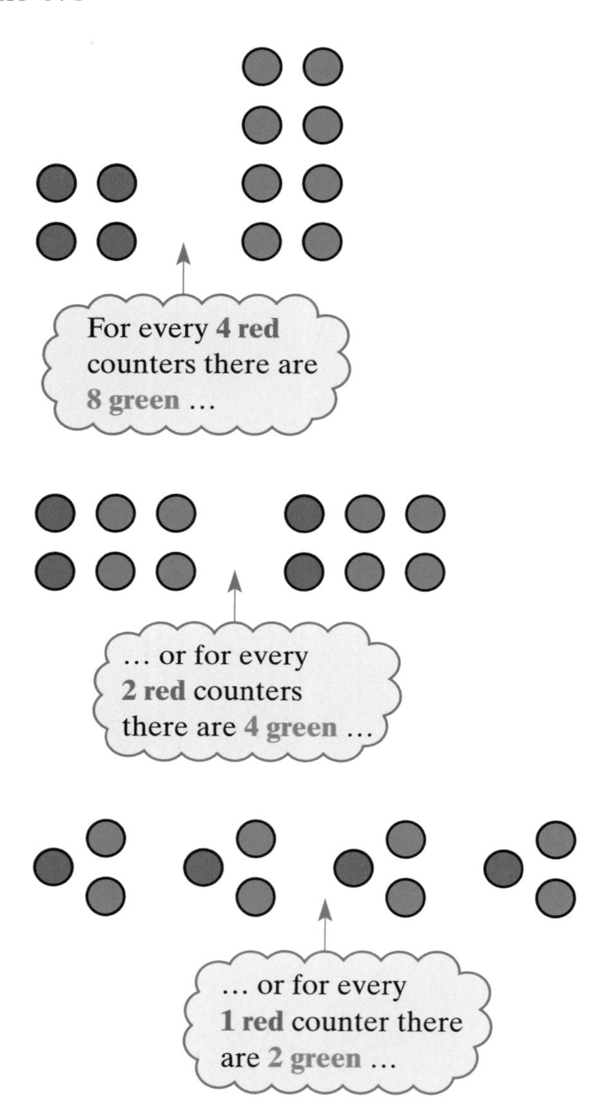

For every 4 red counters there are 8 green ...
... or for every 2 red counters there are 4 green ...
... or for every 1 red counter there are 2 green ...

The ratios 4 : 8, 2 : 4 and 1 : 2 are **equivalent**.
The ratios 4 : 8 and 2 : 4 can be **simplified** to 1 : 2

EXAMPLE

Simplify the following ratios:

a) 12 : 4 **b)** 15 : 18

SOLUTION

a) Divide both sides of the ratio by 4:

4 is the largest number that divides into both 12 and 4

12 : 4

÷4 ÷4

<u>3 : 1</u>

b) Divide both sides of the ratio by 3:

3 is the largest number that divides into both 15 and 18

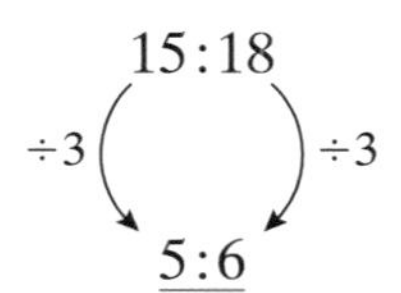

EXERCISE 2.1

1 Simplify the following ratios:

a) 3 : 6 **b)** 8 : 12 **c)** 21 : 15 **d)** 24 : 36
e) 18 : 9 **f)** 27 : 36 **g)** 8 : 24 **h)** 72 : 24

2 Complete these equivalent ratios:

a) 1 : 2 = 2 : ? **b)** 2 : 3 = 10 : ? **c)** 5 : 7 = 10 : ?
d) 8 : 9 = ? : 18 **e)** 5 : 8 = ? : 40 **f)** 4 : 5 = ? : 40

3 Match together the equivalent ratios:

12:18	6:18	4:12
3:2	6:9	2:3
1:3	2:6	12:8

2.2 Dividing a quantity in a given ratio

EXAMPLE

Ayesha, Kassim and Hussein share £120 in the ratio 2 : 3 : 5
How much should they each receive?

SOLUTION

There are 2 + 3 + 5 = 10 shares.

Add up all the numbers in the ratio to find the total number of shares.

So 10 shares = £120

÷10 ÷10

1 share = £12

Work out what 1 share is worth.

So 2 shares = £24
3 shares = £36
5 shares = £60

So Ayesha gets £24

Kassim gets £36

and Hussein gets £60

Check: £24 + £36 + £60 = £120 ✓

EXAMPLE

Myra, Kate and Natasha buy a car for £4320
Myra pays £1240 and Kate pays £2000
They sell the car for £3000
How much should Natasha receive?

SOLUTION

You can think of 1 share equalling £1.

Natasha pays: £4320 − £2000 − £1240 = £1080

There are 4320 shares in the car in the ratio 1240 : 2000 : 1080

So 4320 shares = £3000

÷4320 ÷4320

1 share = £0.694 444 444

Don't round this answer.

×1080 ×1080

1080 shares = £750

So Natasha receives £750

EXAMPLE

The ratio of boys to girls in a school is 8 : 9
There are 720 boys.
How many girls are at the school?

SOLUTION

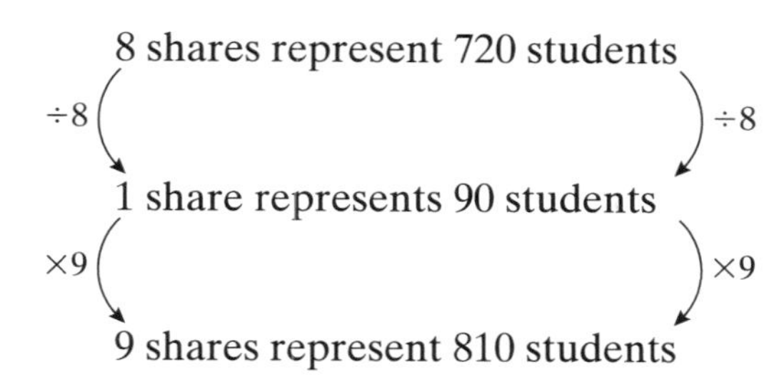

8 shares represent 720 students
÷8
÷8
1 share represents 90 students
×9
×9
9 shares represent 810 students

There are 810 girls at the school.

EXERCISE 2.2

Do not use your calculator for Questions 1–6.

1 Divide £240 in the following ratios:
a) 1 : 2 **b)** 3 : 5 **c)** 1 : 2 : 3 **d)** 3 : 4 : 5

2 Divide £360 in the following ratios:
a) 1 : 5 **b)** 4 : 5 **c)** 1 : 3 : 5 **d)** 1 : 4 : 7

3 Ben and Eliza share a bag of sweets.
Ben takes three sweets for every two that Eliza has.
There are 40 sweets in the bag.
How many sweets does Ben have?

4 The ratio of students to teachers on a school trip is 8 : 1
120 students go on the trip.
How many teachers go?

5 To make an orange drink you need to mix 5 parts water with 2 parts orange juice.
Work out how many litres of orange juice are needed to make:
a) 7 litres **b)** 14 litres **c)** 35 litres of orange drink

6 The ratio of CDs to DVDs sold in a music shop is 4 : 3
The shop sells 220 CDs one Saturday.
How many DVDs does it sell?

You can use your calculator for Question 7.

7 Magda and Janek buy a bicycle for £360
Magda pays £200 towards it.
They sell the bicycle for £126
How much money should they each receive?

2.3 Unitary method

EXAMPLE

Which toothpaste is the best buy?

SOLUTION

Method 1

Work out how much toothpaste you can buy for 1p

Small: 65p buys 110 ml

$\div 65$ $\div 65$

1p buys 1.6923… ml

It is easier to work in pence.

Large: 140p buys 240 ml

$\div 140$ $\div 140$

1p buys 1.7142… ml

You get more toothpaste for 1p

So the large toothpaste is the best buy.

Method 2

Work out how much 1 ml of toothpaste costs.

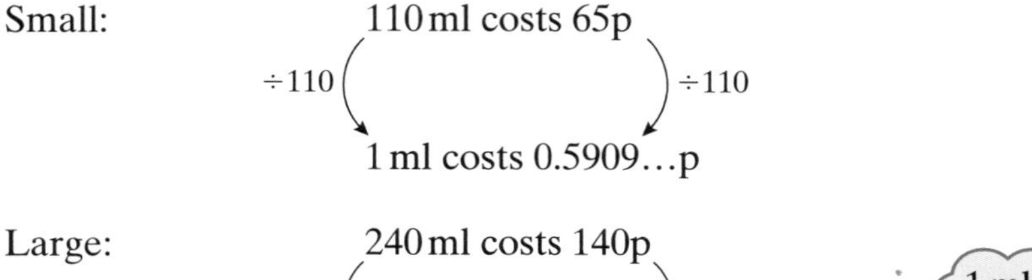

Small: 110 ml costs 65p

$\div 110$ $\div 110$

1 ml costs 0.5909…p

Large: 240 ml costs 140p

$\div 240$ $\div 240$

1 ml costs 0.5833…p

1 ml of toothpaste is cheaper.

So the large toothpaste is the best buy.

EXAMPLE

To make enough pastry for 12 small pies you need:

240 g flour   120 g butter
4 tablespoons water $\frac{1}{2}$ teaspoon salt.

How much flour and butter do you need to make 9 pies?

Keep the ingredients in the same **proportions**.

SOLUTION

12 pies need 240 g flour
$\div 12$ $\div 12$
1 pie needs 20 g flour
$\times 9$ $\times 9$
9 pies need 180 g flour

12 pies need 120 g butter
$\div 12$ $\div 12$
1 pie needs 10 g butter
$\times 9$ $\times 9$
9 pies need 90 g butter

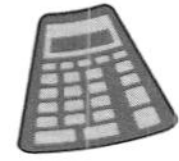

EXAMPLE

Leila is going on holiday to the USA.
She needs to buy some foreign currency.
The exchange rate is £1 = \$1.5

a) How many dollars can she buy for £300?

When Leila comes back from her holiday she has \$69 left.
The bank uses the same exchange rate.

b) How many pounds can Leila buy for \$69?

SOLUTION

a) Using the exchange rate of £1 = \$1.5

£1 = \$1.5
$\times 300$ $\times 300$
£300 = \$450

b) Using the exchange rate of \$1.5 = £1

Write the exchange rate with 'dollars on the left-hand side.

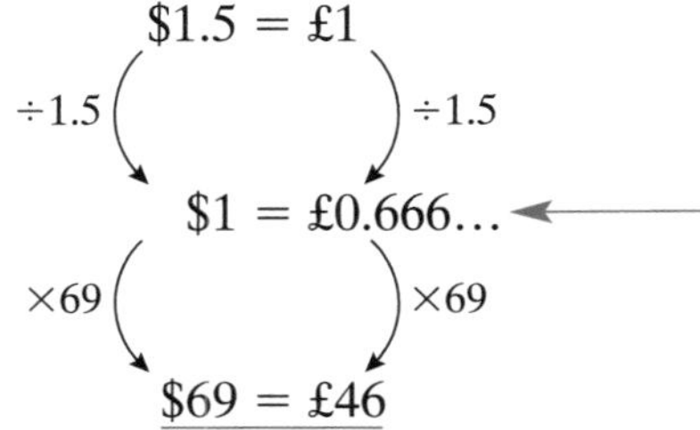

Find what \$1 is worth. Don't round the answer.

EXERCISE 2.3

Do not use your calculator for Questions 1–4.

1 Phil buys three CDs for £22.50
How much would five CDs cost?

MegaMusic
Special Offer
on *all* CDs

2 Alfie is charged £2.25 for using his mobile phone for $1\frac{1}{2}$ hours.
Find the cost of phone calls lasting:
a) 30 minutes **b)** 2 hours

3 Here are the ingredients for making 24 flapjacks:

120 g butter

100 g sugar

200 g oats

1 tablespoon golden syrup

Find how much **(i)** butter **(ii)** sugar **(iii)** oats are needed to make:
a) 12 flapjacks **b)** 6 flapjacks **c)** 30 flapjacks

4 Ben is going on holiday to Australia.
He needs to buy some foreign currency.
The exchange rate is £1 = A$3
a) How many Australian dollars can he buy for £900?

When Ben comes back from his holiday he has A$120 left.
The bank uses the same exchange rate.
b) How many pounds can Ben buy for A$120?

While on holiday, Ben buys some trainers for A$180
He sees an identical pair in his home town in Wales for £56
c) In which country were the trainers cheaper, and by how much?
Give your answer in pounds.

You can use your calculator for Questions 5–8.

5 Which of these bags of crisps is the best buy?

6 100 ml of lemonade contains 50 calories.
Find how many calories there are in:
a) a 330 ml can
b) a 1 litre (1000 ml) bottle.

7 Fiona is going on holiday to France.
She needs to buy some foreign currency.
The exchange rate is £1 = 1.65 euros.
a) How many euros can she buy for £460?

Fiona's hotel bill comes to 363 euros.
b) Work out the cost of Fiona's hotel bill in pounds.

8 Here are the ingredients for a blackberry and almond tart recipe:

300 g pastry
140 g butter
160 g sugar
100 g almonds
500 g blackberries
Makes 12 portions.

Work out how much of each ingredient is needed to make:
a) 9 portions **b)** 30 portions.

2.4 Scales

You will see ratios used on maps.

They tell you what the **scale** of the map is.

EXAMPLE

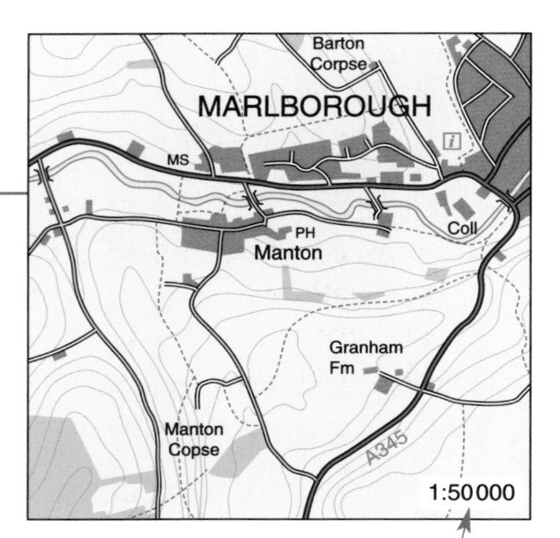

A map has a scale of 1 : 50 000

a) What does each of these distances on the map represent?

(i) 1 cm

(ii) 5 cm

(iii) 4 mm

b) The distance between two towns is 12 km
How far apart are the towns on the map?

This means that 1 unit on the map represents 50 000 units on the ground.
For example, 1 cm on the map represents 50 000 cm on the ground.

SOLUTION

a) (i) 1 cm on the map represents 50 000 cm on the ground.
1 cm represents 500 m

Divide by 100 to change cm to m

(ii) 1 cm represents 500 m
×5 ↓ ×5
5 cm represents 2500 m
or 2.5 km

Divide by 1000 to change m to km

(iii) 1 mm on the map represents 50 000 mm
So 1 mm represents 5000 cm
So 1 mm represents 50 m
So 4 mm represents 4 × 50 m = 200 m

Divide by 10 to change mm to cm

b) 12 km = 12 000 m

Multiply by 1000 to change km to m

Every 50 000 m on the ground is represented by 1 m on the map.
So you need to find how many 50 000's there are in 12 000:

$$\frac{12\,000}{50\,000} = 0.24$$

So the towns are 0.24 m or 24 cm apart on the map.

EXERCISE 2.4

1 An estate agent draws a floor plan of a house.
She uses a scale of 1 : 50

a) Work out the measurements, in centimetres, of the following rooms on her plan:

(i)	living room:	6 m by 5 m
(ii)	dining room:	5 m by 4 m
(iii)	bedroom:	3 m by 3.5 m

Remember: 1 m = 100 cm

b) Work out the measurements, in metres, of the following rooms.

(i)	bathroom:	6 cm by 5 cm on the plan
(ii)	master bedroom:	7.2 cm by 8.4 cm on the plan
(iii)	kitchen:	6.3 cm by 5.8 cm on the plan

Remember: The scale is 1 : 50

2 A model railway is made on a scale of 1 cm represents 12 m

a) Write this scale as a ratio.

The model station platform is 20 cm long.

b) How long would the actual platform be?

The distance between two stations is 3 km

Remember: 1 km = 1000 m

c) What length of track is needed for the model?

3 A map has a scale of 1 : 20 000

a) Find the distance on the ground, in kilometres, represented by:

(i) 2 cm on the map
(ii) 10 cm on the map
(iii) 5 mm on the map
(iv) 8.2 cm on the map

b) Find how far apart on the map, in centimetres, are towns:

(i) 10 km apart on the ground
(ii) 6 km apart on the ground
(iii) 4.2 km apart on the ground

REVIEW EXERCISE 2

Do not use your calculator for Questions 1–6.

1 Simplify the following ratios:

a) 6 : 12	**b)** 8 : 20	**c)** 15 : 40	**d)** 25 : 60
e) 14 : 42	**f)** 5 : 10 : 25	**g)** 21 : 35 : 49	**h)** 8 : 12 : 32

2 Ingrid and Hilmar share £420 in the ratio 3 : 4
How much do they get each?

3 The ratio of male to female workers in a company is 5 : 4
There are 200 men working at the company.
How many female employees are there?

4 The ratio of rap to pop to rock music in John's CD collection is 3 : 4 : 5
John has 720 CDs.
State how many of the following CDs John has got:
a) rap CDs **b)** pop CDs **c)** rock CDs

5 Here is a recipe for breakfast muesli.

For 20 servings (3200 calories), mix:

600 g wheatflakes
240 g mixed fruit
100 g mixed nuts
60 g oatflakes

a) How much of each ingredient is needed for: **(i)** 10 servings, **(ii)** 30 servings?
b) How many calories are there in: **(i)** 10 servings, **(ii)** 30 servings?

6 Here are the ingredients needed to make 500 ml of custard.

400 ml milk
3 large egg yolks
50 g sugar
2 teaspoons of cornflour

a) Work out the amount of sugar needed to make 2000 ml custard.
b) Work out the amount of milk needed to make 750 ml custard. **[Edexcel]**

You can use your calculator for Questions 7–15.

7 An architect draws the following floor plan using a scale of 1 : 100

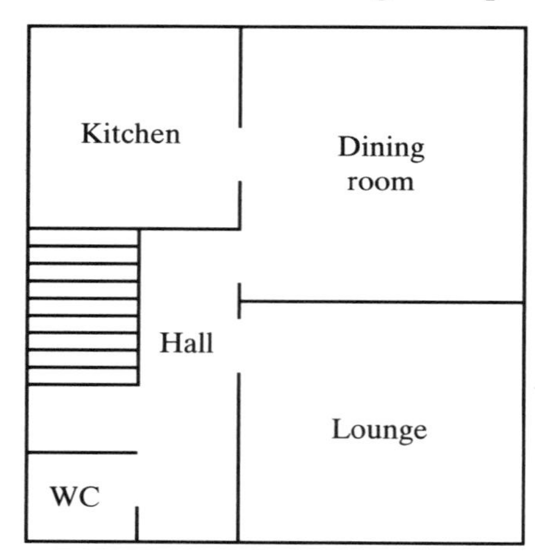

Scale 1:100

a) Find the actual dimensions, in metres, of:
(i) the kitchen, **(ii)** the dining room, **(iii)** the lounge

The architect decides to add a conservatory measuring 4.2 m by 5.1 m
b) What are the measurements on the plan of the conservatory, in centimetres?

8 Marie bought four identical computer discs for £3.60
Work out the cost of nine of these computer discs. **[Edexcel]**

9 Barry has a list of ingredients to make 'chicken surprise' for 8 people.

CHICKEN SURPRISE
for 8 people
40 ml tomato ketchup
16 ml dark brown sauce
16 ml mild mustard
32 ml dark soy sauce
8 chicken legs

Barry wants to make a chicken surprise meal for 20 people.
a) Work out how much of each ingredient Barry will need.

Barry has 0.5 litres of tomato ketchup.
He has plenty of the other ingredients.
b) For how many people could he make chicken surprise? **[Edexcel]**

1 litre = 1000 ml

10 Sarah went to Germany.
She changed £300 into euros.
The exchange rate was £1 = 1.64 euros.
a) Work out the number of euros Sarah got.

Sarah came home, she had 119 euros left.
The new exchange rate was £1 = 1.50 euros.
b) Work out how much Sarah got in pounds for 119 euros **[Edexcel]**

11 This is a list of ingredients for making a pear and almond crumble for 4 people.

80 g plain flour
60 g ground almonds
90 g soft brown sugar
60 g butter
4 ripe pears

Work out the amount of each ingredient needed to make a pear and almond crumble for 10 people. **[Edexcel]**

12 Michael buys three files.

The total cost of these is £5.40
Work out the cost of 7 of these files.

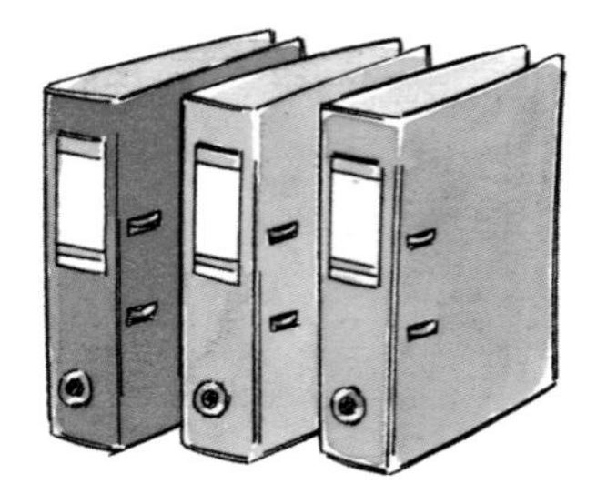

[Edexcel]

13 This is a map of part of northern England.
Scale: 1 cm represents 10 km

A plane flies in a straight line from Preston to Stoke-on-Trent.

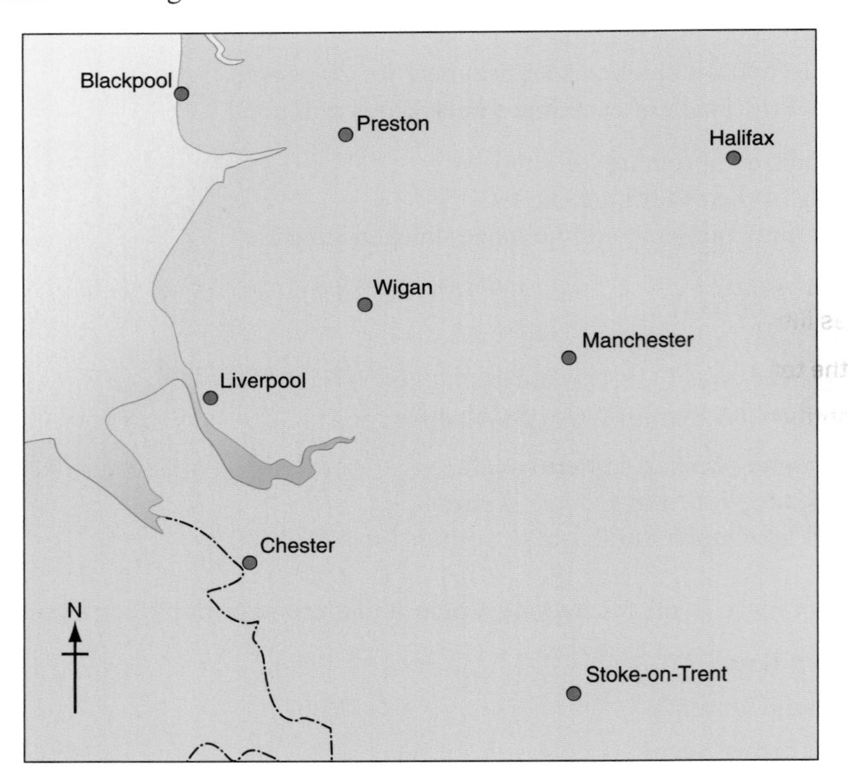

How far does it fly?
Give your answer in kilometres.

[Edexcel]

14 The length of a coach is 15 metres.
Jonathan makes a model of the coach.
He uses a scale of 1 : 24
Work out the length, in centimetres, of the model coach.

[Edexcel]

15 Margaret goes on holiday to Switzerland.
The exchange rate is £1 = 2.10 francs

She changes £450 into francs.

a) How many francs should she get?

In Switzerland, Margaret buys a railway ticket.
The cost of the railway ticket is 63 francs.

b) Work out the cost of the ticket in pounds.

[Edexcel]

KEY POINTS

1. A ratio can be simplified by dividing both sides of the ratio by the largest number that goes into both sides.
2. To find the total number of 'shares' add together the numbers in the ratio.
3. Two quantities are in proportion when one increases as the other increases.
4. On a map the scale 1 : 10 000 means that 1 unit on the map represents 10 000 of the same unit on the ground.

CHAPTER 3

Decimals

In this chapter you will learn how to:

- order decimals
- round decimals to a given degree of accuracy
- add and subtract decimal numbers
- multiply and divide decimal numbers
- read scales
- solve money problems.

Starter: Getting the point

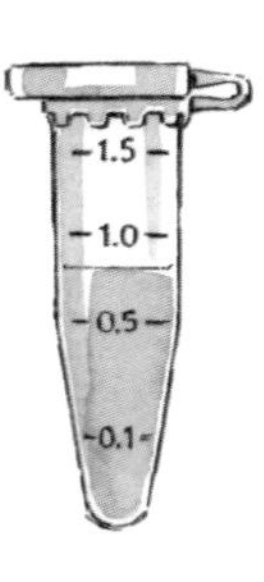

These are all examples of using **decimal numbers**.
Any number which has digits after the decimal point is a decimal.

0.25 has the same value as 0.250 or 0.2500, and so on.

Hundreds	Tens	Units	.	Tenths	Hundredths	Thousandth
		0	.	2	5	
	1	5	.	4	3	
		1	.	6	7	
	1	0	.	1	9	8

The value of the digit 8 in 10.198 is 8 thousandths or $\frac{8}{1000}$

To order decimals it helps to have the same number of digits after the decimal point.

EXAMPLE

Put these numbers in descending order:

0.205 0.316 0.21 0.3 0.39

> Descending order means put the largest number first.

SOLUTION

Write the numbers on a place value diagram:

Units	.	Tenths	Hundredths	Thousandth
0	.	2	0	5
0	.	3	1	6
0	.	2	1	**0**
0	.	3	**0**	**0**
0	.	3	9	**0**

> Fill in **0**'s so that all the numbers have 3 decimal places.
> This doesn't change their value.

> Look for the largest number in the tenths column first, then the hundredths and so on.

In descending order:

0.390 0.316 0.300 0.210 0.205

So the required order is <u>0.39 0.316 0.3 0.21 0.205</u>

Put these numbers in descending order:

1 1.2, 1.204, 1.32, 1.29

2 3.2, 3.54, 3.239, 3.47, 3.4

3 5.3, 5.299, 5.29, 5.31, 5.307

4 0.039, 0.13, 0.207, 0.1004, 0.099

5 0.004, 0.4291, 0.0444, 0.47, 0.4, 0.04

6 3.3, 3.33, 3.03, 3.333, 3.003, 3.303

3.1 Rounding to a given number of decimal places

EXAMPLE

Round these numbers to 1 decimal place:

a) 1.849 **b)** 2.351 **c)** 3.04 **d)** 5.97

SOLUTION

> Look at the digit in the 2nd decimal place.
> If it is **4 or less**, then leave the digit in the 1st decimal place **as it is**.
> If it is **5 or more**, then round the digit in the 1st decimal place **up**.

'd.p.' stands for decimal place.

a) 1 . 8 | 4 9 = <u>1.8</u> to 1 d.p.

b) 2 . 3 | 5 1 = <u>2.4</u> to 1 d.p.

c) 3 . 0 | 4 = <u>3.0</u> to 1 d.p.

You need to write in the **0** as '3' doesn't have 1 decimal place.

d) 5 . 9 | 7 = <u>6.0</u> to 1 d.p.

The '9 tenths' needs to round up to '10 tenths' as 7 is more than 5 but '10 tenths' is the same as '1 unit'. So the 5 rounds up to a 6

EXAMPLE

Round these numbers to 2 decimal places:

a) 2.831 **b)** 4.6893 **c)** 7.003 **d)** 6.395

SOLUTION

> Look at the digit in the 3rd decimal place.
> If it is **4 or less**, then leave the digit in the 2nd decimal place **as it is**.
> If it is **5 or more**, then round the digit in the 2nd decimal place **up**.

a) 2 . 8 3 | 1 = <u>2.83</u> to 2 d.p.

b) 4 . 6 8 | 9 3 = <u>4.69</u> to 2 d.p.

c) 7 . 0 0 | 3 = <u>7.**00**</u> to 2 d.p.

You need to put in the two **0**'s as '7' doesn't have 2 decimal places.

d) 6 . 3 9 | 5 = <u>6.40</u> to 2 d.p.

EXERCISE 3.1

1 Round these numbers to 1 decimal place:

a) 4.32 **b)** 5.619 **c)** 7.238 **d)** 8.561
e) 9.02 **f)** 100.04 **g)** 9.99 **h)** 7.954

2 Round these numbers to 2 decimal places:

a) 1.111 **b)** 7.654 **c)** 3.2771 **d)** 3.0129
e) 3.104 **f)** 2.302 **g)** 5.199 **h)** 9.9999

3 Round 3.04896 to:

a) 1 decimal place **b)** 2 decimal places
c) 3 decimal places **d)** 4 decimal places

3.2 Rounding to a given number of significant figures

The 1st significant figure of a number is the first **non-zero digit** in that number (moving from left to right).
The 2nd significant figure is the digit (including zero) immediately to the right of the first significant figure, and so on.

In the following numbers the 1st significant figure is **blue** and the 2nd significant figure is **green**:

4.329   2.014   0.00 001 234   0.000 305 67

EXAMPLE

Round these numbers to 2 significant figures:

a) 2.137 **b)** 0.0143 **c)** 0.0003021 **d)** 0.00199

SOLUTION

> Look at the 3rd significant figure.
> If it is **4 or less**, **leave** the 2nd significant figure **as it is.**
> If it is **5 or more**, then **round** the 2nd significant figure **up**.

a) 2.1|3 7 = 2.1 to 2 s.f.

's.f.' stands for significant figures.

b) 0.014|3 = 0.014 to 2 s.f.

c) 0.00030|21 = 0.000 30 to 2 s.f.

You need to put in a **0** after the 3 as '0.0003' has only 1 significant figure.

d) 0.0019|9 = 0.0020 to 2 s.f.

EXERCISE 3.2

1 Round these numbers to 2 significant figures:

a) 5.12 **b)** 7.619 **c)** 3.038 **d)** 0.0561
e) 0.000124 **f)** 0.003081 **g)** 0.000496 **h)** 11.729

2 Round these numbers to 3 significant figures:

a) 3.123 **b)** 1.759 **c)** 0.1234 **d)** 0.6789
e) 0.02199 **f)** 0.00200187 **g)** 0.003199 **h)** 9.9999

3 Round 1.04396 to:

a) **(i)** 1 significant figure **(ii)** 1 decimal place
b) **(i)** 2 significant figures **(ii)** 2 decimal places
c) **(i)** 3 significant figures **(ii)** 3 decimal places
d) **(i)** 4 significant figures **(ii)** 4 decimal places

3.3 Adding and subtracting decimal numbers

EXAMPLE

Work out 2.34 + 3.5 + 4.78

SOLUTION

Method 1

```
    2 . 3 4
    3 . 5 0
+   4 . 7 8
-----------
  1 0 . 6 2
    1   1
```

3.5 and 3.5**0** are the same. It is easier to add together numbers which have the same number of decimal places.

Keep the decimal points lined up.

So 2.34 + 3.5 + 4.78 = 10.62

Method 2

Add together 2.34 and 3.5**0** first:

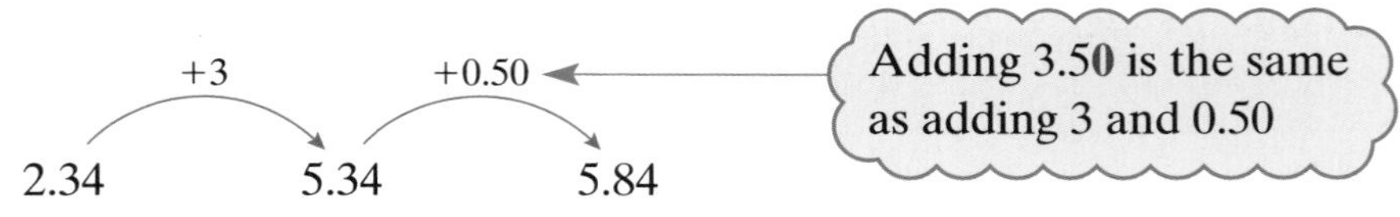

Adding 3.5**0** is the same as adding 3 and 0.50

Now add 5.84 and 4.78

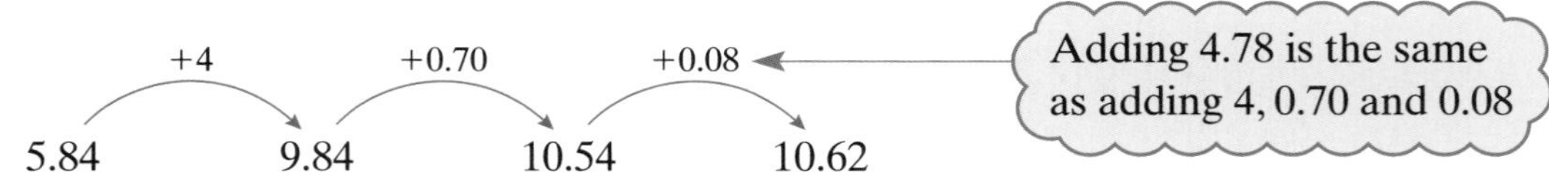

Adding 4.78 is the same as adding 4, 0.70 and 0.08

So 2.34 + 3.5 + 4.78 = 10.62

EXAMPLE

Work out 6.3 − 4.49

SOLUTION

Method 1

$$\begin{array}{r} {}^{5}\not{6}\,.\,{}^{1}{}^{2}\not{3}\,{}^{1}0 \\ -\;4\,.\,4\;9 \\ \hline 1\,.\,8\;1 \end{array}$$

6.3 and 6.3**0** are the same. It is easier to subtract numbers which have the same number of decimal places.

Keep the decimal points lined up.

So 6.3 − 4.49 = <u>1.81</u>

Method 2

Work out what you need to add to 4.49 to get 6.30

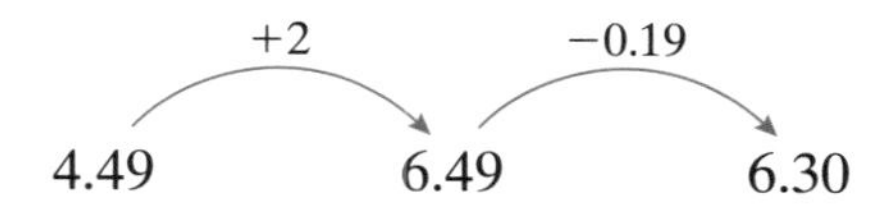

2 − 0. 19 = 1.81

So 6.3 − 4.49 = <u>1.81</u>

EXERCISE 3.3

1 Work out:

a) 0.1 + 0.2 + 0.3 **b)** 0.7 + 0.5 **c)** 1.4 + 2.8 + 3.9

d) 4.43 + 7.24 **e)** 1.87 + 8.11 **f)** 2.46 + 3.27

2 Work out:

a) 0.23 + 0.4 **b)** 1.7 + 2.35 **c)** 3.44 + 2.98

d) 5.06 + 5.7 **e)** 8.29 + 3.452 **f)** 2.4 + 3.77 + 2.98

3 Work out:

a) 0.5 − 0.4 **b)** 1.8 − 1.2 **c)** 3.26 − 2.14

d) 7.36 − 5.72 **e)** 7.79 − 6.41 **f)** 12.42 − 7.76 − 3.15

4 Work out:

a) 4.61 − 3.2 **b)** 9.82 − 2.4 **c)** 8.67 – 2.9

d) 7.16 − 5.7 **e)** 6.2 − 4.83 **f)** 15.7 − 12.84

3.4 Multiplying decimal numbers

When you multiply two decimal numbers, count how many digits there are after the decimal point.

There will be the same number of digits after the decimal point in the answer.

EXAMPLE

Work out 1.2×0.8

SOLUTION

Work out the sum as though there were no decimal point:

$12 \times 8 = 96$

There are two numbers after the decimal point in the question (2 and 8), so there will be two numbers after the decimal point in the answer.

So $1.2 \times 0.8 = \underline{0.96}$

EXAMPLE

Work out 1.35×2.4

SOLUTION

Work out the sum as though there were no decimal point:

135×24

×	100	30	5	Total
20	2000	600	100	2700
4	400	120	20	540
				3240

See page 11 for different ways of working this out.

$135 \times 24 = 3240$

There are three numbers after the decimal point in the question (3, 5 and 4), so there will be three numbers after the decimal point in the answer.

$1.35 \times 2.4 = 3.240$

You don't need to write the final 0

So $1.35 \times 2.4 = \underline{3.24}$

EXERCISE 3.4

1 Work out:

a) 5×0.3 **b)** 0.2×8 **c)** 0.4×7 **d)** 6×0.8
e) 0.3×0.8 **f)** 0.7×0.9 **g)** 0.04×0.3 **h)** 0.06×0.07
i) 0.02×0.01 **j)** 1.1×0.5 **k)** $0.6 \times 0.2 \times 0.1$ **l)** $2.5 \times 0.4 \times 0.2$

2 Work out:

a) 1.53×0.2 **b)** 2.68×0.4 **c)** 3.4×7.8
d) 4.6×3.8 **e)** 2.93×1.2 **f)** 5.21×8.9

3 Use the fact that $173 \times 29 = 5017$ to work out:

a) 17.3×29 **b)** 1.73×29 **c)** 1.73×2.9 **d)** 17.3×2.9

4 Use the fact that $283 \times 46 = 13\,018$ to work out:

a) 28.3×46 **b)** 2.83×46 **c)** 2.83×4.6 **d)** 28.3×4.6

3.5 Dividing by decimal numbers

EXAMPLE

Work out $5 \div 0.2$

SOLUTION

Step 1: Write the sum as a fraction.

$$5 \div 0.2 = \frac{5}{0.2}$$

Step 2: Multiply both the numerator and denominator of the fraction by 10

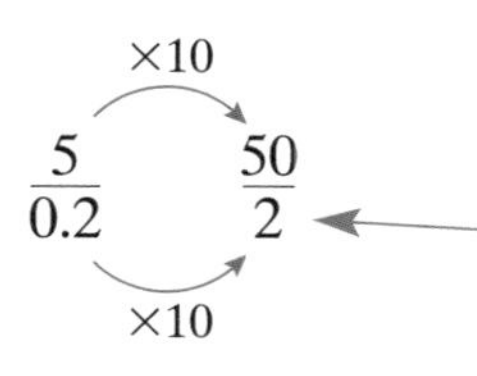

This makes the bottom of the fraction a whole number without changing the value of the fraction (see page 65).

$$\frac{50}{2} = 25$$

So $5 \div 0.2 = \underline{25}$

EXAMPLE

Work out $1.2 \div 0.03$

SOLUTION

Step 1: Write the sum as a fraction.

$$1.2 \div 0.03 = \frac{1.2}{0.03}$$

Step 2: Multiply both the top and bottom of the fraction by 100

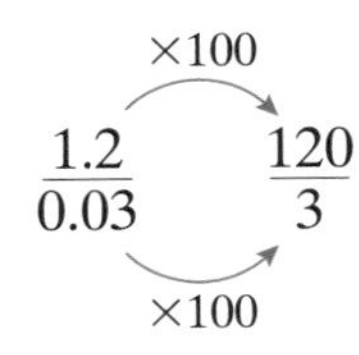

$$\frac{120}{3} = 40$$

So $1.2 \div 0.03 = \underline{40}$

EXERCISE 3.5

1 Work out:

a) $8 \div 0.2$ **b)** $6 \div 0.3$ **c)** $4 \div 0.4$ **d)** $12 \div 0.06$
e) $15 \div 0.03$ **f)** $21 \div 0.07$ **g)** $16 \div 0.08$ **h)** $200 \div 0.05$

2 Work out:

a) $1.6 \div 0.2$ **b)** $0.6 \div 0.2$ **c)** $0.12 \div 0.03$ **d)** $1.4 \div 0.7$
e) $1.8 \div 0.06$ **f)** $3.5 \div 0.7$ **g)** $4.8 \div 0.06$ **h)** $0.72 \div 0.09$

3 Use the fact that $304 \div 19 = 16$ to work out:

a) $304 \div 1.9$ **b)** $304 \div 0.19$ **c)** $30.4 \div 1.9$ **d)** $30.4 \div 19$

4 Use the fact that $14 \times 17 = 238$ to work out:

a) **(i)** $238 \div 17$ **(ii)** $23.8 \div 1.7$ **(iii)** $23.8 \div 0.17$
b) **(i)** $238 \div 14$ **(ii)** $23.8 \div 1.4$ **(iii)** $2.38 \div 1.4$

3.6 Reading scales

EXAMPLE

What numbers are the arrows **a**, **b**, **c** and **d** pointing to?

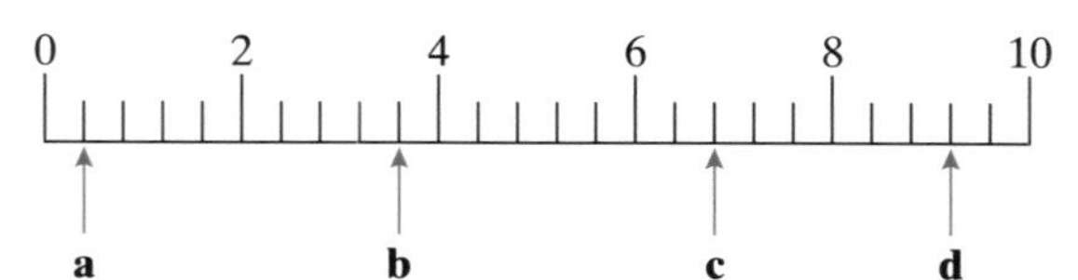

SOLUTION

First you need to work out what one division is worth.

5 divisions = 2 units

$\div 5$ $\qquad$ $\div 5$

1 division = 0.4 units

So arrow **a** is pointing to 0.4

arrow **b** is pointing to 3.6

arrow **c** is pointing to 6.8

arrow **d** is pointing to 9.2

2 plus 4 lots of 0.4
$= 2 + 4 \times 0.4$
$= 2 + 1.6$
$= 3.6$

EXERCISE 3.6

Don't use your calculator for Questions 1–3.

1 What numbers are the arrows **a**, **b**, **c** and **d** pointing to?

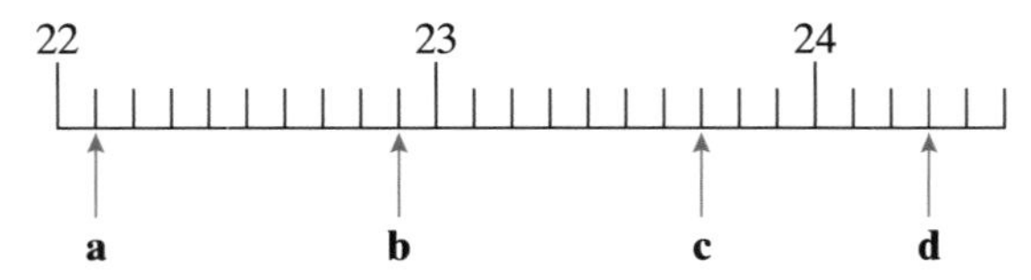

2 What numbers are the arrows **a**, **b**, **c** and **d** pointing to?

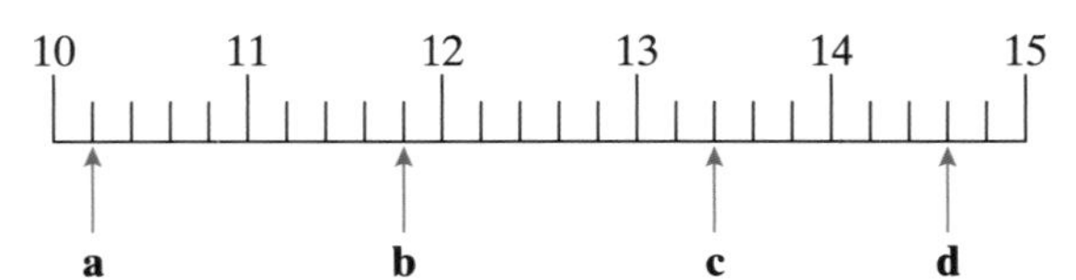

3 a) Write down the number marked with an arrow.

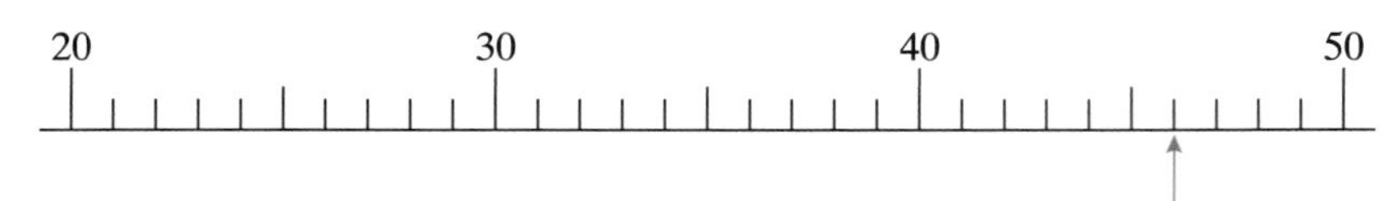

b) Write down the number marked with an arrow.

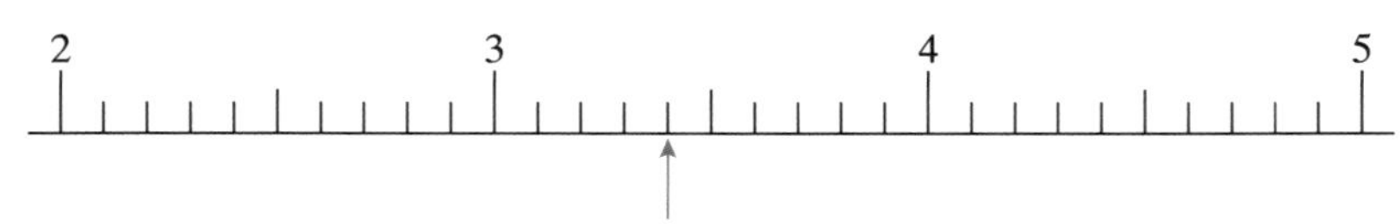

c) Find the number 430 on the number line.
On a copy mark it with an arrow (↑).

d) Find the number 3.7 on the number line.
On a copy mark it with an arrow (↑).

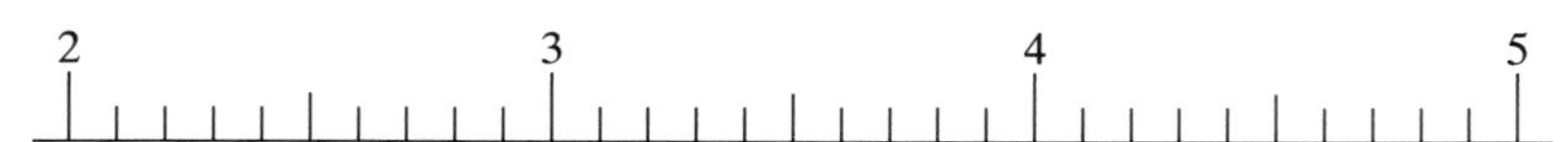

[Edexcel]

You can use your calculator for Question 4.

4 a) Write down the number marked with an arrow.

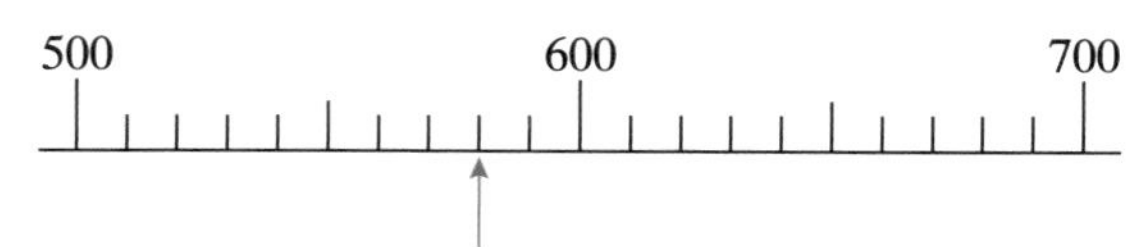

b) Write down the number marked with an arrow.

c) Find the number 48 on the number line.
On a copy mark it with an arrow (↑).

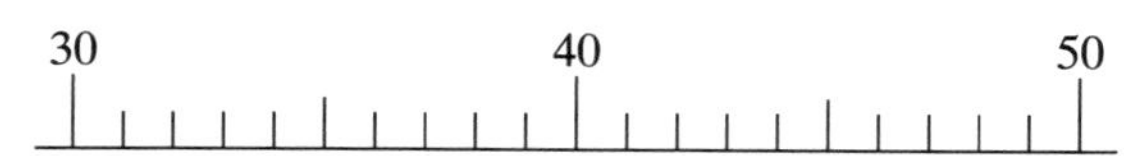

d) Find the number 6.7 on the number line.
On a copy mark it with an arrow (↑).

[Edexcel]

3.7 Solving money problems

EXAMPLE

A calculator costs seven pounds and five pence.

a) Write down the cost of the calculator in figures.

b) Work out the cost of 16 calculators.

SOLUTION

a) £7.05

b) You need to work out 16 × £7.05

It is easier to work in pence: £7.05 = 705p

$$\begin{array}{r} 7\ 0\ 5 \\ \times \quad 1\ 6 \\ \hline 7\ 0\ 5\ 0 \\ \text{Add} \quad 4\ 2\ 3_3\ 0 \\ \hline 1\ 1\ 2\ 8\ 0 \end{array}$$

See page 11 for a reminder of long multiplication.

So 16 calculators cost 11280p = £112.80

EXAMPLE

Tom goes shopping.
He buys:

2 bottles of lemonade at 57p each
3 packets of crisps at 24p each
1 chocolate bar for 49p

How much change does Tom get from £5?

SOLUTION

You need to work out how much Tom has spent first:

Lemonade 2 × 57p = 114p
Crisps 3 × 24p = 72p
Chocolate 1 × 49p = 49p

Keep answers in pence.

The total cost is 114p + 72p + 49p = 235p

Now work out 500p − 235p = 265p

Tom gets £2.65 change.

EXAMPLE

Beth is an electrician.
She charges:

£47.50 for the first hour she works
plus £34.60 for every additional hour.

She charges one customer £255.10

How many hours has she worked?

SOLUTION

£255.10 − £47.50 = £207.60

Work out how much Beth charges for any work after the first hour.

You need to work out how many times £34.60 goes into £207.60

$$207.60 \div 34.60 = 6$$

So Beth has worked for 7 hours

6 hours plus the first hour

EXERCISE 3.7

Don't use your calculator for Questions 1–5.

1 Shannon buys 3 bags of crisps for 45p each.
  a) How much does she spend?
  b) How much change does she get from £2?

2 David buys 3 CDs and 2 DVDs.
How much does he spend?

3 Kim has seven pounds and three pence.
  a) Write down the amount of money Kim has in figures.

  Kim spends £3.50
  b) How much money does Kim have left?

4 A book costs £5.99
Work out the cost of 6 of these books.

5 Tim is taking his family to the cinema.
Tim buys cinema tickets for 3 adults and 4 children.

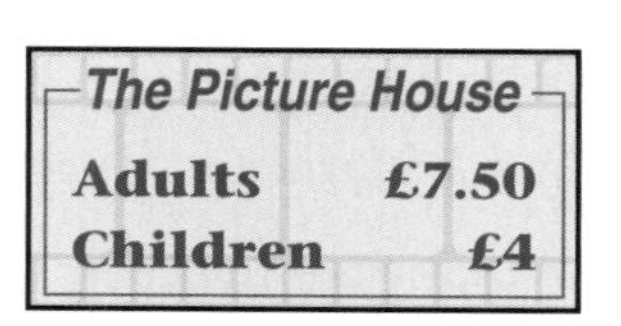

a) How much does he spend on tickets?

Cola	£1.60
Large popcorn	£1.80
Ice cream	£1.25

Tim buys 3 colas, 2 large popcorns and 4 ice-creams.
b) How much does Tim spend on food and drinks?
c) How much should the cinema trip cost altogether?

You can use your calculator for Questions 6–8.

6 Fred is a plumber.
He charges:

£85.60 for the first hour he works and

£60.35 for each additional hour.

a) How much should Fred charge for one job lasting 6 hours?

Fred charges £266.65 for another job.
b) How many hours did this job take Fred?

7 Sabra is decorating.
She has hired a steamer to help her remove the wallpaper.
The cost of hiring the steamer was:

£30.50 for the first day and

£22.25 for each extra day.

Sabra's total cost of hiring the steamer was £119.50.
a) For how many days did Sabra hire the steamer?

Amrik hires the steamer for 8 days.
b) How much should it cost to hire the steamer for 8 days?

8 Will has been shopping.
He bought:

2 loaves of bread for 65p each

$\frac{1}{2}$ kilogram of apples for 30p per kilogram

3 tins of baked beans for 35p per tin

4 bars of chocolate for 27p each

How much change should Will get from a £5 note?

REVIEW EXERCISE 3

Don't use your calculator for Questions 1–12.

1 Round the following numbers to:
(i) 1 decimal place **(ii)** 2 decimal places
a) 2.345 **b)** 5.029 **c)** 4.321 **d)** 6.897

2 Place the following numbers in descending order:
a) 5.5, 5.05, 5.55, 5.005, 5.055
b) 4.1, 4.11, 4.21, 4.01, 4.2, 4.12
c) 3.7, 3.8, 3.78, 3.88, 3.07, 3.08

3 Work out:
a) $0.2 + 0.3 - 0.4$
b) $2.89 + 4.7 + 3.24$
c) 5×0.04
d) 3.6×0.02
e) $1 \div 0.02$
f) $0.6 \div 0.3$
g) 3.2×2.55
h) $4.8 \div 0.12$
i) 1.28×3.4
j) 9.6×0.42

4 Ken had one thousand and twenty pounds.
Lisa had eight pounds and six pence.
Write down, in figures, how much money Ken and Lisa each had. **[Edexcel]**

5 Nick takes 26 boxes out of his van.
The weight of each box is 32.9 kg
a) Work out the total weight of the 26 boxes.
Then Nick fills the van with large wooden crates.
The weight of each crate is 69 kg
The greatest weight the van can hold is 990 kg.
b) Work out the greatest number of crates that the van can hold. **[Edexcel]**

6 Natasha has one pound sixty pence.
Her friend, Kelly, has two pounds five pence.
Write down, in figures, how much money Kelly and Natasha each have. **[Edexcel]**

7 The cost of a calculator is £6.79
Work out the cost of 28 of these calculators. **[Edexcel]**

8 Tanya goes shopping.
She buys:

$\frac{1}{2}$ kg of apples at 72p per kg

4 bananas at 24p each

5 kg of potatoes at 25p per kg

She pays with a £5 note. Work out how much change she should get. **[Edexcel]**

9 Fatima bought 48 teddy bears at £9.55 each.
a) Work out the total amount she paid.

Fatima sold all the teddy bears for a total of £696
She sold each teddy bear for the same price.
b) Work out the price at which Fatima sold each teddy bear.

[Edexcel]

10 Round the following numbers to:
(i) 1 significant figure **(ii)** 2 significant figures
a) 4.125 **b)** 0.034 81 **c)** 0.0865 **d)** 0.000 439

11 Use the information that

$13 \times 17 = 221$

to write down the value of:
a) 1.3×1.7 **b)** $22.1 \div 1700$ **[Edexcel]**

12 Using the information that

$97 \times 123 = 11\,931$

write down the value of:
a) 9.7×12.3 **b)** $0.97 \times 123\,000$ **c)** $11\,931 \div 9.7$ **[Edexcel]**

You can use your calculator for Questions 13–16.

13 What is the reading on each of these scales?

a)

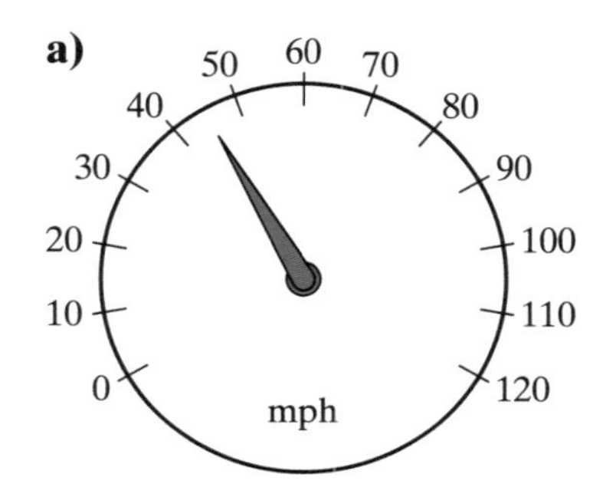

b)

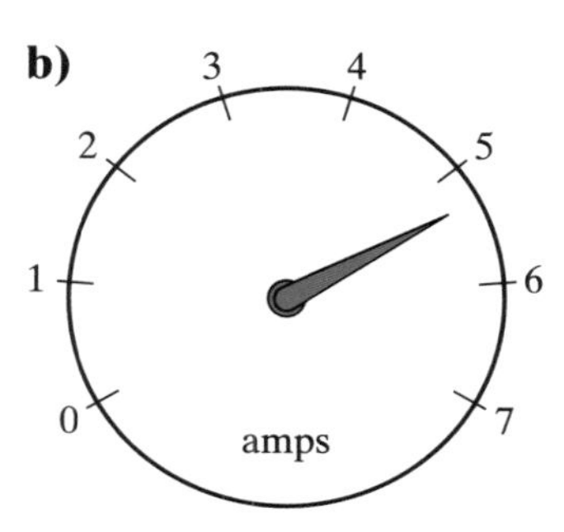

[Edexcel]

14 Christine buys:

a calculator costing £5.95
a pencil case costing £1.62
a ruler costing 25p
two pens costing 48p each

She pays with a £10 note.
a) How much change should she get from her £10 note?

Christine needs 160 tiles for a room.
Tiles are sold in boxes.
There are 12 tiles in each box.
b) Work out the least number of boxes that Christine needs.

Each box of tiles costs £12.20
c) Work out the total cost of the boxes of tiles that Christine needs. **[Edexcel]**

15 Martin cleaned his swimming pool.
He hired a cleaning machine to do this job.
The cost of hiring the cleaning machine was:

£35.50 for the first day
then £18.25 for each extra day.

Martin's total cost of hiring the machine was £163.25
a) For how many days did Martin hire the machine?

Martin filled the pool with 54 000 gallons of water.
He paid £2.38 for each 1000 gallons of water.
b) Work out the total amount he paid for 54 000 gallons of water. **[Edexcel]**

16 The table below shows the cost of each of three calculators.

Compact	£2.30
Studio	£2.15
Basic	£2.80

Barbara buys one Studio calculator and one Compact calculator.
She pays with a £10 note.
a) How much change should she get?

Mrs Brown wants to buy some Basic calculators.
She has £60 to spend.
b) Work out the greatest number of Basic calculators she can buy. **[Edexcel]**

KEY POINTS

1 When ordering decimals add 0's to make all the numbers have the same number of decimal places.

2 When rounding a number to 2 decimal places:
- look at the digit in the 3rd decimal place
- if it is 4 or less then leave the digit in the 2nd decimal place as it is
- if it is 5 or more then round the digit in the 2nd decimal place up.

3 The 1st significant figure of a number is the first non-zero digit.

4 When rounding a number to 2 significant figures:
- underline the first 2 significant figures
- look at the 3rd significant figure
- if it is 4 or less leave the 2nd significant figure as it is
- if it is 5 or more then round the 2nd significant figure up.

5 When adding/subtracting decimal numbers add 0's to make all the numbers have the same number of decimal places.

6 When you multiply 2 decimal numbers, find how many digits there are after the decimal point – there will be the same number of digits after the decimal point in the answer.

7 When you divide a number by a decimal:
- write the sum as a fraction
- multiply the numerator and denominator (top and bottom) by 10, 100... to make the denominator a whole number.

8 When you read a scale, make sure you work out the value of one division.

9 When you are solving money problems, it is often easier to work in pence.
Always write your answers to 2 decimal places.
For example, £4.60 not £4.6

CHAPTER 4

Fractions

In this chapter you will **learn how to:**

- write a fraction as a decimal
- find equivalent fractions
- simplify a fraction
- write a terminating decimal as a fraction
- add, subtract, multiply and divide fractions
- change an improper, or top heavy, fraction to a mixed number
- find a fraction of an amount.

You will also be **challenged to:**

- investigate Egyptian Fractions

Starter: Fractions of shapes

1 What fraction of the following shapes is **(i)** shaded, **(ii)** unshaded?

a) 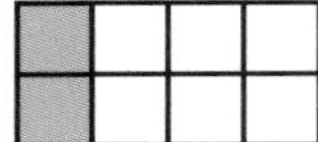

b) 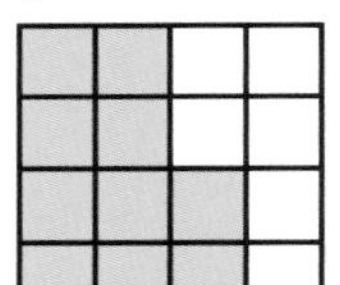

c) 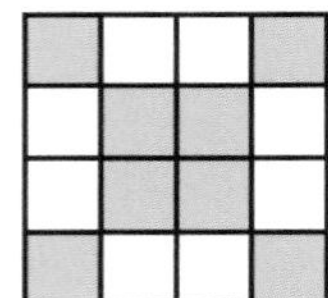

d) 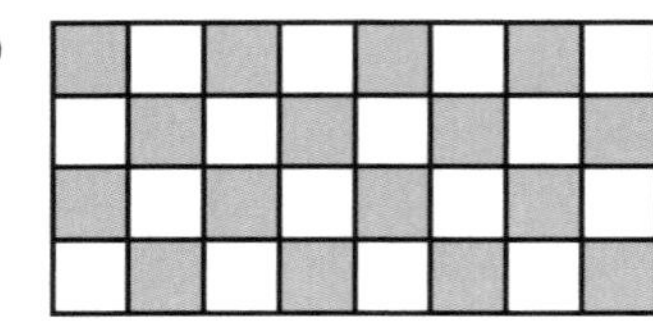

e) 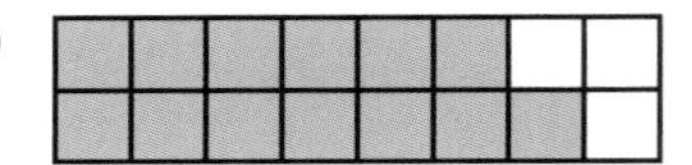

f) 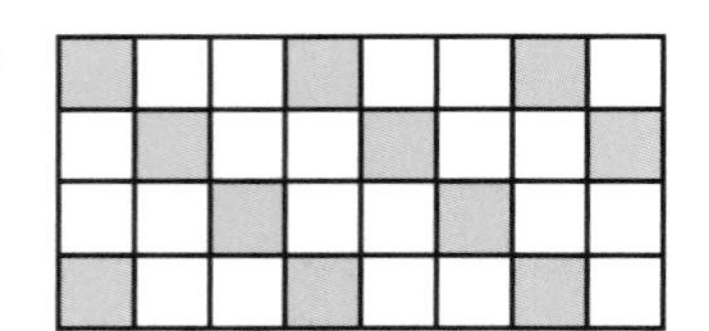

2 Find three different ways of shading in the following fractions of this shape.

a) $\frac{1}{2}$ **d)** $\frac{1}{3}$

$\frac{1}{2}$ is the same as the ratio 1 : 1

c) $\frac{3}{4}$ **d)** $\frac{1}{6}$

$\frac{1}{3}$ is the same as the ratio 1 : 2

e) $\frac{3}{8}$ **f)** $\frac{5}{8}$

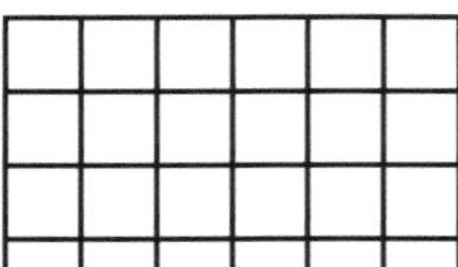

3 $\frac{1}{4}$ of this shape is shaded red.
What is the ratio of red to white squares?

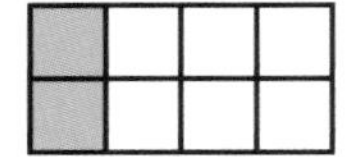

4.1 Writing a fraction as a decimal

$\frac{4}{5}$ means '4 out of 5'.

The top line of a fraction is called the **numerator**.

The bottom line of a fraction is called the **denominator**.

To write a fraction as a decimal, find the:

numerator ÷ denominator

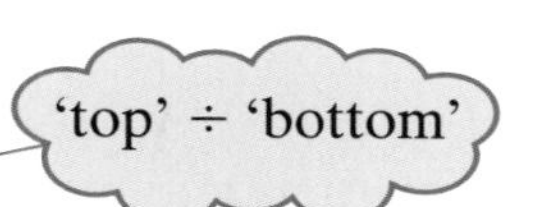

EXAMPLE

Write these fractions as decimals:

a) $\frac{4}{5}$ **b)** $\frac{1}{3}$

SOLUTION

a) $\frac{4}{5} = 4 \div 5 = \underline{0.8}$

b) $\frac{1}{3} = 1 \div 3 = \underline{0.\dot{3}}$

The 'dot' above the 3 shows that the 3's carry on forever. This is called a recurring decimal – because the 3 keeps re-occurring!

EXAMPLE

Write these fractions as decimals:

a) $\frac{3}{5}$ **b)** $\frac{5}{8}$

SOLUTION

a) Use short division: $5\overline{)3.0}$ giving 0.6

5 doesn't go into 3, so put a '0' and a decimal point.

5 goes into 30 **6 times**, so 5 goes into 3.0 **0.6 times.**

So $\frac{3}{5} = \underline{0.6}$

b) $8\overline{)5.0^{2}0^{4}0}$ giving $0.6\ 2\ 5$

8 goes into 50 **6 times, remainder 2**

8 goes into 20 **2 times, remainder 4**

8 goes into 40 **5 times.**

So $\frac{5}{8} = \underline{0.625}$

EXAMPLE

Write these numbers in order of size.
Write the smallest number first.

$\frac{3}{7}$ 0.42 $\frac{2}{5}$ 0.44 $\frac{4}{9}$

SOLUTION

You need to write the fractions as decimals so that you can compare the numbers.

$\frac{3}{7} = 3 \div 7 = 0.428\ldots$

$\frac{2}{5} = 2 \div 5 = 0.4$

$\frac{4}{9} = 4 \div 9 = 0.4\dot{4}$

Use '…' to show that the decimal carries on.

$\frac{1}{7} = 0.142\,857\,142\,857\ldots$ is a recurring decimal.

Now you need to order:

0.428 0.42**0** 0.4**00** 0.44**0** 0.444

Remember: Adding in **0**'s after the decimal point doesn't change the value of the number.

In order of size:

0.400 0.420 0.428 0.440 0.444

So in order of size:

$\frac{2}{5}$ 0.42 $\frac{3}{7}$ 0.44 $\frac{4}{9}$

Write the numbers as they appear in the question.

EXERCISE 4.1

You can use your calculator for questions 1 and 2.

1 Write the following fractions as decimals:

a) $\frac{1}{2}$ **b)** $\frac{3}{4}$ **c)** $\frac{7}{8}$ **d)** $\frac{7}{10}$

e) $\frac{2}{3}$ **f)** $\frac{2}{9}$ **g)** $\frac{8}{15}$ **h)** $\frac{17}{20}$

2 Write the following numbers in order of size.
Write the largest number first:

a) $\frac{4}{5}$ 0.7 $\frac{3}{4}$ $\frac{7}{8}$ 0.72

b) 0.33 $\frac{3}{10}$ $\frac{3}{8}$ $\frac{1}{3}$ 0.303

c) $\frac{11}{15}$ 0.8 0.73 $\frac{7}{10}$ 0.78

d) 0.91 $\frac{6}{7}$ 0.8 $\frac{9}{10}$ $\frac{8}{9}$

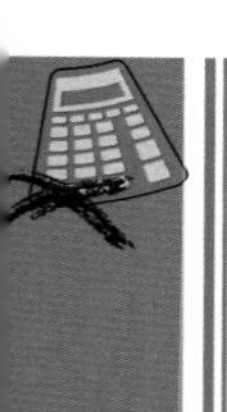

3 Write the following fractions as decimals:

a) $\frac{4}{10}$ **b)** $\frac{1}{5}$ **c)** $\frac{6}{12}$

d) $\frac{1}{8}$ **e)** $\frac{3}{8}$ **f)** $\frac{3}{25}$

g) $\frac{7}{25}$ **h)** $\frac{19}{20}$ **i)** $\frac{9}{20}$

4.2 Equivalent fractions

Two fractions are **equivalent** if they have the same decimal value.

For example, $\frac{3}{5}$ and $\frac{18}{30}$ are equivalent because:

$$\frac{3}{5} = 3 \div 5 = 0.6 \quad \text{and} \quad \frac{18}{30} = 18 \div 30 = 0.6$$

So they both equal 0.6

EXAMPLE

Which of these fractions are equivalent?

$$\frac{3}{5} \quad \frac{7}{10} \quad \frac{12}{20} \quad \frac{21}{30} \quad \frac{24}{40}$$

Work out the decimal value of each fraction.

SOLUTION

$\frac{3}{5} = 3 \div 5 = 0.6$ $\quad$ $\frac{7}{10} = 7 \div 10 = 0.7$

$\frac{12}{20} = 12 \div 20 = 0.6$ $\quad$ $\frac{21}{30} = 21 \div 30 = 0.7$

$\frac{24}{40} = 24 \div 40 = 0.6$

They all equal 0.6

So $\frac{3}{5}, \frac{12}{20}$ and $\frac{24}{40}$ are equivalent.

They both equal 0.7

Also $\frac{7}{10}$ and $\frac{21}{30}$ are equivalent.

numerator

When you multiply the 'top' and 'bottom' of a fraction by the same number, you do not change the value of the fraction.

denominator

Look at the fractions used in the last example.

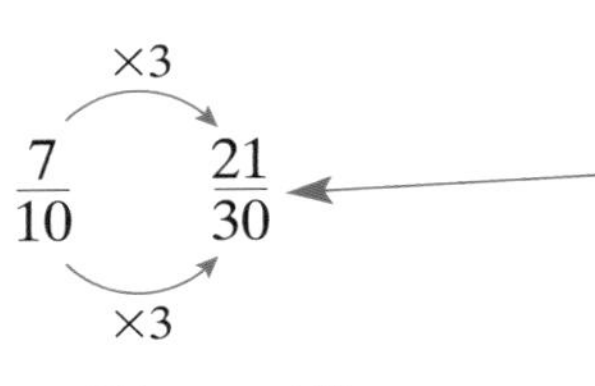

Really you are multiplying by $\frac{3}{3}$ which is 1
When you multiply by 1 the number stays the same.

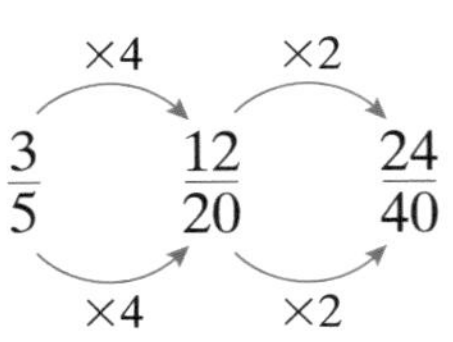

This means they cannot be **simplified**.

$\frac{3}{5}$ and $\frac{7}{10}$ are examples of fractions in their **lowest terms**.

You can **simplify** $\frac{21}{30}$ to $\frac{7}{10}$ by dividing 'top' and 'bottom' of the fraction by 3.

You can also **simplify** $\frac{24}{40}$ and $\frac{12}{20}$ to $\frac{3}{5}$

Sometimes this is called **cancelling**.

EXAMPLE

Simplify:

a) $\frac{36}{48}$ **b)** $\frac{12}{54}$

SOLUTION

a) $\frac{36}{48} \xrightarrow{\div 2} \frac{18}{24} \xrightarrow{\div 2} \frac{9}{12} \xrightarrow{\div 3} \frac{3}{4}$

To simplify the fraction in one step:
Divide both 'top' and 'bottom' by 12
12 is the largest number that divides into both 36 and 48

$\frac{3}{4}$ can't be simplified any further as there is no number (other than 1) that divides exactly into both 3 and 4

So $\frac{36}{48}$ simplifies to $\underline{\frac{3}{4}}$

b)

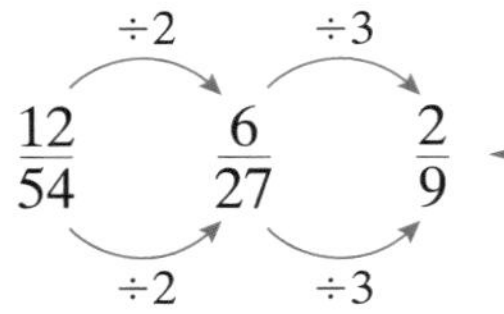

To simplify the fraction in one step: divide both 'top' and 'bottom' by 6.
6 is the largest number that divides into both 12 and 54

So $\frac{12}{54}$ simplifies to $\underline{\frac{2}{9}}$

EXAMPLE

Write these fractions in order of size:

$\frac{7}{12}$ $\frac{1}{2}$ $\frac{2}{3}$ $\frac{5}{6}$ $\frac{3}{4}$

SOLUTION

Write all fractions over the same denominator:

'bottom'

$$\frac{7}{12} \quad \frac{1}{2} = \frac{6}{12} \ (\times 6) \quad \frac{2}{3} = \frac{8}{12} \ (\times 4) \quad \frac{5}{6} = \frac{10}{12} \ (\times 2) \quad \frac{3}{4} = \frac{9}{12} \ (\times 3)$$

2, 3, 4, 6 and 12 all go into 12

So order of size: $\frac{6}{12}$ $\frac{7}{12}$ $\frac{8}{12}$ $\frac{9}{12}$ $\frac{10}{12}$

$\frac{1}{2}$ $\frac{7}{12}$ $\frac{2}{3}$ $\frac{3}{4}$ $\frac{5}{6}$

Write these fractions as they appear in the question.

EXERCISE 4.2

1 Match together equivalent fractions.

$\frac{2}{3}$ $\frac{8}{12}$ $\frac{3}{8}$ $\frac{3}{7}$ $\frac{60}{140}$ $\frac{21}{56}$ $\frac{12}{32}$ $\frac{15}{35}$ $\frac{10}{15}$

Don't use your calculator for Questions 2 to 4.

2 Write these fractions in their simplest form:

a) $\frac{25}{50}$ **b)** $\frac{15}{20}$ **c)** $\frac{4}{16}$ **d)** $\frac{3}{27}$

e) $\frac{30}{35}$ **f)** $\frac{16}{20}$ **g)** $\frac{135}{200}$ **h)** $\frac{250}{400}$

3 Find the missing values in these equivalent fractions:

a) $\frac{1}{3} = \frac{?}{6} = \frac{?}{9} = \frac{4}{?} = \frac{5}{?}$ **b)** $\frac{3}{8} = \frac{9}{?} = \frac{15}{?} = \frac{?}{56}$

c) $\frac{4}{5} = \frac{?}{10} = \frac{?}{25} = \frac{28}{?} = \frac{40}{?}$ **d)** $\frac{2}{3} = \frac{?}{12} = \frac{?}{21} = \frac{20}{?} = \frac{30}{?}$

4 Write these fractions in order of size:

a) $\frac{2}{3}$ $\frac{5}{6}$ $\frac{4}{12}$

b) $\frac{5}{8}$ $\frac{1}{2}$ $\frac{3}{4}$ $\frac{3}{8}$

c) $\frac{7}{10}$ $\frac{1}{5}$ $\frac{3}{10}$ $\frac{4}{5}$

d) $\frac{7}{18}$ $\frac{1}{3}$ $\frac{1}{6}$ $\frac{5}{9}$

e) $\frac{3}{4}$ $\frac{16}{20}$ $\frac{3}{5}$ $\frac{7}{10}$ $\frac{1}{2}$

f) $\frac{7}{12}$ $\frac{3}{4}$ $\frac{2}{3}$ $\frac{19}{24}$ $\frac{5}{6}$

4.3 Writing a terminating decimal as a fraction

0.4, 0.825 and 0.1234 are all examples of **terminating decimals**. Any terminating decimal can be written as a fraction.

They 'end' – they have a certain number of decimal places and don't continue forever.

EXAMPLE

Write 0.235 as a fraction in its simplest form.

SOLUTION

Write 0.235 as a fraction: $0.235 = \frac{0.235}{1}$

You can divide a number by 1 without changing its value.

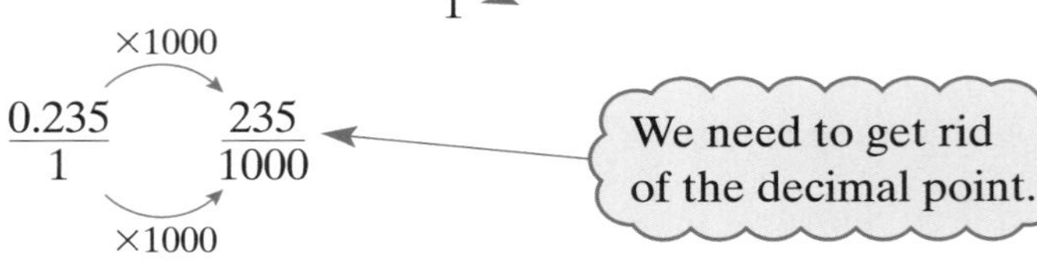

×1000
0.235 / 1
235 / 1000
×1000
We need to get rid of the decimal point.

Simplifying:

$$\frac{235}{1000} \xrightarrow{\div 5} \frac{47}{200}$$

Check: 47 ÷ 200 = 0.235 ✓

So $0.235 = \frac{47}{200}$

EXERCISE 4.3

Write the following decimals as fractions.
Write each fraction in its simplest form:

1 0.5	**2** 0.75	**3** 0.2
4 0.9	**5** 0.95	**6** 0.35
7 0.825	**8** 0.455	**9** 0.128

4.4 Adding and subtracting fractions

To add or subtract fractions with the same (or **common**) denominator you simply add together the numerators of both fractions.

So $\frac{\mathbf{5}}{8} + \frac{\mathbf{2}}{8} = \frac{\mathbf{7}}{8}$

Or in words:

'five eighths plus two eighths equals seven eighths'.

Or in a diagram:

Also $\frac{7}{12} - \frac{5}{12} = \frac{2}{12}$

Or in words:

'seven twelfths minus five twelfths equals two twelfths'.

Or in a diagram:

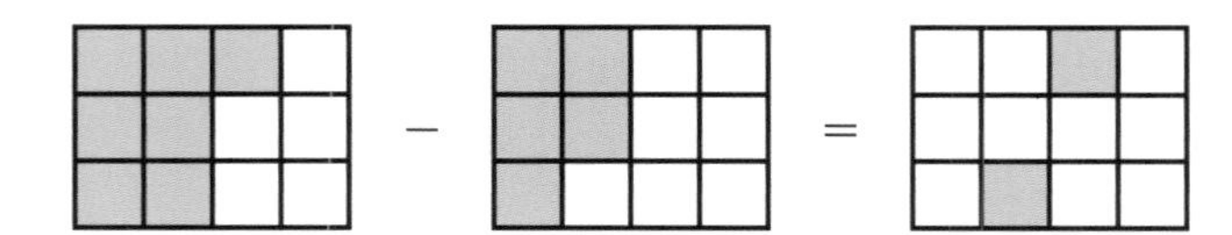

To add or subtract fractions without the same denominator you need to find equivalent fractions with the same denominator.

'bottom'

EXAMPLE

Work out $\frac{2}{3} + \frac{1}{5}$

$3 \times 5 = 15$
So 3 and 5 go into 15

SOLUTION

Find a number that both 3 and 5 go into.

Now rewrite both fractions so they are both 'some number' over 15

$\frac{2}{3} = \frac{?}{15}$ so $\frac{2}{3} \xrightarrow{\times 5} \frac{10}{15}$ (numerator $\times 5$, denominator $\times 5$)

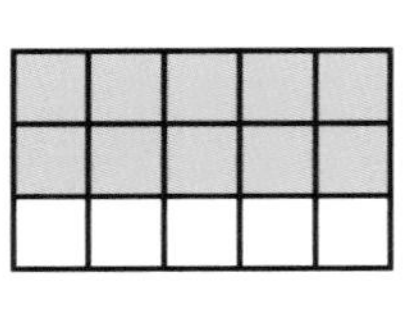

+

$\frac{1}{5} = \frac{?}{15}$ so $\frac{1}{5} \xrightarrow{\times 3} \frac{3}{15}$ (numerator $\times 3$, denominator $\times 3$)

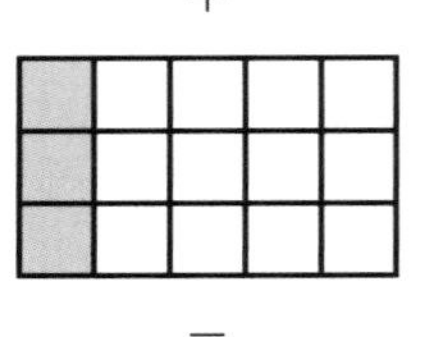

=

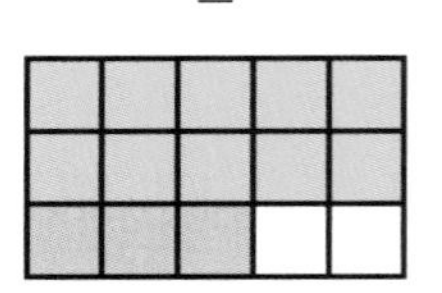

$\frac{10}{15} + \frac{3}{15} = \frac{13}{15}$

So $\frac{2}{3} + \frac{1}{5} + \underline{\frac{13}{15}}$

EXAMPLE

Work out $\frac{5}{6} - \frac{2}{3}$

6 and 3 both go into 6

SOLUTION

Find a number that both 6 and 3 go into.

Now rewrite both fractions so they are both 'some number' over 6.

For $\frac{2}{3} = \frac{?}{6}$ so $\frac{2}{3} \xrightarrow{\times 2} \frac{4}{6}$ (numerator $\times 2$, denominator $\times 2$)

$\frac{5}{6} - \frac{4}{6} = \frac{1}{6}$

So $\frac{5}{6} - \frac{2}{3} = \underline{\frac{1}{6}}$

EXAMPLE

Work out $\frac{1}{2} + \frac{1}{4} - \frac{2}{5}$

SOLUTION

Find a number that 2, 4 and 5 all go into.

2, 4 and 5 go into 20

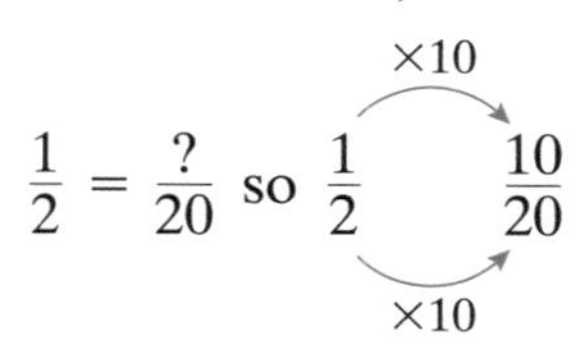

$\frac{1}{2} = \frac{?}{20}$ so $\frac{1}{2} \xrightarrow{\times 10} \frac{10}{20}$ (×10)

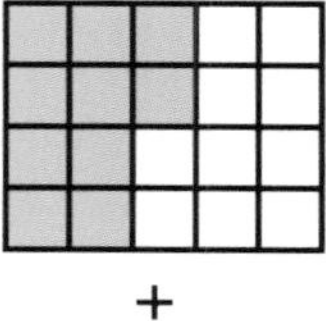

+

$\frac{1}{4} = \frac{?}{20}$ so $\frac{1}{4} \xrightarrow{\times 5} \frac{5}{20}$ (×5)

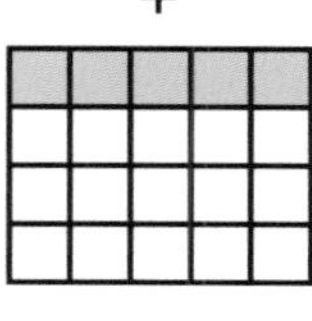

−

$\frac{2}{5} = \frac{?}{20}$ so $\frac{2}{5} \xrightarrow{\times 4} \frac{8}{20}$ (×4)

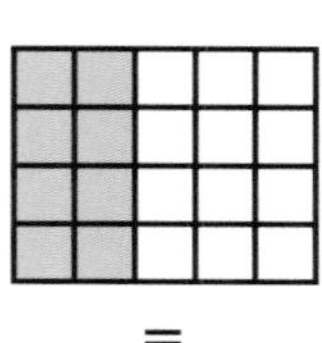

=

$\frac{10}{20} + \frac{5}{20} - \frac{8}{20} = \frac{7}{20}$ so $\frac{1}{2} + \frac{1}{4} - \frac{2}{5} = \underline{\frac{7}{20}}$

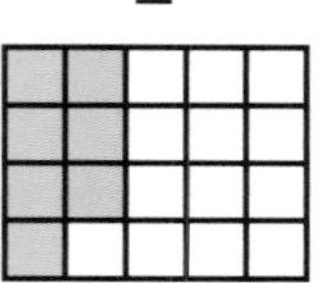

EXERCISE 4.4

1 Work out the following – give your answers as a fraction in its simplest form:

a) $\frac{1}{3} + \frac{1}{3}$ **b)** $\frac{3}{7} + \frac{2}{7}$ **c)** $\frac{11}{17} + \frac{3}{17} + \frac{1}{17}$

d) $\frac{2}{4} + \frac{1}{4}$ **e)** $\frac{1}{8} + \frac{5}{8}$ **f)** $\frac{7}{12} + \frac{1}{12}$

2 Work out the following – give your answers as a fraction in its simplest form:

a) $\frac{4}{5} - \frac{1}{5}$ **b)** $\frac{8}{9} - \frac{3}{9}$ **c)** $\frac{19}{24} - \frac{5}{24} - \frac{3}{24}$

d) $\frac{3}{4} - \frac{1}{4}$ **e)** $\frac{8}{10} - \frac{2}{10}$ **f)** $\frac{7}{18} - \frac{1}{18}$

3 Work out the following – give your answers as a fraction in its simplest form:

a) $\frac{7}{10} + \frac{1}{5}$ **b)** $\frac{3}{4} - \frac{1}{2}$ **c)** $\frac{1}{4} + \frac{1}{12}$

d) $\frac{7}{15} - \frac{1}{5}$ **e)** $\frac{7}{20} + \frac{2}{5}$ **f)** $\frac{14}{18} - \frac{1}{9}$

g) $\frac{1}{2} - \frac{1}{6}$ **h)** $\frac{3}{4} - \frac{5}{12}$ **i)** $\frac{3}{7} + \frac{5}{14}$

4 Work out the following – give your answers as a fraction in its simplest form:

a) $\frac{1}{2}+\frac{1}{3}$ **b)** $\frac{3}{4}-\frac{1}{5}$ **c)** $\frac{1}{6}+\frac{2}{5}$ **d)** $\frac{1}{2}-\frac{1}{3}$

e) $\frac{3}{7}+\frac{1}{8}$ **f)** $\frac{2}{3}-\frac{1}{4}$ **g)** $\frac{7}{8}-\frac{3}{5}$ **h)** $\frac{1}{4}+\frac{1}{3}$

i) $\frac{2}{5}-\frac{1}{4}$ **j)** $\frac{1}{2}+\frac{1}{3}-\frac{1}{6}$ **k)** $\frac{2}{5}+\frac{1}{3}-\frac{1}{4}$ **l)** $\frac{1}{2}+\frac{1}{3}-\frac{1}{4}$

5 Maria spends $\frac{2}{5}$ of her pocket money on a CD.

She spends $\frac{1}{5}$ of her pocket money on make-up.

Maria saves the rest of her pocket money.

a) What fraction of her pocket money does Maria spend?

b) What fraction of her pocket money does Maria save?

If you have a fraction button a^b/c on your calculator, you can use it to check your answers.

6 Phil spends $\frac{1}{8}$ of his pocket money on magazines.

He spends $\frac{2}{3}$ of his pocket money on a computer game.

Phil saves the rest of his pocket money.

a) What fraction of his pocket money does Phil spend?

b) What fraction of his pocket money does Phil save?

4.5 Improper fractions and mixed numbers

$\frac{11}{6}$ is an **improper** or **top heavy fraction** because the numerator is larger than the denominator.

The 'top' (11) is larger than the 'bottom' (6).

$1\frac{5}{6}$ is called a **mixed number** because it has a whole number and a fractional part.

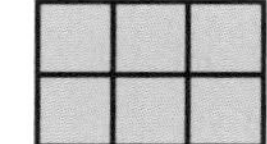 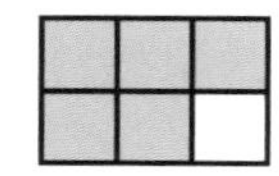

This represents $1\frac{5}{6}$ or $\frac{11}{6}$

EXAMPLE

Change $2\frac{3}{5}$ to an improper fraction.

SOLUTION

$2\frac{3}{5} = 2 + \frac{3}{5}$

$2\frac{3}{5}$ means 2 wholes and 3 fifths.

We need to write the '2' as a fraction, so $2 = \frac{10}{5}$

There are 5 fifths in 1 whole, so there are 10 fifths in 2 wholes.

$\frac{10}{5}+\frac{3}{5}=\frac{13}{5}$

So $2\frac{3}{5} = \frac{13}{5}$

EXAMPLE

Write $\frac{19}{8}$ as a mixed number.

SOLUTION

$$\frac{19}{8} = \frac{8}{8} + \frac{8}{8} + \frac{3}{8}$$

$$= 1 + 1 + \frac{3}{8}$$

$$= \underline{2\frac{3}{8}}$$

$8 + 8 + 3 = 19$

Or work out $19 \div 8 = 2$ remainder 3. So there are 2 wholes and 3 eighths.

EXAMPLE

Work out $2\frac{3}{4} - 1\frac{2}{3}$

SOLUTION

Write both fractions as top heavy fractions.

$$2\frac{3}{4} = 2 + \frac{3}{4} = \frac{8}{4} + \frac{3}{4} = \frac{11}{4}$$

$$1\frac{2}{3} = 1 + \frac{2}{3} = \frac{3}{3} + \frac{2}{3} = \frac{5}{3}$$

So you need to work out $\frac{11}{4} - \frac{5}{3}$

Now 4 and 3 both go into 12.

$$\frac{11}{4} = \frac{?}{12} \text{ so } \frac{11}{4} \xrightarrow{\times 3} \frac{33}{12}$$

$$\frac{5}{3} = \frac{?}{12} \text{ so } \frac{5}{3} \xrightarrow{\times 4} \frac{20}{12}$$

To solve this problem, you could find $2 - 1 = 1$ and $\frac{3}{4} - \frac{2}{3} = \frac{1}{12}$ and add the answers to get $1\frac{1}{12}$

So $\frac{11}{4} - \frac{5}{3} = \frac{33}{12} - \frac{20}{12} = \frac{13}{12}$

and $\frac{13}{12} = \frac{12}{12} + \frac{1}{12} = 1\frac{1}{12}$

Write the fraction as a **mixed number**.

So $2\frac{3}{4} - 1\frac{2}{3} = \underline{1\frac{1}{12}}$

EXERCISE 4.5

1 Write the following as top heavy fractions:

a) $1\frac{1}{4}$ **b)** $1\frac{2}{3}$ **c)** $1\frac{2}{7}$ **d)** $5\frac{1}{2}$

e) $3\frac{3}{4}$ **f)** $4\frac{5}{6}$ **g)** $2\frac{5}{8}$ **h)** $4\frac{5}{9}$

If you have a fraction button a^b/c on your calculator, you can use it to check your answers.

2 Write the following as mixed numbers:

a) $\frac{7}{2}$ **b)** $\frac{7}{3}$ **c)** $\frac{9}{4}$ **d)** $\frac{13}{7}$

e) $\frac{8}{3}$ **f)** $\frac{22}{9}$ **g)** $\frac{25}{4}$ **h)** $\frac{17}{4}$

3 Work out the following – give your answers as a fraction in its simplest form:

a) $1\frac{1}{3} + 3\frac{1}{3}$ **b)** $5\frac{1}{2} - 3\frac{1}{2}$ **c)** $2\frac{3}{7} + 1\frac{1}{7}$

d) $2\frac{3}{4} - 1\frac{1}{4}$ **e)** $3\frac{2}{5} + 1\frac{4}{5}$ **f)** $3\frac{1}{5} - 2\frac{3}{5}$

4 Work out the following – give your answers as a fraction in its simplest form:

a) $1\frac{1}{6} + 1\frac{1}{3}$ **b)** $1\frac{3}{10} - \frac{1}{5}$ **c)** $3\frac{3}{8} - 2\frac{1}{4}$

d) $2\frac{3}{4} - 1\frac{1}{2}$ **e)** $2\frac{1}{2} + 1\frac{1}{3}$ **f)** $1\frac{1}{2} + 2\frac{2}{3}$

g) $3\frac{3}{4} - 2\frac{1}{3}$ **h)** $4\frac{3}{7} - 3\frac{2}{5}$ **i)** $5\frac{1}{3} - 3\frac{1}{4}$

4.6 Multiplying and dividing fractions

EXAMPLE

Work out **a)** $\frac{3}{4} \times \frac{5}{7}$ **b)** $1\frac{2}{5} \times \frac{2}{3}$

SOLUTION

a) $\frac{3}{4} \times \frac{5}{7} = \frac{3 \times 5}{4 \times 7} = \underline{\frac{15}{28}}$

Multiply the 'tops' and then multiply the 'bottoms'.

b) $1\frac{2}{5} = \frac{7}{5}$

$\frac{7}{5} \times \frac{2}{3} = \frac{7 \times 2}{5 \times 3}$

$= \underline{\frac{14}{15}}$

Sometimes you can cancel the fractions before you multiply them.

EXAMPLE

Work out:

a) $\frac{2}{3} \times \frac{3}{5}$ **b)** $\frac{5}{12} \times \frac{8}{9}$

SOLUTION

a) $\frac{2}{3} \times \frac{3}{5} = \frac{2}{\not{3}} \times \frac{\not{3}}{5} = \underline{\frac{2}{5}}$

The 3's cancel as you are multiplying by 3 and then dividing by 3

b) $\frac{5}{12} \times \frac{8}{9} = \frac{5 \times \overset{2}{\not{8}}}{{}_{3}\not{12} \times 9} = \frac{5 \times 2}{3 \times 9} = \underline{\frac{10}{27}}$

You can divide both the 8 and the 12 by 4

EXAMPLE

Work out: **a)** $\frac{3}{8} \div \frac{2}{5}$ b) $2\frac{2}{3} \div \frac{3}{4}$

SOLUTION

a) $\frac{3}{8} \div \frac{2}{5} = \frac{3}{8} \times \frac{5}{2}$

Turn the second fraction upside down and multiply. You have found the reciprocal of $\frac{2}{5}$ see page 115

$\frac{3}{8} \times \frac{5}{2} = \frac{3 \times 5}{8 \times 2} = \frac{15}{16}$

$\frac{3}{8} \div \frac{2}{5} = \underline{\frac{15}{16}}$

b) $2\frac{2}{3} = \frac{8}{3}$

Change to improper fraction.

$\frac{8}{3} \div \frac{3}{4} = \frac{8}{3} \times \frac{4}{3} = \frac{32}{9}$

$\frac{32}{9} = \underline{3\frac{5}{9}}$

EXERCISE 4.6

1 Work out the following – give your answers as a fraction in its simplest form:

a) $\frac{2}{3} \times \frac{4}{5}$ **b)** $\frac{2}{9} \times \frac{5}{7}$ **c)** $\frac{1}{2} \times \frac{1}{3}$

d) $\frac{2}{5} \times \frac{3}{10}$ **e)** $1\frac{3}{4} \times \frac{5}{8}$ **f)** $2\frac{2}{7} \times \frac{7}{9}$

If you have a fraction button a^b/c on your calculator, you can use it to check your answers.

2 Work out the following – give your answers as a fraction in its simplest form:

a) $\frac{2}{3} \div \frac{4}{5}$ **b)** $\frac{2}{7} \div \frac{3}{4}$ **c)** $\frac{1}{3} \div \frac{1}{2}$

d) $1\frac{1}{2} \div \frac{1}{3}$ **e)** $\frac{4}{5} \div \frac{5}{4}$ **f)** $2\frac{3}{4} \div \frac{3}{5}$

3 Work out the following – give your answers as a fraction in its simplest form:

a) $\frac{1}{2} \times \frac{1}{4}$ **b)** $\frac{2}{3} \times \frac{5}{9}$ **c)** $\frac{3}{4} \div \frac{1}{2}$ **d)** $\frac{5}{9} \div \frac{1}{3}$

e) $\frac{4}{5} \times \frac{10}{12}$ **f)** $\frac{1}{5} \div \frac{1}{2}$ **g)** $\frac{3}{12} \div \frac{3}{4}$ **h)** $\frac{4}{12} \times \frac{8}{20}$

i) $1\frac{1}{2} \times \frac{2}{3}$ **j)** $1\frac{1}{2} \div \frac{2}{3}$ **k)** $2\frac{1}{2} \div \frac{1}{3}$ **l)** $1\frac{3}{5} \div \frac{1}{2}$

4.7 Finding a fraction of an amount

EXAMPLE

Find:

a) $\frac{1}{7}$ of 63

b) $\frac{4}{9}$ of 54

SOLUTION

a) $\frac{1}{7}$ of $63 = \frac{1}{7} \times 63$

$= \underline{9}$

'of' means '×'

$\frac{1}{7} \times 63 = \frac{63}{7} = 9$

b) Find $\frac{1}{9}$ of 54 first.

$\frac{1}{9}$ of $54 = \frac{1}{9} \times 54 = 6$

So $\frac{4}{9}$ of $54 = \underline{24}$

$\frac{4}{9}$ is 4 times more than $\frac{1}{9}$ and $4 \times 6 = 24$

EXAMPLE

Work out $\frac{3}{5}$ of £72

SOLUTION

$\frac{3}{5} \times 72 = \frac{3 \times 72}{5}$

$= \frac{216}{5}$

$= £43.2$

$\frac{3}{5}$ of £72 = $\underline{£43.20}$

Always give money amounts to two decimal places.

EXAMPLE

a) The normal price of a television is £650
What is the sale price?

b) The sale price of a DVD player is £96.66
What is its normal price?

Electric Ed's
MASSIVE SALE!
EVERYTHING
$\frac{1}{3}$ OFF!

SOLUTION

a) **Method 1**

Work out $\frac{1}{3}$ of £650

$\frac{1}{3} \times 650 = 216.666\ldots$

$650 - 216.66\ldots = 433.333\ldots$

So the sale price is £433.33

Find $\frac{1}{3}$ of £650 and then take it away from £650.

Don't round until your final answer.

Method 2

The sale price is $\frac{2}{3}$ of the normal price.

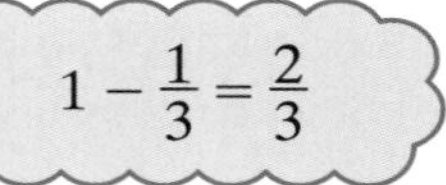

So work out $\frac{2}{3}$ of £650

$\frac{2}{3} \times 650 = 433.333\ldots$

So the sale price is £433.33

This method works out the sale price without the subtraction.

b) The sale price is $\frac{2}{3}$ of the normal price.

So £96.66 = $\frac{2}{3}$ of the normal price

(÷2 on both sides)

£48.33 = $\frac{1}{3}$ of the normal price

(×3 on both sides)

£144.99 = the normal price

So the normal price is £144.99

EXERCISE 4.7

Don't use your calculator for Questions 1–3.

1 Work out:

a) $\frac{1}{3}$ of 27 **b)** $\frac{1}{9}$ of 36 **c)** $\frac{1}{4}$ of 12

d) $\frac{1}{5}$ of 100 **e)** $\frac{1}{6}$ of 60 **f)** $\frac{1}{12}$ of 96

2 Work out:

a) $\frac{3}{5}$ of 20 **b)** $\frac{2}{7}$ of 14 **c)** $\frac{6}{11}$ of 44

d) $\frac{11}{20}$ of 80 **e)** $\frac{7}{9}$ of 81 **f)** $\frac{2}{3}$ of 63

g) $\frac{11}{15}$ of 60 **h)** $\frac{9}{17}$ of 51 **i)** $\frac{5}{12}$ of 144

3 At Northfield School $\frac{1}{20}$ of the students were absent on the last Monday of term. There are 1120 students at the school.

a) How many students were absent on that Monday?
b) What fraction of the students were present on that Monday?
c) How many students were present on that Monday?

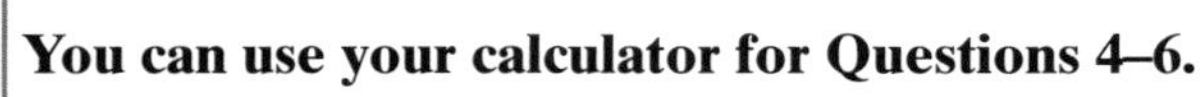

You can use your calculator for Questions 4–6.

4

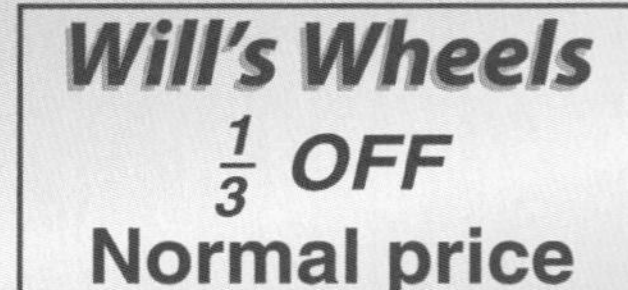

Which bicycle is the cheapest?
Show all your working.

5 a) A CD normally costs £9.60
How much does it cost in the sale?
b) Harry buys a double CD for £10.
What is the normal price of the CD?

6 a) A shirt normally costs £22
How much does it cost in the sale?
b) Jess buys a pair of jeans for £48.75
What is the normal price of the jeans?

Clothing Sale
$\frac{1}{4}$ off all
normal prices

REVIEW EXERCISE 4

Don't use your calculator for Questions 1–16.

1 Write these fractions in their simplest forms:

a) $\frac{20}{40}$ **b)** $\frac{15}{20}$ **c)** $\frac{5}{15}$

d) $\frac{22}{55}$ **e)** $\frac{20}{36}$ **f)** $\frac{36}{40}$

2 Write these decimals as a fraction – write your answers as a fraction in its simplest form:

a) 0.125 **b)** 0.8 **c)** 0.7

d) 0.44 **e)** 0.875 **f)** 0.675

3 Work out:

a) $\frac{1}{4}+\frac{1}{4}$ **b)** $\frac{2}{5}+\frac{1}{5}$ **c)** $\frac{3}{7}+\frac{2}{7}+\frac{1}{7}$

d) $\frac{8}{9}-\frac{4}{9}$ **e)** $\frac{11}{17}-\frac{5}{17}$ **f)** $\frac{13}{15}-\frac{4}{15}-\frac{7}{15}$

4 Work out the following – write your answers as a fraction in its simplest form:

a) $\frac{3}{4}-\frac{1}{3}$ **b)** $\frac{2}{9}+\frac{1}{3}$ **c)** $\frac{2}{5}-\frac{3}{8}$

d) $\frac{3}{7}+\frac{1}{4}$ **e)** $\frac{5}{6}-\frac{1}{2}$ **f)** $\frac{7}{9}-\frac{2}{3}$

5 Work out the following – write your answers as a fraction in its simplest form:

a) $\frac{3}{4}\times\frac{2}{3}$ **b)** $\frac{2}{7}\times\frac{3}{4}$ **c)** $\frac{3}{5}\times\frac{5}{6}$

d) $\frac{3}{8}\times\frac{1}{4}$ **e)** $\frac{1}{6}\div\frac{1}{2}$ **f)** $\frac{7}{12}\div\frac{2}{3}$

6 Write the following as mixed numbers:

a) $\frac{7}{4}$ **b)** $\frac{9}{5}$ **c)** $\frac{12}{7}$ **d)** $\frac{18}{5}$

7 Write the following as improper fractions:

a) $1\frac{4}{9}$ **b)** $2\frac{2}{3}$ **c)** $3\frac{1}{7}$ **d)** $2\frac{5}{6}$

8 Work out the following – write your answers as a fraction in its simplest form:

a) $1\frac{1}{3}+1\frac{1}{3}$ **b)** $2\frac{3}{4}-1\frac{1}{2}$ **c)** $2\frac{3}{4}-1\frac{1}{5}$

9 Work out:

a) $\frac{1}{5}$ of 30 **b)** $\frac{3}{5}$ of 80 **c)** $\frac{3}{4}\times 24$

d) $\frac{5}{9}\times 45$ **e)** $24\times\frac{1}{8}$ **f)** $72\times\frac{5}{12}$

10 Write these numbers in order of size – start with the smallest number:

a) 75, 56, 37, 9, 59 **b)** 0.56, 0.067, 0.6, 0.65, 0.605

c) 5, −6, −10, 2, −4 **d)** $\frac{1}{2}, \frac{2}{3}, \frac{2}{5}, \frac{3}{4}$ [Edexcel]

11 a) Write down the fraction of this shape that is shaded.
Write your fraction in its simplest form.

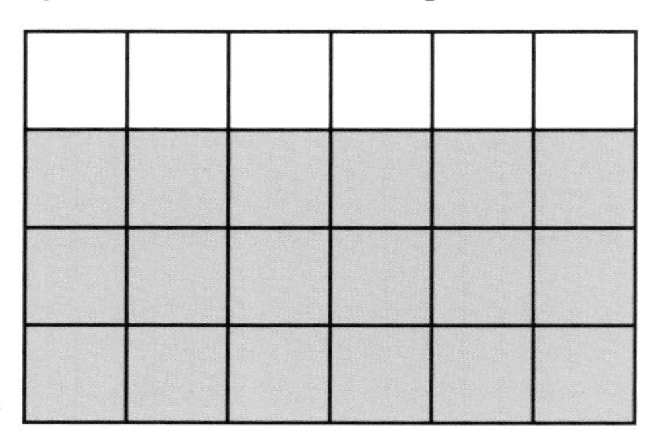

b) On a copy, shade $\frac{2}{3}$ of this shape.

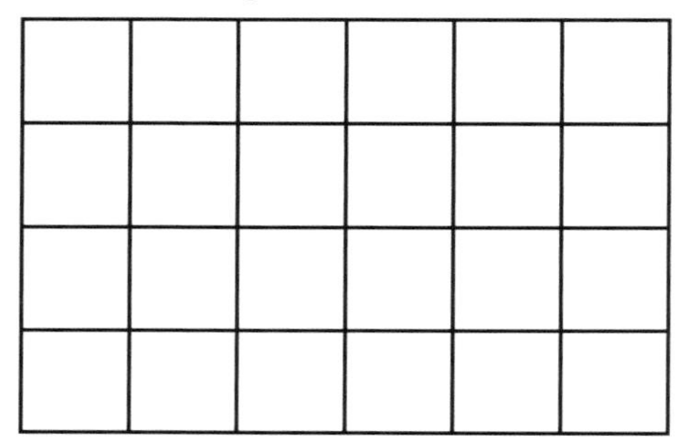

[Edexcel]

12 Here are two fractions: $\frac{3}{5}$ and $\frac{2}{3}$
Explain which is the larger fraction.
You may use a copy of the grids to help with your explanation.

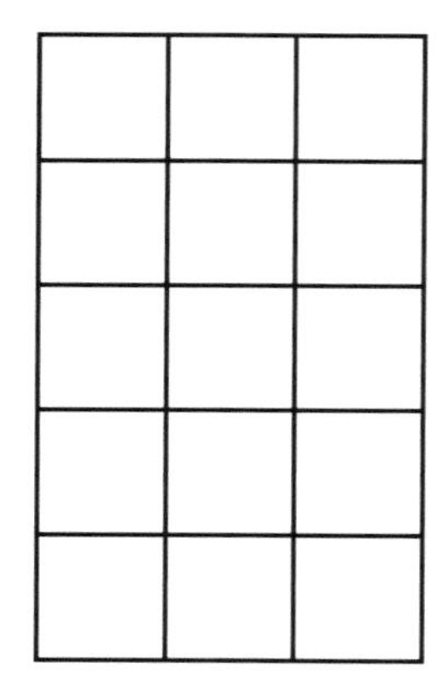

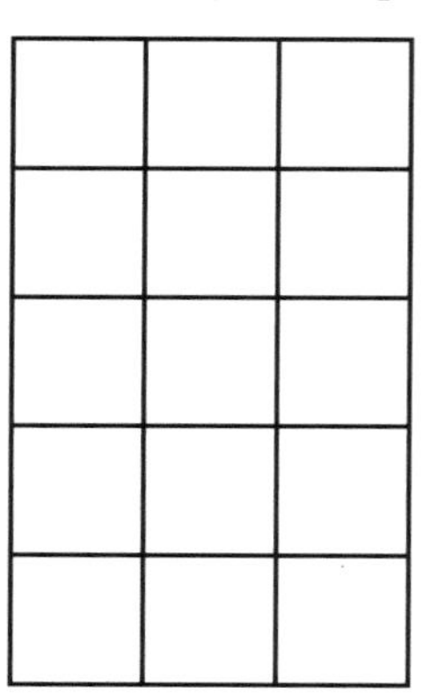

[Edexcel]

13

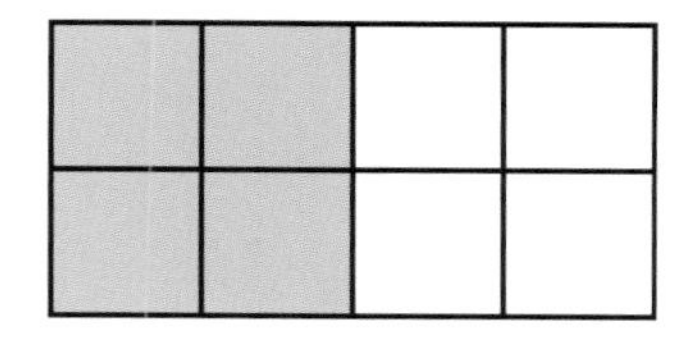

a) i) Write down the fraction of this shape that is shaded.
Write your fraction in its simplest form.

ii) On a copy shade $\frac{1}{4}$ of this shape.

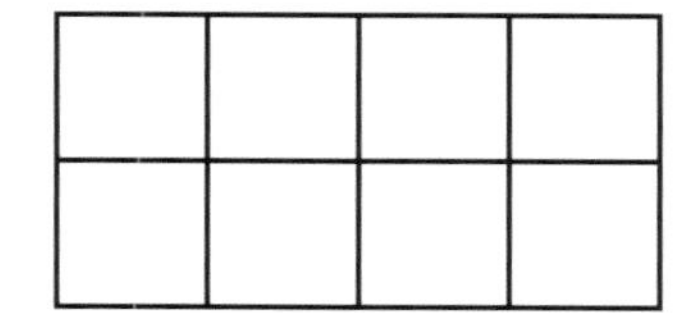

9 is the number that is half way between 6 and 12.

6……9……12

b) Work out the number which is halfway between:

i) 20 …………….. 60

ii) 100 000 ……………… 200 000

iii) 6.5 ………………… 6.6

iv) $\frac{1}{4}$ …………………… $\frac{1}{2}$ **[Edexcel]**

14 Simon spent $\frac{1}{3}$ of his pocket money on a computer game.

He spent $\frac{1}{4}$ of his pocket money on a ticket for a football match.

Work out the fraction of his pocket money that he had left. **[Edexcel]**

15 a) Work out $\frac{11}{12} - \frac{5}{6}$

b) Estimate the value of $\frac{68 \times 401}{198}$ **[Edexcel]**

16 a) Work out the value of $\frac{2}{3} \times \frac{3}{4}$

Give your answer as a fraction in its simplest form.

b) Work out the value of $1\frac{2}{3} + 2\frac{3}{4}$

Give your answer as a fraction in its simplest form. **[Edexcel]**

You can use your calculator for Questions 17–20.

17 Write these fractions as decimals:

a) $\frac{1}{2}$ **b)** $\frac{1}{4}$ **c)** $\frac{3}{4}$ **d)** $\frac{3}{8}$ **e)** $\frac{2}{3}$ **f)** $\frac{5}{9}$

18

0 $\frac{1}{4}$ $\frac{1}{2}$ $\frac{3}{4}$ Full

The diagram shows the measuring scale on a petrol tank.

a) What fraction of the tank is empty?

The petrol tank holds 28 litres when full.
A litre of petrol cost 74p

b) Work out the cost of the petrol which has to be added to the tank so that it is full.

[Edexcel]

19 a) Write down these fractions in order of size – start with the smallest fraction:

$$\frac{3}{4} \quad \frac{1}{2} \quad \frac{3}{8} \quad \frac{2}{3} \quad \frac{1}{6}$$

b) Write down these numbers in order of size – start with the smallest number:

$$0.65 \quad \frac{3}{4} \quad 0.72 \quad \frac{2}{3} \quad \frac{3}{5}$$

[Edexcel]

20 Alison travels by car to her meetings.
Alison's company pay her 32p for each mile she travels.

One day Alison writes down the distance readings from her car:

Start of the day: *2430 miles*
End of the day: *2658 miles*

a) Work out how much the company pays Alison for her day's travel.

The next day Alison travelled a total of 145 miles.

She travelled $\frac{2}{5}$ of this distance in the morning.

b) How many miles did she travel during the rest of the day?

[Edexcel]

KEY POINTS

1 To write a fraction as a decimal you divide the 'top' of the fraction by the 'bottom'.

2 To compare decimals and fractions:

- change all the fractions to decimals, *or*
- write all the fractions over a common denominator ('bottom').

3 Any terminating decimal $\left(\text{e.g. } 0.23 = \frac{23}{100}\right)$ can be written as a fraction.

4 When you multiply or divide the 'top' *and* 'bottom' of a fraction by a number you don't change the value of the fraction.

5 Equivalent fractions have the same decimal value.

6 To simplify a fraction:

- find the largest number that goes into *both* the 'top' *and* the 'bottom' of the fraction
- divide both the 'top' and 'bottom' of the fraction by this number.

7 To add/subtract fractions:

- make sure they have the same number at the 'bottom' (common denominator)
- add/subtract the 'top' numbers.

8 To multiply fractions, multiply the 'top' numbers together and then the 'bottom' numbers.

9 To divide fractions:

- turn the 2nd fraction upside down
- multiply.

10 Mixed numbers can be written as top-heavy fractions.

11 To find a fraction of an amount:

- divide by the 'bottom' of the fraction, and
- multiply by the 'top'.

Internet Challenge 4 

Egyptian Fractions

It is thought that the ancient Egyptians only used unit fractions like $\frac{1}{2}, \frac{1}{3}, \frac{1}{4}$, etc.
– in other words, in the form $\frac{1}{n}$

They wrote other fractions as sums of (different) unit fractions.

For example, we can write $\frac{5}{6}$ as $\frac{1}{2} + \frac{1}{3}$ using Egyptian Fractions.

Answer the following questions about Egyptian Fractions.

Use the internet where appropriate.

1 Work out the value of $\frac{1}{8} + \frac{1}{24}$

2 Work out the value of $\frac{1}{4} + \frac{1}{5} + \frac{1}{10}$

3 Find two (different) Egyptian Fractions that add up to $\frac{3}{4}$

4 Find three (different) Egyptian Fractions that add up to 1

5 Find three (different) Egyptian Fractions that add up to $\frac{7}{9}$

6 Can *every* ordinary fraction be written using (different) Egyptian Fractions in this way?

7 Simon and Sarah have five sacks of corn, and they want to share them out between eight chicken coops, so that each coop gets (about) the same amount of corn.
They do not have any weighing or measuring equipment available.

We'll just have to judge five-eighths of a sack as best we can.
Simon
I know a better way, using Egyptian Fractions.
Sarah

What method might Sarah be planning to use?

CHAPTER 5

Percentages

In this chapter you will **learn how to:**

- convert between decimals, fractions and percentages
- write one amount as a percentage of another
- find a percentage of an amount
- work out percentages without using a calculator
- increase and decrease a number by a given percentage
- solve problems using percentages.

You will also be **challenged to:**

- investigate inflation.

Starter: How many per cent?

Look at the hexagonal honeycomb shape below.

How many different paths can you find that spell 'PERCENT'?

The paths must be continuous.

One path has been shown for you.

What strategies could you use?

Hint: Start at the middle 'T' and look outwards.

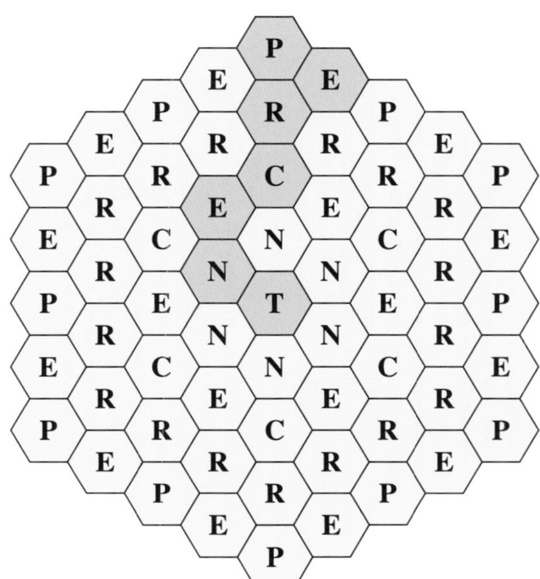

5.1 Changing between decimals, fractions and percentages

Per cent means 'for every 100'.
We use the symbol '%' to show that a number is a percentage.

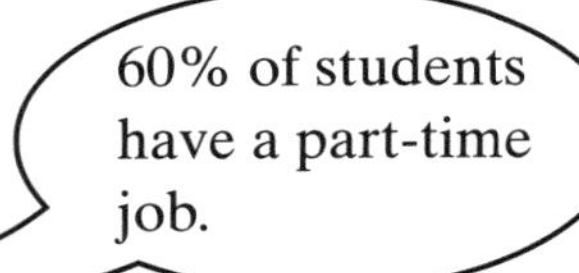
60% of students have a part-time job.

So '60% of students' means 60 out of every 100 students.

So $\frac{60}{100}$ of students have a part-time job.

0.6 of students have a part-time job.

To change a percentage to a decimal you **divide by 100**

Move the decimal point two places to the left.

To change a decimal to a percentage you **multiply by 100**

Move the decimal point two places to the right.

EXAMPLE

Change these percentages to decimals:

a) 5% **b)** 17.5% **c)** 0.1%

SOLUTION

a) $5. \div 100 = \mathbf{.\overset{\frown}{0}\overset{\frown}{5}} = \mathbf{0}.05$

So $5\% = \underline{0.05}$

Don't forget to put in a '0' before the decimal point.

b) $17.5 \div 100 = \mathbf{.\overset{\frown}{1}\overset{\frown}{7}5} = \mathbf{0}.175$

So $17.5\% = \underline{0.175}$

c) $0.1 \div 100 = \mathbf{.\overset{\frown}{0}\overset{\frown}{0}1} = \mathbf{0}.001$

So $0.1\% = \underline{0.001}$

EXAMPLE

Change these decimals to percentages:

a) 0.32 **b)** 0.02 **c)** 0.125

SOLUTION

a) $0.32 \times 100 = \mathbf{.\overset{\frown}{3}\overset{\frown}{2}} = 32\%$

So $0.32 = \underline{32\%}$

b) $0.02 \times 100 = \mathbf{.\overset{\frown}{0}\overset{\frown}{2}} = 2\%$

So $0.02 = \underline{2\%}$

c) $0.125 \times 100 = \mathbf{\overset{\frown}{1}\overset{\frown}{2}.5} = 12.5\%$

So $0.125 = \underline{12.5\%}$

EXAMPLE

Change these percentages to fractions:

a) 80% **b)** 37.5%

SOLUTION

a) 80% means 80 out of every 100

So $80\% = \frac{80}{100}$

This can be simplified.

$$\frac{80}{100} \xrightarrow{\div 20} \frac{4}{5}$$

So $80\% = \frac{4}{5}$

b) 37.5% means 37.5 out of every 100

So $37.5\% = \frac{37.5}{100}$

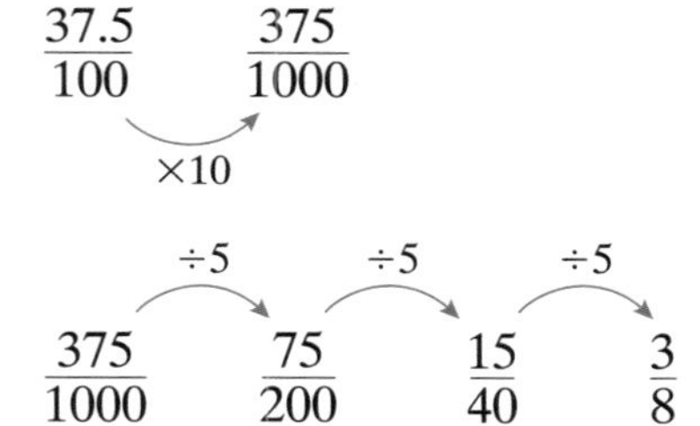

$$\frac{37.5}{100} \xrightarrow{\times 10} \frac{375}{1000}$$

Multiply the 'top' and 'bottom' of the fraction by 10 to get rid of the decimal.

$$\frac{375}{1000} \xrightarrow{\div 5} \frac{75}{200} \xrightarrow{\div 5} \frac{15}{40} \xrightarrow{\div 5} \frac{3}{8}$$

You may spot that you can simplify this fraction in one step by dividing top ' and 'bottom' by 125

So $37.5\% = \frac{3}{8}$

EXAMPLE

Change $\frac{7}{25}$ to a percentage.

SOLUTION

Per cent means 'per hundred' so we need to find an equivalent fraction which has denominator of 100.

'bottom'

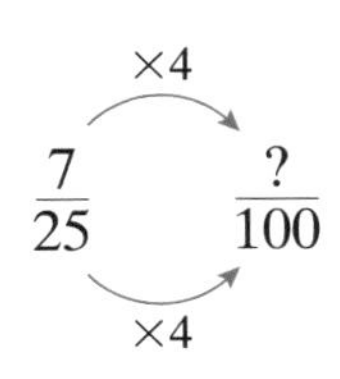

So $\frac{7}{25} = \frac{28}{100} = \underline{28\%}$

EXERCISE 5.1

Don't use your calculator for Questions 1–4.

1 Change these percentages to decimals:

a) 23% **b)** 54% **c)** 64% **d)** 7%
e) 27% **f)** 10.5% **g)** 113% **h)** 0.5%

2 Change these decimals to percentages:

a) 0.65 **b)** 0.7 **c)** 0.41 **d)** 0.01
e) 0.9 **f)** 0.03 **g)** 1.2 **h)** 1.37

3 Change these fractions to percentages:

a) $\frac{34}{100}$ **b)** $\frac{22}{100}$ **c)** $\frac{32}{50}$ **d)** $\frac{13}{50}$

e) $\frac{8}{25}$ **f)** $\frac{6}{20}$ **g)** $\frac{9}{10}$ **h)** $\frac{4}{5}$

4 Change these percentages to fractions – write each fraction in its simplest form:

a) 70% **b)** 35% **c)** 84% **d)** 99%
e) 2% **f)** 17.5% **g)** 115% **h)** 120%

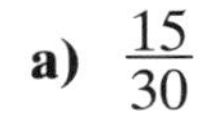

You can use your calculator for Questions 5 and 6.

Hint: Write the fractions as a decimal first.

5 Change these fractions to percentages:

a) $\frac{15}{30}$ **b)** $\frac{18}{30}$ **c)** $\frac{10}{40}$ **d)** $\frac{8}{40}$

e) $\frac{11}{55}$ **f)** $\frac{18}{80}$ **g)** $\frac{27}{60}$ **h)** $\frac{14}{80}$

6 Copy and complete the following table.

Percentage	Decimal	Fraction
1%		
	0.1	
	0.125	
		$\frac{1}{5}$
25%		
33.33%		$\frac{1}{3}$
50%		
		$\frac{2}{3}$
75%		
	0.8	
90%		

You need to learn these.

5.2 Writing one quantity as a percentage of another

This can be a useful way of comparing two quantities.

It is also used to calculate percentage profit or loss.

EXAMPLE

Felix gets the following scores in his end of year exams:

French: 64 out of 80
Maths: 74 out of 95
Science: 53 out of 65

Which subject has he done best in?

SOLUTION

'out of' means divide.

As all the marks are out of different totals we need to write them as percentages to make a fair comparison.

French: $\frac{64}{80} \times 100 = 80\%$

Maths: $\frac{74}{95} \times 100 = 77.894\ldots = 78\%$ to the nearest 1%

Science: $\frac{53}{65} \times 100 = 81.538\ldots = 82\%$ to the nearest 1%

So Felix has done best in Science.

EXAMPLE

Scott buys a car for £8000 and sells it one year later for £6000
What is his percentage loss?

SOLUTION

Scott has made a loss of £8000 − £6000 = £2000
We need to work out what percentage £2000 is of the original £8000:

$$\frac{2000}{8000} \times 100 = \frac{2}{8} \times 100 = \frac{200}{8} = 25\%$$

You can cancel by dividing 'top' and 'bottom' by 1000

Scott has made a loss of 25%

For **percentage profit** or **loss**, you can use the formula:

$$\text{Percentage profit or loss} = \frac{\text{Difference in price}}{\text{Original price}} \times 100\%$$

EXERCISE 5.2

Don't use your calculator for Questions 1–3.

1 Natasha surveys 400 students from her school to find out what end-of-year celebration they would like:

200 students would like a trip to a theme park
160 students would like a prom
40 students would like a picnic.

Write each of these preferences as a percentage.

2 Meera buys a car for £12 000 and sells it two years later for £9000
a) How much money has she lost?
b) What is her percentage loss?

3 Nell buys a painting for £500 and sells it for £700
a) How much profit has she made?
b) What is her percentage profit?

You can use your calculator for Questions 4 and 5.

4 Aimee gets the following marks in her end-of-year exams.

Subject	Mark
English	10 out of 20
French	50 out of 200
Maths	150 out of 200
Science	32 out of 40
Art	21 out of 60
German	52 out of 80

a) Write each of Aimee's marks as a percentage.
b) Which subject has she done best in?

5 Kieran owns an electrical goods store.
What percentage profit or loss has he made on each of these items?
a) Digital camera bought for £300, sold for £330
b) DVD player bought for £300, sold for £270
c) Computer bought for £800, sold for £980
d) Printer bought for £60, sold for £57

5.3 Working out a percentage of an amount

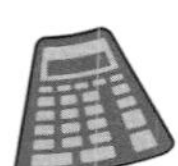

EXAMPLE

A shop offers a 15% discount to all its customers.
What is the discount on a pair of jeans costing £50?

SOLUTION

You need to find 15% of £50

$\frac{15}{100} \times 50$

$0.15 \times 50 = 7.5$

So the discount is £7.50

Step 1 Write the percentage as a decimal.
Step 2 Multiply.

Always give money amounts to 2 decimal places.

EXERCISE 5.3

1 Work out:

a) 25% of 600
b) 32% of 60
c) 5% of 90
d) 20% of 160
e) 70% of 49
f) 80% of 80
g) 8% of 24
h) 1% of 56
i) 17.5% of 500
j) 1.5% of 150
k) 120% of 60
l) 115% of 900

2 John buys a television.
How much VAT does he pay?

Special Offer!
Plasma TV
only
£595
plus 17.5% VAT

VAT stands for Value Added Tax. It is an extra tax you pay when buying many goods and services.

3 At Allstars School, 47% of the students are girls.
There are 900 students at Allstars.

a) What percentage of the students are boys?

b) How many boys are there at the school?

4

I beat you in the maths test – I got 87%

Emily

No – you didn't. I got 120 out of 140, so I did better.

Simon

Who is right?
Explain why.

5 A salesman gets 3% commission on all of his sales.
Find how much commission he earns when his monthly sales are worth:

a) £50 000 **b)** £34 000 **c)** £62 500

5.4 Working out percentages without a calculator

EXAMPLE

Josie buys a stereo for £240 plus VAT.
How much VAT does Josie pay?

SOLUTION

You need to work out 17.5% of £240

Find 10%: 10% of £240 = £24 ← Find 10% by dividing by 10

\+

5% of £240 = £12 ← 5% is half of 10%

\+

2.5% of £240 = £6 ← 2.5% is half of 5%

Total: 17.5% of £240 = £42 ← 17.5% is 10% + 5% + 2.5%

So Josie pays £42 in VAT.

EXERCISE 5.4

1 Find **(i)** 10% **(ii)** 5% **(iii)** 15% of:

a) 50 **b)** 700
c) 1200 **d)** 82

2 Find 25% of:

25% is the same as $\frac{1}{4}$, so divide by 4

a) 200 **b)** 60
c) 4200 **d)** 88

3 Find:

a) 50% of 62 **b)** 75% of 140 **c)** 15% of 60
d) 2.5% of 360 **e)** 5% of 90 **f)** 20% of 70
g) 30% of 180 **h)** 45% of 1200 **i)** 80% of 600

4 Work out the VAT (at 17.5%) that needs to be paid on the following items.

a)

b)

c)

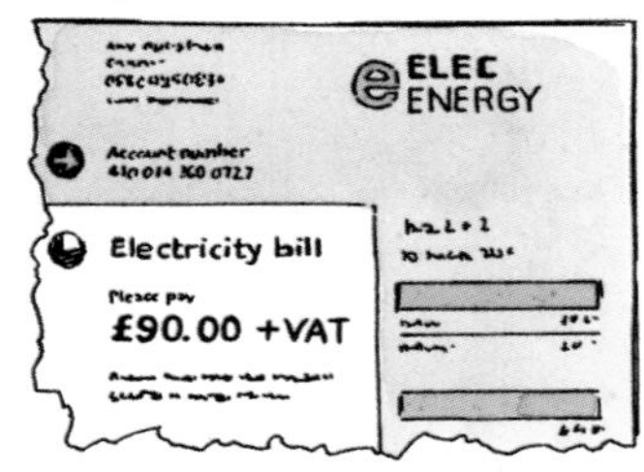

5 A shop offers a discount of 15% to all of its customers.
How much is the discount on the following items?

a)

b)

c)

6 **a)** Find 10% of 800
b) Find 10% of your answer to **a)**.
c) State which of these is the same as your answer to **b)**:
1% of £800 or 20% of £800

7 **a)** Find 20% of £60
b) Find 50% of your answer to **a)**.
c) State which of these is the same as your answer to **b)**:
10% of £60 or 70% of £60

5.5 Increasing and decreasing an amount by a percentage

When you have worked out a given percentage, you may need to:

- add it to (increase) or
- subtract it from (decrease)

the original amount.

EXAMPLE

How much would a £60 pair of jeans cost in the sale?

All jeans 20% off marked price

SOLUTION

Method 1

First you need to find 10% of £60, and then double that to find 20%.

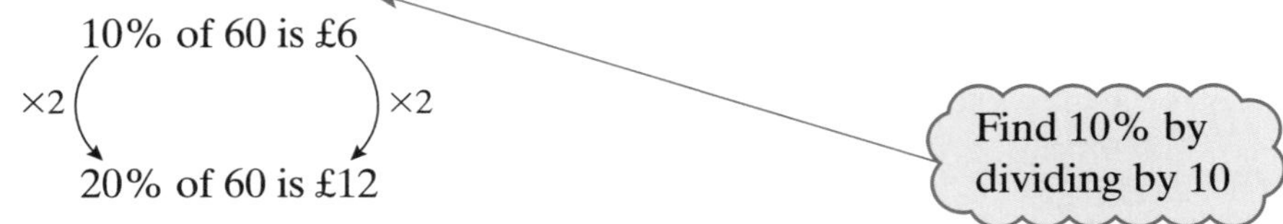

10% of 60 is £6
×2
×2
20% of 60 is £12
Find 10% by dividing by 10

So the price is reduced by £12

£60 − £12 = £48

The jeans cost £48

Method 2

The original price of the jeans is 100%.
There is a 20% decrease so you are paying 80% of the original price.
So you need to find 80% of £60

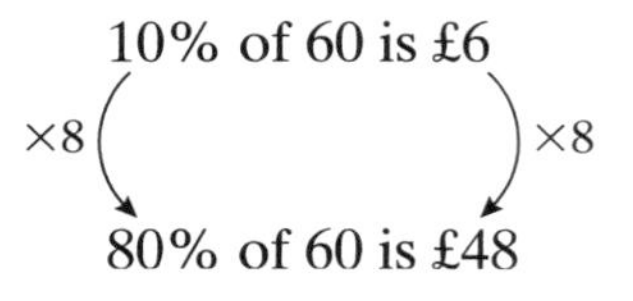

100% − 20% = 80%

The jeans cost £48

EXAMPLE

Tom buys a mobile phone.
How much does it cost?

Mobile Phones
£90 plus 17.5% VAT

SOLUTION

Method 1

$$\frac{17.5}{100} \times 90 = 15.75$$

The price has increased by £15.75

$$90 + 15.75 = 105.75$$

The mobile phone costs £105.75

Method 2

The original price of the mobile phone is 100%.

There is a 17.5% increase, so you are paying 117.5% of the original price.

So you need to find 117.5% of £90:

$1.175 \times £90 = £105.75$

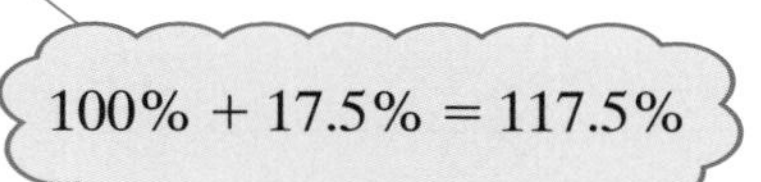
100% + 17.5% = 117.5%

The mobile phone costs £105.75

EXAMPLE

Isobel invests £5000 for 1 year at 6% simple interest.

How much does she have at the end of one year?

SOLUTION

Method 1

$\frac{6}{100} \times 5000 = 300$

$5000 + 300 = 5300$

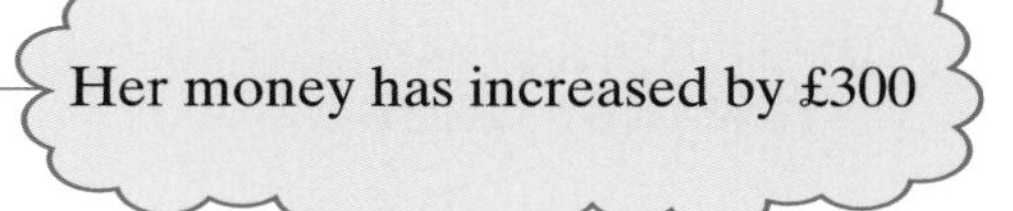
Her money has increased by £300

Isobel has £5300 after one year.

Method 2

The original investment is 100%

There is a 6% increase after 1 year to 106%

100% + 6% = 106%

So you need to find 106% of £5000

$\frac{106}{100} \times 5000 = 5300$

Isobel has £5300 after one year

EXERCISE 5.5

Don't use your calculator for Questions 1–3.

1 1250 students attend Allstars School.
One Monday 4% of the students were absent.
How many students were present?

2 Simon has been on a diet and has lost 15% of his body weight.
His original weight was 90 kg
How much does he weigh now?

3 Vijitha invests £6600 at 4% simple interest for one year.
How much does she have after one year?

You can use your calculator for Questions 4–6.

4 Liam needs to make a 12% profit on all the items he sells in his shop. His accounts book shows the buying price of some of his stock. What should the selling price for each item be?

Item	Buying price (per unit)
Widescreen televisions	£650
DVD players	£75
Stereos	£1100
Laptops	£800

5 Leela's mobile phone bill is £65 plus VAT at 17.5%
What is the total cost of the phone bill?

6 Matthew invests £4300 at 3.5% simple interest for one year.
How much does he have after one year?

5.6 Interest rates

When you leave money in a bank or building society for more than one year you earn interest on any interest you have already earned.

EXAMPLE

Tosin invests £600 at an interest rate of 5%.
How much does he have after three years?

SOLUTION

Interest in Year 1 $\frac{5}{100} \times 600 = 30$

Tosin has earned £30 in interest – this £30 will earn interest from now on.

At the end of Year 1 $600 + 30 = 630$

Interest in Year 2 $\frac{5}{100} \times 630 = 31.50$

At the end of Year 2 $630 + 31.50 = 661.50$

Interest in Year 3 $\frac{5}{100} \times 661.50 = 33.075$

At the end of Year 3 $661.50 + 33.075 = 694.575$
$= £694.58$

Always give money amounts to two decimal places.

Tosin has £694.58 after 3 years

EXERCISE 5.6

1 Lucas invests £8000 at an interest rate of 5%.
Find how much he has after:
a) 1 year **b)** 2 years **c)** 3 years

2 A £14 000 car depreciates by 15% every year.
So its value **decreases** by 15% every year.
Work out how much the car is worth after:
a) 1 year **b)** 2 years **c)** 3 years

3 Mysa invests £4000 at an interest rate of 7%.
How much does she have after three years?

4 The value of Mr Jones's house has increased by 15% each year for the last three years.
It was worth £180 000 three years ago.
How much is it worth today?
Give your answer to the nearest thousand pounds.

REVIEW EXERCISE 5

Don't use your calculator for Questions 1–8.

1 At a pet rescue centre, 60% of the animals are dogs.
a) What percentage are not dogs?
b) There are 350 animals at the centre.
How many dogs are there?

2 a) Write 87% as a decimal. **b)** Write $\frac{2}{5}$ as a percentage.
c) Write 60% as a fraction.
Give your fraction in its simplest form.
d) Write $5\frac{1}{2}$ million in figures.
e) 55% of the students in a school are female.
What percentage are male? [Edexcel]

3

> **Cat facts**
> - *40% of people named cats as their favourite pet.*
> - *98% of women said they would rather go out with someone who liked cats.*
> - *About $7\frac{1}{2}$ million families have a cat.*
> - *$\frac{1}{4}$ of cat owners keep a cat because cats are easy to look after.*

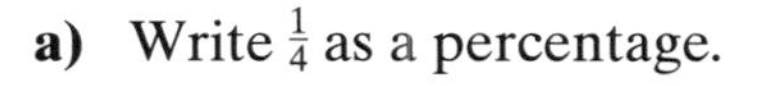

a) Write $\frac{1}{4}$ as a percentage. **b)** Write 40% as a fraction.
Give your fraction in its simplest form.
c) Write 98% as a decimal. **d)** Write $7\frac{1}{2}$ million in figures. [Edexcel]

4 Write these numbers in order of size:

a) 76, 103, 13, 130, 67

b) −3, 5, 0, −7, −1

c) 0.72, 0.7, 0.072, 0.07, 0.702

d) 70%, $\frac{3}{4}$, 0.6, $\frac{2}{3}$ **[Edexcel]**

5 A shop makes 4 of its 80 employees redundant.
What percentage of the staff are made redundant?

6 'Top Togs' is having a summer sale.
How much would each of the following items cost?

a)

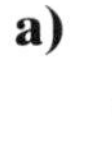

b)

c)

7 There are 800 students at Prestfield School.
144 of these were absent on Wednesday.

a) Work out how many students were *not* absent on Wednesday.

Trudy says that more than 25% of the 800 students were absent on Wednesday.

b) Is Trudy right?
Explain your answer.

45% of these 800 students are girls.

c) Work out 45% of 800

There are 176 students in Year 10

d) Write 176 out of 800 as a percentage. **[Edexcel]**

8

Ben bought a car for £12 000.
Each year the value of the car depreciated by 10%.
Work out the value of the car two years after he bought it.
[Edexcel]

You can use your calculator for Questions 9–16.

9 Copy and complete the following table.

Percentage	Decimal	Fraction
65%		
	0.95	
		$\frac{17}{20}$
62.5%		
	0.3	
		$\frac{9}{40}$

10 In a survey, some families were asked to name their favourite supermarket.
Some of the results are shown in the diagram.

a) Write as a fraction the percentage whose favourite supermarket was Montrose.
b) Write as a decimal the percentage whose favourite supermarket was Salisbury.

200 families took part in the survey.
c) Work out the number of families whose favourite supermarket was Tresco. **[Edexcel]**

11 Three women earned a total of £36
They shared the £36 in the ratio 7 : 3 : 2
Donna received the largest amount.
a) Work out the amount Donna received.

A year ago, Donna weighed 51.5 kg
Donna now weighs $8\frac{1}{2}$%less.
b) Work out how much Donna weighs now.
Give your answer to an appropriate degree of accuracy. **[Edexcel]**

12 Alistair sells books.
He sells each book for £7.60 plus VAT at $17\frac{1}{2}$%
He sells 1650 books.
Work out how much money Alistair receives. **[Edexcel]**

13 Simon repairs computers.
He charges:

£56.80 for the first hour he works on a computer, and
£42.50 for each extra hour's work.

Yesterday Simon repaired a computer and charged a total of £269.30

a) Work out how many hours Simon worked yesterday on this computer.

Simon reduces his charges by 5% when paid promptly.
He was paid promptly for yesterday's work on the computer.

b) Work out how much he was paid. **[Edexcel]**

14 Martin bought some cleaning materials.
The cost of the cleaning materials was £64.00 plus VAT at $17\frac{1}{2}\%$
Work out the total cost of the cleaning materials. **[Edexcel]**

15 In a sale, a supermarket took 20% off its normal prices.
On Fun Friday, it took 30% off its sale prices.
Joe is wrong.
Explain why. **[Edexcel]**

That means there was 50% off the normal prices.

KEY POINTS

1. Per cent means parts per 100
2. To change a decimal to a percentage you multiply by 100
3. To change a percentage to a decimal you divide by 100
4. To change a fraction to a percentage:
   - write the fraction as a decimal ('top' of fraction ÷ 'bottom' of fraction)
   - multiply by 100
5. To change a percentage to a fraction write it over 100 and then simplify the fraction.
6. When you are working out percentages without a calculator find 10% first (by dividing by 10).
7. To work out a percentage of an amount:
   - write the percentage as a decimal
   - multiply the decimal by the amount.
8. Percentage profit or loss $= \dfrac{\text{difference in price}}{\text{original price}} \times 100\%$

Internet Challenge 5

Investigating inflation

Each year the price of things goes up, and so does the amount people earn.
This is called **inflation**, and is usually measured using percentages to give an annual rate of inflation.

Here are some questions about inflation.
You will need internet access to help you research the answers.

1 What is the meaning of the retail prices index (RPI)?

2 How often is it calculated?

3 What is the value of the present UK yearly rate of inflation, based on the RPI?

4 What is the meaning of the Bank of England 'base rate'?

5 How often is it calculated?

CHAPTER 6

Powers and roots

In this chapter you will learn how to:

- find the factors and multiples of a number
- find the least common multiple of two numbers
- find the highest common factor of two numbers
- check whether a number is prime
- work out squares and square roots
- work out cubes and cube roots
- use the laws of indices
- write a number as a product of prime factors
- work out the reciprocal of a number
- use BIDMAS correctly
- use powers of 10
- use your calculator efficiently

You will also be challenged to:

- identify astronomical numbers.

Starter: Off and on

Holly works in a factory testing light bulbs.
She has 100 light bulbs to test – they are numbered 1 to 100.

Holly starts by switching all the bulbs *on*.
She then presses the switch on every 2nd light bulb (all the multiples of 2).

So light bulbs 2, 4, 6, 8 … 98, 100 are now *off*.

Holly then presses the switch on every 3rd light bulb (all the multiples of 3).
Then she presses the switch on every 4th light bulb, and so on.

She presses the switch on every 5th, 6th, … 100th light bulb.

Which light bulbs will be on when Holly has finished?
What is special about these numbers?

You will find a table like this will help.

Round \ Bulb	1	2	3	4	5	6	7	8	9	10	…
1	on	on	on	on	on	on	on	on	on	on	…
2	–	off	–	off	–	off	–	off	–	off	…
3	–	–	off	–	–	on	–	–	off	–	…
4											
…											

6.1 Factors, multiples and primes

A **multiple** of a number is in the 'times tables' for that number.

EXAMPLE

Write down the first five multiples of 8

SOLUTION

$$1 \times 8 = 8, 2 \times 8 = 16, 3 \times 8 = 24, 4 \times 8 = 32, 5 \times 8 = 40$$

So the first five multiples of 8 are 8, 16, 24, 32 and 40

EXAMPLE

Find the **least common multiple** of 6 and 10

The smallest number that is in both the 6 and 10 'times tables'.

SOLUTION

List the first few multiples of 6 and 10

Look for the smallest number that is a multiple of 6 and 10:

6: 6, 12, 18, 24, **30**, 36, 42, ...

10: 10, 20, **30**, 40, 50, 60, 70, ...

The LCM of 6 and 10 is 30

LCM stands for **l**east **c**ommon **m**ultiple.

A **factor** (divisor) is number that divides exactly into another number.

EXAMPLE

Write down all the factors of 72

SOLUTION

Find pairs of numbers that multiply together to give 72:

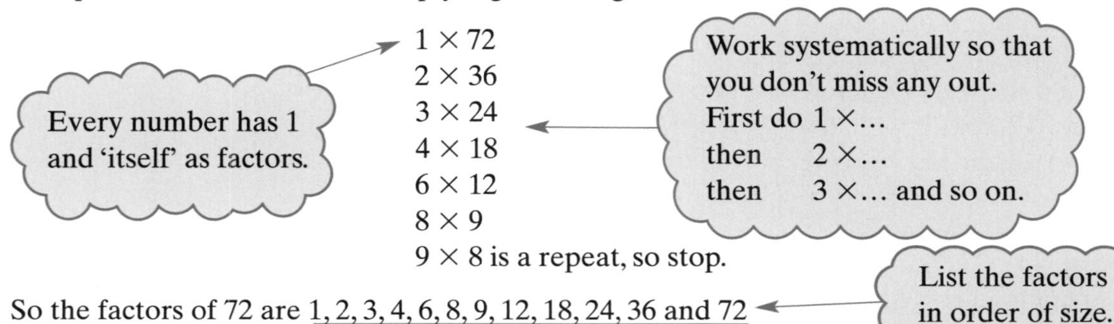

Every number has 1 and 'itself' as factors.
Work systematically so that you don't miss any out. First do 1 ×... then 2 ×... then 3 ×... and so on.
List the factors in order of size.

1×72
2×36
3×24
4×18
6×12
8×9
9×8 is a repeat, so stop.

So the factors of 72 are 1, 2, 3, 4, 6, 8, 9, 12, 18, 24, 36 and 72

EXAMPLE

Find the **highest common factor** of 36 and 48

The largest number that divides into both 36 and 48

SOLUTION

Find the factors of 36 and 48:

36:	1×36	48:	1×48
	2×18		2×24
	3×12		3×16
	4×9		4×12
	6×6		6×8

The factors of 36 are 1, 2, 3, 4, 6, 9, **12**, 18, and 36

The factors of 48 are 1, 2, 3, 4, 6, 8, **12**, 16, 24 and 48

The HCF of 36 and 48 is 12

HCF stands for **H**ighest **C**ommon **F**actor.

A **prime number** is a number which has exactly two factors – '1' and 'itself'.

1 is not prime because it has only one factor (namely 1).

EXAMPLE

Is 91 prime?

SOLUTION

If 91 is prime it will only be divisible by 1 and 91.

Check:	$91 \div 2 = 45.5$ ✗	$91 \div 3 = 30.333\ldots$ ✗
	$91 \div 4 = 22.75$ ✗	$91 \div 5 = 18.2$ ✗
	$91 \div 6 = 15.166\ldots$ ✗	$91 \div 7 = 13$ ✓

No, 91 is not prime because 7 and 13 are factors.

EXERCISE 6.1

1 a) Write down the first six multiples of:
(i) 12 **(ii)** 20 **(iii)** 24 **(iv)** 36 **(v)** 60

b) Find the **least common multiple** of:
(i) 12 and 20 **(ii)** 24 and 36 **(iii)** 20 and 24
(iv) 24 and 60 **(v)** 20 and 60 **(vi)** 36 and 60

2 a) Find the factors of:
(i) 12 **(ii)** 20 **(iii)** 72 **(iv)** 180

b) Find the **highest common factor** of:
(i) 12 and 20 **(ii)** 72 and 180 **(iii)** 20 and 180

3 Which of the following numbers are prime?
Give a reason for your answer.

63 21 11 51 9
23 27 7 61

4 *Eratosthenes' Sieve* is a method for finding prime numbers.

Step 1 Ring the number 2.
Shade in all the multiples of 2 except 2

Step 2 Ring the number 3
Shade in all the multiples of 3 (except 3) that aren't already shaded.

Step 3 Ring the next number that hasn't been shaded.
Shade in all the multiples of this number.

Step 4 Carry on until every number is either ringed or shaded.
All the ringed numbers are prime.

The next number is 5

(2)	(3)	4	5	6	7
8	9	10	11	12	13
14	15	16	17	18	19
20	21	22	23	24	25
26	27	28	29	30	31
32	33	34	35	36	37
38	39	40	41	42	43
44	45	46	47	48	49
50	51	52	53	54	55
56	57	58	59	60	61
62	63	64	65	66	67
68	69	70	71	72	73
74	75	76	77	78	79
80	81	82	83	84	85
86	87	88	89	90	91
92	93	94	95	96	97

These shaded numbers can't be prime as they have 2 or 3 (or both) as factors.

a) Use Eratosthenes' Sieve to find all the prime numbers between 1 and 100.

b)

2 is the only even prime number.

Is Mel right?
Explain why.

c)

Every prime number except 2 and 3 is either one before or one after a multiple of 6

Is Simon right?
Explain why.

6.2 Finding squares and square roots

When you multiply a number by itself, you are finding its **square**.

So '3 squared' is 9 because $3 \times 3 = 9$
You can write this as $3^2 = 3 \times 3 = 9$

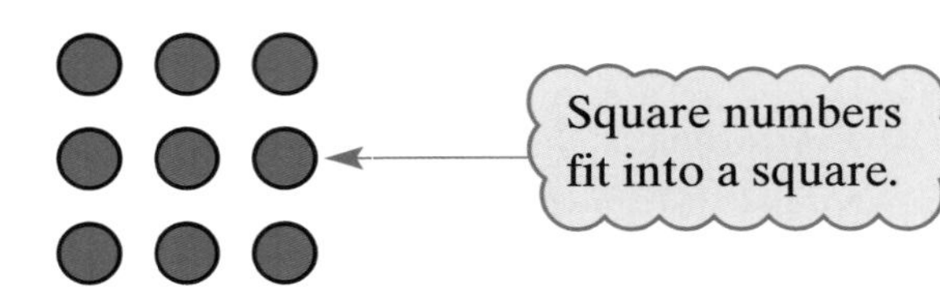
Square numbers
fit into a square.

9 is a **square number**.

What is the first square number?
You need to learn the first 15 square numbers.

The opposite of squaring is **square rooting**.
The square roots of 9 are 3 and -3 because:

$$3 \times 3 = 9$$

and $$-3 \times -3 = 9$$

Every positive number has two square roots.

You can write the square roots of 9 as ± 3:

3 is the **positive square root** of 9 -3 is the **negative square root** of 9
$\sqrt{\ }$ is a square root sign.
It means 'the positive square root' only.
So $\sqrt{9} = 3$

Can a negative number have a square root?

EXAMPLE

Estimate $\sqrt{40}$

SOLUTION

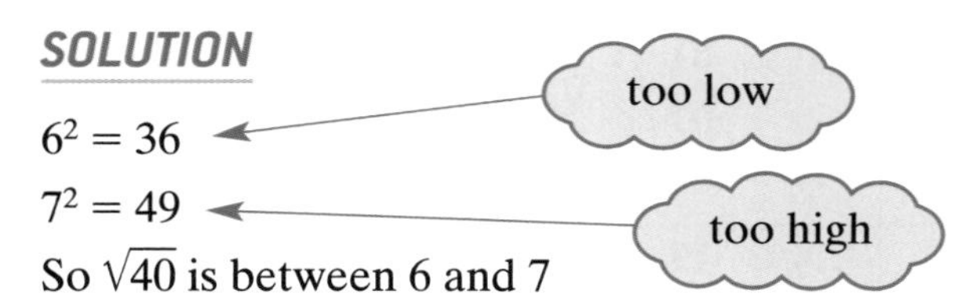
too low
too high

$6^2 = 36$

$7^2 = 49$

So $\sqrt{40}$ is between 6 and 7

40 is nearer to 6^2 than to 7^2, so 6.3 is a reasonable estimate.

EXERCISE 6.2

1 Square: **a)** 5 **b)** 6 **c)** 9 **d)** 12 **e)** 15

2 Work out: **a)** 2^2 **b)** 4^2 **c)** 7^2 **d)** 1^2 **e)** 13^2

3 Write down the first 15 square numbers.

4 Work out: **a)** $\sqrt{1}$ **b)** $\sqrt{49}$ **c)** $\sqrt{64}$ **d)** $\sqrt{225}$

5 10 000 counters are arranged in a square.
How many counters are along one side of the square?

6 **a)** **(i)** Square root 9 and then square your answer.
(ii) Square root 25 and then square your answer.
(iii) What do you notice?
b) Work out: **(i)** $(\sqrt{5})^2$ **(ii)** $(\sqrt{10})^2$ **(iii)** $(\sqrt{17})^2$

7 Write down the whole number that is closest in value to:
a) $\sqrt{10}$ **b)** $\sqrt{15}$ **c)** $\sqrt{70}$ **d)** $\sqrt{50}$

8 Estimate: **a)** $\sqrt{45}$ **b)** $\sqrt{150}$ **c)** $\sqrt{90}$ **d)** $\sqrt{60}$

9 Find the square roots of: **a)** 25 **b)** 16 **c)** 100 **d)** 144

6.3 Using your calculator to find squares and square roots

You can work out squares and square roots using your calculator.

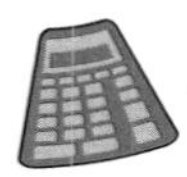

EXAMPLE

Work out 3.1^2

SOLUTION

Use these calculator keys:

So $3.1^2 = \underline{9.61}$

EXAMPLE

Find the square roots of 6.32

Give your answer correct to 3 significant figures.

SOLUTION

Use these calculator keys:

So $\sqrt{6.32} = 2.513\ldots$

So the square root of 6.32 is $\underline{\pm 2.51}$ (to 3 s.f.)

Your calculator will only give you the positive square root.

EXERCISE 6.3

1 Work out: **a)** $\sqrt{289}$ **b)** $\sqrt{86.49}$ **c)** $\sqrt{0.01}$ **d)** $\sqrt{51.84}$

2 Work out the following – give your answers to 1 decimal place:
a) 2.3^2 **b)** 4.7^2 **c)** 12.8^2 **d)** 32.1^2

3 Work out the following – give your answers to 2 decimal places:
a) $25.2^2 - \sqrt{123}$ **b)** $3.1^2 - \sqrt{10}$ **c)** $\sqrt{40} - \sqrt{30}$

4 Work out the following – give your answers to 2 decimal places:
a) $\dfrac{7.3^2}{\sqrt{2.8}}$ **b)** $\dfrac{9.5^2}{\sqrt{12.7}}$ **c)** $\dfrac{\sqrt{15.6}}{1.2^2}$

5 Find the square roots of:
a) 625 **b)** 0.25 **c)** 132.25 **d)** 17.64

6.4 Cubes and cube roots

When you multiply a number by itself twice, you are finding its **cube**.

So '**2 cubed**' is 8 because $2 \times 2 \times 2 = 8$
You can write this as $2^3 = 2 \times 2 \times 2 = 8$

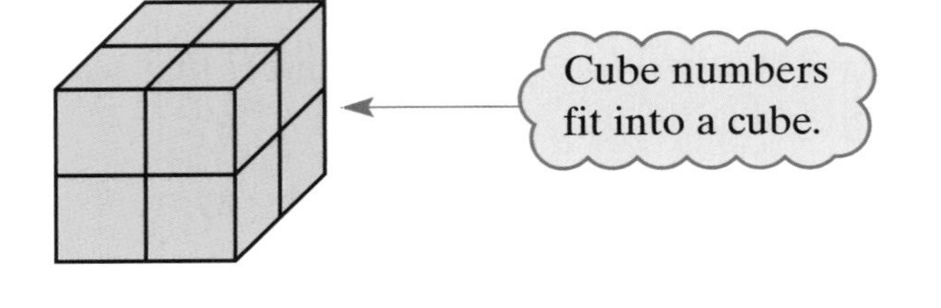

8 is a **cube number**.

What is the first cube number?
You need to learn the first five cube numbers.
The opposite of cubing is **cube rooting**.

The **cube root** of 8 is 2

You can write this as: $\sqrt[3]{8} = 2$

You can work out cubes and cube roots using your calculator.

EXAMPLE

Work out 4.1^3

SOLUTION

Use these calculator keys:

So $4.1^3 = \underline{68.921}$

EXAMPLE

Work out $\sqrt[3]{343}$

SOLUTION

You may have a cube root key $\sqrt[3]{\ }$ on your calculator.

Use these calculator keys:

So $\sqrt[3]{343} = \underline{7}$

EXERCISE 6.4

1 Find the first five cube numbers.

2 Work out:

a) 10^3 **b)** 7^3 **c)** 8^3 **d)** 0.1^3

3 Work out:

a) $\sqrt[3]{729}$ **b)** $\sqrt[3]{216}$ **c)** $\sqrt[3]{3.375}$ **d)** $\sqrt[3]{0.008}$

6.5 Indices

$4 \times 4 \times 4 \times 4 \times 4$ can be written as 4^5

5 is called the **index** or **power**. There are **5** fours being multiplied together.

You say '4 to the **power** of 5'

You can work out 4^5 on your calculator like this:

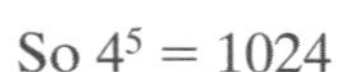

So $4^5 = 1024$

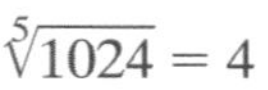

You can also write: $\sqrt[5]{1024} = 4$

You say 'the fifth root of 1024 is 4'

You can work this out on your calculator like this:

EXERCISE 6.5

1 Write the following using indices:

a) $4 \times 4 \times 4 \times 4 \times 4 \times 4$

b) $3 \times 3 \times 3 \times 3 \times 3$

c) $9 \times 9 \times 9 \times 9$

d) $7 \times 7 \times 7 \times 7 \times 7 \times 7 \times 7$

e) $2.1 \times 2.1 \times 2.1 \times 2.1 \times 2.1 \times 2.1$

f) $3.7 \times 3.7 \times 3.7 \times 3.7$

You can use your calculator for Questions 2–4.

2 Use your calculator to work out:

a) 5^7 **b)** 3^9 **c)** 2^{14}

d) 4^4 **e)** 2^5 **f)** 7^5

3 Use your calculator to work out:

a) $\sqrt[6]{729}$ **b)** $\sqrt[4]{4096}$ **c)** $\sqrt[7]{128}$

d) $\sqrt[6]{1}$ **e)** $\sqrt[4]{6561}$ **f)** $\sqrt[5]{7776}$

4 a) Match together cards with the same answer.

$2^4 \times 2^6$	$2^9 \div 2^5$	2^2
2^4	$2^3 \times 2^4$	$2^6 \div 2$
2^7	2^{10}	$2^2 \times 2^4$
$2^7 \div 2^5$	2^6	2^5

b) What do you notice?

6.6 The laws of indices

Make sure you have answered Question 4 in Exercise 6.5 above.

You will have noticed that $2^3 \times 2^4 = 2^7$

This is because:

$$2^3 \times 2^4 = (2 \times 2 \times 2) \times (2 \times 2 \times 2 \times 2)$$

…and $(2 \times 2 \times 2) \times (2 \times 2 \times 2 \times 2) = 2^7$

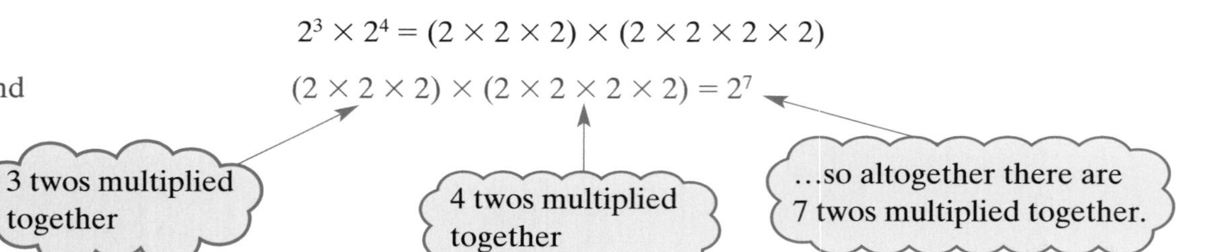
3 twos multiplied together
4 twos multiplied together
…so altogether there are 7 twos multiplied together.

In general:

$$2^a \times 2^b = 2^{a+b}$$

You add the indices.

In fact this rule works for any number.

We can use the letter n to represent 'any number' and write:

$$n^a \times n^b = n^{a+b}$$

EXAMPLE

Write $3^4 \times 3^5$ as a single power of 3

SOLUTION

$3^4 \times 3^5 = 3^{4+5} = \underline{3^9}$

You will have noticed that $2^7 \div 2^5 = 2^2$

This is because:

$$2^7 \div 2^5 = \frac{2 \times 2 \times \cancel{2} \times \cancel{2} \times \cancel{2} \times \cancel{2} \times \cancel{2}}{\cancel{2} \times \cancel{2} \times \cancel{2} \times \cancel{2} \times \cancel{2}} = 2 \times 2$$

Multiplying by 2 and then dividing by 2 cancels out.

$$2 \times 2 = 2^2$$

So altogether there are 2 twos multiplied together.

In general: $2^a \div 2^b = 2^{a-b}$

You **subtract** the indices.

In fact this rule works for any number.

We can use the letter n to represent 'any number' and say:

$$n^a \div n^b = n^{a-b}$$

EXAMPLE

Write $7^9 \div 7^4$ as a single power of 7

SOLUTION

$7^9 \div 7^4 = 7^{9-4} = \underline{7^5}$

The laws of indices are:

$$n^a \times n^b = n^{a+b}$$

$$n^a \div n^b = n^{a-b}$$

EXERCISE 6.6

Don't use your calculator for Questions 1 and 2.

1 Write each of the following as a single power:

a) $5^6 \times 5^8$ **b)** $4^3 \times 4^6$ **c)** $12^9 \div 12^3$

d) $3^{20} \div 3^6$ **e)** $11^3 \times 11^9$ **f)** $6^2 \times 6^7$

g) $8^{100} \div 8^{96}$ **h)** $20^{11} \div 20^4$ **i)** $9^{10} \times 9^{17}$

j) $16^3 \times 16^7$ **k)** $\frac{7^{200}}{7^{100}}$ **l)** $\frac{5^{100}}{5^{50}}$

2 Write each of the following as a single power:

a) $\dfrac{5^7 \times 5^6}{5^4}$ **b)** $\dfrac{4^2 \times 4^4}{4^3}$ **c)** $\dfrac{6^{12} \times 6^9}{6^6}$

You can use your calculator for Questions 3 and 4.

3 a) Write down the following as a single power:

(i) $\dfrac{7^7}{7^6}$ **(ii)** $\dfrac{4^9}{4^8}$ **(iii)** $\dfrac{2^{12}}{2^{11}}$

b) Write down the value of:

(i) $\dfrac{7^7}{7^6}$ **(ii)** $\dfrac{4^9}{4^8}$ **(iii)** $\dfrac{2^{12}}{2^{11}}$

c) How else can you write n^1?

4 a) Write the following as a single power:

(i) $\dfrac{3^4}{3^4}$ **(ii)** $\dfrac{6^7}{6^7}$ **(iii)** $\dfrac{8^5}{8^5}$

b) Write down the value of:

(i) $\dfrac{3^4}{3^4}$ **(ii)** $\dfrac{6^7}{6^7}$ **(iii)** $\dfrac{8^5}{8^5}$

b) What does n^0 always equal?

6.7 Writing a number as a product of prime factors

Every number can be written as a **product of prime factors**.

For example: $15 = 3 \times 5$

$60 = 2 \times 2 \times 3 \times 5$

or $60 = 2^2 \times 3 \times 5$

EXAMPLE

Write 180 as a product of prime factors.

SOLUTION

You can make a prime factor tree like this:

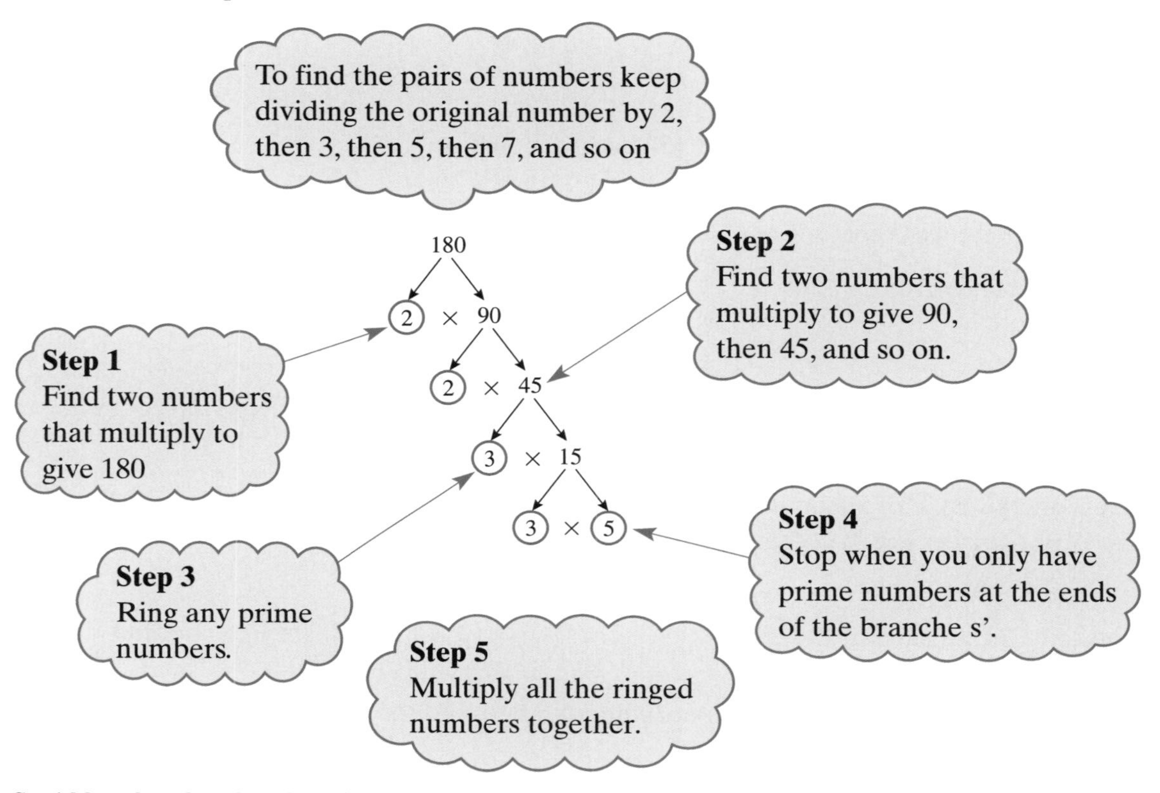

To find the pairs of numbers keep dividing the original number by 2, then 3, then 5, then 7, and so on
180
2 × 90
2 × 45
3 × 15
3 × 5
Step 1 Find two numbers that multiply to give 180
Step 2 Find two numbers that multiply to give 90, then 45, and so on.
Step 3 Ring any prime numbers.
Step 4 Stop when you only have prime numbers at the ends of the branche s'.
Step 5 Multiply all the ringed numbers together.

So $180 = 2 \times 2 \times 3 \times 3 \times 5$

You can write this as $\underline{180 = 2^2 \times 3^2 \times 5}$

EXAMPLE

Using the information that $108 = 2^2 \times 3^3$ and that $180 = 2^2 \times 3^2 \times 5$, find the:

a) highest common factor, HCF **b)** least common multiple, LCM

of 108 and 180

SOLUTION

a) **Step 1**

Find the **prime factors** that 108 and 180 have in **common**.

Step 2

Multiply them together.

Prime factors of 108: **2** **2** **3** **3** 3

Prime factors of 180: **2** **2** **3** **3** 5

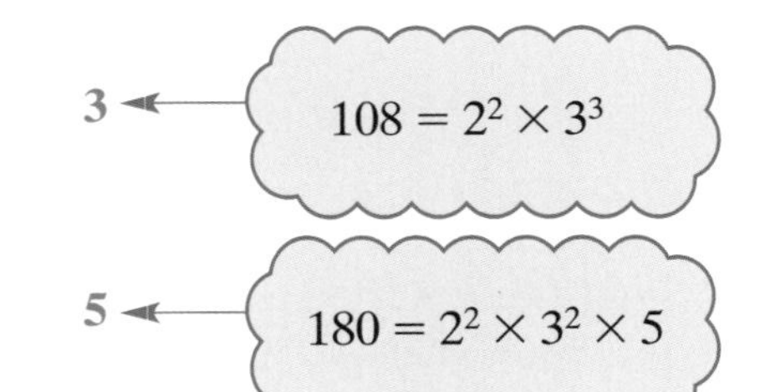

So the HCF of 108 and 180 is $\mathbf{2 \times 2 \times 3 \times 3} = \underline{36}$

b) Multiply the HCF of 108 and 180 by the prime factors **not** in common:

$36 \times 5 \times 3 = 540$

$5 \times (2 \times 2 \times 3 \times 3 \times 3) = 5 \times \mathbf{108} = 540$

and

$3 \times (2 \times 2 \times 3 \times 3 \times 5) = 3 \times \mathbf{180} = 540$

So the LCM of 108 and 180 is $\underline{540}$

Or you can use the method shown in Exercise 6.1 to find the LCM.

EXERCISE 6.7

1 Write the following numbers as a product of prime factors:

a) 36	**b)** 80	**c)** 72
d) 150	**e)** 140	**f)** 198
g) 52	**h)** 360	**i)** 396

2 Find the

(i) highest common factor(HCF); **(ii)** least common multiple (LCM) of:

a) 36 and 80	**b)** 80 and 140	**c)** 72 and 150
d) 198 and 396	**e)** 36 and 52	**f)** 150 and 360
g) 36 and 360	**h)** 72 and 396	**i)** 140 and 360

6.8 Reciprocals

To find the reciprocal of a number you find 1 divided by that number.

So the reciprocal of 3 is $\frac{1}{3}$

You can't find the reciprocal of 0 because $1 \div 0$ is undefined.

EXAMPLE

Find the reciprocal of:

a) 4 **b)** 0.5 **c)** $\frac{1}{5}$ **d)** $\frac{3}{7}$

SOLUTION

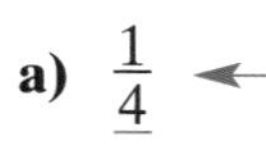

a) $\underline{\frac{1}{4}}$ ← You can leave the answer as a fraction.

b) $\frac{1}{0.5} = 1 \div 0.5 = \underline{2}$ ← There are 2 'halves' in 1 'whole'.

c) $\frac{1}{\frac{1}{5}} = 1 \div \frac{1}{5} = \underline{5}$ ← There are 5 'fifths' in 1 'whole'.

d) $\frac{1}{\frac{3}{7}} = 1 \div \frac{3}{7} = 1 \times \frac{7}{3} = \underline{\frac{7}{3}}$ ← To find the reciprocal of a fraction you invert the fraction (turn it 'upside down').

The reciprocal button on your calculator looks like $\boxed{\frac{1}{x}}$ or $\boxed{x^{-1}}$ so, to find the reciprocal of 4, press  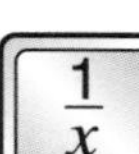 

EXERCISE 6.8

1 Write down the reciprocal of:

a) 1 **b)** 7 **c)** 8 **d)** 0.1

e) 0.25 **f)** $\frac{1}{6}$ **g)** $\frac{3}{5}$ **h)** $\frac{2}{9}$

2 a) Write down the reciprocal of:

(i) 2; **(ii)** 0.2; **(iii)** $\frac{3}{4}$

b) Write down the reciprocal of each of your answers to **a)**.

c) What do you notice?

3 a) Multiply each number in question **1** by its reciprocal.

b) What do you notice?

6.9 BIDMAS

It would be very confusing if there were two answers to questions like this.

So mathematicians have agreed on the order in which operations should be done.

This is the order:

	Brackets	work out any brackets first
	Indices	then any indices or powers
	Divide	then division and multiplication
and	**M**ultiply	
	Add	then addition and subtraction
and	**S**ubtract	

So $2 + 4 \times 5 = 2 + 20 = 22$

EXAMPLE

Work out:

a) $12 - 8 \div 4$ **b)** $(2 + 4^2) - 3$ **c)** $(\sqrt{9} - 1)^3$

SOLUTION

a) $12 - 8 \div 4 = 12 - 2$
$= \underline{10}$

b) $(2 + 4^2) - 3 = (2 + 16) - 3$
$= 18 - 3$
$= \underline{15}$

c) $(\sqrt{9} - 1)^3 = (3 - 1)^3$
$= 2^3$
$= \underline{8}$

EXAMPLE

Use your calculator to work out:

a) $(\sqrt{80} - 4.1) \times 2$ **b)** $\dfrac{4.6 + 3.7}{2.1 \times 1.4}$

Give your answers to 2 significant figures.

SOLUTION

a) Press the following keys on your calculator:

Make sure you know how to use brackets on your calculator.

The answer is 9.68854382

So $(\sqrt{80} - 4.1) \times 2 = \underline{9.7}$ to 2 significant figures

b) Press the following keys on your calculator:

You need to write brackets around the top and bottom of the fraction $\dfrac{(4.6 + 3.7)}{(2.1 \times 1.4)}$

The answer is 2.823129252

So $\dfrac{4.6 + 3.7}{2.1 \times 1.4} = \underline{2.8}$ to 2 significant figures

EXERCISE 6.9

1 Work out:

a) $3 \times 4 + 2$ **b)** $5 \times 4 - 2 \times 3$ **c)** $15 - 4 \div 2$ **d)** $(4^2 - \sqrt{9}) \times 5$

e) $\dfrac{25 + 5}{2 \times 3}$ **f)** $\dfrac{4 + 4 \times 2^3}{4^2 - 7} + 3$ **g)** $\dfrac{2 \times (5^2 - 4)}{3 + 4}$ **h)** $\dfrac{(2 \times 3 + 7^2) \times 2 - 10}{2 + 3 \times 6}$

2 Use your calculator to work out the following – give your answers to 1 decimal place:

a) $\sqrt{56} + 3$ **b)** $\sqrt{(3.2 + 4.7)}$ **c)** $\dfrac{13.6 - 0.7}{4.1 + 1.4}$ **d)** $\sqrt{95} + 5 \times 3.4^2$

e) $(\sqrt{200} + 3) \times 2.4^3$ **f)** $\dfrac{23.7 - 4.91}{3.8 + 4 \times 1.11}$ **g)** $\dfrac{2.1^2}{\sqrt{4.2}}$ **h)** $\dfrac{6.3^2 - 3.1}{\sqrt{8}}$

3 Insert brackets into the following to make them correct.

a) $3.2 + 2.4 \times 4.1 - 1.1 = 21.86$

b) $8.5 - 3.6 \times 2.9 + 1.7 = 22.54$

c) $5.1 + 5.6 - 2.9 \times 1.6 + 2.9 \times 7.2 = 30.3$

6.10 Powers of 10

I earn 25 thousand pounds a year.

Lottery Jackpot
£17.4 million

There are over 6 billion people on earth

These are all examples of using powers of 10

1 thousand = 1000 = $10 \times 10 \times 10 = 10^3$

1 million = 1 000 000 = $10 \times 10 \times 10 \times 10 \times 10 \times 10 = 10^6$

1 billion = 1 000 000 000 = $10 \times 10 \times 10 \times 10 \times 10 \times 10 \times 10 \times 10 \times 10 = 10^9$

It is often quicker to write down a number using powers 10 rather than writing down lots of zeros.

Sometimes your calculator will give you a large answer using powers of 10

This is called **standard index form**.

For example:

2.5 04 means 2.5×10^4 = 25000. = 25 000

1.74 07 means 1.74×10^7 = 17400000. = 17 400 000

6 09 means 6×10^9 = 6000000000. = 6 000 000 000

This means:
$2.5 \times 10 \times 10 \times 10 \times 10$
So you need to move the decimal point 4 places to the right.

EXERCISE 6.10

1 Work out:

a) 10^4 **b)** 10^7 **c)** 10^8

2 Write the following as a power of 10:

a) 100 **b)** 100 000 **c)** 100 000 000 000 000

3 10^{100} is called a 'googol'.
A googol is written as 1 followed by how many zeros?

4 Write the following as ordinary numbers:

a) 4 03 **b)** 3.2 05 **c)** 2.6 08
d) 1.78 05 **e)** 3.24 10 **f)** 5.1298 07

5 Use your calculator to work out the following, giving your answer in standard index form:

a) $(3.2 \times 10^6) \times (4 \times 10^3)$ **b)** $(5 \times 10^{17}) \div (2 \times 10^4)$
c) $(7.4 \times 10^4) \times (2.3 \times 10^9)$ **d)** $(8.2 \times 10^{25}) \div (2.05 \times 10^{14})$

You can use the power button y^x on your calculator, or ask your teacher how to use the **Exp** button.

REVIEW EXERCISE 6

Don't use your calculator for Questions 1–10.

1 Write down the reciprocal of:

a) 9 **b)** $\frac{1}{12}$ **c)** 0.4 **d)** $\frac{5}{8}$

2

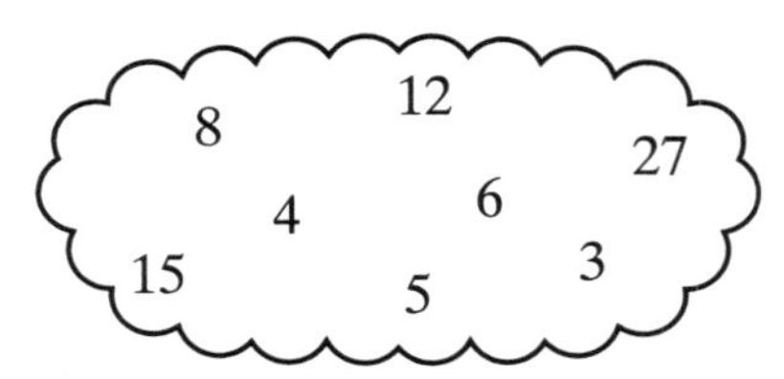

Using only the numbers in the cloud, write down:

a) all the multiples of 6 **b)** all the square numbers
c) all the factors of 12 **d)** all the cube numbers. **[Edexcel]**

3 Here is a list of eight numbers:

5 6 12 20 25 26 28 33

a) From the list write down:

(i) a square number
(ii) a number that is a multiple of 7
(iii) two numbers that are factors of 40
(iv) two numbers with a sum of 59

b) Tony says '6 is a cube number because $2^3 = 6$'
Tony is wrong.
Explain why. **[Edexcel]**

4 Here is a list of eight numbers:

11 16 18 36 68 69 82 88

a) Write down two numbers from the list with a sum of 87
b) Write down a number from the list which is:
(i) a multiple of 9 **(ii)** a square number.

cube	**multiple**	**factor**	**product**

c) Use a word from the list above to complete this sentence correctly:
11 is a .... of 88 **[Edexcel]**

5 Write as a power of 7:
a) $7^3 \times 7^4$ **b)** $7^{11} \div 7^5$ **[Edexcel]**

6 **a)** Express 108 as a product of its prime factors.
b) Find the highest common factor of 108 and 24 **[Edexcel]**

7 **a)** Work out the value of $2^4 \times 3^2$
b) Write down the whole number that is closest in value to $\sqrt{40}$ **[Edexcel]**

8 **a)** Write as a power of 3: **(i)** $3^7 \times 3^2$ **(ii)** $3^8 \div 3^3$
b) $3^x \div 3^6 = 3^3$
and $3^x \times 3^y = 3^{12}$
(i) Work out the value of x
(ii) Work out the value of y

9 The number 40 can be written as $2^n \times m$ where m and n are prime numbers.
Find the value of m and the value of n. **[Edexcel]**

10 Work out the value of:
a) $(2^2)^3$ **b)** $(\sqrt{3})^2$ **c)** $\sqrt{2^4 \times 9}$ **[Edexcel]**

You can use your calculator for Questions 11–18.

11 a) Work out the value of $3.8^2 - \sqrt{75}$
Write down all the figures on your calculator display.
b) Write your answer to part **a)** correct to 1 significant figure. [Edexcel]

12 Every day, a quarter of a million babies are born in the world.
a) Write a quarter of a million using figures.
b) Work out the number of babies born in 28 days.
Give your answer in millions. [Edexcel]

13 Joe can do, on average, four calculations on his calculator every minute.
a) How many calculations, on average, can he do in $7\frac{1}{2}$ minutes?
b) Use your calculator to work out the value of $\sqrt{(15 + 27.25)}$ [Edexcel]

14 a) Use your calculator to work out $(2.3 + 1.8)^2 \times 1.07$
Write down all the figures on your calculator display.
b) Put brackets in the expression below so its value is 45.024
$1.6 + 3.8 \times 2.4 \times 4.2$ [Edexcel]

15 Use your calculator to work out the value of:
$$\frac{6.27 \times 4.52}{4.81 + 9.63}$$
a) Write down all the figures on your calculator display.
b) Write your answer to part **a)** to an appropriate degree of accuracy. [Edexcel]

16 Use your calculator to work out the value of:
$$\frac{\sqrt{(1.3^2 + 4.2)}}{5.1 - 2.02}$$
a) Write down all the figures on your calculator display.
b) Write your answer to part **a)** to an appropriate degree of accuracy. [Edexcel]

17 a) Express the following numbers as products of their prime factors:
(i) 60 **(ii)** 96
b) Find the highest common factor of 60 and 96
c) Work out the least common multiple of 60 and 96 [Edexcel]

18 Work out:
a) 7^2 **b)** $\sqrt{121}$ **c)** 10^4 **d)** $\sqrt[3]{0.001}$

KEY POINTS

1 A multiple of a number is in the 'times table' for that number.

2 A factor (divisor) is a number that divides exactly into another number.

3 The least common multiple (LCM) is the lowest multiple that two numbers have in common.

4 The highest common factor (HCF) is the highest factor that two numbers have in common.

5 A prime number has exactly *two* factors.

6 The first 15 square numbers are:
1, 4, 9, 16, 25, 36, 49, 64, 81, 100, 121, 144, 169, 196, 225

7 To square a number you have to multiply the number by itself.

8 The opposite of squaring is square rooting.

9 Every positive number has two square roots – one positive and one negative.

10 The $\sqrt{\ }$ sign means just give the positive square root.

11 The first five cube numbers are:
1, 8, 27, 64, 125

12 $10^3 = 1000$

13 The opposite of cubing is cube rooting.

14 The laws of indices are:

- $n^a \times n^b = n^{a+b}$
- $n^a \div n^b = n^{a-b}$
- $n^1 = n$
- $n^0 = 1$

15 To find the reciprocal of a number you divide 1 by that number (1 'over' the number). To find the reciprocal of a fraction, turn the fraction upside down.

16 BIDMAS gives you the order in which operations should be carried out:

Brackets
Indices
[Divide
[Multiply
[Add
[Subtract

17 Sometimes your calculator will use standard index form (powers of 10) to display large numbers:

2.36^{14} means 2.36×10^{14} or 236 000 000 000 000

Internet Challenge 6

Astronomical numbers

Astronomers work with very large numbers, so they often use powers of 10

Here are some astronomical statements with missing values.

The values are given, in jumbled-up order, in the right-hand column.

Use the internet to help you decide which value belongs to which statement.

1 Astronomers have calculated that the mass of the Sun is about ...... kg	5×10^9 5 thousand million
2 The Sun is thought to have formed about ...... years ago.	6×10^3 6 thousand
3 Each second the Sun's mass decreases by about ...... tonnes.	10^{-9} one billionth
4 The surface temperature of the Sun has been measured to be about ...... degrees C	94×10^6 94 million
5 It takes our Solar System about ...... years to make one revolution around the Milky Way galaxy.	2.8×10^6 2. 8 million
6 Light travels through space at a speed of ...... metres per second.	2.25×10^8 2.25 hundred million
7 The Sun is about ...... miles away.	10^{11} 1 hundred thousand million
8 X-rays can have wavelengths as short as ...... metres.	2×10^{30} 2 billion, billion, billion, billion, billion
9 The Andromeda galaxy is so remote that light from it takes ...... years to reach us.	3×10^8 3 hundred million
10 It is thought that the Universe contains about ...... individual galaxies.	4×10^6 4 million

CHAPTER 7

Working with algebra

In this chapter you will **learn how to**:

- use a formula
- write down an expression
- substitute numbers into formulae and expressions
- simplify expressions
- work with indices
- expand brackets and collect like terms
- factorise algebraic expressions
- use identities
- generate formulae
- change the subject of a formula.

You will also be **challenged to**:

- investigate the language of algebra.

Starter: Right or wrong?

Each question is followed by two possible answers.
Identify the correct one.

Explain what slip might have caused the wrong one in each case.

1	3×0	3	or	0
2	$6 - 5 + 1$	0	or	2
3	5^2	10	or	25
4	$2 + 3 \times 4$	20	or	14
5	$4 + 10 \div 2$	9	or	7
6	$(-3)^2$	9	or	-9

7.1 Using formulae

A **formula** is a rule that is used to work something out.

You can write down a formula in words:

distance = speed × time

Or you can use letters – in other words, **algebra**:

$d = s \times t$

where d is the distance travelled, s is the speed and t is the time taken.

When you use letters you don't need to write the '×' sign.

So you can write:

$d = st$

In this formula d, s and t represent numbers (or **variables**) which can change (**vary**).

You can **substitute** values into a formula.
This means that you replace the letters with numbers.

EXAMPLE

A printing company works out the cost of making some party invitations using this formula:

cost (in £) = 7 + 0.25 × number of cards.

Work out the cost of printing 50 cards.

SOLUTION

Cost (in £) = 7 + 0.25 × number of cards
= 7 + 0.25 × 50
= 7 + 12.5
= 19.5

Cost = £19.50

The number of cards = 50

Always write money amounts to 2 decimal places.

EXAMPLE

A car hire company works out the cost of hiring a car using the formula:

$$C = 150 + 30d + 0.3m$$

where: C is the cost in pounds of hiring a car
d is the number of days the car is hired for
m is the number of miles the car is driven.

Remember d and m represent *numbers* not words – they do not stand for 'days' and 'miles'.

Work out the cost of hiring a car for 7 days driving a total of 360 miles.

SOLUTION

Substitute $d = 7$ and $m = 360$:

$$\begin{aligned} C &= 150 + 30(7) + 0.3(360) \\ &= 150 + 210 + 108 \\ &= 468 \end{aligned}$$

Replace d with 7 and m with 360

So the cost of hire is £468

EXERCISE 7.1

Don't use your calculator for Questions 1 and 2.

1 Use the formulae:

area = length × width
perimeter = (2 × length) + (2 × width)

to work out the area and perimeter of these rectangles:

a) 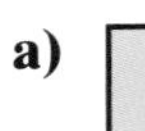
5 cm, 3 cm

b) 
6 cm, 4 cm

c) 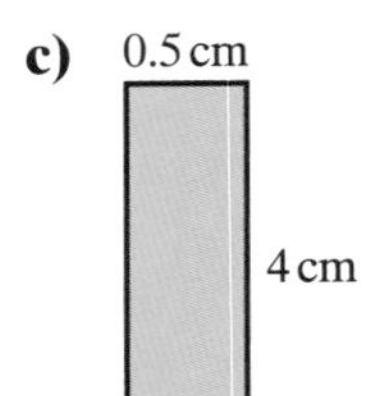
0.5 cm, 4 cm

2 Use the formula:

$$d = st$$

to work out the distance travelled by the following:

a) a bicycle travelling at 15 km/h for 2 hours
b) a car travelling at 80 km/h for 3 hours
c) a man walking at 5 km/h for 4 hours
d) a woman jogging at 8 km/h for 30 minutes.

You can use your calculator for Questions 3–5.

3 A phone company uses the following formula to work out their customers' phone bills.

Total charges (in £)	=	Line rental	+	0.1 × number of off-peak minutes	+	0.2 × number of peak rate minutes

The line rental for Tom's phone is £10 per month.
In May, Tom made 210 minutes of off-peak calls and 150 minutes of peak rate calls.

a) How much is Tom's phone bill for May?

The line rental for Sam's phone is £12 per month.
In May, Sam made 330 minutes of off-peak calls and 90 minutes of peak rate calls.

b) What is the difference between Sam's phone bill and Tom's phone bill for May?

4 The voltage V in an electronic circuit is given by the formula:

$$V = IR$$

Find the value of V when:

a) $I = 13$ and $R = 20$
b) $I = 7$ and $R = 15$
c) $I = 15$ and $R = 22$
d) $I = 7.5$ and $R = 14.7$

5 The final velocity, v, of a particle is found using the formula:

$$v = u + at$$

where u is the initial velocity, a is the acceleration and t is the time taken.

Find v when:

a) $u = 3, a = 10$ and $t = 2$
b) $u = 1, a = 5$ and $t = 7$
c) $u = 0, a = 12$ and $t = 5$
d) $u = 20, a = 0$ and $t = 11$

7.2 Using expressions

An **expression** is like a formula, but without the '=' sign.
Any letters in an expression represent numbers (**variables**).

Here are some examples of expressions:

$n + 4$ means '4 more than n' or 'n add 4' or '4 add n'

$p - 5$ means '5 less than p' or 'p take away 5'

$5 - p$ means 'p less than 5' or '5 take away p'

$7y$ means '$7 \times y$' or '7 times y' or 'y multiplied by 7'

ab means '$a \times b$' or 'a lots of b' or 'b multiplied by a'

$\frac{9}{d}$ means '$9 \div d$' or '9 divided by d'

c^2 means 'c squared' or '$c \times c$'

e^3 means 'e cubed' or '$e \times e \times e$'

$3c^2$ means '3 lots of c squared' or '$3 \times c \times c$'

$(3c)^2$ means '3 lots of c all squared' or '$(3 \times c) \times (3 \times c)$'

$\sqrt{x}$ means 'the (positive) square root of x'

$4 + n$ is the same as $n + 4$ as you can add in any order.

$5 - p$ is **not** the same as $p - 5$

Write numbers first then letters – so $7y$ and not $y7$

Write letters in alphabetical order – so ab and not ba

$3c^2$ is **not** the same as $(3c)^2$

EXAMPLE

Alan has n pounds.
Claudia has £5 more than Alan.
Richard has twice as much money as Claudia.

a) Write down an expression, in terms of n, for the amount of money that:

(i) Claudia has, **(ii)** Richard has.

b) How much money does Richard have when Alan has £42?

SOLUTION

a) **(i)** $n + 5$

(ii) Richard has $2 \times (n + 5)$

Richard has $2(n + 5)$

b) n represents the amount of money Alan has.

Substitute $n = 42$ into $2(n + 5)$:

$$2(n + 5) = 2 \times (42 + 5)$$
$$= 2 \times 47 = 94$$

So Richard has £94

You need brackets so that the whole expression '$n + 5$' is multiplied by 2

You don't need to write the '×' sign.

Replace n with 42 and write the × signs in.

EXERCISE 7.2

1 Write down expressions for the following:

a) 7 more than p
b) 10 less than y
c) n less than 4
d) 5 lots of g
e) h multiplied by 10
f) 18 divided by k
g) m divided by n
h) c lots of d
i) r multiplied by s
j) t squared
k) 7 lots of t squared
l) $7t$ all squared
m) $3a$ multiplied by b
n) x cubed multiplied by y squared

2 Elliot has c CDs.
Petra has three times as many CDs as Elliot.

a) Write down an expression, in terms of c, for the number of CDs that Petra has.

Mysa has 10 CDs fewer than Petra has.

b) Write down an expression, in terms of c, for the number of CDs that Mysa has.

Jamie has twice as many CDs as Mysa.

c) Write down an expression, in terms of c, for the number of CDs that Jamie has.

d) Elliot has 50 CDs.
How many CDs does Jamie have?

3 The length of a rectangle is 5 cm more than its width, w.

a) Write down an expression, in terms of w, for the length of the rectangle.

b) Write down an expression, in terms of w, for the area of the rectangle.

Remember that the area of a rectangle is length × width.

c) Work out the area of the rectangle when:
i) $w = 3$ cm **ii)** $w = 2$ cm **iii)** $w = 5$ cm.

4 a) Write down an expression for the area of this square.

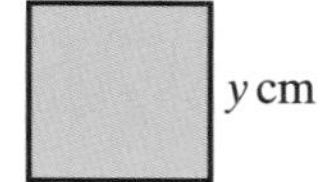

Four of these squares are put together to make a larger square.

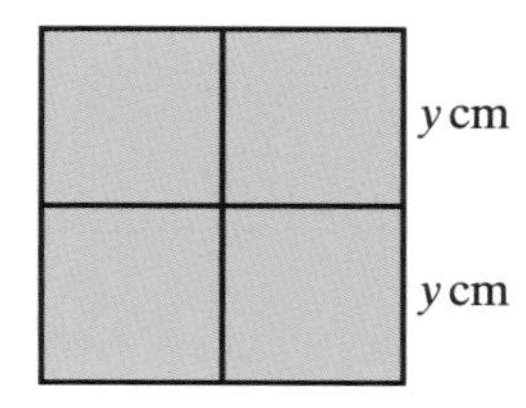

b) Write down an expression for the area of the large square.

c) What is the length of one side of the large square?

d) Explain how this shows that $4y^2 = (2y)^2$

7.3 Substituting values into an expression

When you substitute values into an expression or formula, remember the order of operations described by **BIDMAS**.

Work out	**B**rackets first
then	**I**ndices (squares, cubes, etc.)
next come	**D**ivision **M**ultiplication
and then	**A**ddition **S**ubtraction

EXAMPLE

Work out the values of:

a) $3a + 4b$ **b)** $2a^2$ **c)** $ab - 5c$

when $a = 5, b = 2$ and $c = -4$

See page 18 to remind yourself about negative numbers.

SOLUTION

a) $3a + 4b = 3(\mathbf{5}) + 4(\mathbf{2})$
$= 15 + 8$
$= \underline{23}$

Replace a with 5 and b with 2

b) $2a^2 = 2 \times (\mathbf{5})^2$
$= 2 \times 25$
$= \underline{50}$

Remember this means two lots of a^2 and **I**ndices come before **M**ultiplication.

c) $ab - 5c = (\mathbf{5})(\mathbf{2}) - 5(\mathbf{-4})$
$= 10 - -20$
$= 10 + 20$
$= \underline{30}$

EXAMPLE

Work out the values of:

a) $3(x + 2z)^2$ **b)** $\dfrac{2x + 3y}{z + y}$

when $x = 4, y = -1$ and $z = 3$

SOLUTION

a) $3(x + 2z)^2 = 3[(4) + 2(3)]^2$
$= 3 \times (4 + 6)^2$
$= 3 \times (10)^2$
$= 3 \times 100$
$= \underline{300}$

Work out the brackets first – **Brackets, Indices, Multiplication…**

b) $\dfrac{2x + 3y}{z + y} = \dfrac{2(4) + 3(-1)}{(3) + (-1)}$
$= \dfrac{(8 + -3)}{(3 + -1)}$
$= \dfrac{5}{2}$
$= \underline{2\tfrac{1}{2}}$

Work out the top of the fraction and the bottom of the fraction separately.

EXERCISE 7.3

Don't use your calculator for Questions 1–3.

1 When $a = 3, b = 5, c = 4$ and $d = 0$, find the value of:

a) $a + 5$ **b)** $10 - b$ **c)** $3c$ **d)** $3a + b$
e) $a + b - c$ **f)** $10c - 4b$ **g)** ab **h)** $ac - 5.3d$
i) $ab + 2ac$ **j)** a^2 **k)** c^3 **l)** abc

2 When $r = 2, s = 6$ and $t = 1$, find the value of:

a) $2r^2$ **b)** $(2r)^2$ **c)** $3(r + s)$
d) $2(3r + 2s)$ **e)** $4rst$ **f)** $2r(s - t)$
g) $s^2 - rt$ **h)** $s^2 - r^2$ **i)** $\dfrac{s}{r}$

3 When $e = 8, f = 2$ and $g = -2$, find the value of:

a) $f + g$ **b)** $f^2 + g$ **c)** $f^2 - g$
d) $\dfrac{e}{2f}$ **e)** $\dfrac{e + 4}{g + 4}$ **f)** $\dfrac{2e + f}{g + 5}$

You can use your calculator for Questions 4–8.

4 When $p = 6, q = 5$ and $r = 2$, find the value of:

a) $3p - 2q$ **b)** $5pq + 10q$ **c)** $r^2 - 3p$ **d)** $(2p - 3r)^2$

5 When $x = 4, y = 6$ and $z = -2$, find the value of:

a) $x^2 + y^2$ **b)** $3x - z$ **c)** $2z - 3y$ **d)** $(y - 2z)^2$

6 When $f = -3$, $g = -2$ and $h = 2$, find the value of:

a) f^2 **b)** $5(f + g)$ **c)** $f^2 + gh$ **d)** h^3

e) $3h^3$ **f)** g^3 **g)** $2g^3$ **h)** $2f^2$

i) $3f^2 + 2f$ **j)** $3f^2 - 2f$ **k)** $\frac{g}{h}$ **l)** $\frac{4g + h^2}{h - g}$

7 The distance, s, travelled by a particle is given by the formula:

$$s = ut + \tfrac{1}{2}at^2$$

where u is the initial velocity, t is the time and a is the acceleration.

a) Use your calculator to work out the value of s when $a = 9.8$, $t = 3.5$ and $u = 2.4$
Write down all the figures on your calculator display.

b) Write your answer to part **a)** to the nearest integer.

8 The number of bacteria, N, in a colony is modelled by the formula:

$$N = 2500(1 + kt)$$

where t is the time and k is a growth factor.

a) Find the value of N when $k = 0.3$ and $t = 1.5$

b) Write down the number of bacteria at time $t = 0$

7.4 Simplifying expressions

Look at this expression:

$$4a + 3b - 5a + 5b$$

$4a$ and $-2a$ are **like terms**.

$+\,3b$ and $+\,5b$ are also like terms.

Like terms have the same letters or combination of letters.

Look at this expression:

$$5ab + 7b + 2ab - 6b$$

$5ab$ and $+\,2ab$ are like terms.

$+\,7b$ and $-\,6b$ are also like terms.

Keep the '+' or '−' sign *and* the term together.

When you **simplify** an expression you combine **like terms**.

First group like terms: $4a + 3b - 2a + 5b = 4a - 2a + 3b + 5b$

Then combine like terms: $= 2a + 8b$

This can't be simplified any further – you can't combine unlike terms.

So to simplify $5ab + 7b + 2ab - 6b$ we:

First group like terms: $5ab + 7b + 2ab - 6b = 5ab + 2ab + 7b - 6b$

Then combine like terms: $= 7ab + 1b$

$= 7ab + b$

You don't need to write the '1' – so '$1b$' is written as b

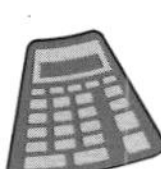

EXAMPLE

Simplify the following expressions:

a) $b + b + b + b$
b) $5x + 2x - x$
c) $4s - 2t + 5s + 3t - s$
d) $4a + 2 - 3a - 1$

SOLUTION

a) $b + b + b + b = \underline{4b}$

b) $5x + 2x - x = \underline{6x}$

c) $4s - 2t + 5s + 3t - s = 4s + 5s - s - 2t + 3t$
$= \underline{8s + t}$

d) $4a + 2 - 3a - 1 = 4a - 3a + 2 - 1$
$= \underline{a + 1}$

EXAMPLE

Simplify $4xy + x - 2xy + 2x - 3x$

SOLUTION

$4xy + x - 2xy + 2x - 3x = 4xy - 2xy + x + 2x - 3x$
$= \underline{2xy}$

The 'x' terms have cancelled out – 'zero lots of x' = 0

EXERCISE 7.4

1 Simplify:

a) $a + a + a + a$ **b)** $3b + 2b + 5b$ **c)** $7c + 3c - 5c$
d) $4d + 3d + 2d + d$ **e)** $6e - 3e - 2e$ **f)** $5f - 3f - 2f$
g) $5g - 7g$ **h)** $4h - 2h - 3h$ **i)** $7i - 6i - 2i$

2 Simplify:

a) $a + 3 + a + 2$ **b)** $b + 5 + 3b - 4$ **c)** $3c + 4 - 2c - 5$
d) $7d + 2 - 8d + 1$ **e)** $5 + 3e - 2e - e$ **f)** $8 - 5f - 2f - 2$

3 Simplify:

a) $a + b + a + b$ **b)** $2a + 3a + 2b - b$ **c)** $3c + 2c - 4d + 3d$
d) $2c + 5d - c + 3d$ **e)** $4d - 3e + 2e - 3d - d$ **f)** $2f - 3f - 2g + 3g$
g) $7g + 4h - 6g - g - 3h$ **h)** $4g - 2h - 3h + g$ **i)** $3i - 2h + 2i - 2h$

4 Simplify:

a) $a^2 + a^2 + a^2 + a^2$ **b)** $3b^2 + 2b^2$ **c)** $3c^3 + 3c^3$
d) $3d^3 + 2d^3 - d^3$ **e)** $5e^2 + 3e^2 + 3e - 7e^2 - 2e$ **f)** $7fg - 4fg - 2fg$
g) $5g^2 - 7g^2 - 4g^2 + 6g^2$ **h)** $3h^2 - 2h^2 - 2h^2 + 4h^2$ **i)** $2hi + 3hi - 4hi$

5 The length of this rectangle is x cm
The width is 3 cm less than the length.

x cm

a) Write down an expression for the width of the rectangle.

The perimeter of the rectangle is found using the formula:

perimeter = width + length + width + length

b) Write down an expression, in terms of x, for the perimeter of the rectangle.
Simplify your expression.

c) What is the perimeter of the rectangle when $x = 6$?

6 Here are three bags of counters.

The medium bag contains n counters.
The large bag contains twice as many counters as the medium bag.

a) Write down an expression, in terms of n, for the number of counters in the large bag.

The small bag contains 5 counters fewer than the medium bag.

b) Write down an expression, in terms of n, for the number of counters in the small bag.

c) Write down an expression, in terms of n, for the total number of counters in all three bags.
Simplify your expression.

d) What is the total number of counters when $n = 8$?

7.5 Working with indices

You have already met indices – they are powers such as squares or cubes – in Chapter 6.

Remember, $7 \times 7 \times 7 \times 7 \times 7 = 7^5$

'5' is called the power or index – it tells you how many 7's are multiplied together.

Also, using the laws of indices:

$$7^4 \times 7^2 = 7^{4+2} = 7^6$$
$$7^5 \div 7^2 = 7^{5-2} = 7^3$$

The laws of indices also work with algebra (letters).

Here are the rules for multiplication and division.

Multiplication	Division
$x^a \times x^b = x^{a+b}$	$x^a \div x^b = x^{a-b}$

EXAMPLE

Simplify these expressions:

a) $x^2 \times x^3$ **b)** $x^7 \div x^3$

SOLUTION

a) $x^2 \times x^3 = x^{2+3} = \underline{x^5}$

b) $x^7 \div x^3 = x^{7-3} = \underline{x^4}$

EXAMPLE

Simplify these expressions:

a) $x^4 \div x^3$ **b)** $x^5 \div x^5$

SOLUTION

a) $x^4 \div x^3 = x^{4-3} = x^1 = \underline{x}$

You don't need to write the '1'

b) $x^5 \div x^5 = x^{5-5} = x^0 = \underline{1}$

Any number divided by itself = 1
e.g. $9 \div 9 = 1$ and $103 \div 103 = 1$
So $x^0 = 1$

Sometimes the algebraic terms will have whole number multiples in front of them – these are called **coefficients**.

When these are present, you multiply or divide the coefficients *as well as* the algebraic terms (the letters).

EXAMPLE

Simplify these expressions:

a) $5x^2 \times 4x^3$ **b)** $12x^7 \div 4x^3$

SOLUTION

a) $5x^2 \times 4x^3 = 5 \times 4 \times x^{2+3}$
$= 20 \times x^5$
$= 20x^5$

Simplify the numbers first and then the letters.

b) $12x^7 \div 4x^3 = (12 \div 4) \times x^{7-3}$
$= 3 \times x^4$
$= 3x^4$

EXERCISE 7.5

1 Simplify:

a) $a \times a \times a$ **b)** $b \times b \times b \times b \times b \times b$ **c)** $c \times c \times c \times c$
d) $4 \times d \times d \times d$ **e)** $7 \times e \times e$ **f)** $9 \times f \times f \times f \times f$

2 Simplify:

a) $x^3 \times x^5$ **b)** $x^3 \times x^2$ **c)** $y^7 \times y^3$ **d)** $y^9 \times y^4$
e) $z^5 \times z$ **f)** $z^4 \times z^3 \times z^2$ **g)** $y^{20} \div y^9$ **h)** $y^{10} \div y^3$
i) $x^7 \div x^6$ **j)** $y^{200} \div y^{199}$ **k)** $\frac{x^4}{x}$ **l)** $\frac{z^{12}}{z^{12}}$

3 Simplify:

a) $2x^3 \times 5x^4$ **b)** $3x^2 \times 5x^3$ **c)** $4x \times 6x^5$
d) $4x^3 \times 10x^2$ **e)** $4y^2 \times 2y^4$ **f)** $10x^7 \times 10x^3$
g) $4x^4 \times 3x$ **h)** $2x^2 \times 3x^3 \times x^4$ **i)** $5x^2 \times 2x^3 \times 4x^4$

4 Simplify:

a) $12y^6 \div 6y^3$ **b)** $18z^6 \div 3z$ **c)** $48x^4 \div 16x$
d) $40z^6 \div 20z^4$ **e)** $12y^{10} \div 12y^9$ **f)** $4y^2 \div 2y$
g) $\frac{4x^4}{2x^4}$ **h)** $\frac{2x^{10}}{2x^9}$ **i)** $\frac{5x^2}{5x^2}$

7.6 Expanding brackets

EXAMPLE

Expand $3(2x + 5)$

Expand means 'clear away the brackets'.

SOLUTION

$$\mathbf{3}(2x + 5) = \mathbf{3} \times 2x + \mathbf{3} \times 5$$
$$= \underline{6x + 15}$$

The bracket contains two terms, namely $2x$ and $+\ 5$; each term gets multiplied by the **3**

Many exam questions ask you to expand two sets of brackets, and then collect like terms to write the result in a neater form.

In such a case you will be told to **'expand and simplify'**.

EXAMPLE

Expand and simplify $4(3x + 7) + 5(x + 2)$

SOLUTION

First you multiply out the brackets …

$$\mathbf{4}(3x + 7) + \mathbf{5}(x + 2) = \mathbf{4} \times 3x + \mathbf{4} \times 7 + \mathbf{5} \times x + \mathbf{5} \times 2$$
$$= 12x + 28 + 5x + 10$$
$$= 12x + 5x + 28 + 10$$
$$= \underline{17x + 38}$$

… then collect up $12x$ and $5x$ to make $17x$ …
and collect up $+\ 28$ and $+\ 10$ to make $+\ 38$

Sometimes there are minus signs inside one or more of the brackets.

See page 18 for a reminder on negative numbers.

EXAMPLE

Expand and simplify $4(3x - 7) + 5(2 - x)$

SOLUTION

$$\begin{aligned} 4(3x - 7) + 5(2 - x) &= 4 \times 3x - 4 \times 7 + 5 \times 2 - 5 \times x \\ &= 12x - 28 + 10 - 5x \\ &= 12x - 5x - 28 + 10 \\ &= \underline{7x - 18} \end{aligned}$$

Again, multiply out the brackets …

$(+5) \times (-x) = -5x$

… then collect up $+12x$ and $-5x$ to make $7x$ … and -28 and $+10$ to make -18

Watch carefully when there is a minus sign *in front of* one of the brackets, because the multiplication is much more tricky.

EXAMPLE

Expand and simplify $5(4x + 3) - 2(x + 3)$

SOLUTION

$$\begin{aligned} 5(4x + 3) - 2(x + 3) &= 5 \times 4x + 5 \times 3 - 2 \times x - 2 \times 3 \\ &= 20x + 15 - 2x - 6 \\ &= 20x - 2x + 15 - 6 \\ &= \underline{18x + 9} \end{aligned}$$

The terms in the second bracket are multiplied by -2

When you multiply by a negative number the signs change.

Finally, watch for a double minus multiplying up to give a positive term, as in this example.

EXAMPLE

Expand and simplify $2(4x - 1) - 3(x - 2)$

SOLUTION

$$\begin{aligned} 2(4x - 1) - 3(x - 2) &= 2 \times 4x - 2 \times 1 - 3 \times x - 3 \times -2 \\ &= 8x - 2 - 3x + 6 \\ &= 8x - 3x - 2 + 6 \\ &= \underline{5x + 4} \end{aligned}$$

The -3 multiplies with -2 to give $+6$

EXERCISE 7.6

1 Expand the following brackets:

a) $2(a + 5)$ **b)** $3(4 + b)$ **c)** $5(3 - c)$ **d)** $7(d - 1)$
e) $3(e + 4)$ **f)** $12(3 - f)$ **g)** $14(g + 1)$ **h)** $-2(h + 3)$
i) $-5(i - 3)$ **j)** $-2(3 + j)$ **k)** $-7(2 - k)$ **l)** $-3(4 - l)$

2 Expand the following brackets:

a) $2(3a + 7)$ **b)** $4(2b + 1)$ **c)** $5(7c - 2)$ **d)** $4(3 + 9d)$
e) $13(2e - 3)$ **f)** $7(3 - 2f)$ **g)** $5(2g - 1)$ **h)** $-2(3h + 1)$
i) $-3(2i + 5)$ **j)** $-2(3 + 4j)$ **k)** $-2(2k - 3)$ **l)** $-3(7 - 5l)$

3 Multiply out the brackets and simplify the results:

a) $2(x + 5) + 5(x + 2)$ **b)** $3(2x + 1) + 2(x + 5)$
c) $4(2x + 5) + 3(x + 3)$ **d)** $2(x + 5) + 3(2x + 5)$
e) $10(x + 1) + 6(2x + 1)$ **f)** $3(x + 5) + 4(x - 1)$
g) $2(x - 1) + 7(2x + 3)$ **h)** $3(x + 2) + 2(2x - 1)$
i) $5(x + 1) + 4(3x - 1)$ **j)** $6(3x - 2) + 5(4x + 3)$

4 Multiply out the brackets and simplify the results.
Take special care when there is a negative number in front of the second bracket.

a) $6(2x - 1) + 3(3x - 1)$ **b)** $4(x + 3) - 2(x + 1)$
c) $6(2x + 1) - 3(3x + 1)$ **d)** $8(2x - 5) + 5(3x - 2)$
e) $16(x - 1) + 5(3x - 2)$ **f)** $12(x + 2) - 3(2x + 4)$
g) $5(2x + 5) - 4(x - 2)$ **h)** $6(x + 1) - 2(x - 3)$
i) $7(x - 1) - 2(2x + 1)$ **j)** $4x + 3(2x - 1) - 5x$

7.7 More brackets

EXAMPLE

Expand:

a) $a(3a - 2)$
b) $2b(5a - 3b)$

Remember that both terms inside a bracket need to be multiplied by the term outside the bracket.

SOLUTION

a) $a(3a - 2) = \mathbf{a} \times 3a - \mathbf{a} \times 2$
$= \underline{3a^2 - 2a}$

$a \times 3a = 3 \times a \times a = 3a^2$

b) $2b(5a - 3b) = \mathbf{2b} \times 5a - \mathbf{2b} \times 3b$
$= \underline{10ab - 6b^2}$

$2b \times 5a = 2 \times 5 \times b \times a = 10ba = 10ab$
Write letters in alphabetical order.

You need to know how to expand two brackets multiplied together.

Here are three methods you could use.

EXAMPLE

Expand and simplify $(x + 3)(x + 5)$

SOLUTION

$$(x + 3)(x + 5) = x^2 + 3x + 5x + 15$$
$$= x^2 + 8x + 15$$

Each term in the first bracket is multiplied by each term in the second one – the '**smiles**' and '**eyebrows**' show which pairs of terms are being multiplied at each stage.

EXAMPLE

Expand and simplify $(x + 3)(x - 1)$

SOLUTION

$$(x + 3)(x - 1) = x^2 - x + 3x - 3$$
$$= x^2 + 2x - 3$$

Here we are using **FOIL**:
First x times x gives x^2
Outside: x times -1 gives $-x$
Inside: $+3$ times x gives $+3x$
Last: $+3$ times -1 gives -3

EXAMPLE

Expand and simplify $(x - 1)(x - 5)$

SOLUTION

	x	**-1**
x	x^2	$-x$
-5	$-5x$	$+5$

$$(x - 1)(x - 5) = x^2 - x - 5x + 5$$
$$= x^2 - 6x + 5$$

The two terms from the first bracket are written along one side of the grid, and the terms from the other bracket down the other side.

The grid is then filled in by multiplying corresponding pairs of terms, for example -1 times -5 gives $+5$

You may use whichever of these methods you prefer.

EXERCISE 7.7

1 Expand:

a) $a(a + 2)$ **b)** $b(b + 4)$ **c)** $c(c + 1)$
d) $d(3 - d)$ **e)** $e(7 - e)$ **f)** $f(4f - g)$
g) $g(3g + h)$ **h)** $h(3h - 2g)$ **i)** $i(6h - i)$

2 Expand:

a) $2a(a + 3)$ **b)** $3b(b + 2)$ **c)** $4c(c - 1)$
d) $2d(1 - d)$ **e)** $5e(2e - 1)$ **f)** $3f(5f - g)$
g) $2g(3g + 3h)$ **h)** $2h(3g - h)$ **i)** $7i(3h - i)$

3 Expand and simplify these brackets.
You must show all the steps in your working.

a) $(x + 2)(x + 3)$ **b)** $(x + 2)(x + 4)$ **c)** $(x + 3)(x + 4)$
d) $(x + 2)(x - 7)$ **e)** $(x + 2)(x - 1)$ **f)** $(x - 2)(x - 5)$
g) $(x - 3)(x + 2)$ **h)** $(x + 3)(x + 1)$ **i)** $(x - 4)(x - 2)$
j) $(x + 4)(x + 1)$ **k)** $(x - 1)(x - 3)$ **l)** $(x + 7)(x + 2)$

4 The length of one side of this square is $(b + 2)$ cm
The square has been divided up into four shapes.

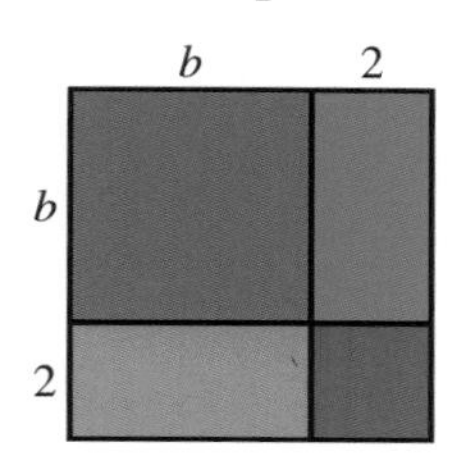

a) What is the area of the **red square**?
b) Write down an expression in terms of b for the area of:
 (i) the **green rectangle**
 (ii) the **pink rectangle**
 (iii) the **blue square**.
c) Write down an expression for the total area of the large square.
Simplify your expression.
d) How does the expression $(b + 2)^2$ relate to the large square?

$(b + 2)^2 = (b + 2)(b + 2)$

5 Expand:

a) $(a + 3)^2$ **b)** $(b + 1)^2$
c) $(c - 1)^2$ **d)** $(d - 2)^2$

7.8 Identities

The identity sign '≡' is used to show that two expressions are always equal to each other.

For example, $3x + 2x$ is always equal to $5x$ no matter what value of x you choose.
So you can write:

$3x + 2x \equiv 5x$

$3x + 2x \equiv 5x$ is an **identity**.

However, $3x + 2x = 5$, is only true when $x = 1$, so this is an **equation** (see Chapter 8).

EXAMPLE

a) Show that $(x + 1)^2 = x^2 + 2x + 1$ when $x = 2$

b) By expanding $(x + 1)^2$ show that you can write:
$(x + 1)^2 \equiv x^2 + 2x + 1$

So it is true for all values of x

SOLUTION

a) When $x = 2$ $\quad (x + 1)^2 = (2 + 1)^2$
$= 3^2$
$= 9$

Substitute $x = 2$ into both expressions.

When $x = 2$ $\quad x^2 + 2x + 1 = 2^2 + 2 \times 2 + 1$
$= 4 + 4 + 1$
$= 9$

Since both expressions equal 9 then $(x + 1)^2 = x^2 + 2x + 1$ when $x = 2$

b) $(x + 1)^2 = (x + 1)(x + 1)$

Using the grid method for expanding brackets:

×	**x**	**1**
x	x^2	x
1	x	1

So $\quad (x + 1)(x + 1) = x^2 + x + x + 1$
$= x^2 + 2x + 1$

So $\quad (x + 1)^2 \equiv x^2 + 2x + 1$

EXERCISE 7.8

1 Which of the following statements are correct?

a) $x + x + x \equiv 3x$ **b)** $x + 2x + 3x \equiv 6x$

c) $3(a + b) \equiv 3a + b$ **d)** $4x + 3x \equiv 7x$

2 Which of the following expressions are always equal to $6n$?

$10n - 4n$	$3 + 3n$	$2n \times 3n$
$3n + 3n$	$6 + n$	$3(n + n)$

3 a) Show that $(x + 5)^2 = x^2 + 10x + 25$ when $x = 3$

b) By expanding $(x + 5)^2$ show that you can write $(x + 5)^2 \equiv x^2 + 10x + 25$

7.9 Factorising

Sometimes you are given an expanded expression to factorise.

When you **factorise** an expression you write it using brackets – it is the opposite of expanding.

EXAMPLE

Factorise $15a + 6$

SOLUTION

Step 1 Find the highest common factor of both terms.

The largest number which goes into both 15 and 6 is 3

$15a + 6 = 3(\ldots + \ldots)$

Keep the same sign inside the brackets.

Step 2 We need to find the two terms to go inside the brackets.

There are two terms in the expression $15a + 6$ so there have to be two terms inside the brackets.

Find what needs to go in the brackets so that when you expand you get $15a + 6$

So $15a + 6 = 3(5a + 2)$

To get $15a$ you have to multiply 3 by $5a$

To get 6 you have to multiply 3 by 2

EXAMPLE

Factorise $5ab - a$

SOLUTION

Step 1 Find the highest common factor of both terms.

Both terms have an 'a' in them.
So 'a' can go outside the brackets: $a(\ldots - \ldots)$

Step 2 Find what needs to go in the brackets so that when you expand you get $5ab - a$

$$5ab - a = \underline{a(5b - 1)}$$

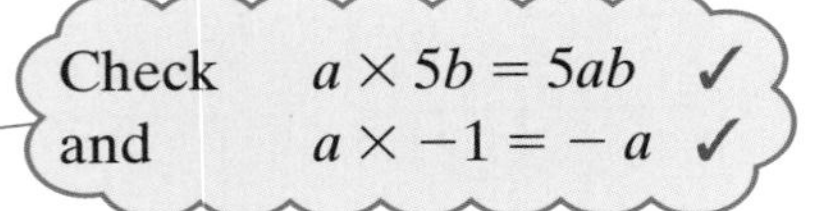
Check $a \times 5b = 5ab$ ✓
and $a \times -1 = -a$ ✓

EXAMPLE

Factorise $3x + x^2$

SOLUTION

$$3x + x^2 = \underline{x(3 + x)}$$

Check $x \times 3 = 3x$ ✓
and $x \times x = x^2$ ✓

EXERCISE 7.9

1 Factorise these expressions:

a) $3a + 6$ **b)** $9b + 12$
c) $15 + 10c$ **d)** $2d - 4$
e) $3 - 9e$ **f)** $5f - 5$

2 Factorise these expressions:

a) $8a - 12$ **b)** $30b - 24$
c) $12 - 24c$ **d)** $24d + 32$
e) $8e - 4$ **f)** $4 - 12f$

3 Factorise these expressions:

a) $2ab + 3a$ **b)** $ab + a$
c) $12cd - 6c$ **d)** $8d - 16cd$
e) $24ef - 30f$ **f)** $42f - 21ef$

4 Factorise these expressions:

a) $a^2 + 6a$ **b)** $b^2 - 10b$
c) $6c + c^2$ **d)** $2d^2 - d$
e) $12e^2 + e$ **f)** $f^2 + 12f$
g) $2x^2 + xy$ **h)** $y^2 + 8y$
i) $fg - 3g^2$ **j)** $5jk - 6k^2$

7.10 Writing formulae

Formulae can be generated from information given in words.

You can also generate them from information given by a diagram, or even another formula.

EXAMPLE

Write down a formula for the cost, C, of n people to go bowling.

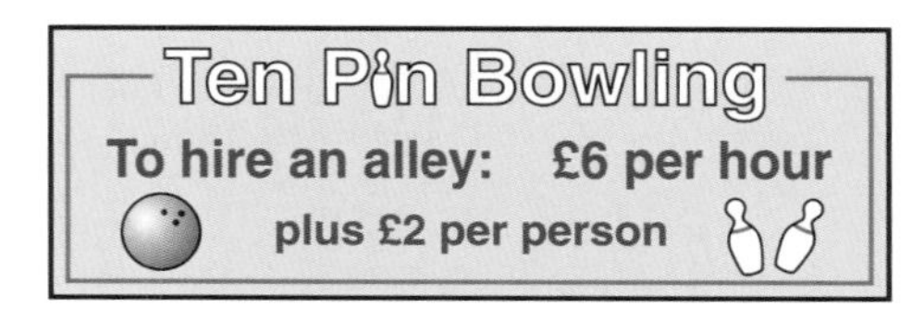

SOLUTION

Total cost (in pounds) $= £6 + £2 \times$ number of people

So $\underline{C = 6 + 2n}$

Don't put the sign in the formula.

EXAMPLE

A factory produces handbags by cutting and shaping rectangles of material.

To make one bag, a rectangle of material x cm by y cm is needed.

y cm

x cm

The factory finds that to make n bags without wastage they need a piece of material of area A

Obtain a formula for A in terms of x, y and n

SOLUTION

The area required for one bag is xy

Area of a rectangle is length $\times$ width

For n such bags, the area of material must be n times bigger, that is nxy

So the required formula is:

$\underline{A = nxy}$

EXERCISE 7.10

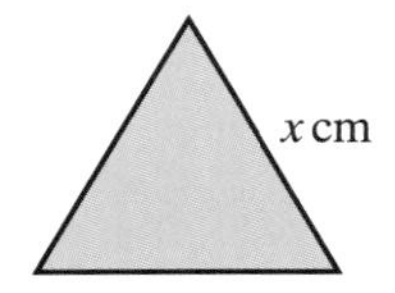

1 An equilateral triangle has sides of length x cm
Write down a formula for the perimeter, P, of the triangle.

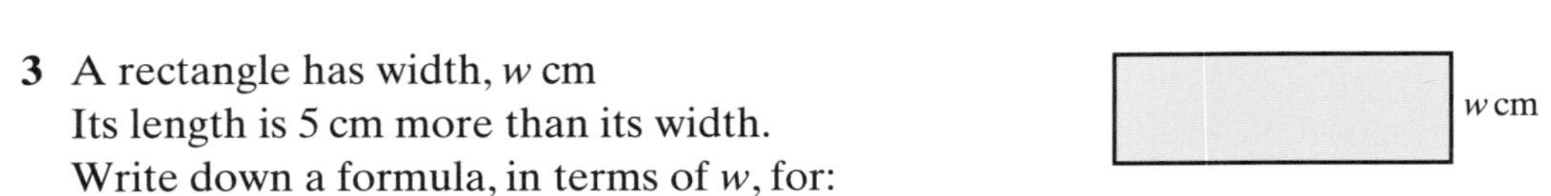

2 A rectangle has width, w cm, and length, l cm
Write down a formula in terms of w and l for:

a) the perimeter of the rectangle
b) the area of the rectangle.

3 A rectangle has width, w cm
Its length is 5 cm more than its width.
Write down a formula, in terms of w, for:

a) the length of the rectangle
b) the perimeter of the rectangle
c) the area of the rectangle.

4 A helicopter consumes n litres of fuel per minute.
Obtain a formula for the total number of litres, T, of fuel consumed during a flight lasting half an hour.

5 Each month Hugo pays £x for his house mortgage.
Each year Hugo pays £y for buildings insurance.

a) Write down an expression, in terms of x, for the cost of Hugo's mortgage for one year.
b) Find a formula for the total, £T, Hugo pays in mortgage and buildings insurance in one year.

6 A rectangle of area a has length l cm
Find a formula for its width, w cm, in terms of a and l.

7 Pencils cost 15 pence each, and pens 25 pence each.

a) Write an expression for the cost of x pencils.
b) Write a formula for the total cost, T pence, of x pencils and y pens.

8 The cost of hiring a bicycle is £5 plus a daily charge of £2 per day.

a) Find the cost of hiring the bicycle for 5 days.
b) Obtain a formula for the cost, £C, of hiring the bicycle for n days.

7.11 Changing the subject of a formula

A formula usually has a single letter term on the left-hand side of the equals sign – this is called the **subject** of the formula.

For example, the formula $C = 2\pi r$ has C as its subject.

Sometime you will want to rearrange the formula so that one of the other letters becomes the subject instead.

EXAMPLE

Make b the subject of: $a = b + c$

SOLUTION

$$a = b + c$$
$$b + c = a$$

Write the formula so that the letter you want is on the left-hand side.

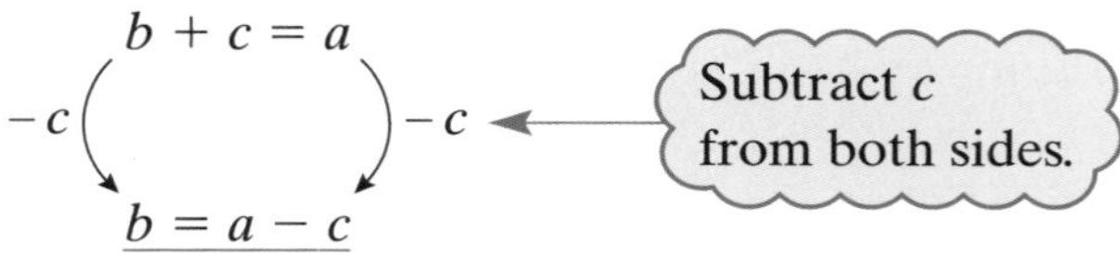

Subtract c from both sides.

EXAMPLE

Make r the subject of $C = 2\pi r$

SOLUTION

$$C = 2\pi r$$
$$2\pi r = C$$

First, rewrite the original equation with the left- and right-hand sides swapped over …

$$2\pi r = C$$
$$\div 2\pi \qquad \div 2\pi$$
$$r = \frac{C}{2\pi}$$

… then divide both sides by 2π

Some problems require a mixture of addition/subtraction and multiplication/division.

You need to think carefully about the appropriate order in which to do these.

EXAMPLE

Make x the subject of $y = mx + c$

SOLUTION

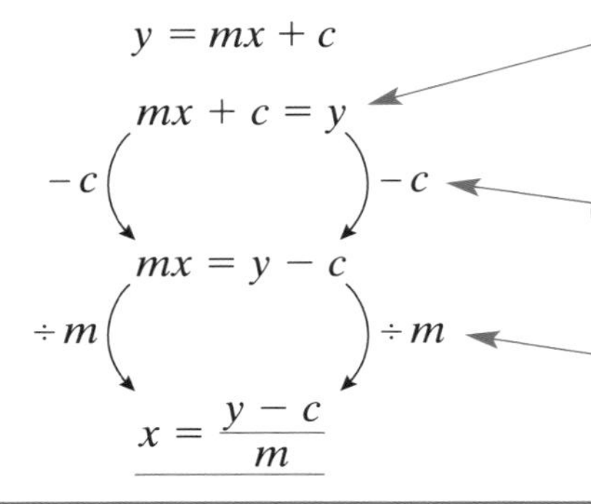

Again, begin by swapping the left- and right-hand sides over …

… next, subtract c from both sides …

… and finish off by dividing both sides by m

EXERCISE 7.11

1 Rearrange these formula to make a the subject:

a) $b = a + 3$ **b)** $b = 7 + a$ **c)** $b = a - 1$

d) $b = 2a$ **e)** $b = \dfrac{a}{4}$ **f)** $b = 3 - a$

2 Rearrange these formula to make x the subject:

a) $y = 2x + 1$ **b)** $y = 3 + 2x$ **c)** $y = 5x - 1$

d) $y = \dfrac{x}{3}$ **e)** $y = \dfrac{x}{4} + 5$ **f)** $y = \dfrac{x}{2} - 1$

3 Rearrange these formula to make c the subject:

a) $y = abc$ **b)** $y = a + 2c$ **c)** $y = ac - b$

d) $y = \dfrac{a}{c}$ **e)** $y = 2a + bc$ **f)** $y = \dfrac{ac}{b}$

4 Rearrange these formulae so that the bracketed letter becomes the subject:

a) $A = lw$ (make l the subject) **b)** $v = u + at$ (u)

c) $v = u + at$ (a) **d)** $E = mc^2$ (m)

e) $y = 4x + 3$ (x) **f)** $V = abc$ (b)

g) $A = \frac{1}{2}bh$ (h) **h)** $y = m(x - a)$ (m)

i) $y = mx + c$ (m) **j)** $y = mx + c$ (c)

REVIEW EXERCISE 7

Don't use your calculator for Questions 1–29.

1 When $a = 3, b = 4$ and $c = -2$, find the value of:

a) $a + b$ **b)** $2b$ **c)** $7 - b$ **d)** $b + 3a$
e) $b + c$ **f)** $b - c$ **g)** b^2 **h)** $2c^2$

2 When $a = 4, b = 2$ and $c = -5$, find the value of:

a) $3ab$ **b)** $2a^2$ **c)** $4a - 3b$ **d)** abc

3 When $x = 3, y = -2$ and $z = 10$, find the value of:

a) $2x^2$ **b)** y^3 **c)** $3z - xy$ **d)** $z(x + y)$

4 When $s = 1, t = 4$ and $u = -1$, find the value of:

a) su **b)** $t^2 + 3u$ **c)** $2s + 3t + 4u$ **d)** u^2

5 Look at these algebra cards.

$2n + n$	$2(n + n)$	$2n + 2n$
$2 \times 2n$	$2 + 2n$	$2n \times 2n$

Which of the expressions always have the same value as $4n$?

6 The cost, in pounds, of buying time on a satellite link can be worked out using this rule:

> Add three to the number of hours of time bought.
> Multiply your answer by 1000

a) Work out the cost of buying 4 hours of satellite time.

Julian bought some satellite time.
The cost was £12 000

b) Work out the number of hours of satellite time that Julian bought.

The cost of buying n hours of satellite time is C pounds.

c) Write down a formula for C in terms of n **[Edexcel]**

7 Andrew, Brenda and Callum each collect football stickers.
Andrew has x stickers.
Brenda has three times as many stickers as Andrew.

a) Write down an expression for the number of stickers that Brenda has.

Callum has nine fewer stickers than Andrew.

b) Write down the number of stickers that Callum has. **[Edexcel]**

8 Lisa packs pencils in boxes.
She packs 12 pencils in each box.
Lisa packs x boxes of pencils.
a) Write an expression, in terms of x, for the number of pencils Lisa packs.

Lisa also packs pens in boxes.
She packs 10 pens into each box.
Lisa packs y boxes of pens.
b) Write an expression, in terms of x and y, for the *total* number of pens and pencils that Lisa packs. **[Edexcel]**

9 Simplify the following:
a) $a + a + a$ **b)** $5b - 3b + 2b$ **c)** $10c + 3 - 4c - 5$
d) $4c - 3d + 5c + 4d$ **e)** $5e^2 + 3e^2 - e^2$ **f)** $4f^2 + 2f^2 - f^2$

10 Expand:
a) $4(2a + 3)$ **b)** $b(4a - 1)$ **c)** $c(3c - 4)$ **d)** $4(2d - 3)$

11 Expand and simplify:
a) $5(x + 2) + 2(x + 3)$ **b)** $2(y + 5) + 3(y + 1)$ **c)** $7(x - 1) + 6(x + 2)$
d) $4(x - 2) - 2(x + 4)$ **e)** $3(z + 1) + 5(z - 2)$ **f)** $4(2x - 2) + 2(x - 3)$

12 Factorise:
a) $4a + 2$ **b)** $8b - 4$ **c)** $12c + 8$
d) $8cd - 20c$ **e)** $12ef - 32e$ **f)** $4f^2 - 3f$
g) $8g^2 + 6g$ **h)** $15h + 30h^2$ **i)** $24hi - 18i$

13 Simplify:
a) $a + a + a + a + a$ **b)** $b \times b \times b \times b \times b$ **c)** $3c + 4c - 2c$ **d)** $2c \times 3d$

14 Simplify:
a) $7x + 4y - 6x + 2y$ **b)** $3x + 7y - 2x - 5y - x$ **c)** $6c^2 - 2c^2$

15 **a)** Simplify: **(i)** $e + f + e + f + e$ **(ii)** $e^2 + e^2 + e^2 + e^2$
b) Work out the value of $5e + 1$ when $e = -3$
c) Factorise $6ab - 3a$

16 Simplify:
a) $3x + 2y - 2x + 2y$
b) $5n^2 - 4n^2$
c) $5r + 4s - 4r - 2s$
d) $6a \times 2b$

17 **a)** Simplify: **(i)** $3g + 5g$ **(ii)** $2r \times 5p$
b) Expand $5(2y - 3)$
c) Expand and simplify $2(3x + 4) - 3(4x - 5)$ [Edexcel]

18 Tayub said, 'When $x = 3$, then the value of $4x^2$ is 144.'
Bryani said, 'When $x = 3$, then the value of $4x^2$ is 36.'
a) Who is right?
Explain why.
b) Work out the value of $4(x + 1)^2$ when $x = 3$ [Edexcel]

19 **a)** Work out the value of $2a + ay$ when $a = 5$ and $y = -3$
b) Work out the value of $5t^2 - 7$ when $t = 4$ [Edexcel]

20 Simplify:
a) $a \times a \times a$ **b)** $b \times b \times b \times b \times b$ **c)** $c \times c \times c \times c$
d) $d^3 \times d^2$ **e)** $e^7 \times e^9$ **f)** $f^9 \times f$
g) $g^8 \div g^6$ **h)** $\frac{h^7}{h^6}$ **i)** $\frac{i^6}{i^2}$

21 Simplify:
a) $3x^4 \times x^2$ **b)** $10y^8 \div 2y^5$ **c)** $8z^5 \div 2z^4$
d) $5y^3 \times 3y^4$ **e)** $4x^3 \times 3x^2$ **f)** $12y^8 \div 4y^5$

22 Expand and simplify:
a) $(x + 5)(x + 1)$ **b)** $(y + 5)(y + 7)$
c) $(x - 5)(x + 4)$ **d)** $(z + 4)^2$

23 A post office sells x 26 pence stamps and y 19 pence stamps during one day.
The total income from the stamps is T pence.
Write a formula expressing T in terms of x and y.

24 Rearrange the formula $C = 2\pi r$ to make r the subject.

25 Rearrange the formula $P = 3l$ to make l the subject.

26 Rearrange the formula $P = 2w + 2l$ to make w the subject.

27 **a)** Factorise: **(i)** $6x + 4$ **(ii)** $3x^2 - 6x$ **(iii)** $8xy + 4x$
b) Simplify: **(i)** $x^5 \div x^2$ **(ii)** $x^6 \times x^4$ **(iii)** $5x \times 3x$
c) Expand and simplify: **(i)** $4(x + 5) + 3(x - 7)$ **(ii)** $(x + 3)(x + 2)$ **(iii)** $(x - 3)^2$

28 **a)** Factorise $a^2 + 2a$

b) Make a the subject of $b = 2a - 1$

c) Expand and simplify $(x + 3)(x - 2)$

29 **a)** Simplify $a^6 \div a^4$

b) Simplify $3y \times 5y^3$

c) Make a the subject of $v = u + at$

You can use your calculator for Questions 30–37.

30 A school has a photocopier and a printing machine.
The cost of using the photocopier is given by the rule:

The cost of one copy is 4 pence.
Geoff makes 96 copies.

a) Work out the cost of using the photocopier to make 96 copies.

The cost of using the printing machine is given by the following rule:

Cost of using the printing machine = Copy fee + Number of copies × Cost of one copy

The cost of one copy is 3 pence.
The copy fee is 40 pence.
Charlotte makes 96 copies using the printing machine.

b) Work out the difference in their costs between Geoff and Charlotte. **[Edexcel]**

31 Audrey sells packets of sweets.
There are three sizes of packets.

There are n sweets in the small packet.
There are twice as many sweets in the medium packet as there are in the small packet.

a) Write down an expression, in terms of n, for the number of sweets in the medium packet.

There are 15 more sweets in the large packet than in the medium packet.

b) Write down an expression, in terms of n, for the number of sweets in the large packet.

A small packet of sweets costs 20p
Sebastian buys q small packets of sweets.

c) Write down an expression, in terms of q, for the cost in pence of the sweets. [Edexcel]

32 You can use this rule to work out the total number of points a football team got last season.

Multiply the number of wins by 3 and then add the number of draws.

Last season Rovers had 10 wins and 0 draws.

a) Use the rule to work out the total number of points Rovers got last season.

Last season United had 20 wins and 5 draws.

b) Use the rule to work out the total number of points United got last season. [Edexcel]

33 Eggs are sold in boxes.
A small box holds 6 eggs.
A large box holds 12 eggs.

Hina buys x small boxes of eggs.

a) Find, in terms of x, the total number of eggs in these small boxes.

Hina also buys 4 fewer of the large boxes of eggs than the small boxes.

b) Find, in terms of x, the total number of eggs in the *large* boxes that Hina buys.

c) Find, in terms of x, the *total* number of eggs that Hina buys.
Give your answer in its simplest form. [Edexcel]

34 a) Simplify $y^3 \times y^4$

b) Expand and simplify $5(2x + 3) - 2(x - 1)$

c) Factorise $4a + 6$

d) Factorise completely $6p^2 - 9pq$ [Edexcel]

35 a) Work out the value of $2n$ when n is:

(i) 1 **(ii)** 2 **(iii)** 32 **(iv)** 100

b) When n is an integer (whole number), what sort of number is:.

(i) $2n$ **(ii)** $2n + 1$?

36 Sam says that the expression $n^2 + n + 11$ always produces a prime number when n is an integer.

a) Work out the value of $n^2 + n + 11$ when n is:

(i) 1 **(ii)** 2 **(iii)** 3 **(iv)** 6 **(v)** 10

b) Is Sam right?
Give a reason for your answer.

37 The diagram represents a field in the shape of a rectangle.
All measurements are given in metres.
The field has a barn in one corner.
The barn is a square of side x.

a) Write down an expression, in terms of x, for the short side of the field.

b) Find a formula for the perimeter, P, of the field in terms of x.
Give your answer in its simplest form.

c) Make x the subject of your formula.

The perimeter of the field is 90 metres.

d) Find the value of x.

Diagram *not* accurately drawn

KEY POINTS

1 When substituting numbers into expressions, remember the BIDMAS sequence:

Work out	Brackets first
then	Indices (squares, cubes, etc.)
Next come	Division
	Multiplication
and then	Addition
	Subtraction

2 Remember that, for example, the value of $2x^2$ when $x = 3$ is

$2 \times 3^2 = 2 \times 9 = 18$

and *not* $6^2 = 36$

- The square must be done before the multiplication by 2

3 The laws of indices are:

$x^a \times x^b = x^{a+b}$

$x^a \div x^b = x^{a-b}$

4 When expanding brackets, watch for a minus sign in front of a bracket – this will change the sign of all the terms inside the bracket, for example:

$-3(2x - 5) = -6x + 15$

5 Factorising, in algebra, means the reverse of expanding.

There will be the same number of terms inside the bracket as there were in the original expression.

6 To change the subject of a formula, rewrite it with the letter you want on the left-hand side first.

Internet Challenge 7

The language of algebra

The wordsearch below contains 20 algebraic words for you to find.
When you have located them, use the internet to check the precise meaning of each word.

T	E	Q	U	A	T	I	O	N	N	D	P	O	R	I
V	X	S	P	G	F	E	N	Y	M	Z	S	E	R	Y
B	P	O	W	E	R	F	T	L	A	C	I	D	A	R
Q	R	X	O	Y	U	O	M	K	P	P	M	S	T	L
U	E	S	N	E	O	A	N	T	P	N	P	C	I	H
O	S	L	D	R	U	S	A	O	I	S	L	D	O	S
T	S	A	O	D	E	R	L	U	N	F	I	V	N	J
I	I	N	D	E	X	Y	E	A	G	I	F	A	A	Y
E	O	J	E	L	N	C	D	U	H	C	Y	R	L	T
N	N	D	N	O	I	T	C	N	U	F	E	I	Q	I
T	E	R	M	G	E	O	W	L	A	L	H	A	F	T
A	R	I	L	T	C	U	D	O	R	P	N	B	A	N
V	A	B	R	A	C	K	E	T	W	T	X	L	B	E
L	Y	T	S	F	A	C	T	O	R	I	S	E	L	D
C	I	T	A	R	D	A	U	Q	F	V	H	I	Y	I

Here are the words to find.
They may run left, right, up, down or diagonally.

EQUATION	EXPAND	EXPRESSION	FACTORISE
FUNCTION	IDENTITY	INDEX	MAPPING
POLYNOMIAL	POWER	PRODUCT	QUADRATIC
QUOTIENT	BRACKET	RATIONAL	ROOT
SIMPLIFY	SURD	TERM	VARIABLE

CHAPTER 8

Equations

In this chapter you will **revise and extend earlier work on:**

- the language of algebra.

You will **learn how to:**

- solve simple equations
- solve harder linear equations
- solve equations involving brackets
- set up an equation in order to solve a problem
- find approximate solutions by trial and improvement
- solve inequalities.

You will also be **challenged to:**

- investigate the mathematics of Carl Friedrich Gauss.

Starter: Triangular arithmagons

In a triangular arithmagon, the number along each side of the triangle is obtained by adding up the numbers in the adjacent corners.

The first arithmagon has been filled in, to show you how this works.

Complete the other three arithmagons.

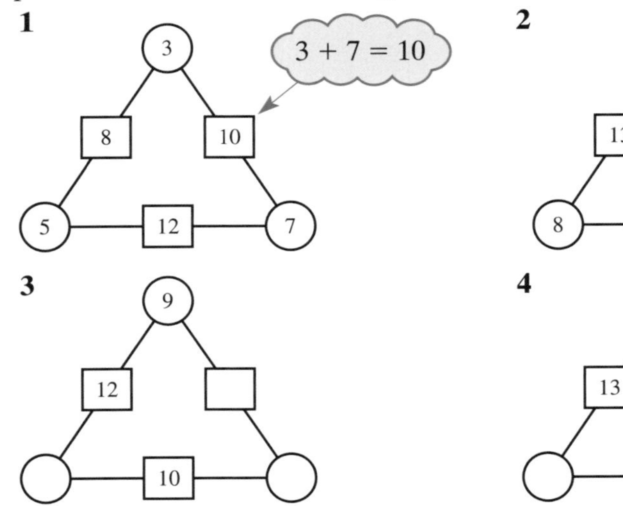

2 Top: 5; sides: 13, [ ]; bottom corners: 8, 9; bottom: [ ]

4 Top: ( ); sides: 13, 17; bottom corners: ( ), ( ); bottom: 16

Now make some arithmagon puzzles of your own.

8.1 Expressions and equations

In your work on algebra, you have met the words **expression**, **equation** and **formula**.

They all mean similar, but slightly different, things.

An **expression** is a piece of algebra usually containing one or more letters:

$6x + 1$ is an expression.

An **equation** also contains letters and numbers, but has an equals sign:

$6x + 1 = 13$ is an equation.

A **formula** is designed to solve a problem:

$V = IR$ is a formula used to calculate electrical voltage, V

You can think of equations as being statements that are true for some values of the quantities involved, and false for others.

For example, the equation $6x + 1 = 13$ is:

true when $x = 2$, since $6 \times 2 + 1 = 13$, but
false when $x = 1$, since $6 \times 1 + 1 = 7$

Some equations, however, are true for *all* values of x and these are called **identities**:

$3(x + 5) = 3x + 15$ is an identity.

Strictly speaking, identities should be written using a special symbol, $3(x + 5) \equiv 3x + 15$, but in practice this is often overlooked.

EXERCISE 8.1

Look at the various algebraic statements labelled **A** to **J**.

A $1 + x$ **B** $A = \pi r^2$ **C** $10x - 3$ **D** $x + x = 2x$
E $x + 4 = 12$ **F** $x + 5 = 17$ **G** $5x = 20$ **H** $x + x + x = 3x$
I $x^2 = 9$ **J** $C = 2\pi r$

1 Which ones are expressions?

2 Which ones are equations?

3 Which ones would you call formulae?

4 Pick out any identities, and rewrite them using the identity symbol '$\equiv$'.

8.2 Balancing equations

Some simple algebraic equations can be solved in just one step, or even **by inspection**.

Just by looking.

For example, if $x + 2 = 9$ you can spot, by inspection, that x must be 7

A more formal approach would be to subtract 2 from both sides:

$$x + 2 = 9$$
$$-2 \qquad -2$$
$$x = 7$$

You must do the same thing to both sides of the equation – this is called **balancing**.

You say that you have **solved** the equation $x + 2 = 9$
The solution is $x = 7$

In this section you will be solving simple equations of this kind.

Although you may well be able to spot some of the solutions by inspection, it is better to solve them formally, since this equips you with the skills needed for harder equations where the solutions cannot be spotted by inspection.

In the following examples we are solving equations by **balancing**.

This means you must apply the same **operation** to both sides of the equation.

$+$, $-$, $\times$ or $\div$

EXAMPLE

Solve these equations:

a) $3x = 12$

b) $x + 3 = 7$

c) $x - 3 = 2$

d) $\frac{x}{3} = 8$

e) $\frac{15}{x} = 5$

SOLUTION

a)

$$3x = 12 \quad (\div 3 \text{ both sides})$$

$$\underline{x = 4}$$

To undo multiply by 3 you need to divide by 3

b)

$$x + 3 = 7 \quad (-3 \text{ both sides})$$

$$\underline{x = 4}$$

To undo add 3 you need to subtract 3

c)

$$x - 3 = 2 \quad (+3 \text{ both sides})$$

$$\underline{x = 5}$$

To undo subtract 3 you need to add 3

d)

$$\frac{x}{3} = 8 \quad (\times 3 \text{ both sides})$$

$$\underline{x = 24}$$

$\frac{x}{3}$ means x divided by 3'
To undo divide by 3 you need to multiply by 3
Note: $\frac{x}{3}$ is the same as $\frac{1}{3}x$

e)

$$\frac{15}{x} = 5 \quad (\times x \text{ both sides})$$

$$15 = 5x \quad (\div 5 \text{ both sides})$$

$$3 = x$$

So $\underline{x = 3}$

The solution to an equation isn't always an **integer** (whole number). It may be a negative number or a fraction.

EXAMPLE

Solve: **a)** $10a = 5$ **b)** $b + 3 = 1$ **c)** $c - 3 = -5$ **d)** $\frac{1}{4}d = -1$

SOLUTION

a) $10a = 5$

$\div 10$ on both sides

$a = \frac{5}{10}$

$\underline{a = 0.5 \text{ or } \frac{1}{2}}$

b) $b + 3 = 1$

-3 on both sides

$\underline{b = -2}$

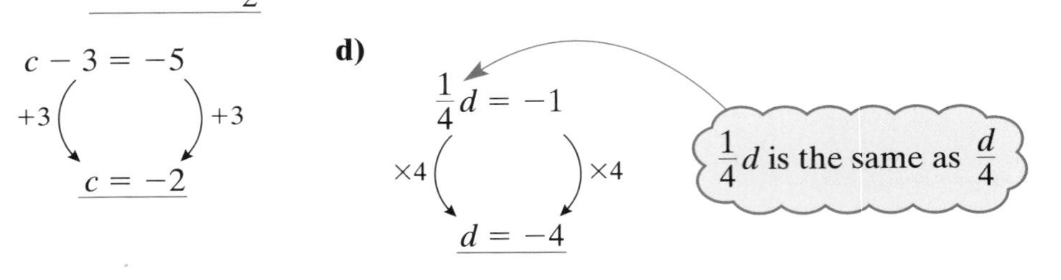

EXERCISE 8.2

Don't use your calculator for Questions 1–4.

Show all your working when answering the questions in this exercise.

1 Solve: **a)** $a + 4 = 10$ **b)** $b + 20 = 80$ **c)** $c + 13 = 20$

2 Solve: **a)** $a - 4 = 4$ **b)** $b - 10 = 20$ **c)** $c - 17 = 0$

3 Solve: **a)** $2a = 16$ **b)** $4b = 40$ **c)** $7c = 21$

4 Solve: **a)** $\frac{a}{3} = 5$ **b)** $\frac{1}{6}b = 1$ **c)** $\frac{c}{9} = 2$

You can use your calculator for Questions 5 and 6.

5 Solve:

a) $x + 3 = 10$ **b)** $x - 2 = 7$ **c)** $x + 13 = 4$

d) $\frac{x}{10} = 3$ **e)** $2x = 4$ **f)** $\frac{1}{5}x = 2$

g) $x - 3 = 1$ **h)** $\frac{x}{2} = 6$ **i)** $3x = 24$

6 Solve:

a) $3t = 48$ **b)** $u + 7 = 4$ **c)** $8p = 4$

d) $21g = 7$ **e)** $4q = 1$ **f)** $w + 6 = 4$

g) $5y = 2$ **h)** $\frac{12}{h} = 4$ **i)** $z + 10 = 3$

8.3 Further solving equations

You often have to solve an equation which involves more than one step.

Always 'undo' addition and subtraction first, and then multiplication and division.

EXAMPLE

Solve:

a) $5x - 1 = 14$ **b)** $3x + 8 = 2$

SOLUTION

a)

$$5x - 1 = 14$$

$+1$ (both sides)

$$5x = 15$$

$\div 5$ (both sides)

$$\underline{x = 3}$$

b)

$$3x + 8 = 2$$

-8 (both sides)

$$3x = -6$$

$\div 3$ (both sides)

$$\underline{x = -2}$$

EXAMPLE

Solve:

a) $10 - a = 15$ **b)** $12 - 2b = 11$

SOLUTION

a)

$$10 - a = 15$$

-10 (both sides)

$$-a = 5$$

The '−' sign stays with the '*a*'

$$\underline{a = -5}$$

You can work this out by inspection. Or by multiplying both sides by −1

b)

$$12 - 2b = 11$$

-12 (both sides)

$$-2b = -1$$

$-2b$ means $-2 \times b$

$\div -2$ (both sides)

$$\underline{b = \frac{1}{2} \text{ or } 0.5}$$

EXAMPLE

Solve:

a) $\dfrac{2x-1}{3} = 7$ b) $\dfrac{2x}{3} - 1 = 7$

SOLUTION

a)

$$\frac{2x-1}{3} = 7$$

$\times 3$ (both sides)

$$2x - 1 = 21$$

$+1$ (both sides)

$$2x = 22$$

$\div 2$ (both sides)

$$\underline{x = 11}$$

> $2x - 1$ has all been divided by 3 so you need to 'undo' divide by 3 first.

b)

$$\frac{2x}{3} - 1 = 7$$

> Undo '-1' first.

$+1$ (both sides)

$$\frac{2x}{3} = 8$$

$\times 3$ (both sides)

$$2x = 24$$

$\div 2$ (both sides)

$$\underline{x = 12}$$

> You can check you have the right answer by substituting back into the original equation …
>
> … $\dfrac{2(12)}{3} - 1 = \dfrac{24}{3} - 1$
> $= 8 - 1 = 7$ ✓

EXERCISE 8.3

Don't use your calculator for Questions 1 and 2.

1 Solve:

a) $3a + 4 = 16$ **b)** $2b + 1 = 3$ **c)** $8c + 3 = 27$
d) $5d - 1 = 4$ **e)** $10 + 3e = 16$ **f)** $12 + 4f = 32$
g) $6g - 10 = 2$ **h)** $15h - 8 = 7$ **i)** $3i - 6 = 6$

2 Solve:

a) $10 - a = 4$ **b)** $20 - b = 11$ **c)** $18 - c = 12$
d) $5 - 2d = 1$ **e)** $8 - 3e = 2$ **f)** $12 - 4f = 0$
g) $10 - 5g = 0$ **h)** $15 - 3h = 3$ **i)** $8 - 7i = 8$

You can use your calculator for Questions 3 and 4.

3 Solve:

a) $4a - 1 = 1$ **b)** $8b + 6 = 12$ **c)** $10c + 5 = 5$
d) $10 + 5d = 5$ **e)** $18 - e = 20$ **f)** $12 - 2f = 16$
g) $4 + 3g = -5$ **h)** $5h - 1 = 1$ **i)** $8i - 4 = -2$

4 Solve:

a) $\frac{a-1}{4} = 2$ **b)** $\frac{b+3}{5} = 1$ **c)** $\frac{3c}{4} + 2 = 8$
d) $\frac{2+d}{7} = 3$ **e)** $\frac{2e-8}{2} = 1$ **f)** $\frac{f}{3} - 1 = 1$
g) $\frac{g}{2} + 4 = 3$ **h)** $10 - \frac{h}{3} = 6$ **i)** $\frac{3-2i}{2} = 1$

8.4 Solving an equation with brackets

EXAMPLE

Solve $4(3x - 2) = 10$

SOLUTION

Multiply out any brackets first:

$$4(3x - 2) = 4 \times 3x - 4 \times 2$$
$$= 12x - 8$$

So you need to solve

$$12x - 8 = 10$$
$+8$ $+8$
$$12x = 18$$
$\div 12$ $\div 12$
$$x = 1.5$$

Remember multiply out the brackets first …
… then undo + and − …
… and lastly undo × and ÷

Check: $4(3 \times 1.5 - 2) = 10$
$4 \times 2.5 = 10$ ✓

EXERCISE 8.4

Solve:

1 $4(2a + 1) = 12$ **2** $3(2b - 1) = 9$ **3** $5(2c - 1) = 55$ **4** $3(4d - 6) = 18$

5 $6(10 - 3e) = 24$ **6** $2(20 - 2f) = 48$ **7** $8(4 - 4g) = 24$ **8** $9(3 - h) = 54$

9 $2(1 + 3i) = 44$ **10** $4(6j - 1) = 8$ **11** $3(2 + 4k) = 24$ **12** $10(9l - 5) = -50$

8.5 Solving equations with an unknown on both sides

When solving an equation with an unknown (e.g. x) on both sides of the equals sign you have to:

Step 1 Collect all the x terms on one side and all the numbers on the other.

Step 2 Simplify any like terms.

Step 3 Solve the resulting equation.

These examples show you how this works.

EXAMPLE

Solve the equation $5x + 2 = 3x + 16$

SOLUTION

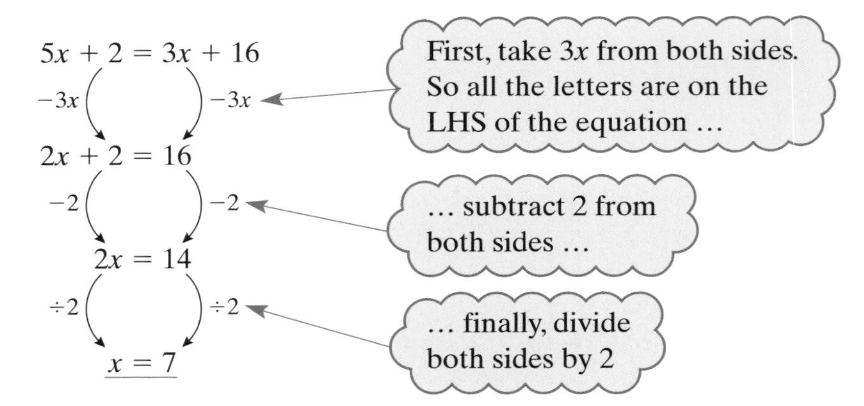

Check: $5 \times 7 + 2 = 3 \times 7 + 16$

$35 + 2 = 21 + 16$

$37 = 37$ ✓

Here is another example, this time with some minus signs.
The principle is exactly the same.

EXAMPLE

Solve the equation $3x - 9 = 3 - x$

SOLUTION

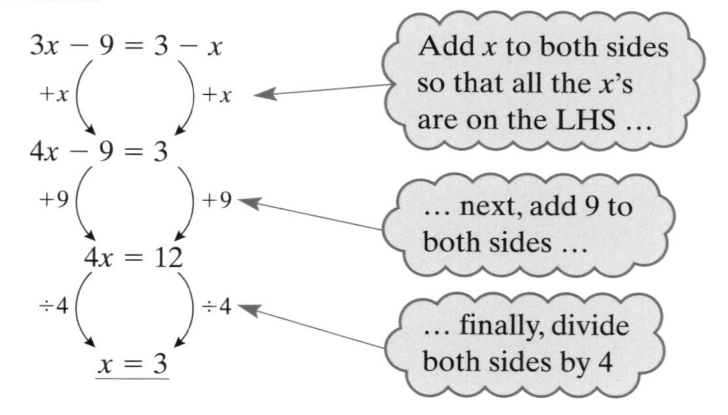

$3x - 9 = 3 - x$
$+x$
$+x$
Add x to both sides so that all the x's are on the LHS …
$4x - 9 = 3$
$+9$
$+9$
… next, add 9 to both sides …
$4x = 12$
$\div 4$
$\div 4$
… finally, divide both sides by 4
$x = 3$

Check: $3 \times 3 - 9 = 3 - 3$

$0 = 0$ ✓

When the overall **coefficient** of x looks like being negative, it may be easier to collect the x terms on the right hand side instead, as in this final example.

The number in front of the x

EXAMPLE

Solve the equation $16 + x = 12 + 5x$

SOLUTION

$16 + x = 12 + 5x$

$-x$ (both sides) — Take x from both sides so that all the x's are on the RHS …

$16 = 12 + 4x$

-12 (both sides) — … next, take 12 from both sides …

$4 = 4x$

$\div 4$ (both sides) — … finally, divide both sides by 4

$x = 1$

Check: $16 + 1 = 12 + 5 \times 1$

$17 = 17$ ✓

EXAMPLE

Solve $7(3x - 1) = 2(3x + 4)$

SOLUTION

First expand the brackets:

$$7(3x - 1) = 2(3x + 4)$$
$$7 \times 3x - 7 \times 1 = 2 \times 3x + 2 \times 4$$
$$21x - 7 = 6x + 8$$

Then solve:

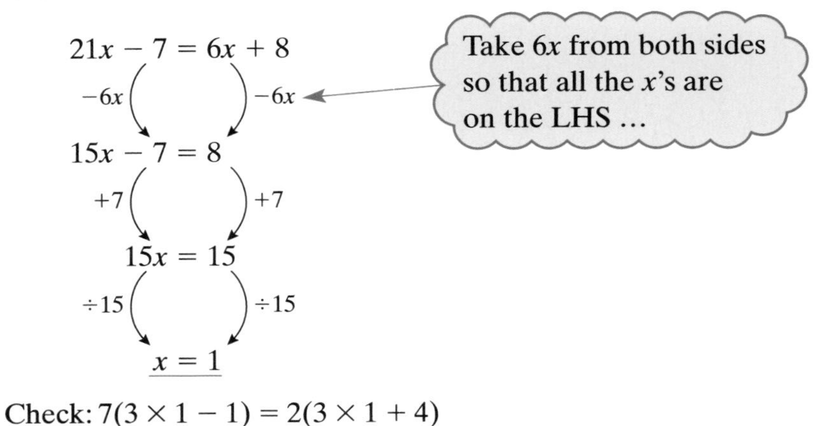

21x − 7 = 6x + 8
−6x
−6x
15x − 7 = 8
+7
+7
15x = 15
÷15
÷15
x = 1
Take 6x from both sides so that all the x's are on the LHS ...

Check: $7(3 \times 1 - 1) = 2(3 \times 1 + 4)$
$7 \times 2 = 2 \times 7 = 14$ ✓

EXERCISE 8.5

Solve these equations, showing the steps of your working clearly.
All the answers are integers (whole numbers), but some may be negative.

1 $4x + 5 = x + 14$

2 $6x + 1 = 8 - x$

3 $10x + 3 = 8x + 11$

4 $7x - 15 = x + 3$

5 $12x + 5 = 44 - x$

6 $15 - 2x = 30 - 5x$

7 $14 + 2x = 9x$

8 $10 + 5x = 3x + 6$

9 $x + 3 = 3 + 2x$

10 $4x = 55 - x$

Expand any brackets and simplify before solving these equations.

11 $5(x - 5) + 3 = 13$

12 $4(x - 1) = 45 - 3x$

13 $3(x + 5) = 23 - x$

14 $2(x - 4) = 4$

15 $3(2x + 7) = 4x + 41$

16 $x + 8 = 6(x - 2)$

17 $5(2x - 1) = 3 + 2(3x + 4)$

18 $4(x - 2) = 4 + 2(x - 3)$

19 $2(3x + 14) = 3(x + 9) - 11$

20 $3(2x + 3) = 2(x + 1) + 23$

8.6 Solving quadratic equations

A quadratic equation has an 'x^2' term in it.

x^2 or 'x squared' means $x \times x$

For example, $x^2 = 9$ and $2x^2 + 1 = 19$ are both quadratic equations.

Remember that $3 \times 3 = 9$ and $-3 = -3 = 9$ so the solution to $x^2 = 9$ is $x = 3$ or -3

You can write this as $x = \pm 3$

'x equals plus or minus 3'

When you solve a quadratic equation you need to rearrange the equation so that you have just 'x^2' on the left-hand side.

You then square root to get rid of the 'square'.

EXAMPLE

Solve $3x^2 + 2 = 50$

SOLUTION

You need to end up with just x^2 on the left-hand side:

$$3x^2 + 2 = 50$$

-2 ( ) -2

$$3x^2 = 48$$

$\div 3$ ( ) $\div 3$

$$x^2 = 16$$

Take the square root of both sides – don't forget the negative square root.

So $x = 4$ or -4

This can be written as $\underline{x = \pm 4}$

Say 'x equals plus or minus 4'

EXERCISE 8.6

Solve the following quadratic equations.

1 $x^2 = 4$ **2** $x^2 = 36$ **3** $x^2 = 144$ **4** $2x^2 = 32$

5 $3x^2 = 675$ **6** $4x^2 = 4$ **7** $x^2 + 1 = 10$ **8** $x^2 + 4 = 125$

9 $x^2 - 6 = 10$ **10** $x^2 - 11 = 70$ **11** $2x^2 - 12 = 6$ **12** $2x^2 + 12 = 350$

13 $3x^2 - 8 = 100$ **14** $\frac{x^2}{2} = 32$ **15** $\frac{x^2}{2} = 72$ **16** $400 - 2x^2 = 8$

8.7 Solving problems

Sometimes you are given a problem that involves setting up an equation and then solving it.

EXAMPLE

I think of a number.
I multiply it by 4 and
subtract 6
My answer is 14

Sarah

What number is Sarah thinking of?

SOLUTION

Let n represent Sarah's 'mystery number'.
So our equation is:

$$4n - 6 = 14$$

$+6$ (both sides)

$$4n = 20$$

$\div 4$ (both sides)

$$n = 5$$

So Sarah's mystery number is 5

Check: $4 \times 5 - 6 = 14$ ✓

Sometimes you need to use brackets when you set up the equation.

EXAMPLE

I think of a number. When I add 20 on to my number I get twice as much as when I add 6

Simon

SOLUTION

Let n represent the number Simon is thinking of.

Then $n + 20$ is twice as much as $n + 6$

So: $n + 20 = 2(n + 6)$

Expand the brackets: $n + 20 = 2n + 12$

Solve:

$$n + 20 = 2n + 12$$

$-n$ (both sides)

$$20 = n + 12$$

-12 (both sides)

$$8 = n$$

So Simon is thinking of the number 8

Check: $8 + 20 = 2(8 + 6)$

$28 = 2 \times 14$ ✓

EXERCISE 8.7

Don't use your calculator for Questions 1–6.

1 A CD costs c pounds.
Tim buys six CDs.

a) Write down an expression, in terms of c, for the cost of Tim's CDs.

He spends £66

b) Write down an equation representing this information.

c) Solve your equation to find the cost of one CD.

2 An equilateral triangle has three sides of length x cm

a) Write down an expression, in terms of x, for the perimeter of the triangle.

The perimeter of the triangle is 48 cm

b) Write down an equation for this information.

c) Solve your equation to find the length of one side of the triangle.

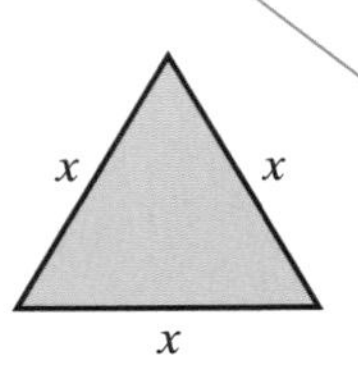

Distance all the way round the outside edge.

3 The cost of a phone call was c pence per minute.
The price has increased by 2p per minute.

a) Write down an expression, in terms of c, for the new cost per minute.

b) Write down an expression, using brackets, for the cost of a 3 minute call.

A 3 minute phone call now costs 12p

c) Write down an equation for this information.

d) Solve your equation to find the value of c

4

I think of a number. I double my number and add 4 The answer is 20

Jamie

I think of a number. I add 4 to my number and multiply the result by 3. The answer is 18

Mel

a) **(i)** Use n to stand for Jamie's mystery number.
Write down an equation for Jamie's mystery number.

(ii) Solve your equation to find Jamie's number.

b) **(i)** Use m to stand for Mel's mystery number.
Write down an expression, in terms of m, for 'add 4 to Mel's number'.

(ii) Use brackets to multiply the whole of your expression by 3

(iii) Write down an equation for Mel's mystery number.

(iv) Solve your equation to find Mel's number.

5 A rectangle has width w centimetres.
The length of the rectangle is 6 cm more than its width.

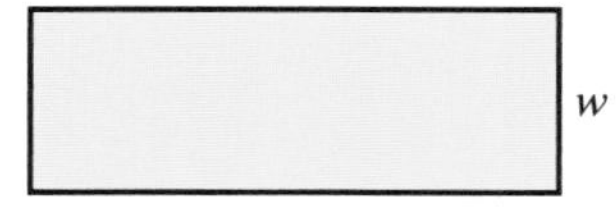

a) Write down an expression for the length of the rectangle.

b) **(i)** Write down an expression for the perimeter of the rectangle.

(ii) Simplify your expression.

The perimeter of the rectangle is 28 cm

c) What is the width of the rectangle?

6 A rectangle has a width of w centimetres.
The length of the rectangle is twice its width.

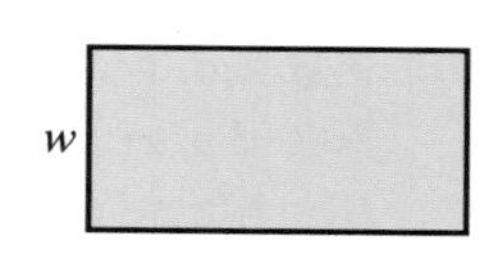

a) Write down an expression for the length of the rectangle.
b) **(i)** Write down an expression for the perimeter of the rectangle.
(ii) Simplify your expression.

The perimeter of the rectangle is 48 cm
c) What is the width of the rectangle?

You can use your calculator for Questions 7–10.

7 **a)** Write down an expression for the area of this square.

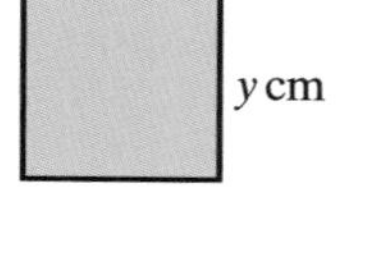

The area of the square is 169 cm^2
b) What is the length of one side of the square?

8 **a)** **(i)** Write down an expression, in terms of x, for the perimeter of this rectangle.
(ii) Simplify your expression.

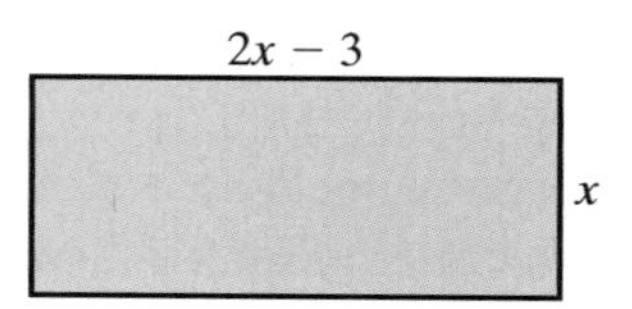

b) **(i)** Write down an expression, in terms of x, for the perimeter of this triangle.
(ii) Simplify your expression.

Not to scale

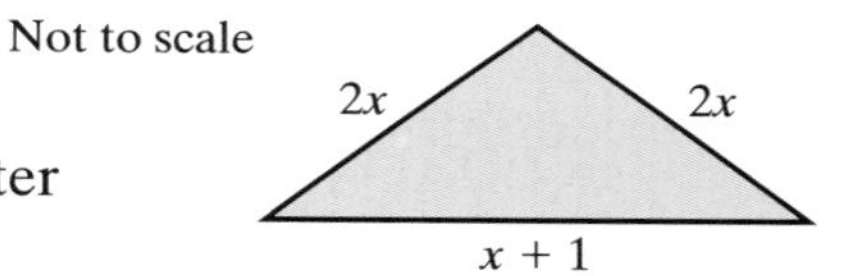

The perimeter of the rectangle is the same as the perimeter of the triangle.
c) **(i)** Write down an equation to show this information.
(ii) Solve your equation to find the width of the rectangle.

You will need to use brackets to solve the next two problems.

9

a) Write down an expression for this information.
Use n for Joe's mystery number.

Joe gets the same answer when he multiplies his mystery number by 4
b) Write down an equation for this information.
c) Solve your equation to find Joe's mystery number.

10 Nat and Marina each think of the same number.
Nat multiplies the number by 7, and then adds 5
Marina adds 7 to the number, and then multiplies by 5
They both end up with the same answer.

a) Write this information as an equation.

b) Solve your equation, to find the number they both thought of.

8.8 Trial and improvement

So far, you have been finding the exact solutions to equations.

Not all equations can be solved like this, however.
So there are ways of using **trial and improvement** to find an approximate solution.

When using such a method you can find the solution correct to 1 decimal place, or 2, and so on – but you will never find the exact solution.

Exam questions will tell you when to use a trial and improvement method and tell you how accurate your final answer needs to be.

EXAMPLE

The equation $x^2 + 4x = 25$ has a solution between 3 and 4
Use trial and improvement to find this solution correct to 1 decimal place.

SOLUTION

We need to find a value of x such that $x^2 + 4x = 25$
You will find a table like this one helpful.

Trial	Working	Result	Comment
$x = 3$	$3^2 + 4 \times 3$	21	Too small
$x = 4$	$4^2 + 4 \times 4$	32	Too big
$x = 3.5$	$3.5^2 + 4 \times 3.5$	26.25	Too big
$x = 3.3$	$3.3^2 + 4 \times 3.3$	24.09	Too small
$x = 3.4$	$3.4^2 + 4 \times 3.4$	25.16	Too big

So x must be between 3 and 4; so try $x = 3.5$ (half way)

So x must be between 3 and 3.5; so try $x = 3.3$

So x must be between 3.3 and 3.5; so try $x = 3.4$

Now we are only asked for one decimal place, so we need to work out whether $x = 3.3$ or 3.4

We use 3.35 as it is half way between 3.3 and 3.4

We do this by checking whether 3.35 is too big or too small.

Trial	Working	Result	Comment
$x = 3.35$	$3.35^2 + 4 \times 3.35$	24.6225	Too small

It is too small, so x must be bigger than 3.35

So $x = 3.4$ to 1 d.p.

If $x = 3.35$ had been too big then x would be smaller than 3.35, so x would be 3.3 to 1 d.p.

In the exam it is really important to show the examiner all your trials – even the ones that missed by a long way. The examiner will want to see the improvement taking place, not just the final result.

For full marks on this question you must test 3.35
Do **not** just take the closest value to 25, that is 3.4, as you will lose marks.

EXERCISE 8.8

Use a table like the one in the example above to help with these questions.
Use trial and improvement to solve these equations correct to 1 decimal place.

1 The equation $x^2 + 7x = 48$ has a solution between $x = 4$ and $x = 5$

2 The equation $x^2 + 5x = 9$ has a solution between $x = 1$ and $x = 2$

3 The equation $x^3 + 2x = 10$ has a solution between $x = 1$ and $x = 2$

4 The equation $x^2 - 2x = 11$ has a solution between $x = 4$ and $x = 5$

5 The equation $x^3 - 10x + 1 = 0$ has a solution between $x = 3$ and $x = 4$

Use trial and improvement to solve these equations correct to 2 decimal places.

6 The equation $x^2 + 7x - 35 = 0$ has a solution between $x = 3$ and $x = 4$

7 The equation $2x^2 + x = 7$ has a solution between $x = 1$ and $x = 2$

8 The equation $x^2 + 2x = \frac{100}{x}$ has a solution between $x = 4$ and $x = 5$

9 The equation $x(x + 1) = 3$ has a solution between 1 and 2

10 **a)** Show that the equation $3x^2 - x - 1 = 0$ has a solution between $x = 0$ and $x = 1$
b) Use trial and improvement to find this solution correct to 2 decimal places.
The same equation has another solution between $x = 0$ and $x = -1$
c) Find this solution, correct to 2 decimal places.

8.9 Inequalities

In mathematics we use an **inequality sign** when two numbers or expressions are not equal.

For example, 4 is less than 5
You can write this using an inequality sign like this:

$4 < 5$

4 is **less than** 5

Or you could write:

$5 > 4$

5 is **greater than** 4

Negative numbers are numbers which are less than 0

Remember: $-5 < -4$

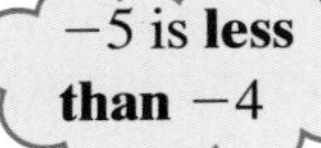

$x < 4$ means x can be any number which is less than 4

$x \geqslant 4$ means x can be any number which is 4 **or more**.

The '**mouth**' opens towards the bigger amount.

You can show these inequalities on a number line like this:

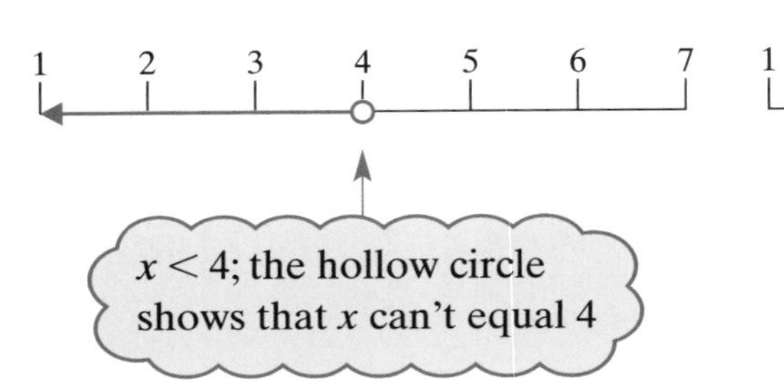

x < 4; the hollow circle shows that x can't equal 4

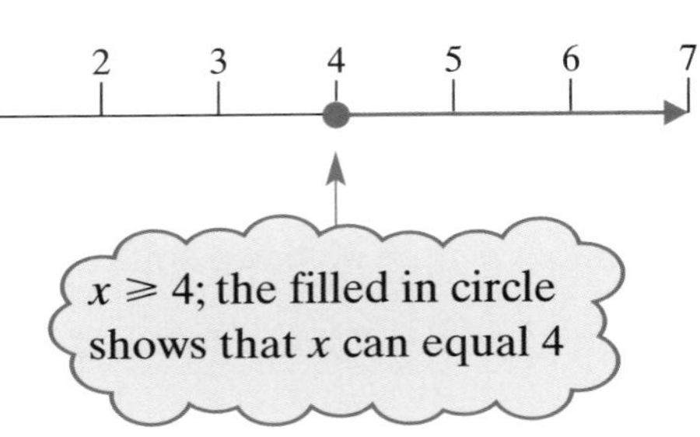

x ⩾ 4; the filled in circle shows that x can equal 4

Look at this notice:

SAFETY NOTICE
No more than 8 people in lift.

You can write this as:

the number of people in the lift is less than or equal to 8

So $p \leqslant 8$, where p stands for 'number of people in the lift'.

p is **less than or equal** to 8

Now you can't have a negative number of people so we can also say that:

***p* must be more than or equal to 0, and less than or equal to 8**

You can write this as:

$0 \leqslant p \leqslant 8$

p is **between** 0 and 8

p must be an integer (whole number), so p could equal 0, 1, 2, 3, 4, 5, 6, 7 or 8

EXAMPLE

For each inequality: **(i)** show the inequality on a number line,
(ii) write down the possible integer values for n

a) $3 \leqslant n < 6$ **b)** $-2 < n \leqslant 3$

SOLUTION

a) $3 \leqslant n < 6$

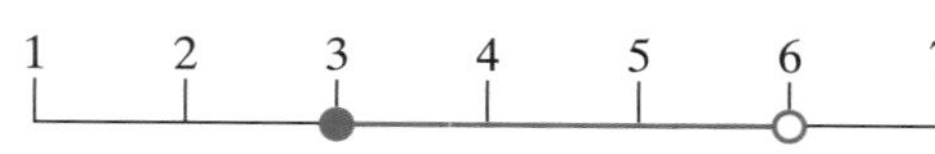

$\underline{n = 3, 4 \text{ or } 5}$

b) $-2 < n \leqslant 3$

n can't equal −2, but can equal 3; remember 0 counts as an integer.

−3 −2 −1 0 1 2 3 4 5

$\underline{n = -1, 0, 1, 2, 3}$

EXAMPLE

Solve these inequalities:

a) $x + 1 < 4$ **b)** $3x \leqslant 18$ **c)** $2x - 3 \leqslant 5$

SOLUTION

Solve the inequalities in the same way that you would solve an equation.
Remember to write an inequality sign not an equals sign.

a)

$x + 1 < 4$

$-1 \quad -1$

$\underline{x < 3}$

When you add 1 to any number less than 3, the answer will always be less than 4

b)

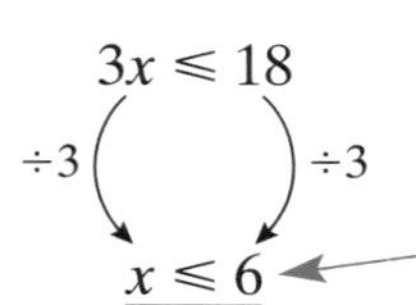

When you multiply any number that is 6 or less by 3, the answer will always be 18 or less

c) $2x - 3 \geqslant 5$

$+3 \quad +3$

$2x \geqslant 8$

$\div 2 \quad \div 2$

$\underline{x \geqslant 4}$

Keep the inequality sign pointing the same way.

EXAMPLE

Solve these inequalities:

a) $-6 \leqslant 2x < 4$ **b)** $-3 < x + 2 < 6$

SOLUTION

a)

$$-6 \leqslant 2x < 4$$

$\div 2$ $\div 2$ $\div 2$

$$\underline{-3 \leqslant x < 2}$$

There are 3 'sides' to this inequality. You must do thesame thing to all 3 'sides'. To undo '×2' you '÷2'

b)

$$-3 < x + 2 \leqslant 6$$

-2 -2 -2

$$\underline{-5 < x \leqslant 4}$$

Keep the inequality signs pointing the same way.

EXERCISE 8.9

1 Write down the correct inequality sign ($<$ or $>$)between these pairs of numbers:

a) 3 … 5 **b)** 2.1 … 2.2 **c)** 7 … 4.1
d) 3.3 … 3.2 **e)** −2 … 0 **f)** −3 … −5
g) −7 … −6 **h)** −2.2 … −2.3 **i)** −4 … −3.9

2 Show these inequalities on a number line:

a) $a \leqslant 3$ **b)** $b > 6$ **c)** $c \geqslant -3$
d) $-4 < d < 3$ **e)** $-2 \leqslant e \leqslant 3$ **f)** $-5 < f \leqslant -1$

3 Write down the possible integer values for n:

a) $-1 \leqslant n \leqslant 5$ **b)** $-6 \leqslant n \leqslant -2$ **c)** $3 \leqslant n < 7$
d) $2 < n < 8$ **e)** $-3 < n < 3$ **f)** $-1 \leqslant n < 4$

4 Solve these inequalities:

a) $x + 1 < 3$ **b)** $x - 3 > 9$ **c)** $4x \geqslant 12$
d) $3x \geqslant 15$ **e)** $3x + 1 \leqslant 10$ **f)** $5x - 7 < 18$

5 Solve these inequalities:

a) $0 < x + 2 < 6$ **b)** $4 \leqslant x - 1 < 10$ **c)** $4 < 2x \leqslant 8$
d) $-6 \leqslant 3x \leqslant 3$ **e)** $-3 \leqslant x + 1 \leqslant 2$ **f)** $-8 \leqslant 4x < 12$

6 Write down the possible integer values for n:

a) $2 \leqslant 2n \leqslant 6$ **b)** $3 \leqslant n + 3 \leqslant 6$ **c)** $-3 \leqslant 3n < 3$
d) $-2 \leqslant n - 2 < 1$ **e)** $-5 < 5n \leqslant 10$ **f)** $-2 \leqslant 2n < 7$

REVIEW EXERCISE 8

Don't use your calculator for Questions 1–9.

1 Insert the best word from this list:

expression   equation   formula   identity

into each of the following sentences.

a) The volume of a cuboid may be found by using the .... $V = abc$

b) My age n years after my fifteenth birthday is given by the .... $15 + n$

c) The .... $3x + 5 = 17$ has a solution of $x = 4$

d) The statement $x(x + 12) = x^2 + 12x$ is an example of an ....

2 Solve these equations:

a) $x + 5 = 11$ **b)** $y + 2 = 0$ **c)** $x - 5 = 2$ **d)** $2y = 1$ **e)** $4x = 20$

f) $\frac{z}{4} = 5$ **g)** $16 - 2x = 8$ **h)** $5x + 1 = 26$ **i)** $3x - 7 = 2$ **j)** $10y - 12 = 8$

3 Solve:

a) $5(a + 1) = 20$ **b)** $7(3b - 4) = 14$ **c)** $-3(4 - 2c) = 6$

4 Solve:

a) $8a + 3 = 6a + 9$ **b)** $6b + 2 = 4(b + 4)$ **c)** $\frac{15}{c} = 3$

5 a) Solve $4(x - 2) = 12 - x$

b) In the inequality $-3 \leqslant y < 3$, y is an integer.
Write down all the possible values of y

6 Write down the possible integer values for n:

a) $4 \leqslant 2n \leqslant 8$ **b)** $-2 \leqslant n - 1 \leqslant 1$ **c)** $-6 \leqslant 3n < 9$ **d)** $-4 \leqslant n - 3 < -1$

7 Solve:

a) $6n \leqslant 12$ **b)** $2n + 3 > 5$ **c)** $-10 \leqslant 5n \leqslant 20$

8 a) Solve the equation $5p - 4 = 11$

b) Solve the equation $7(q + 5) = 21$

c) Solve the equation $21 + \frac{x}{6} = 28$ [Edexcel]

9 The perimeter of this rectangle has to be more than 11 cm and less than 20 cm

(i) Show that $5 < 2x < 14$

(ii) x is an integer.
List all the possible values of x

Diagram *not* accurately drawn

[Edexcel]

You can use your calculator for Questions 10–23.

10 Solve:

a) $y^2 = 144$ **b)** $2y^2 = 72$ **c)** $3y^2 + 3 = 30$

11 Solve these equations – show all the steps in your working:

a) $4x + 3 = x + 15$ **b)** $7x - 5 = 2x + 15$ **c)** $x - 1 = 5 - 2x$

d) $7x + 25 = 13 - x$ **e)** $2x + 6 = 9x - 15$ **f)** $10 - x = 15 - 2x$

g) $16 + 3x = 5x + 20$ **h)** $8 - x = 8 + x$ **i)** $3 - 2x = 2 - x$

12 Expand the brackets and hence solve these equations:

a) $5(y - 1) = 2y + 7$ **b)** $2x - 3 = 5(12 - x)$

c) $9(x + 7) = 7(x + 9)$ **d)** $9z - 3 = 4(z - 2)$

13 The equation $x^2 + 10x = 44$ has a solution between $x = 3$ and $x = 4$
Use the method of trial and improvement to find this solution.
Give your answer correct to 1 decimal place.

14 The equation $x^3 - x = 10$ has a solution between 2 and 3
Use the method of trial and improvement to find this solution.
Give your answer correct to 2 decimal places.

15 a) Show that the equation $x^2 - x = 18$ has a solution between $x = 4$ and $x = 5$

b) Use trial and improvement to find this solution correct to 1 decimal place.

16 a) The equation $x^2 - x - 1 = 0$ has a solution between 1 and 2
Use the method of trial and improvement to find this solution.
Give your answer correct to 2 decimal places.

b) The same equation has another solution, between -1 and 0
Use trial and improvement to find this solution, also correct to 2 decimal places.

17 Glenn has done a homework exercise about solving equations.
He got 19 of the 20 questions right.
Here is the one he got wrong:

$$5(2x - 1) - 2(x + 4) = 19$$
$$10x - 5 - 2x + 8 = 19$$
$$8x + 3 = 19$$
$$8x = 16$$
$$x = 2$$

Look carefully at Glenn's work, and spot what he has done wrong.
Now write out the corrected answer, showing all the lines of working.

18 Simon and Emily are solving the equation $3(2x - 1) + 4(x + 8) = 19$

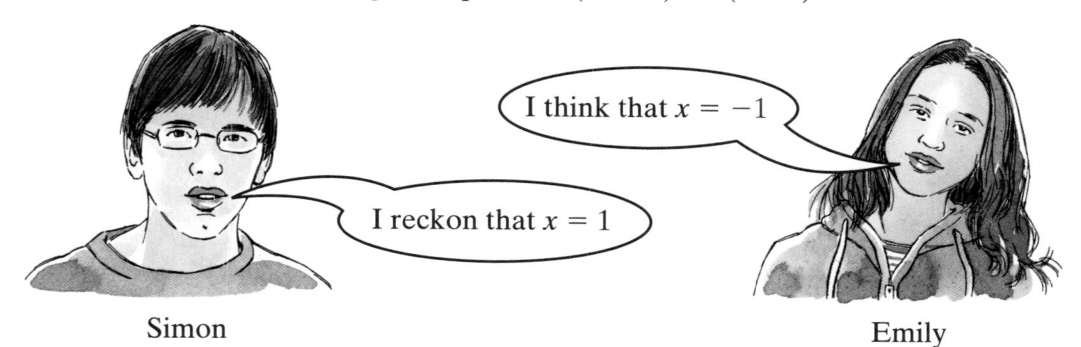
I think that $x = -1$
I reckon that $x = 1$
Simon
Emily

Who is right?
Give a reason for your answer.

19 **a)** Solve $10y - 15 = 8y - 7$

b) Solve $\frac{20 - x}{3} = 4$

20 Sally thinks of a number.
She adds 11 to the number, she then multiplies by 3
Her answer is 60
What number did Sally first think of? **[Edexcel]**

21 The width of a rectangle is x centimetres.
The length of the rectangle is $(x + 4)$ centimetres.

$x + 4$

x

a) Find an expression, in terms of x, for the perimeter of the rectangle.
Give your expression in its simplest form.

b) The perimeter of the rectangle is 54 centimetres.
Work out the width of the rectangle. **[Edexcel]**

22 The equation $x^3 + 3x = 47$ has a solution between 3 and 4
Use a trial and improvement method to find this solution.
Give your answer correct to 1 decimal place.
You must show **all** your working. **[Edexcel]**

23 Nassim thinks of a number.
When he multiplies his number by 5 and subtracts 16 from the result, he gets the same answer as when he adds 10 to his number and multiplies that result by 3
Find the number Nassim is thinking of. **[Edexcel]**

KEY POINTS

1. An expression has no equals sign.
   For example, $5x + 7$
2. An equation has an equals sign and is only true for certain values of x
3. A formula is designed for a purpose, such as to calculate an area.
4. An equation which is always true is called an identity.
5. Simple equations may be solved by inspection.
6. Harder equations may include several stages and, possibly, the expansion of brackets.
   With this kind of problem, you should always show each step of your working carefully.
   Always do the same thing to both sides of the equation.
7. A quadratic equation has an x^2 term.
   It normally has two solutions.
8. When square rooting, remember to write down both the positive and negative square roots.
9. Some equations may be solved using trial and improvement.
   You will always be told when to use trial and improvement.
   You should always include full details of your trials.
10. When you are solving an inequality, use an inequality sign instead of an equals sign.
11. You can check your solution of an equation by substituting your answer back into the original equation.

Internet Challenge 8

Carl Friedrich Gauss

Much pioneering work on the theory of equations was done by Gauss.

Use the internet to help answer these questions about him:

1. What nationality was Carl Friedrich Gauss?
2. When and where was he born?
3. How long did he live?
4. Gauss solved a difficult geometric construction problem while he was still a teenager. What was this?
5. 'Work out 1 + 2 + 3 + … + 100 in your head'
   How did Gauss do this when he was nine years old?
6. Which university did Gauss enter in 1795?
7. What astronomical discovery is jointly credited to Gauss and the Italian astronomer Guiseppe Piazzi?
8. What is the Fundamental Theorem of Algebra?
9. It is sometimes necessary to *degauss* a computer's CRT monitor.
   What does this mean?
10. Gauss allegedly said words to the effect 'Tell her to wait a minute until I've finished' on what occasion?
11. By what regal nickname is Gauss sometimes known?
12. When and where was Gauss buried?
13. What mathematical shape did Gauss want inscribed on his gravestone?
    Was this done?
14. What is the significance of the number 17?

CHAPTER 9

Number sequences

In this chapter you will **learn how to:**

- find the next term in a sequence
- recognise and use common number sequences
- generate a number sequence
- find an expression for the *n*th term of a sequence.

You will also be **challenged to:**

- investigate Fibonacci numbers.

Starter: Circles, lines and regions

Look at the sequence of circles below.

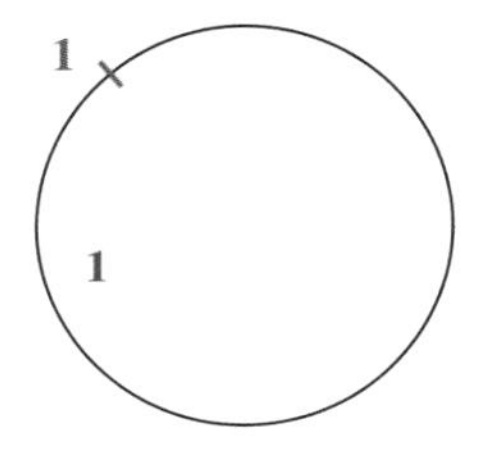

Pattern 1
1 point
0 lines
1 region

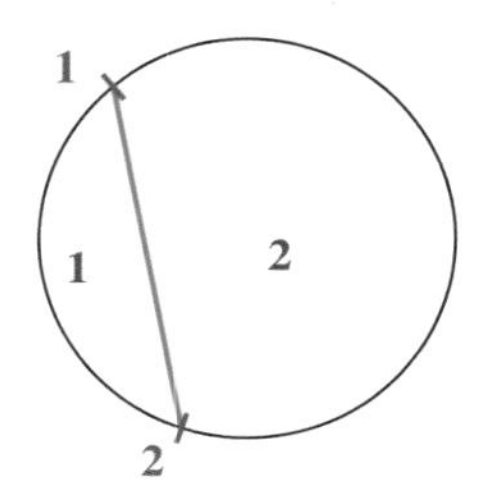

Pattern 2
2 points
1 line
2 regions

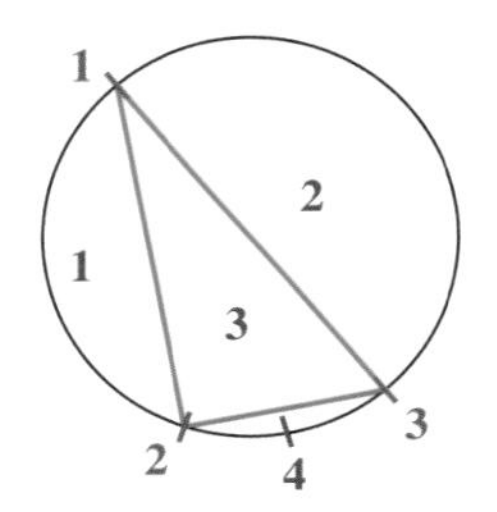

Pattern 3
3 points
3 lines
4 regions

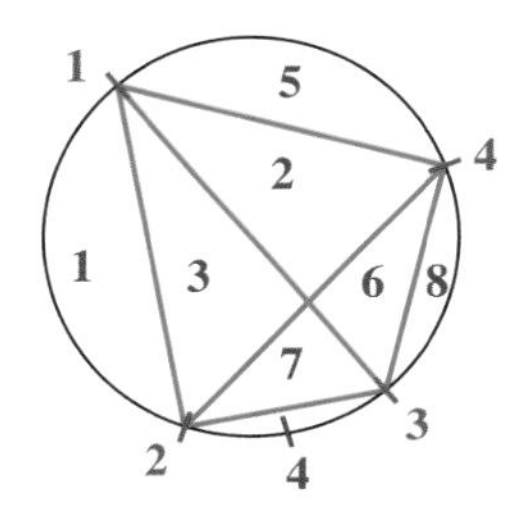

Pattern 4
4 points
6 lines
8 regions

The diagram shows a sequence of circles:

- Each circle has some points marked around its circumference.
- Each point is joined to every other point by a line.
- The lines and regions are then counted.
- The lines and regions are not all the same size.

Task 1 Describe a rule for how the number of points increases in this sequence.

Task 2 Describe a rule for how the number of lines increases.

Task 3 Describe a rule for how the number of regions increases.

Task 4 Now draw Pattern 5 and Pattern 6, and see if your rules seem to be correct.

You should space out the points so that you don't get three lines crossing at the same point, otherwise you lose a region.
For example:

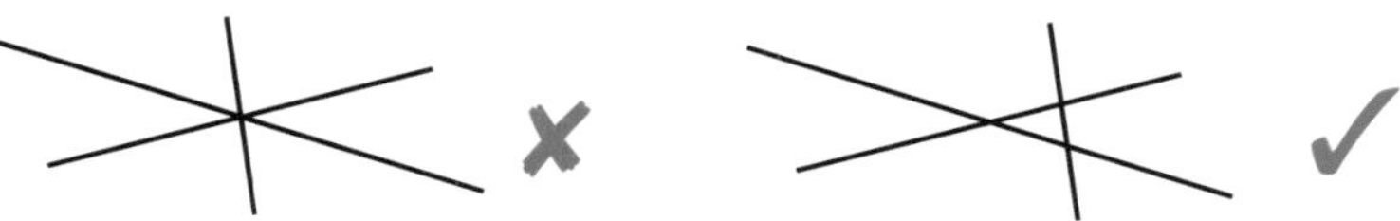

9.1 Number patterns

When you continue a pattern you often find that the numbers relating to the pattern follow a rule.

EXAMPLE

The diagram shows how the tables are arranged in a school canteen.

A × shows where a student can sit.

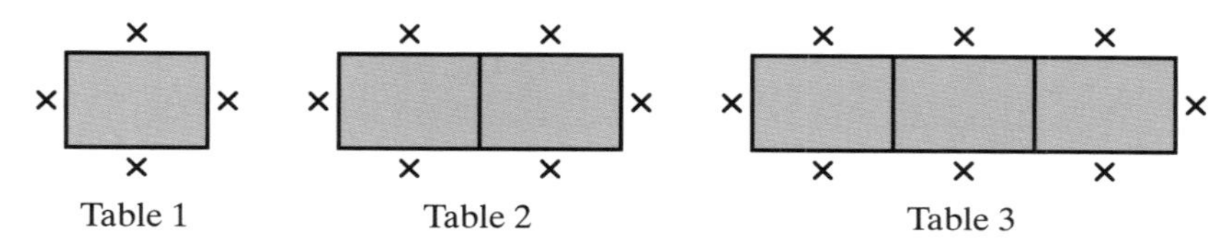

The rest of the tables in the canteen follow the same pattern.

a) Draw table number 4
b) Copy and complete the table.
c) Explain how you would work out how many students can sit at table 6
d) Work out how many students can sit at table 15
e) Can 41 students sit around a table without leaving any spaces?

Table number	1	2	3	4	5
Number of students	4	6	8		

SOLUTION

a)

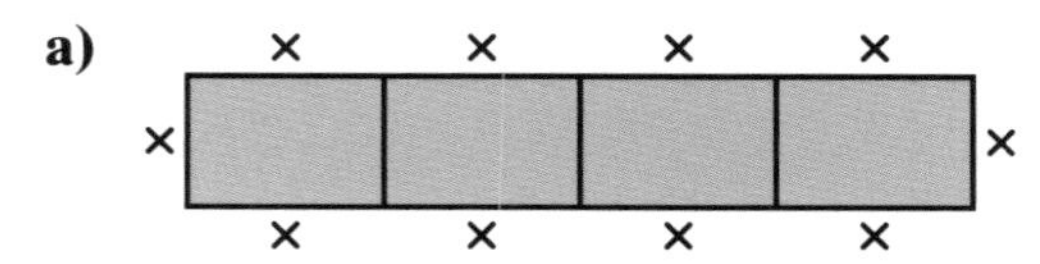

b)

Table number	1	2	3	4	5
Number of students	4	6	8	10	12

c) Add 2 to the number of students who can sit at table 5
So $12 + 2 = 14$

d) $12 + 10 \times 2 = 32$

To find the number for table 15, you need to 'add 2' ten times to the number of students at table 5

e) No. Only an even number of students can sit around a table.

EXERCISE 9.1

1 Here are the first three patterns in a sequence of patterns made from matchsticks.

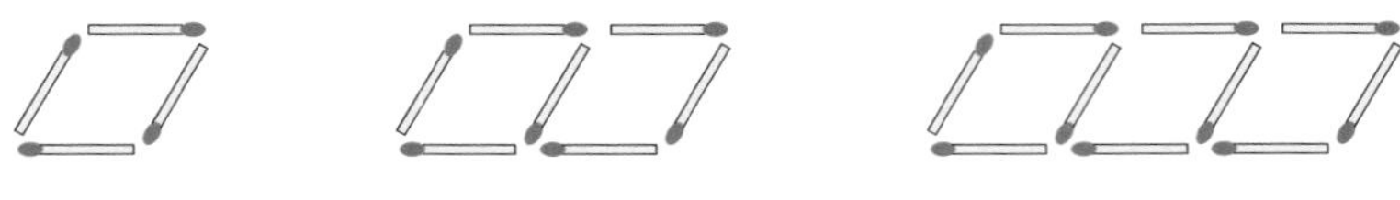

Pattern 1 Pattern 2 Pattern 3

a) Draw pattern number 4
b) Copy and complete the table.
c) Explain how you would work out pattern number 6
d) Work out how many matchsticks in pattern number 15

Pattern number	1	2	3	4	5
Number of matches	4	7	10		

2 Here are the first three patterns in a sequence of patterns made from squares.

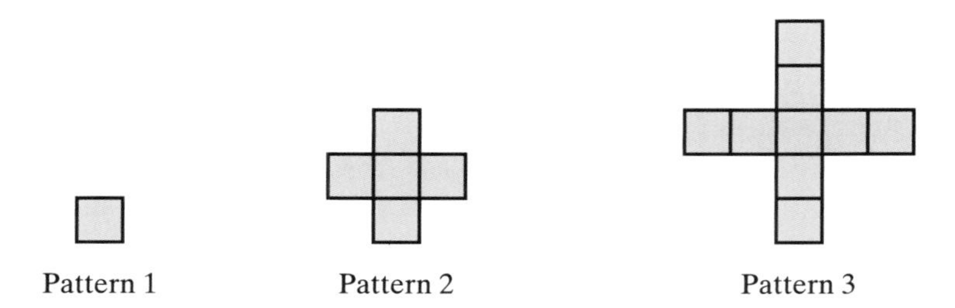

Pattern 1 Pattern 2 Pattern 3

a) Draw pattern number 4
b) Copy and complete the table.
c) Explain how you would work out pattern number 6
d) Work out how many squares are in pattern number 15

Pattern number	1	2	3	4	5
Number of squares	1	5	9		

3 Here are the first three patterns in a sequence of 'huts' made from matchsticks.

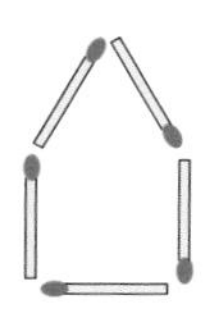

Hut 1

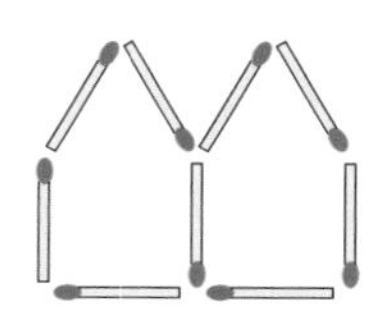

Hut 2

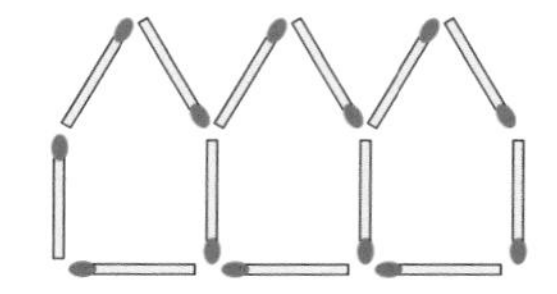

Hut 3

a) Draw hut number 4
b) Copy and complete the table.
c) Work out how many sticks there are in hut number 10
d) Can a hut be made from exactly 50 sticks? Explain your answer.

Hut number	1	2	3	4	5
Number of sticks	5	9	13		

4

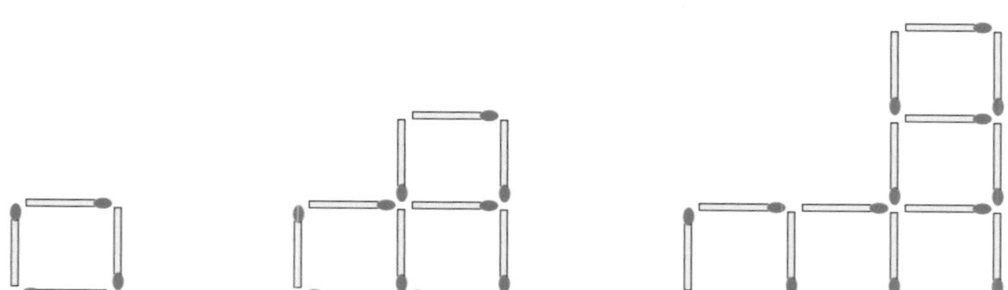

Pattern 1 Pattern 2 Pattern 3

a) Draw pattern number 4.

b) Copy and complete the table.

Pattern number	1	2	3	4	5
Number of matchsticks	4	10	16		
Number of squares	1	3	5		

c) Explain how you would work out the number of:
(i) matchsticks, **(ii)** squares
in pattern number 6

d) Work out how many:
(i) matchsticks, **(ii)** squares
there are in pattern number 20

5 John is building some fences.
The rest of the fences that John builds follow the same pattern.

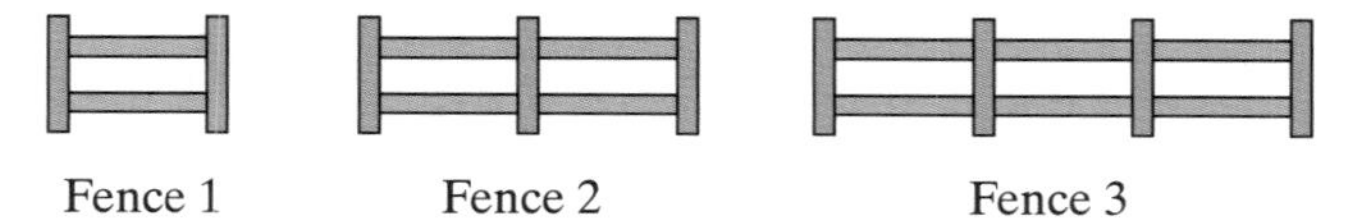

Fence 1 Fence 2 Fence 3

a) Draw fence number 4

b) Copy and complete the table.

Fence number	1	2	3	4	5	10	20
Number of uprights	2	3	4				
Number of crossbars	2	4	6				

c) Which fence number needs 50 uprights?

d) Which fence number needs 50 crossbars?

9.2 Sequences

A **sequence** is a list of numbers which follow a rule. For example:

2, 4, 6, 8, 10, … and
1, 4, 9, 16, 25, …

are both sequences.

The numbers that appear in a sequence are called **terms**.

So in both sequences above, the **2nd term** is **4**

You can work out the pattern or **rule** that the sequence is following by working out the **differences** between the terms:

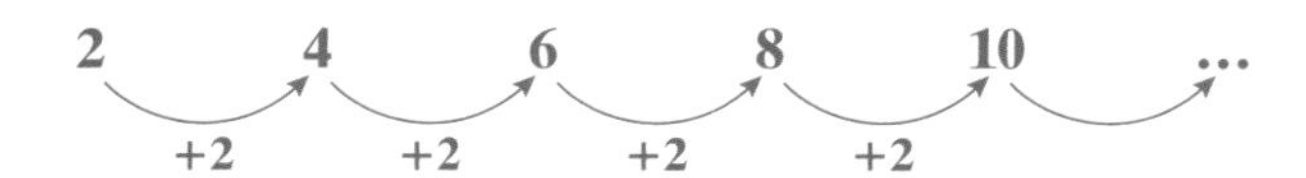

In this sequence, the **difference** is $+2$, so the next term will be $10 + 2 = 12$

EXAMPLE

A number sequence is defined as follows:

- the first term is 3
- each new term is double the previous one.

Use this rule to generate the first five terms of the number sequence.

SOLUTION

Start with 3:

$$3 \times 2 = 6$$
$$6 \times 2 = 12$$
$$12 \times 2 = 24$$
$$24 \times 2 = 48$$

The first five terms of the sequence are $\underline{3, 6, 12, 24, 48}$

EXAMPLE

Write down the next three terms in the following sequence:

2, 5, 10, 17, 26, …

SOLUTION

Look at the **differences** between the terms in the sequence:

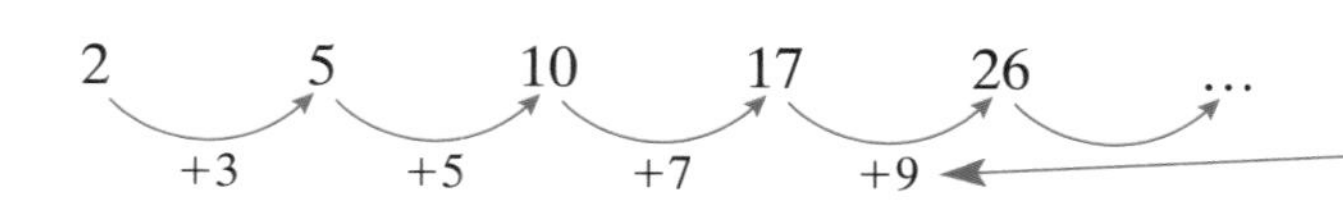

Look for a pattern in the differences to help you find the next term.

The differences are going up by two each time so the next three differences will be 11, 13 and 15

So the sequence will be:

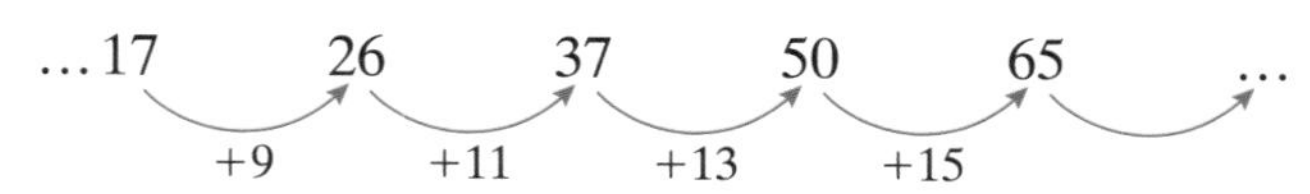

So the next three terms are 37, 50, 65

EXAMPLE

Look at this number sequence: 2, 5, 8, 11, 14, …

a) Write down the next three terms in the sequence.

b) Explain how you found your answer.

The 30th term of the sequence is 89

c) Write down the 31st term.

d) Will 66 be a term in this sequence?
Explain your answer.

SOLUTION

The terms are going up by 3 each time.

a) 17, 20, 23

b) Add 3 to the previous term

c) $89 + 3 = 92$

d) No.

66 is in the 3 times table since $3 \times 22 = 66$

The sequence misses numbers in the 3 times table as it starts at 2 and goes up by 3 each time.

The 3 times table starts at 3 and goes up by 3 each time.

EXERCISE 9.2

1 Write down the first five terms of the number sequences generated by each of these rules.

a) The first term is 7
To find the next term, add 4 to the previous term.

b) The first term is 1
To find the next term, multiply the previous term by 3

c) The first term is 30
To find the next term, subtract 5 from the previous term.

d) The first term is 128
To find the next term, divide the previous term by 2

e) The first term is 8
To find the next term, subtract 3 from the previous term.

f) The first term is 2
To find the next term, double the previous term.

g) The first term is 3
To find the next term, double the previous term and add 1

2 For each sequence:
(i) write down the next two terms,
(ii) explain the rule in words.

a) 3, 5, 7, 9, 11, …
b) 4, 10, 16, 22, 28, …
c) 24, 22, 20, 18, 16, …
d) 42, 36, 30, 24, 18, …
e) 3, 6, 12, 24, …
f) 10, 6, 2, −2, −6, …

3 Find the missing terms in these number sequences:

a) 1, 3, ?, 7, ?, ?
b) ?, 4, 6, 8, ?, ?
c) 23, ?, ?, 32, 35, 38, ?
d) 40, ?, ?, 34, 32, 30, ?
e) ?, ?, ?, 9, 12, 15, 18
f) 12, ?, 4, ?, ?, −8, −12, −16

4 **(i)** Write down the next term in the following sequences.
(ii) Describe how you found the next term.

a) 1, 4, 9, 16, 25, …
b) 1, 2, 4, 8, 16, …
c) 1, 3, 6, 10, 15, …
d) 1, 3, 7, 15, 31, …
e) 1, 1, 2, 3, 5, 8, …
f) 1024, 512, 256, 128, 64, …

5 Look at this number sequence: 1, 6, 11, 16, 21, …

a) Write down the next three terms in the sequence.
b) Explain how you found your answer.

The 50th term of the sequence is 246

c) Write down the 51st term.
d) Will 500 be a term in this sequence?
Explain your answer.

6 Andy has been doing an investigational GCSE coursework task.
He gets this sequence of numbers:

12, 15, 18, 21, 24, …

a) Write down the next three terms in Andy's number sequence.
b) Write a rule for Andy's number sequence.

7 Sarah has been working with even numbers.

The rule for even numbers is:
Add 2 to the pre vious term.'

Sarah is not quite right.
Rewrite Sarah's statement so that it is correct.

8 Look at this number sequence:

1, 4, 7, 10, 13, ….

a) Write down the next three terms in the sequence.
b) Explain how you found your answer.

The 50th term of the sequence is 148
c) Write down the 51st term.
d) Will 99 be a term in this sequence?
Explain your answer.

9 Look at this number sequence:

6, 10, 14, 18, 22, …

a) Write down the next three terms in the sequence.
b) Explain how you found your answer.

The 100th term of the sequence is 402
c) Write down the 101st term.
d) Will 100 be a term in this sequence?
Explain your answer.

9.3 Using an expression for the nth term

Look at these two rules.
They both describe the *same* number sequence.

RULE 1	RULE 2
• The first term is 1 • To find the next term, add 2 to the previous term.	The nth term is found by the expression $2n - 1$
This rule shows how each term is related to the one before.	This rule gives an expression for the term in the nth position.

Both rules describe the odd numbers 1, 3, 5, 7, 9, …

Suppose you want to use these rules to work out the value of the 75th odd number.

To use Rule 1, you would have to work out all of the first 75 terms of the sequence.

Rule 2 is much more efficient because you can calculate the 75th term directly as $2 \times 75 - 1 = 149$

EXAMPLE

The nth term of a number sequence is given by the expression $7n + 3$

a) Work out the first three terms.

b) Find the value of the 10th term.

c) The number 1053 is the kth term of this sequence.
Work out the value of k, solve the equation and find the 150th term.

SOLUTION

a) The first three terms are found by substituting **1**, **2** and **3** into the expression $7n + 3$:

$$7 \times \mathbf{1} + 3 = 10$$
$$7 \times \mathbf{2} + 3 = 17$$
$$7 \times \mathbf{3} + 3 = 24$$

So the first three terms are 10, 17, 24

b) The **10**th term is $7 \times \mathbf{10} + 3 = \underline{73}$

c) If 1053 is the kth term of this sequence, then $7k + 3 = 1053$

This equation may now be solved as follows:

$$7k + 3 = 1053$$
$$-3 \qquad -3$$
$$7k = 1050$$
$$\div 7 \qquad \div 7$$
$$k = \underline{150}$$

So the $\underline{\text{150th term is 1053}}$

EXERCISE 9.3

1 Write down the first five terms of the following sequences:

a) nth term $= 2n$
b) nth term $= 3n$
c) nth term $= 4n$
d) nth term $= n + 1$
e) nth term $= n + 3$
f) nth term $= n - 4$
g) nth term $= 2n - 1$
h) nth term $= 3n + 1$

2 The nth term of a number sequence is given by the expression $8n - 1$

a) Write down the values of the first five terms.
b) Work out the value of the 20th term.
c) 239 is the kth term of the sequence.
Work out the value of k

3 The nth term of a number sequence is given by the expression $\dfrac{3n + 1}{2}$

a) Write down the values of the first six terms.
b) Work out the value of the 23rd term.
c) 26 is the kth term of the sequence.
Work out the value of k

4 Write down the first five terms of the following sequences:

a) nth term $= n^2$
b) nth term $= n^2 - 1$
c) nth term $= n^3$
d) nth term $= n^3 - n^2$

Remember:
n^2 means $n \times n$
n^3 means $n \times n \times n$

5 The nth term of a number sequence is given by the expression $\dfrac{n(n + 1)}{2}$

a) Write down the values of the first five terms.
b) Work out the value of the 30th term.

9.4 Finding an expression for the *n*th term

A number sequence in which the terms go up in equal steps is called an **arithmetic sequence**.

The graph of an arithmetic sequence is a straight line, so it can also be called a **linear sequence**.

The size of the step is called the **common difference**.

You need to be able to find a rule for the *n*th term of a sequence. This is a useful skill for your GCSE coursework.

EXAMPLE

Find an expression for the *n*th terms of the following sequences:

a) 5, 10, 15, 20, 25, …
b) 6, 10, 14, 18, 22, …
c) 20, 18, 16, 14, …

SOLUTION

a) Work out the common difference:

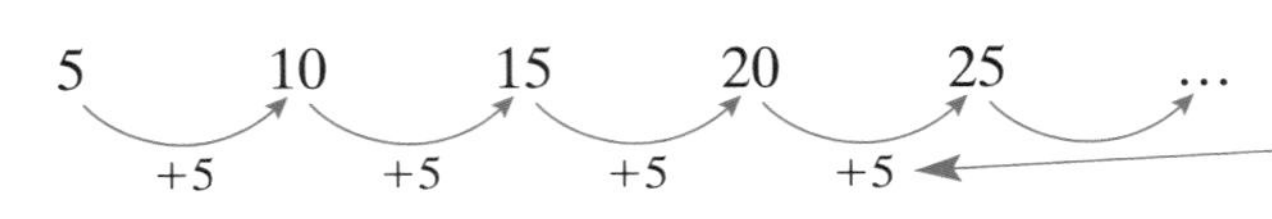

The difference tells you what times table the sequence is based on.

The sequence is the 5 times table so the rule for the *n*th term is $\underline{5n}$

b) Work out the common difference:

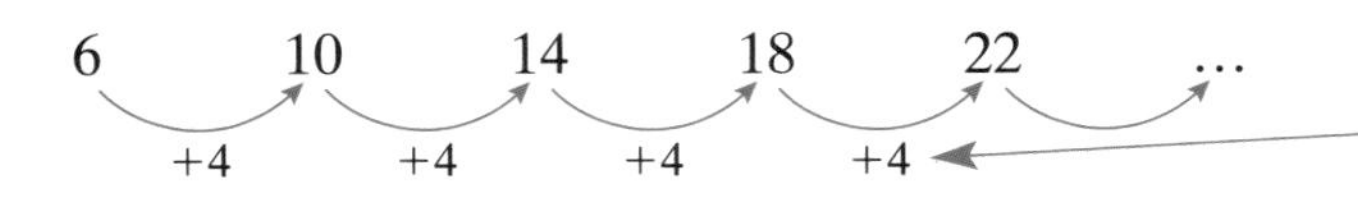

The sequence is based on the 4 times table.

Compare the sequence with the 4 times table:

4	8	12	16	20	$4n$
6	10	14	18	22	$4n + 2$

(+2 from each term in the top row to the term below it)

Each term is 2 more than the terms in the 4 times table.

So the rule for the *n*th term is $\underline{4n + 2}$

c) Work out the common difference:

$$20 \xrightarrow{-2} 18 \xrightarrow{-2} 16 \xrightarrow{-2} 14 \xrightarrow{-2} 12 \longrightarrow \ldots$$

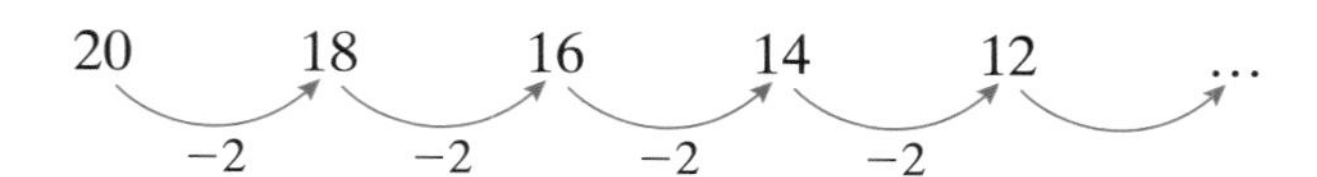

Compare the sequence with the '−2 times table':

$$\begin{array}{cccccc} -2 & -4 & -6 & -8 & -10 & -2n \\ \downarrow{\scriptstyle +22} & & & & & \downarrow{\scriptstyle +22} \\ 20 & 18 & 16 & 14 & 12 & -2n + 22 \end{array}$$

So the rule for the nth term is $\underline{-2n + 22}$

This can be written as $22 - 2n$

The same method can be used for problems set in a more practical context, as in the next example.

EXAMPLE

The table shows the cost of hiring a van.

It is made up of a fixed hire charge plus a daily amount.

a) Work out the cost of hiring the van for 7 days.

b) Express C in terms of n

c) Carlos hired the van and paid £530 For how many days did he hire the van?

Number of days for which the van is hired (n)	Charge for hire (£C)
1	140
2	170
3	200
4	230
5	260
6	290
7	310

SOLUTION

a) Continuing the pattern in the table:

So the cost for 7 days is $\underline{£310}$

b) The common difference is 30, so the formula must be based on the '30 times table'. Compare the sequence with the '30 times table'.

$$\begin{array}{cccccc} 30 & 60 & 90 & 120 & 150 & 30n \\ \downarrow{\scriptstyle +110} & & & & & \downarrow{\scriptstyle +110} \\ 140 & 170 & 200 & 230 & 260 & 30n + 110 \end{array}$$

The formula is therefore $\underline{C = 30n + 110}$

c) Now if $C = 530$, the formula gives:

$$\begin{array}{c} 30n + 110 = 530 \\ \downarrow{\scriptstyle -110} \qquad \downarrow{\scriptstyle -110} \\ 30n = 420 \\ \downarrow{\scriptstyle \div 30} \qquad \downarrow{\scriptstyle \div 30} \\ n = 14 \end{array}$$

So Carlos hired the van for $\underline{\text{14 days}}$

EXERCISE 9.4

1 Find an expression for the nth term of the following sequences:

a) 2, 4, 6, 8, 10, … **b)** 4, 8, 12, 16, 20, …
c) 7, 14, 21, 28, 35, … **d)** 1, 2, 3, 4, 5, …
e) 10, 20, 30, 40, 50, … **f)** 15, 30, 45, 60, 75, …

2 Find an expression for the nth term of the following sequences:

a) 3, 5, 7, 9, 11, … **b)** 1, 3, 5, 7, 9, …
c) 4, 7, 10, 13, 16, … **d)** 5, 9, 13, 17, 21, …
e) 1, 7, 13, 19, 25, … **f)** 2, 10, 18, 26, 34, …

3 Find an expression for the nth term of the following sequences:

a) 8, 6, 4, 2, 0, … **b)** 10, 9, 8, 7, 6, …
c) 15, 12, 9, 6, 3, … **d)** 18, 14, 10, 6, 2, …

4 The first five terms in an arithmetic sequence are:

12, 17, 22, 27, 32, …

a) Find the value of the 10th term.
b) Write down an expression for the nth term.

5 The first four terms in an arithmetic sequence are:

58, 50, 42, 34, …

a) Find the value of the first negative term.
b) Write down an expression for the nth term.

6 Nina has been making patterns with sticks.
Here are her first three patterns.

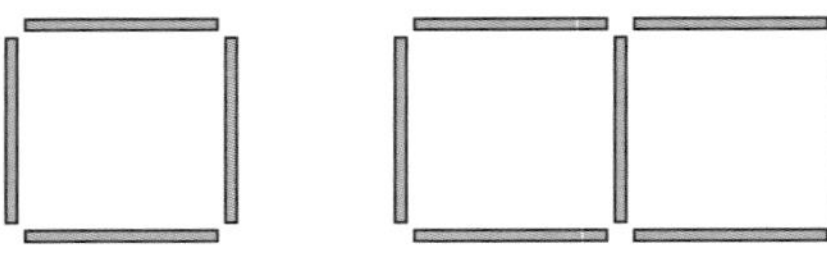

Pattern 1
4 sticks

Pattern 2
7 sticks

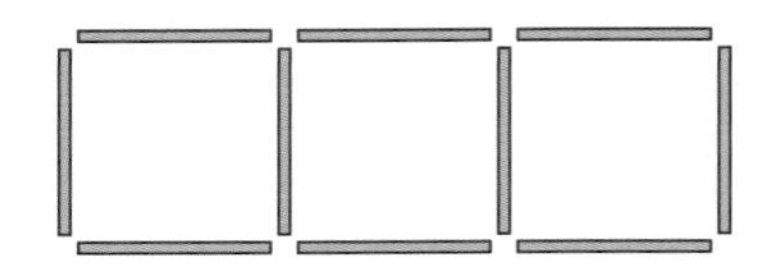

Pattern 3
10 sticks

a) Draw pattern 4
b) Work out the number of sticks in pattern 6
c) Give an expression for the number of sticks, S, in pattern n
d) Explain how your expression relates to the way the sticks fit together.

7 The fifth term of an arithmetic sequence is 30 and the sixth term is 37

a) Write down the value of the common difference for this sequence.
b) Work out the value of the first term.
c) Find an expression for the nth term of the sequence.
d) Check that your expression works when $n = 5$ and $n = 6$

9.5 More number sequences

Many sequences are based on the square numbers.

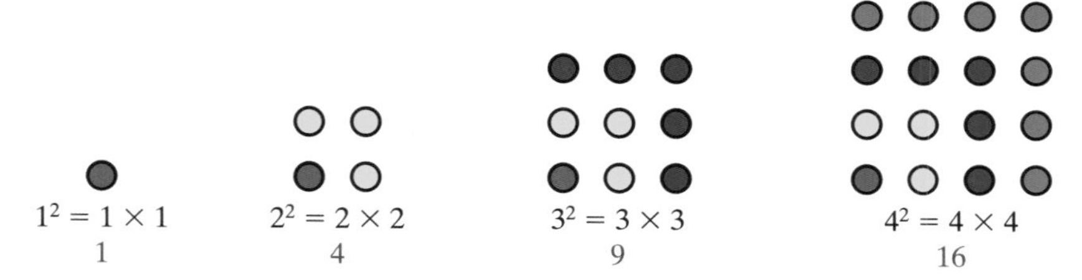

The nth square number is $n \times n = n^2$

Look at the differences in the square numbers:

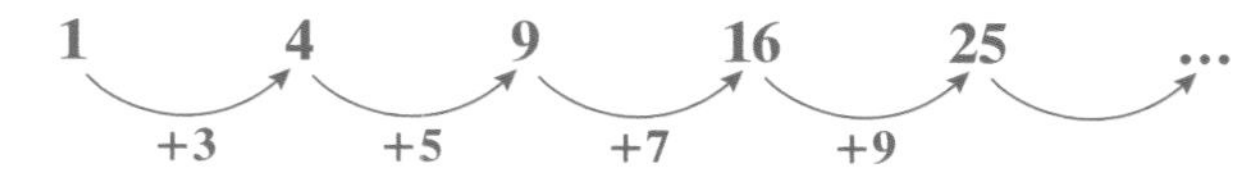

The differences are the odd numbers 3, 5, 7, 9, …

You can see why when you look at the patterns above.
Each time, an odd number of dots is added to get the next pattern.

A sequence where the differences are the odd numbers is based on the square numbers.

EXAMPLE

Find an expression for the nth terms of the sequence 4, 7, 12, 19, 28, …

SOLUTION

Work out the differences:

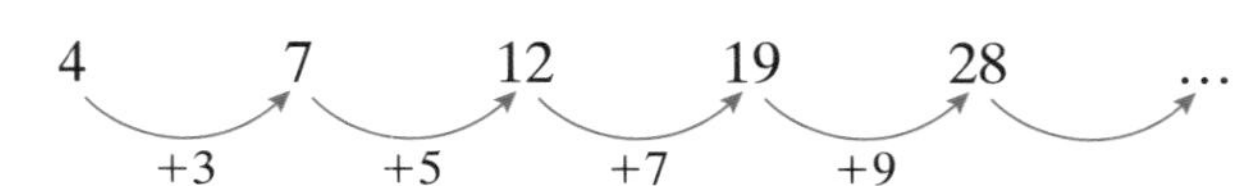

The differences are the odd numbers, so we need to compare the sequence with the square numbers:

1	4	9	16	25	n^2
+3					+3
4	7	12	19	28	$n^2 + 3$

This sequence is 3 more than the square numbers.

So the rule for the nth term is $\underline{n^2 + 3}$

The triangular numbers is another sequence which crops up a lot – especially in coursework.

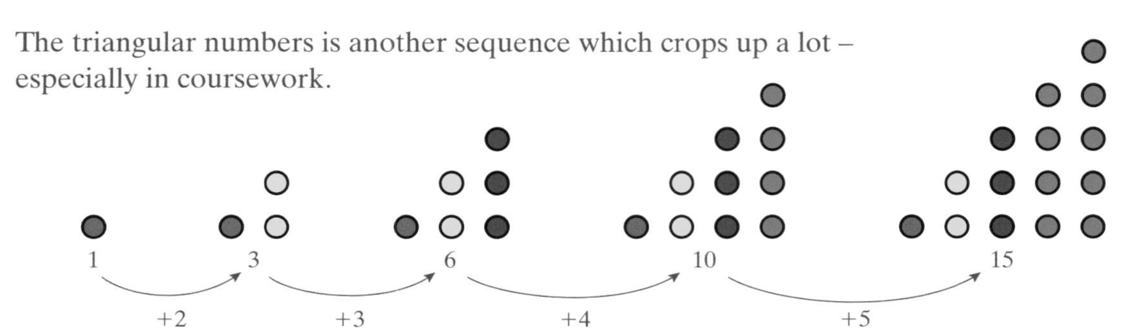

The differences go up by the counting numbers 2, 3, 4, 5, …

Here are some more number sequences that you need to be familiar with.

Name of sequence	First six terms	Formula for the nth term
Positive integers	1, 2, 3, 4, 5, 6, …	n
Even numbers	2, 4, 6, 8, 10, 12, …	$2n$
Odd numbers	1, 3, 5, 7, 9, 11, …	$2n - 1$
Square numbers	1, 4, 9, 16, 25, 36, …	n^2
Cube numbers	1, 8, 27, 64, 125, 216, …	n^3
Powers of 2	2, 4, 8, 16, 32, 64, …	2^n
Powers of 10	10, 100, 1000, 10 000, 100 000, 1 000 000	10^n
Triangular numbers	1, 3, 6, 10, 15, 21, …	$\frac{n(n+1)}{2}$

EXAMPLE

Look at this pattern of squares.

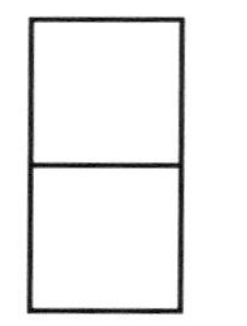
Pattern 1

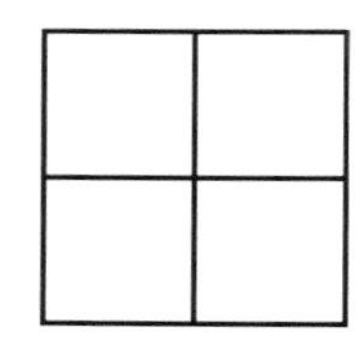
Pattern 2

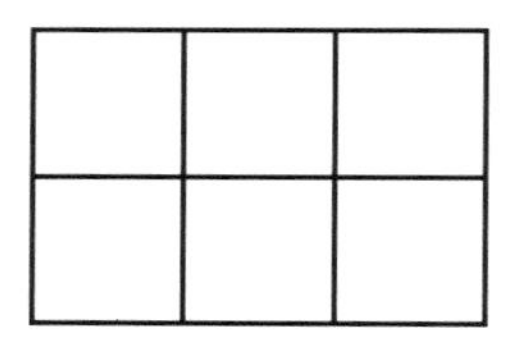
Pattern 3

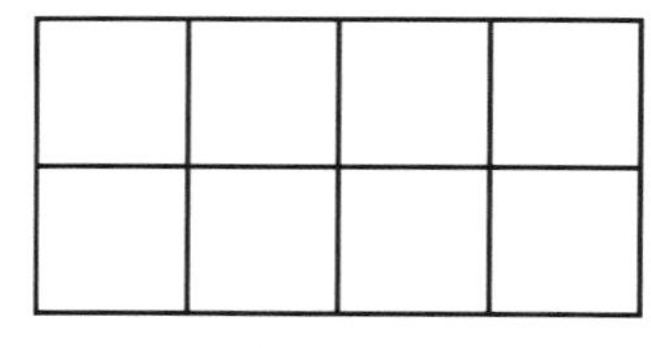
Pattern 4

a) How many squares would there be in pattern 10?
b) Find a formula for the number of squares, S, in pattern n

SOLUTION

The number of squares forms a pattern 2, 4, 6, 8, …, that is, the even numbers.

a) Pattern 10 contains $10 \times 2 =$ 20 squares
b) $S = 2n$

EXERCISE 9.5

Don't use your calculator for Questions 1–3.

1 Look at this pattern of triangles.

Pattern 1

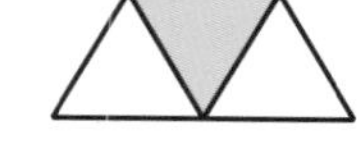
Pattern 2

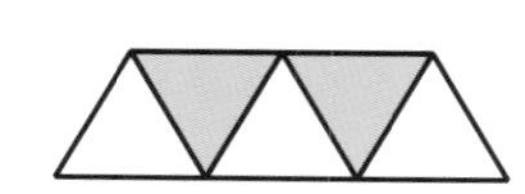
Pattern 3

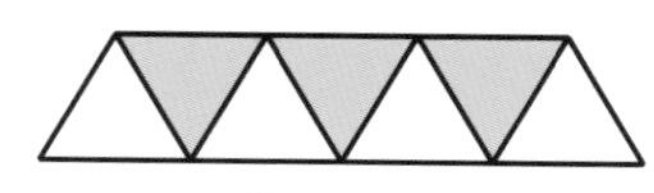
Pattern 4

a) Draw the next pattern.

b) Copy and complete the table.

Pattern number	1	2	3	4	5
Number of triangles	1	3	5		

c) How many triangles would there be in pattern 7?

d) Find a formula for the number of triangles, T, in pattern n

2 **(i)** Write down the next two terms in each of these number sequences.

(ii) Find the 10th term.

(iii) Give a formula for the nth term in each case.

a) 1, 4, 9, 16, 25, …

b) 2, 5, 10, 17, 26, …

c) 0, 3, 8, 15, 24, …

d) 7, 10, 15, 22, 31, …

e) 101, 104, 109, 116, 125, …

f) −2, 1, 6, 13, 22, …

3 Here are the first three patterns in a sequence of patterns made from triangles.

Pattern 1

Pattern 2

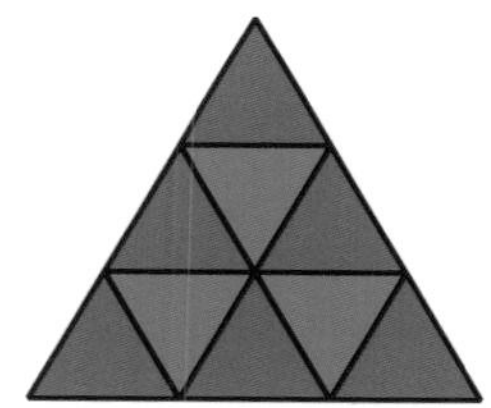
Pattern 3

a) Draw pattern number 4.

b) Copy and complete the table.

Pattern number	1	2	3	4	5
Number of red triangles	1	3	6		
Number of green triangles	0	1	3		
Total number of triangles	1	4	9		

c) Work out the **total** number of triangles in pattern number 10

d) Find a formula for the **total** number of triangles, T, in pattern n

You can use your calculator for Question 4

4 To work out a rule for the nth term for triangular numbers we can put two triangles together to make rectangles like this:

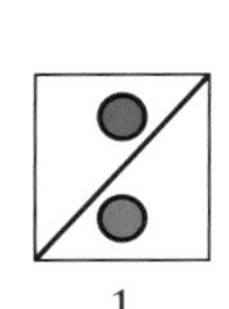

1

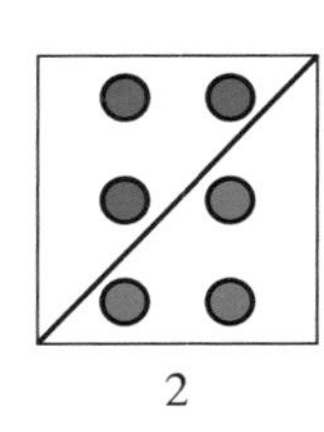

2

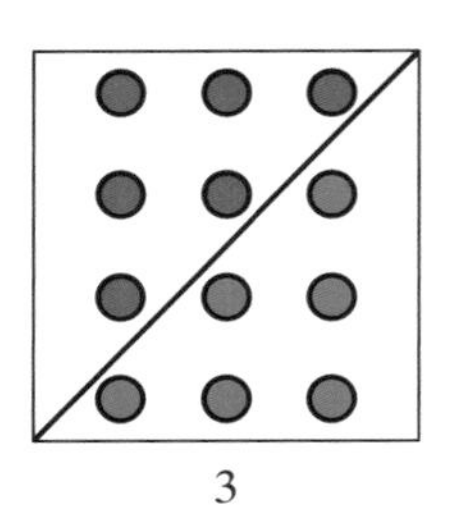

3

4

Pattern number	1	2	3	4
Number of dots in rectangle	$1 \times 2 = 2$	$2 \times 3 = 6$	$3 \times 4 = ?$	$? \times ? = ?$
Number of dots in one triangle	1	3	____	____

a) State how many dots are there in rectangle:
(i) 3, **(ii)** 4, **(iii)** 5

b) State how many dots are there in **triangle**:
(i) 3, **(ii)** 4, **(iii)** 5

c) Fill in the missing values:

(i) The 10th rectangle would have $10 \times (10 + 1) = ?$ dots.

(ii) The 10th **triangle** would have $\dfrac{10 \times (? + ?)}{2}$ dots.

(iii) The nth rectangle would have $? \times (? + ?)$ dots.

(iv) The nth **triangle** would have $\dfrac{? \times (? + ?)}{?}$ dots.

REVIEW EXERCISE 9

Don't use your calculator for Questions 1–10.

1 Find the next three terms in each of these number sequences:

a) 1, 3, 5, 7, 9, …
b) 5, 8, 11, 14, 17, …
c) 1, 2, 4, 8, 16, …
d) 3, 4, 7, 11, 18, 29, …
e) 1, 2, 4, 7, 11, …
f) 40, 38, 36, 34, 32, …
g) 1, 10, 100, 1000, 10 000, …
h) 1, 4, 9, 16, 25, …

2 Write down the first five terms of the following sequences:

a) The first term is 2
To find the next term add 5 to the previous term.

b) The first term is 1
To find the next term add 8 to the previous term.

c) The first term is 15
To find the next term subtract 3 from the previous term.

3 Write down the first five terms of the following sequences:

a) nth term $= 5n$ **b)** nth term $= 7n$
c) nth term $= n - 1$ **d)** nth term $= n + 4$
e) nth term $= n + 5$ **f)** nth term $= n - 2$

4 Write down the first five terms of the following sequences:

a) nth term $= 2n - 1$ **b)** nth term $= 3n + 2$
c) nth term $= 4n - 3$ **d)** nth term $= 5n + 1$
e) nth term $= 6n - 3$ **f)** nth term $= 7n - 7$
g) nth term $= n^2 - 3$ **h)** nth term $= n^2 + 4$

5 Find an expression for the nth term of the following sequences:

a) 3, 6, 9, 12, 15, … **b)** 7, 14, 21, 28, 35, …
c) 8, 16, 24, 32, 40, … **d)** 3, 5, 7, 9, 11, …
e) 10, 16, 22, 28, 34, … **f)** 4, 12, 20, 28, 36, …
g) 22, 33, 44, 55, 66, … **h)** 9, 8, 7, 6, 5, …
i) 1, 4, 9, 16, 25, … **j)** 3, 6, 11, 18, 27, …

6 Look at this number sequence 1, 4, 7, 10, 13, ….

a) Write down the next three terms in the sequence.
b) Explain how you found your answer.

The 50th term of the sequence is 148

c) Write down the 51st term.
d) Will 300 be a term in this sequence?
Explain your answer.

7 Timothy has been drawing patterns.
Here are his first three patterns.

Pattern 1
6 sticks

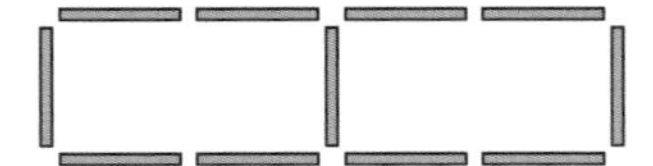
Pattern 2
11 sticks

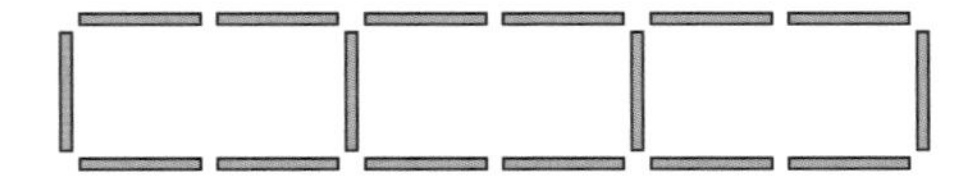
Pattern 3
16 sticks

a) Write down the number of sticks in pattern 5
b) Work out the number of sticks in pattern 12
c) Find an expresson for the number of sticks in pattern n

8 Here are some patterns made from matchsticks.

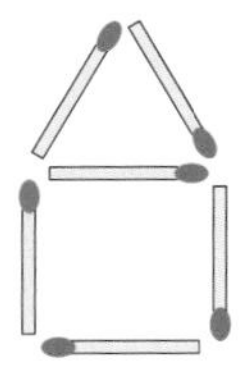
Pattern 1

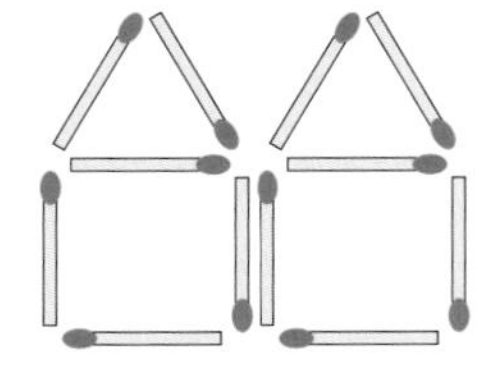
Pattern 2

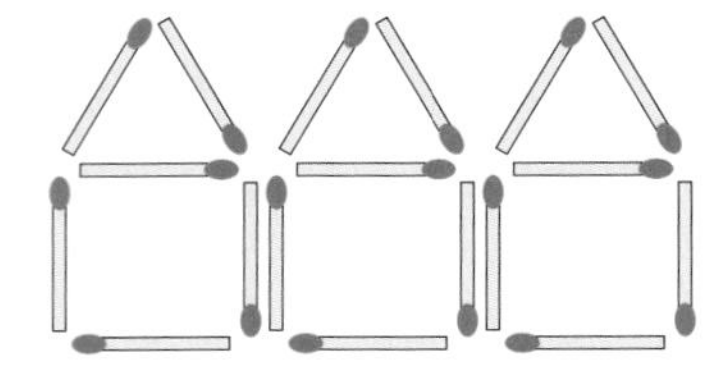
Pattern 3

a) Draw pattern number 4

The graph shows the number of matchsticks m in pattern number n

b) Mark the point which shows the number of matchsticks used in pattern number 4

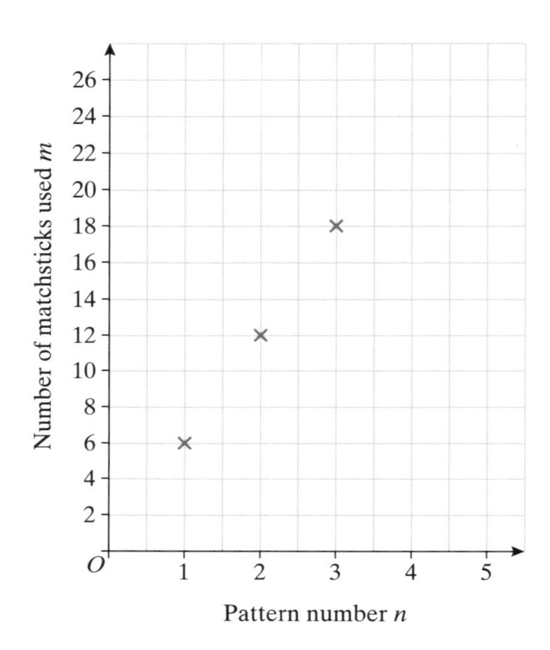

c) How many matchsticks are used in pattern number 10?

d) Write down a formula for m in terms of n **[Edexcel]**

9 Here are the first five terms of a number sequence:

290 284 278 272 266

a) Write down the next two terms of the number sequence.

b) Explain how you found your answer.

The 20th term of the number sequence is 176

c) Write down the 21st term of the number sequence. **[Edexcel]**

10 Here are the first five terms of a number sequence:

3 8 13 18 23

a) Write down the next two terms of the number sequence.
b) Explain how you found your answer.
c) Explain why 387 is not a term of the sequence. [Edexcel]

You can use your calculator for Questions 11–15.

11 a) The first odd number is 1
(i) Find the 3rd odd number. **(ii)** Find the 20th odd number.
b) Write down a method you could use to find the 100th odd number.

Here are some patterns made with dots.

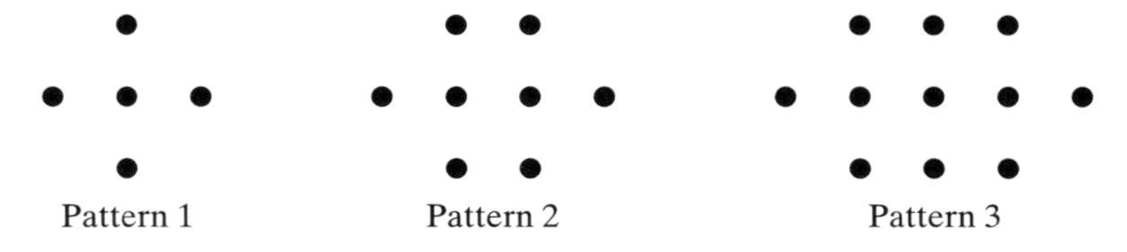

Pattern 1 Pattern 2 Pattern 3

c) Copy and complete pattern number 4

The table shows the number of dots used to make each pattern.

Pattern number	1	2	3	4	5
Number of dots	5	8	11		

d) Copy and complete the table. [Edexcel]

12 a) The first even number is 2

(i) Find the 4th even number.

(ii) Find the 11th even number.

b) Write down a method you could use to find the 200th even number.

Here are some patterns made with crosses.

Pattern 1 Pattern 2 Pattern 3

c) Draw pattern number 4

The table shows the number of crosses used to make each pattern:

Pattern number	1	2	3	4	5
Number of crosses	6	10	14		

d) Copy and complete the table. **[Edexcel]**

13 Here are some patterns made up of dots.

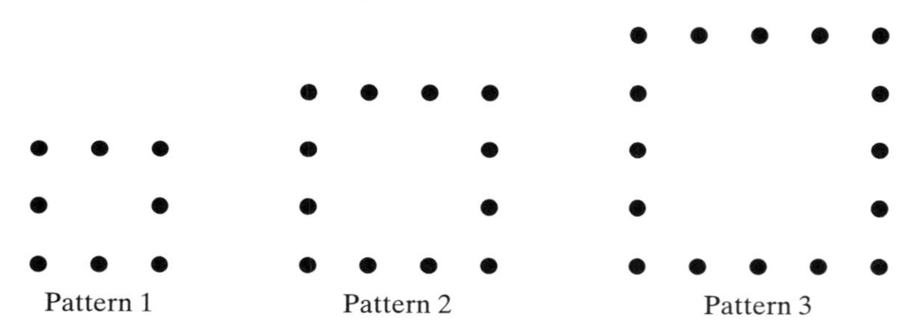

a) Draw pattern number 4

b) Complete the table:

Pattern number	1	2	3	4	5
Number of dots	8	12	16		

c) How many dots are used in pattern number 10?

d) Write down a formula for the number of dots, d, in terms of the pattern number n

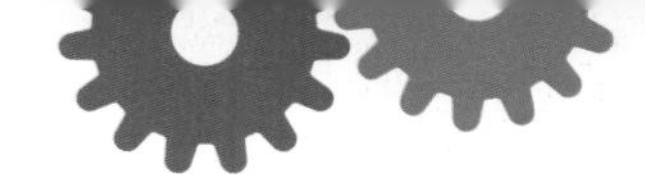

14 Here are the first five terms of a sequence:

30, 29, 27, 24, 20

a) Write down the next two terms in the sequence.

Here are the first five terms of a different sequence:

1, 5, 9, 13, 17

b) Find, in terms of n, an expression for the nth term of the sequence. **[Edexcel]**

15 Here are the first five terms of an arithmetic sequence:

6, 11, 16, 21, 26

Find an expression, in terms of n, for the nth term of this sequence. **[Edexcel]**

KEY POINTS

1 A **sequence** is a list of numbers which follow a rule.

2 A number sequence in which the terms go up in equal steps is called an **arithmetic sequence**.

3 The size of the step is called the **common difference**.

4 The common difference tells you which times table the sequence is based on. You can compare the sequence with the times table to find a rule for the nth term.

5 You can find a rule for a sequence by:

- writing down the first term.
- describing how to find the next term.

6 A number sequence in which the terms go up by adding on the next odd number is based on the square numbers.

7 Number sequences you should learn to recognise are:

Sequence	First 5 terms	Rule
Positive integers	1, 2, 3, 4, 5, ...	n
Even numbers	2, 4, 6, 8, 10, ...	$2n$
Odd numbers	1, 3, 5, 7, 9, ...	$2n - 1$
Square numbers	1, 4, 9, 16, 25, ...	n^2
Cube numbers	1, 8, 27, 64, 125, ...	n^3
Powers of 2	2, 4, 8, 16, 32, ...	2^n
Powers of 10	10, 100, 1000, 10 000, 100 000	10^n
Triangular numbers	1, 3, 6, 10, 15, ...	$\frac{n(n+1)}{2}$

Internet Challenge 9

Fibonacci numbers

Fibonacci numbers are used to model the behaviour of living systems.

The Fibonacci numbers also lead to the 'Golden Ratio', which is widely used in classical art and architecture.

In this challenge you will need to use a spreadsheet at first, before using the internet to complete your work.

Here is the Fibonacci number sequence:

$1, 1, 2, 3, 5, 8, 13, 21, \ldots$

1 Type these numbers into a computer spreadsheet, such as Excel.
(It is a good idea to enter them in a vertical list, rather than a horizontal one.)

2 Each term (apart from the first two) is found by adding together the two previous ones – for example, $13 = 8 + 5$

Use your spreadsheet replicating functions to generate a list of the first 50 Fibonacci numbers automatically.

3 Divide each Fibonacci number by the one before it – for example $8 \div 5 = 1.6$

Set up a column on your spreadsheet to do this up to the 50th Fibonacci number. What do you notice?

The values you found in the last question approach the number called the Golden Ratio.

Now use the internet to help answer the following questions.
Find pictures where appropriate.

4 How was the Golden Ratio used by the builders of the Parthenon in Athens?

5 Whose painting of *The Last Supper* was based on Golden Ratio constructions?

6 Which painter was said to have 'attacked every canvas by the golden section'?

7 When was Fibonacci born?

8 When did Fibonacci die?

9 Is there a formula for finding the nth term for Fibonacci numbers?

10 What sea creature has a spiral shell that is often (mistakenly) said to be based on the Golden Ratio?

CHAPTER 10

Coordinates and graphs

In this chapter you will **revise earlier work on:**

- using coordinates in the first quadrant.

You will also **learn how to:**

- use coordinates in all four quadrants
- find the midpoint of a line segment
- use 3-D coordinates
- plot graphs of linear functions (straight lines)
- find the gradient of a line
- solve simultaneous equations graphically
- plot graphs of quadratic functions (curves).

You will also be **challenged to:**

- investigate parallels.

Starter: Matchstick puzzles

Task 1

Starting with these twelve matches, remove:

a) four matches so that only one square remains

b) four matches so that only two squares remain

c) two matches so that only two squares remain.

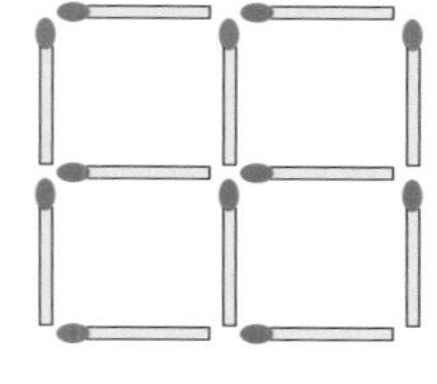

Task 2

These four matches make a cocktail glass containing a cherry. Move:

a) three matches

b) two matches

so that the cherry is outside the glass.

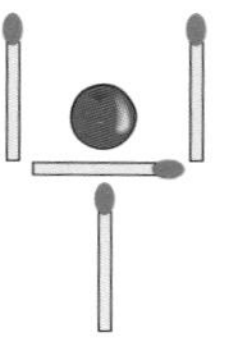

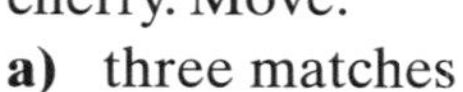

Task 3

These matches make the shape of a fish. Move:

a) four matches

b) three matches

so that the fish swims in the opposite direction.

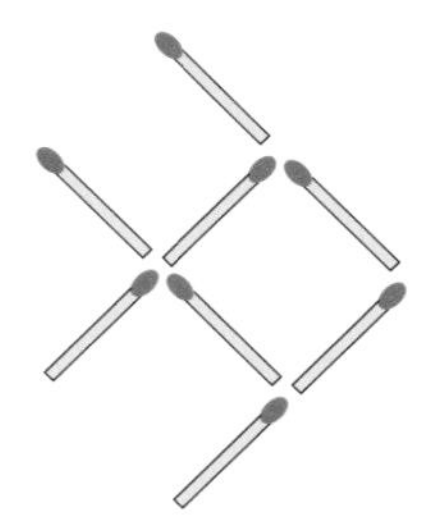

Task 4

Move one match to make a square.

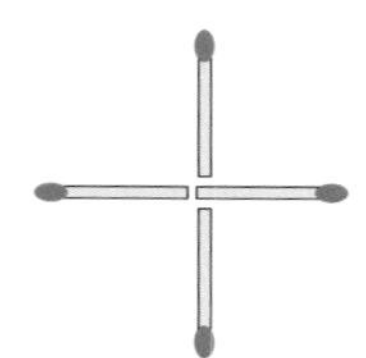

10.1 Coordinates in the first quadrant

The position of a point on a grid can be given by its **coordinates**.

For example, on the grid below the position of the point P is $(3, 2)$

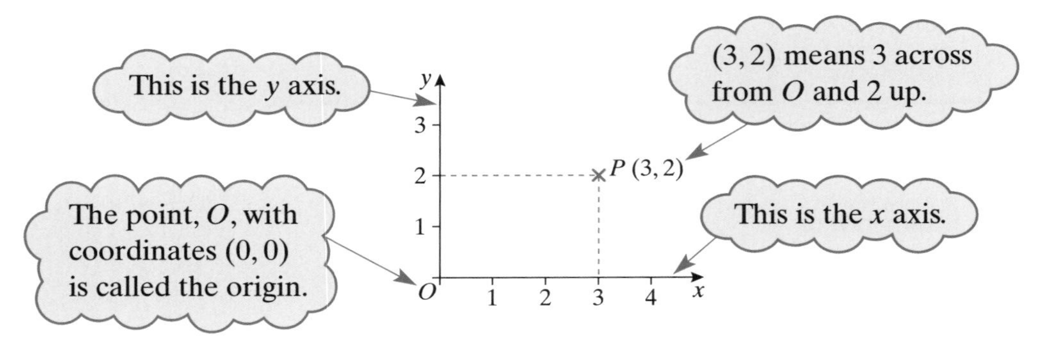
This is the y axis.
The point, O, with coordinates (0, 0) is called the origin.
(3, 2) means 3 across from O and 2 up.
This is the x axis.
P (3, 2)
y
x
O
1
2
3
4

EXAMPLE

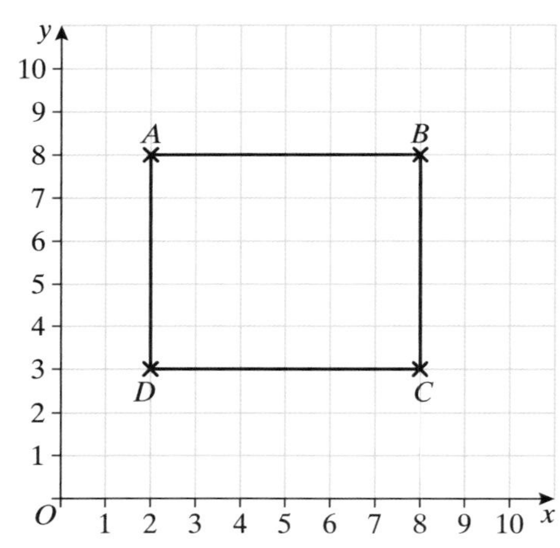
y
x
O
A
B
C
D
1
2
3
4
5
6
7
8
9
10

a) Write down the coordinates of the points A, B, C and D.

b) Write down the coordinates of the midpoint of the line:
(i) AB **(ii)** BC

SOLUTION

a) $A\ (2, 8)$

$B\ (8, 8)$

$C\ (8, 3)$

$D\ (2, 3)$

Remember 'along first then up'.

b) (i) The midpoint of the line AB is the point half way between A and B; the midpoint is at $(5, 8)$

(ii) The midpoint of BC is at $(8, 5\frac{1}{2})$

EXAMPLE

a) Draw a pair of axes on a grid and label both axes from 0 to 10
b) Plot the points A (2, 5) and B (7, 9)
Join the points with a straight line.
c) Mark the midpoint, M, of the line segment AB.
d) Write down the coordinates of M.

SOLUTION

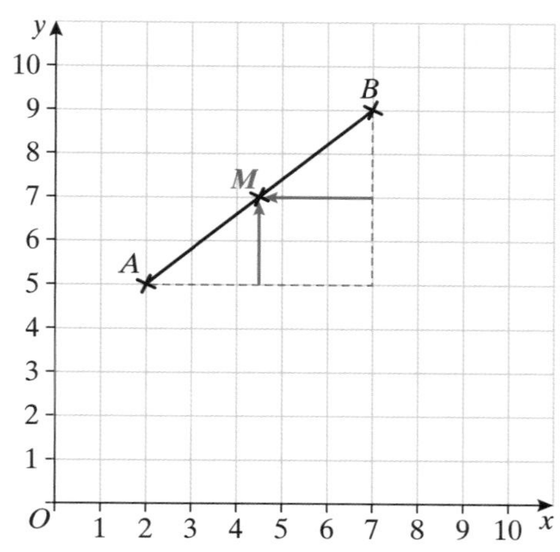

M is at the point $(4\frac{1}{2}, 7)$
You can find the midpoint without a diagram by finding:

- the value halfway between the x coordinates: $\frac{2+7}{2} = 4\frac{1}{2}$
- the value halfway between the y coordinates: $\frac{5+9}{2} = 7$

So the midpoint is at $\underline{(4\frac{1}{2}, 7)}$

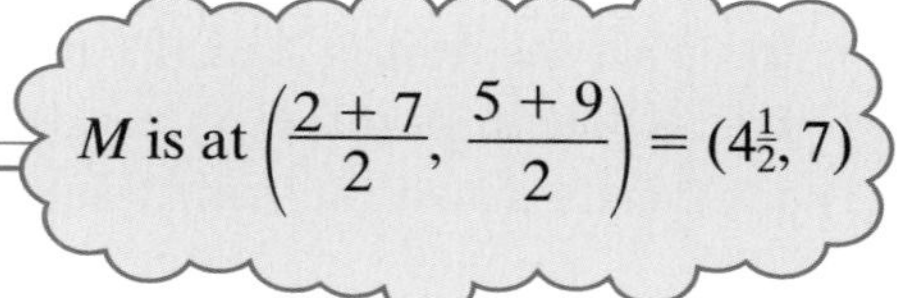

EXERCISE 10.1

1 Write down the coordinates of the points A to H.

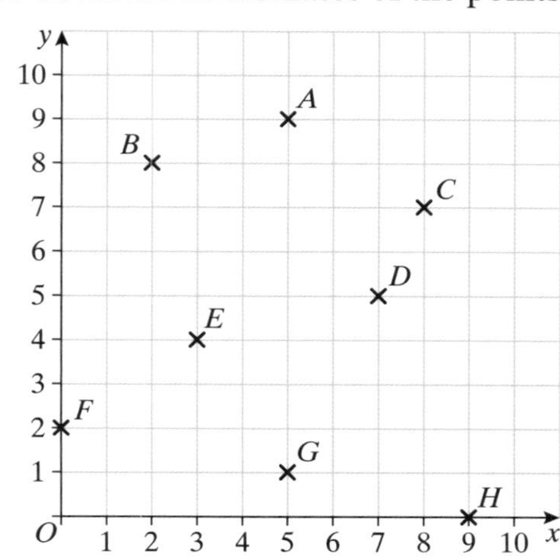

2 Write down the coordinates of the points A to H.

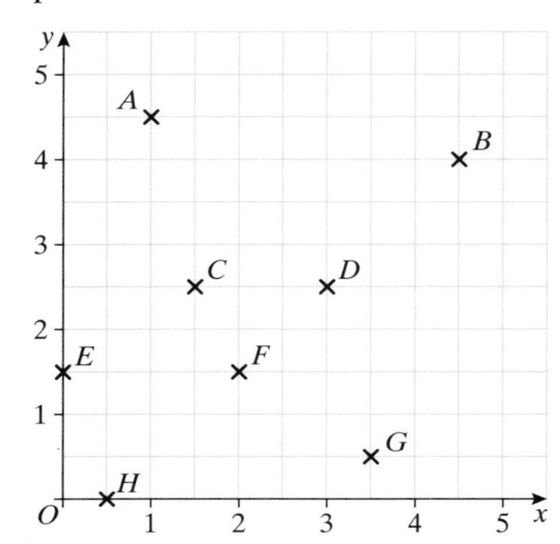

3 a) Write down the coordinates of the points A and B.

b) Write down the coordinates of the midpoint of the line segment AB.

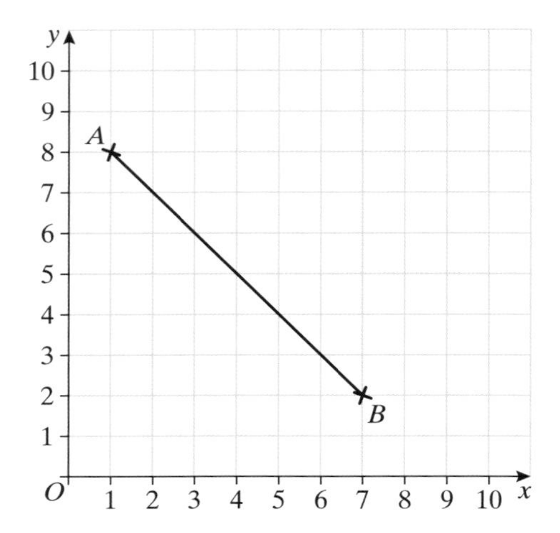

4 Draw a pair of axes on a grid and label both axes from 0 to 10

a) Plot the points $A\ (2, 8)$, $B\ (7, 8)$, $C\ (7, 2)$, and $D\ (2, 2)$

b) Join the points in alphabetical order to form a closed shape.

c) What is the name of the shape?

d) Mark the centre, M, of your shape.
Write down the coordinates of M.

5 Draw a pair of axes on a grid and label both axes from 0 to 6

a) Plot the points $A\ (1, 1)$, $B\ (1, 4)$ and $C\ (4, 4)$

b) A fourth point, D, is needed to form a square.
Write down the coordinates of D.

6 Draw a pair of axes on a grid and label both axes from 0 to 10

a) Plot the points $A\ (1, 9)$ and $B\ (8, 0)$
Join the points with a straight line.

b) Mark the midpoint, M, of the line segment AB.

c) Write down the coordinates of M.

7 The point A has coordinates $(4, 3)$
The point B has coordinates $(10, 7)$
M is the midpoint of the line segment AB.
Find the coordinates of M.

8 The point A has coordinates $(1, 12)$
The point B has coordinates $(6, 2)$
M is the midpoint of the line segment AB.
Find the coordinates of M.

10.2 Coordinates in all four quadrants

You can extend the basic coordinate system into four regions, or **quadrants**, by using negative coordinates, like this:

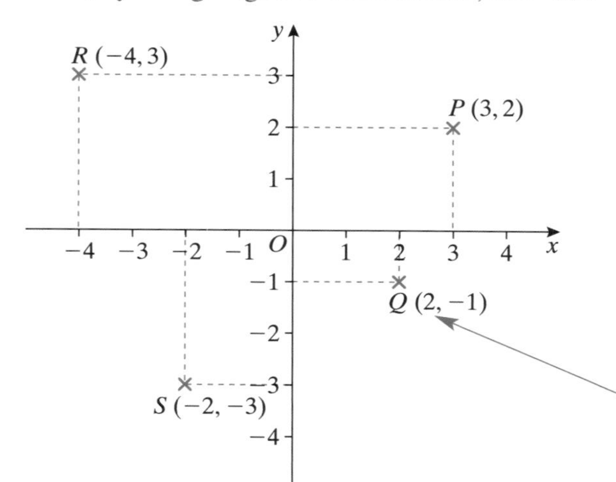

Negative coordinates indicate *left* instead of right (x) …
… or *down* instead of up (y)

EXAMPLE

a) Plot the points $A\,(-1, 2)$, $B\,(5, 1)$, $C\,(2, -1)$, $D\,(-4, 0)$ on a coordinate grid.

b) Join the points up in alphabetical order to form a closed shape. What shape is the result?

c) Find the coordinates of the midpoint, M, of the diagonal BD.

SOLUTION

a)

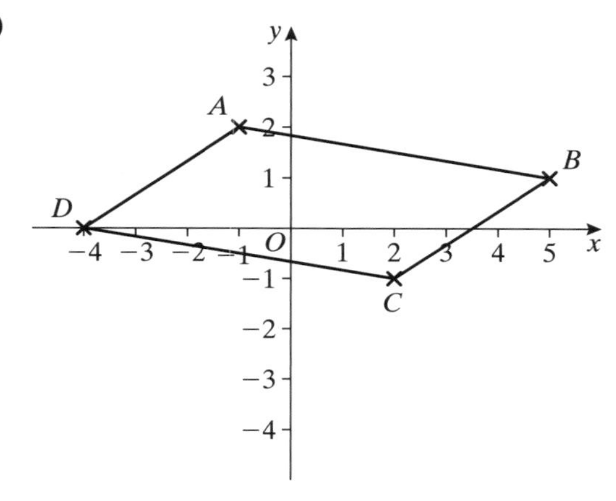

Add together the x coordinates and divide by 2
Do the same for the y coordinates.

b) The shape $ABCD$ is a parallelogram

c) The midpoint of $B\,(5, 1)$ and $D\,(-4, 0)$ is at $\left(\frac{5+-4}{2}, \frac{1+0}{2}\right) = \left(\frac{1}{2}, \frac{1}{2}\right)$

M is at $\left(\frac{1}{2}, \frac{1}{2}\right)$

EXERCISE 10.2

1 Using the diagram below, write down the coordinates of A, B, C, D and E.

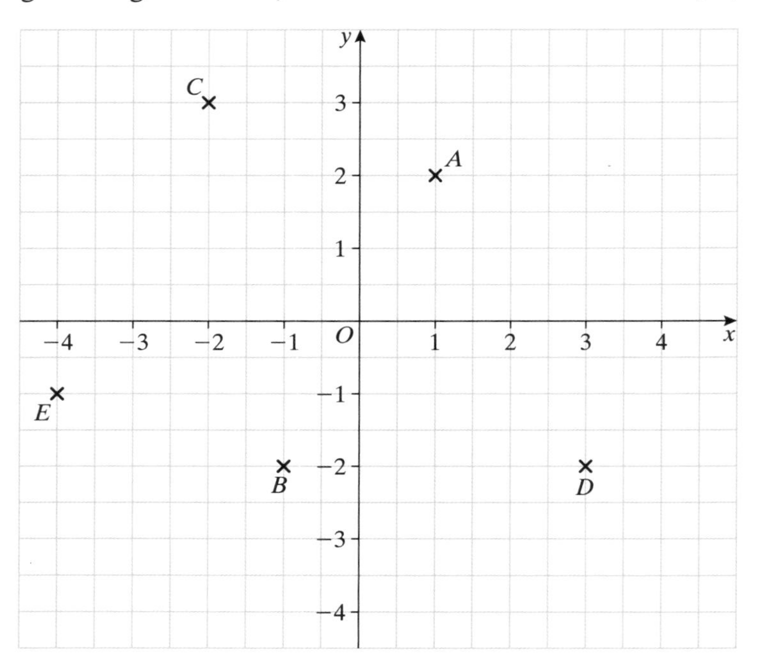

2 a) Which point is at $(-3, 1)$?

b) What are the coordinates of E?

c) Which point is midway between $(-2, 5)$ and $(4, 3)$?

d) What are the coordinates of H?

e) Which point has the **largest** y coordinate?

f) Which point has the **smallest** x coordinate?

g) Which point has the **same** x coordinate and y coordinate?

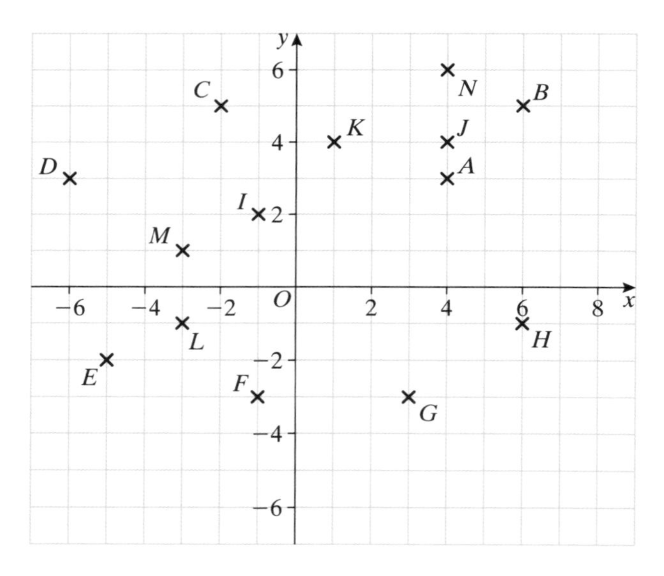

3 Draw a pair of axes on a grid and label both axes from −6 to 6

a) Plot the points A $(-5, 4)$, B $(3, 4)$ and C $(3, -6)$

b) A fourth point, D, is needed to form a rectangle. Write down the coordinates of D.

c) Mark the centre, M, of your shape. Write down the coordinates of M.

4 Follow these instructions carefully.
Draw a coordinate grid and label both axes from -5 to 5
Now draw line segments as follows:

from $(-5, 5)$ to $(5, 5)$
from $(-5, 5)$ to $(-5, -4)$
from $(-5, -5)$ to $(5, -5)$
from $(5, -5)$ to $(5, 4)$
from $(-4, 4)$ to $(2, 4)$
from $(-4, 0)$ to $(2, 0)$
from $(-4, -1)$ to $(-1, -1)$
from $(-1, -1)$ to $(-1, -4)$
from $(-5, -4)$ to $(-1, -4)$
from $(3, 5)$ to $(3, -1)$
from $(4, 4)$ to $(4, -4)$
from $(2, 4)$ to $(2, 0)$
from $(-4, 4)$ to $(-4, 0)$
from $(0, 0)$ to $(0, -5)$
from $(3, -1)$ to $(1, -1)$
from $(1, -4)$ to $(4, -4)$
from $(1, -1)$ to $(1, -4)$

You should find that you have made a maze puzzle.
Enter the maze at the top right corner.
Find a route through the maze to exit at the bottom left corner.

5 Draw a pair of axes on a grid and label both axes from -4 to 4
a) Plot the points $A\ (-3, 2)$ and $B\ (4, -3)$
Join the points with a straight line.
b) Mark the midpoint, M, of the line segment AB.
c) Write down the coordinates of M.

6 Draw a pair of axes on a grid and label both axes from -4 to 4
a) Plot the points $A\ (2, 3)$, $B\ (4, 2)$, $C\ (0, 0)$, and $D\ (-2, 1)$
b) Join the points in alphabetical order to form a closed shape.
c) What is the name of the shape?
d) Write down the coordinates of the midpoint of the diagonals AC and BD.

7 The point A has coordinates $(6, 2)$
The point B has coordinates $(-6, 8)$
M is the midpoint of the line segment AB.
Find the coordinates of M.

8 The point A has coordinates $(4, 9)$
The point B has coordinates $(-2, -3)$
M is the midpoint of the line segment AB.
Find the coordinates of M.

10.3 Coordinates in three dimensions

The position of a point on a line (1-D) is given by one coordinate (x)

The position of a point on a plane (2-D), for example a flat piece of paper, is given by two coordinates (x, y)

The position of a point in a 3-D space, for example a room, is given by three coordinates (x, y, z)

In all three cases the coordinates are given relative to the **origin** – you can think of this as being the starting point.

For example, imagine a spider hanging from the ceiling in your classroom.

To tell someone exactly where the spider is, you would need to say something like this:

Start at the corner of the room.
Walk along the wall 3 m
Walk away from the wall 2 m
The spider is 1 m above this point.

This information can be written as (3, 2, 1), where the origin is the corner of the room.

EXAMPLE

Write down the coordinates of the **vertices** (corners) on this cuboid.

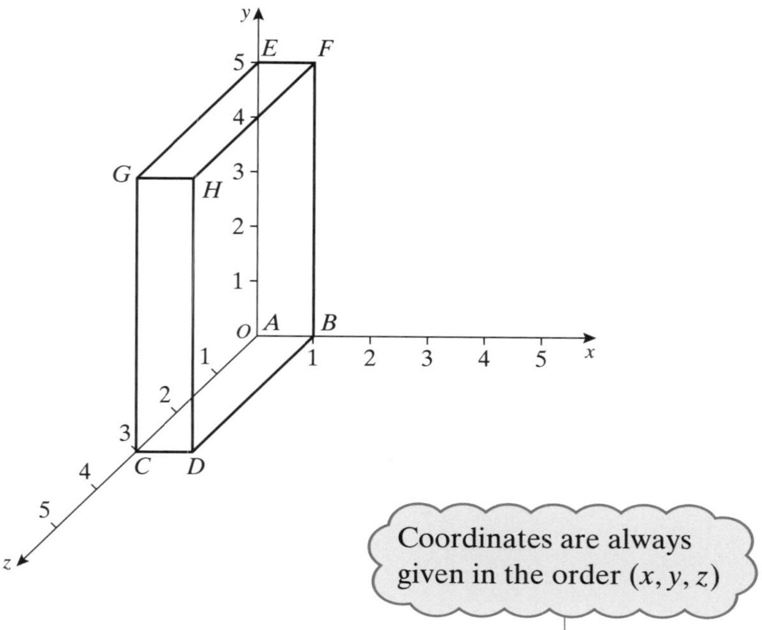

> Coordinates are always given in the order (x, y, z)

SOLUTION

Vertex A is at the origin, so A has coordinates $(0, 0, 0)$

Vertex B is 1 unit to the right of O, so B is at $(1, 0, 0)$

> A, B, C and D are all 'on the floor', so the y coordinate (up) is 0

Vertex C is 3 units out from O, so C is at $(0, 0, 3)$

Vertex D is 1 unit right from D, so D is at $(1, 0, 3)$

Vertex E is 5 units directly above O, so it has coordinates $(0, 5, 0)$

Vertex F is directly above B, so it is at $(1, 5, 0)$

Vertex G is directly above C, so it is at $(0, 5, 3)$

Vertex H is directly above D, so it is at $(1, 5, 3)$

> E, F, G and H are all 5 units up, so the y coordinate (up) is 5

EXERCISE 10.3

Write down the coordinates of the **vertices** (corners) on each of these cuboids.

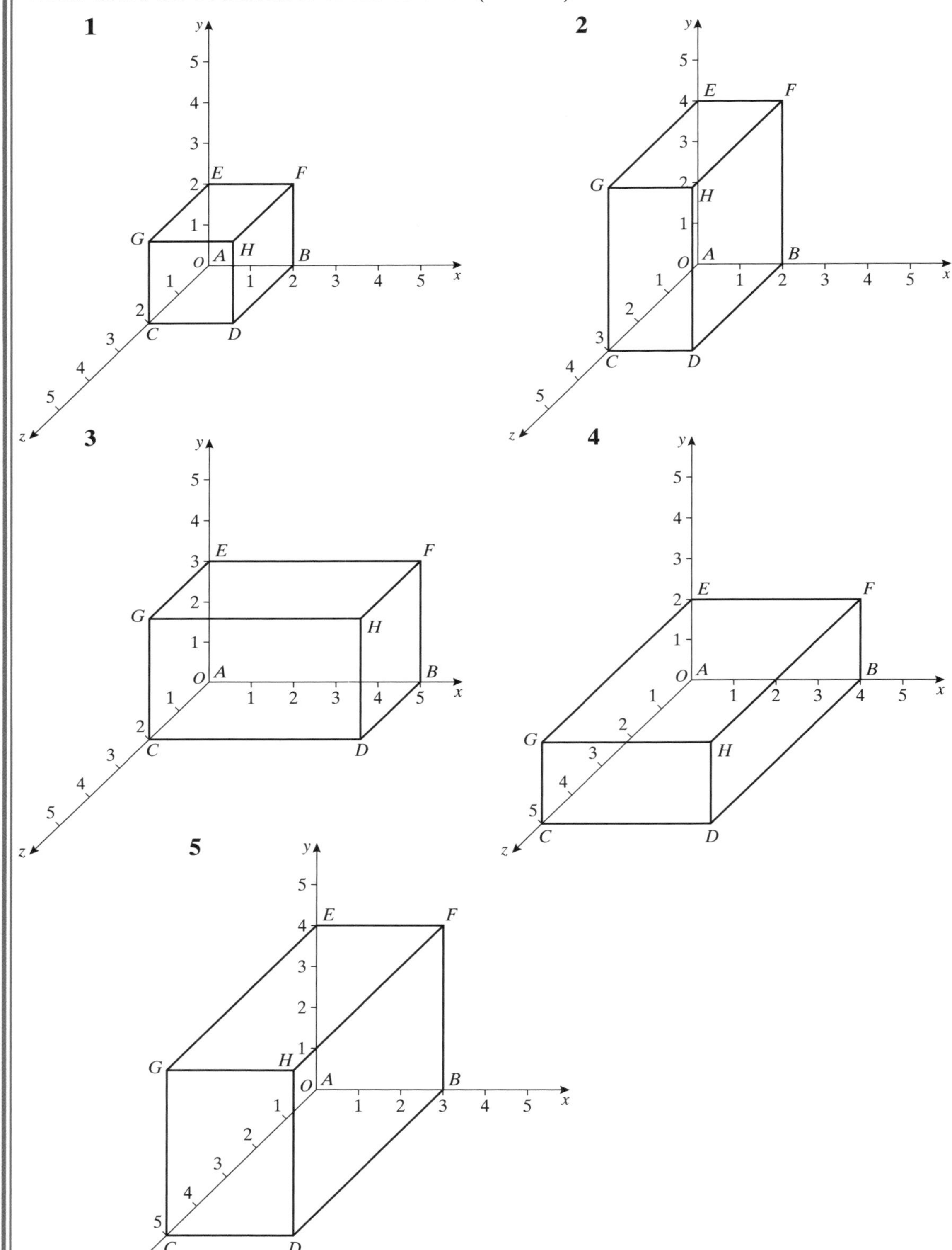

10.4 Straight line graphs

You need to be able to write down the equations of horizontal and vertical lines.

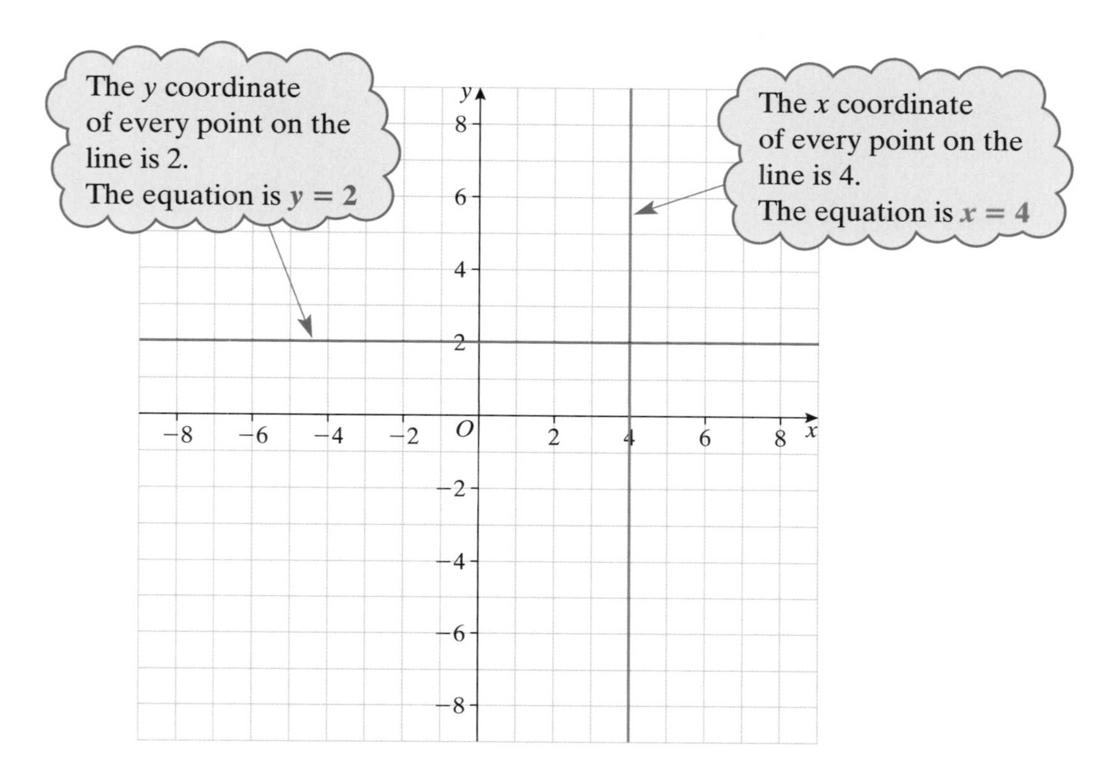

The y coordinate of every point on the line is 2.
The equation is y = 2
The x coordinate of every point on the line is 4.
The equation is x = 4
y
8
6
4
2
O
−2
−4
−6
−8
x
−8
−6
−4
−2
2
4
6
8

Any vertical line has an equation $x = a$, where a is a number.

The y axis is also called $x = 0$

Any horizontal line has an equation $y = b$ where b is a number.

The x axis is also called $y = 0$

You also need to know the following two lines:

For any point on the line,
- the y coordinate has the opposite sign to the x coordinate
- the equation of the line is $y = -x$

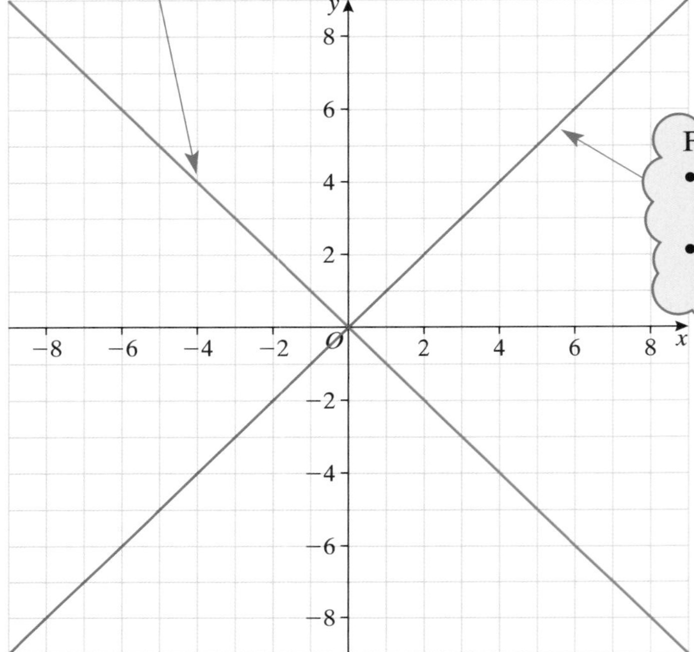

For any point on the line
- the y coordinate is the same as the x coordinate
- the equation of the line is $y = x$

Expressions like $3x + 5$ and $4 - 2x$ are called **linear expressions**.

They do not contain any terms like x^2, x^3, or $\frac{1}{x}$

Linear expressions are always in the form $mx + c$, where m and c are numbers.

An equation of the form:

$$y = mx + c$$

is called a **linear equation**.

This is because when you plot their graphs, the result is a straight line.

EXAMPLE

Draw the graph of $y = 3x - 1$ for $-3 \leqslant x \leqslant 3$

This means use x values between -3 and 3

SOLUTION

Step 1 Make a table:

x	-3	-2	-1	0	1	2	3
y	-10	-7	-4	-1	2	5	8

Multiply each x value by 3 and then subtract 1

Step 2 Draw your axes so that they extend to both the maximum and minimum values of x and y

So the x-axis should go from -3 to 3 and the y axis from -10 to 8

Step 3 Plot the points $(-3, -10), (-2, -7) \dots (3, 8)$

Step 4 Join your points with a straight line.

If one of your points doesn't lie on the line, then check carefully for a mistake.

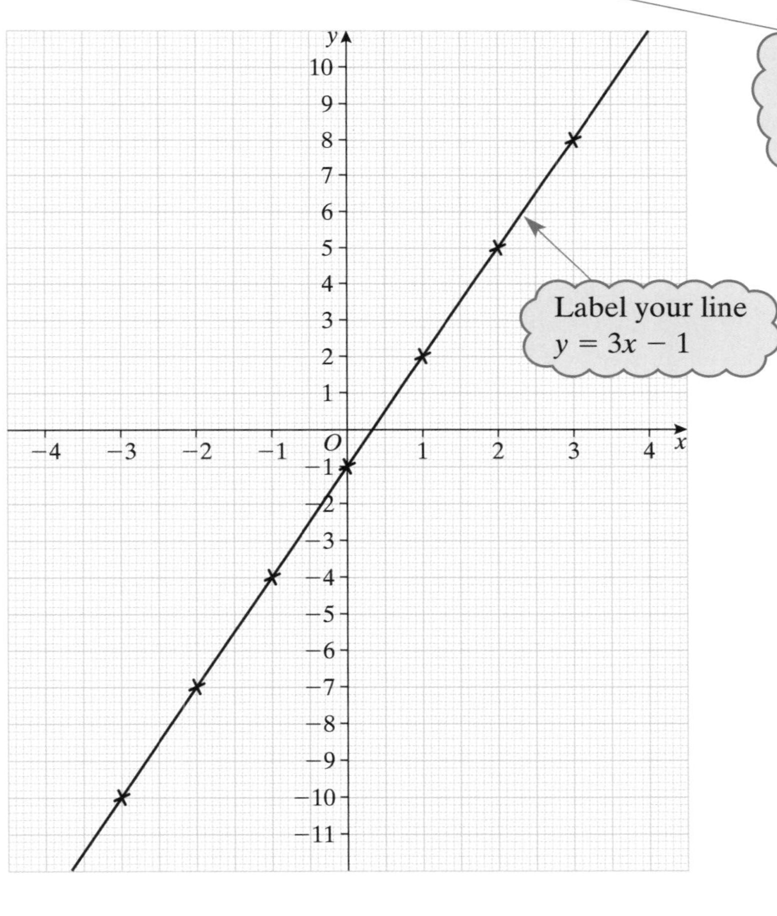

Strictly speaking, the line $y = 3x - 1$ is infinitely long and extends forever in both directions.

The portion of this line cut off between $x = -3$ and $x = 3$ is called a **line segment**.

To draw a graph of a straight line you only need the coordinates of two points.
However, it is best to find the coordinates of at least three points so that you avoid any mistakes.

EXAMPLE

a) Draw the graph of $x + y = 7$ for $0 \leqslant x \leqslant 7$

b) Use your graph to find:

(i) the value of y when $x = 3.7$

(ii) the value of x when $y = 5.8$

This means that the x and y coordinates add to make 7

SOLUTION

a) Step 1 Make a table:

x	0	3	7
y	7	4	0

So the x axis and the y axis both extend from 0 to 7

Step 2 Draw your axes so that they extend to both the maximum and minimum values of x and y

Step 3 Plot the points (0, 7), (3, 4) and (7, 0)

Step 4 Join your points with a straight line.

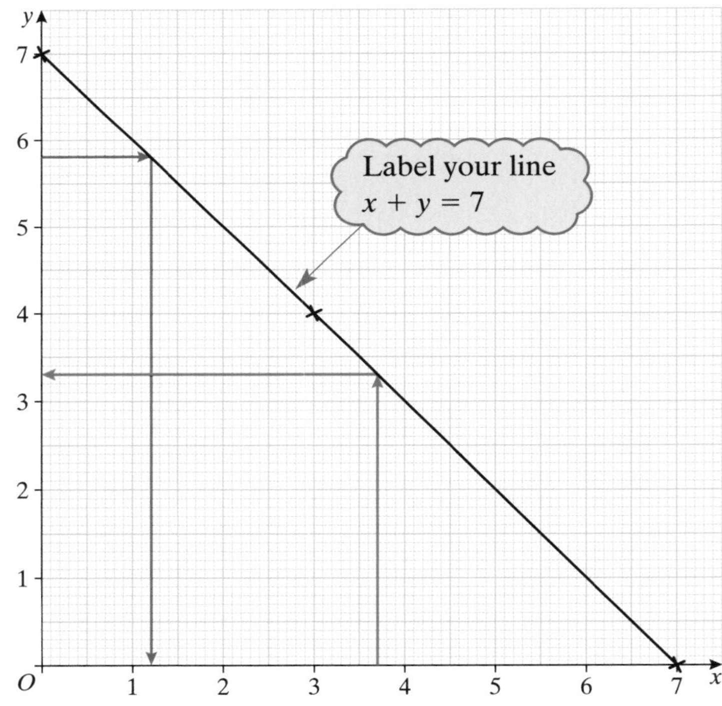

b) (i) Read off the y value at the point where $x = 3.7$

So $y = \underline{3.3}$

(ii) Read off the x value at the point where $y = 5.8$

So $x = \underline{1.2}$

EXERCISE 10.4

1 Write down the equation of each of the following lines:

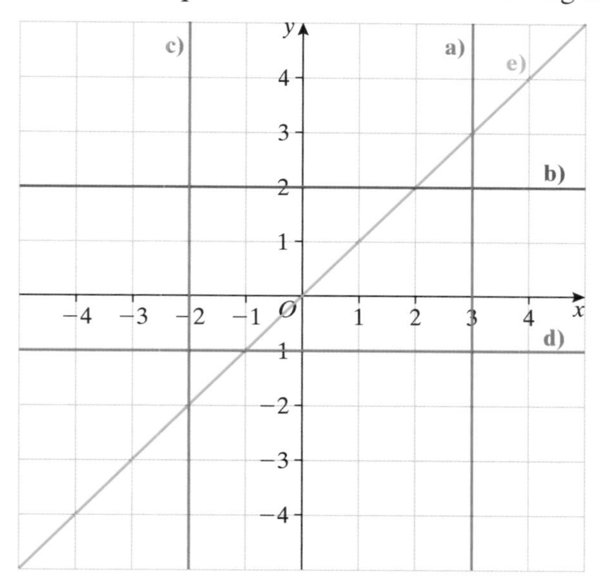

2 Draw a pair of axes on a grid and label both axes from −6 to 6

a) Draw the lines:

(i) $x = 4$ **(ii)** $y = 4$ **(iii)** $x = -3$ **(iv)** $y = -1$ **(v)** $y = -5$ **(vi)** $y = -x$

b) Draw the lines: **(i)** $x = 0$ **(ii)** $y = 0$

c) Write down another name for the lines: **(i)** $x = 0$ **(ii)** $y = 0$

3 a) Complete the table of values for $y = x + 3$

b) Draw the graph of $y = x + 3$

x	0	1	2	3	4	5
y	3		5		7	

4 a) Complete the table of values for $y = 2x$

b) Draw the graph of $y = 2x$

x	0	1	2	3	4	5
y	0	2			8	

5 a) Complete the table of values for $y = 2x + 3$

b) Draw the graph of $y = 2x + 3$

x	0	1	2	3	4	5
y	3	5				13

6 a) Complete the table of values for $y = 5 - x$

b) Draw the graph of $y = 5 - x$

x	0	1	2	3	4	5
y	5	4				0

7 a) Complete the table of values for $y = 2x + 5$

x	-3	-2	-1	0	1	2	3
y	-1		3				11

b) Draw the graph of $y = 2x + 5$

8 a) Complete the table of values for $y = 5x - 7$

x	-3	-2	-1	0	1	2	3
y	-22			-7			

b) Draw the graph of $y = 5x - 7$

9 a) Complete the table of values for $2x + y = 14$

This means 2 times the x coordinate plus the y coordinate make 14

x	0	1	2	3	4	5	6
$2x$	0			6		10	
y	14			8		4	

b) Draw the graph of $2x + y = 14$

10 a) Draw a pair of axes so that:
- the x values start at 0 and finish at 5
- the y values start at 0 and finish at 15

b) Draw the graph of $y = 3x$

c) Use your graph to find:
(i) the value of y when $x = 1.3$ **(ii)** the value of x when $y = 4.8$

11 a) Draw a pair of axes so that:
- the x values start at 0 and finish at 5
- the y values start at 0 and finish at 8

b) Draw the graph of $y = x + 2$

c) Use your graph to find:
(i) the value of y when $x = 2.7$ **(ii)** the value of x when $y = 3.1$

12 a) Draw a pair of axes so that:
- the x values start at 0 and finish at 8
- the y values start at 0 and finish at 8

b) Draw the graph of $x + y = 8$

c) Use your graph to find:
(i) the value of y when $x = 5.9$ **(ii)** the value of x when $y = 1.7$

13 Draw the following graphs:

a) $y = \frac{1}{2}x$ for $0 \leqslant x \leqslant 10$

b) $y = \frac{1}{2}x + 1$ for $0 \leqslant x \leqslant 6$

c) $y = \frac{1}{2}x - 2$ for $0 \leqslant x \leqslant 6$

10.5 Gradients

In question **4** of Exercise 10.4, you were asked to plot the graph of $y = 2x$

Your graph should look like this:

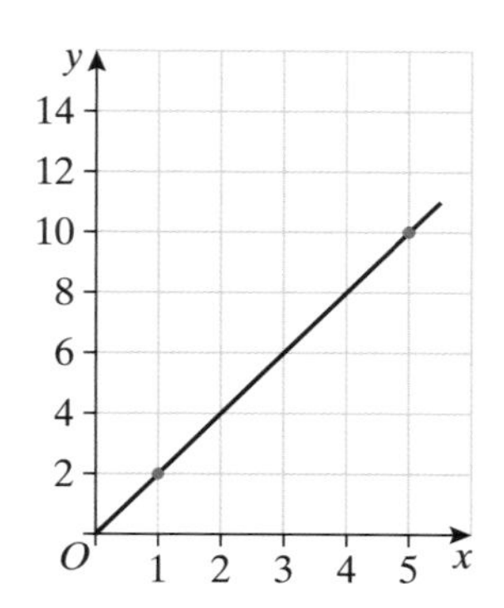

The gradient of a line tells you how steep the line is.

A gradient of 6 is steeper than a gradient of 2

$$\textbf{Gradient} = \frac{\textbf{Rise}}{\textbf{run}}$$

Draw a right-angled triangle (any size will work) under the graph to find the rise and run.

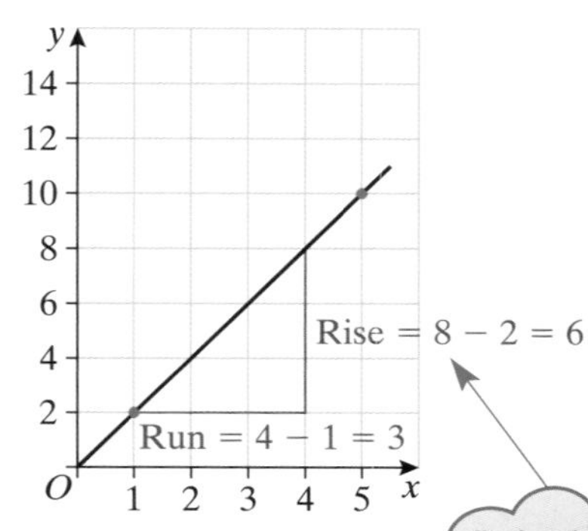

Rise = 8 − 2 = 6

It is dangerous just to count squares. You must read the values off the graph carefully: the x and y axes may have been different scales, as here.

$$\text{So gradient} = \frac{\text{rise}}{\text{run}}$$

$$= \frac{6}{3}$$

$$= 2$$

The gradient of the line $y = 2x$ is 2

The diagram below shows three graphs, all with gradient 2.

Each line crosses the y axis at a different point.
This point is called the ***y* intercept**.

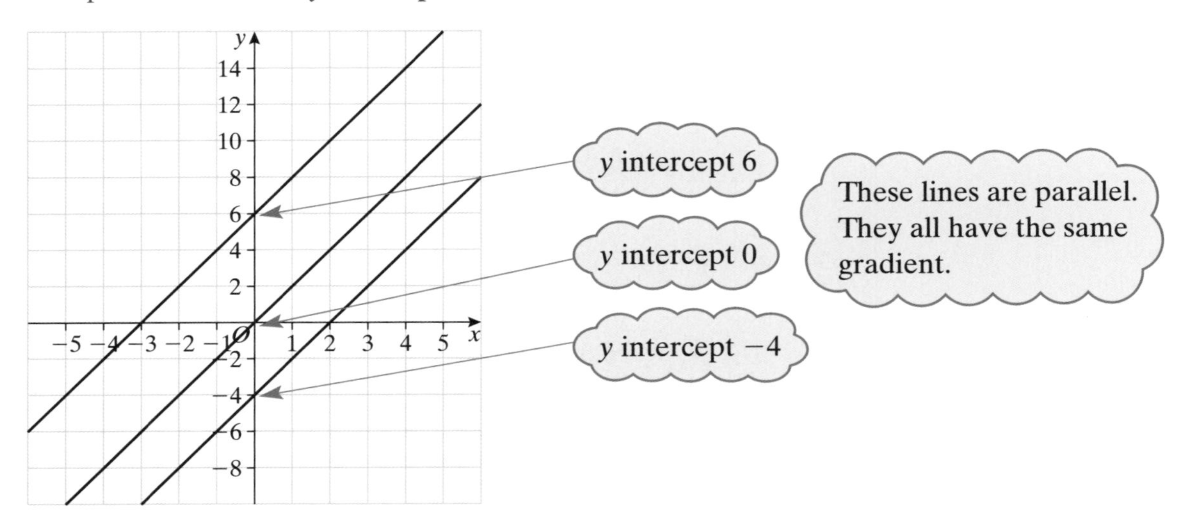

y intercept 6
y intercept 0
y intercept −4
These lines are parallel. They all have the same gradient.

EXAMPLE

A straight line passes through the points (0, 5) and (3, 14)
Find its gradient.

SOLUTION

Rise = 14 − 5 = 9

Run = 3 − 0 = 3

$$\text{Gradient} = \frac{\text{rise}}{\text{run}} = \frac{9}{3} = \underline{3}$$

When the graph slopes downhill, the gradient is negative.

EXAMPLE

Find the gradient of this line.

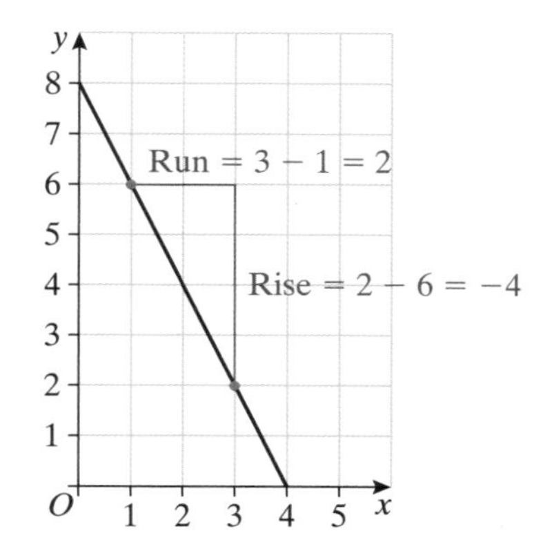

$$\text{Gradient} = \frac{\text{rise}}{\text{run}} = \frac{-4}{2} = \underline{-2}$$

EXERCISE 10.5

Find the gradient of each of the lines marked in questions **1** to **8** below.

1

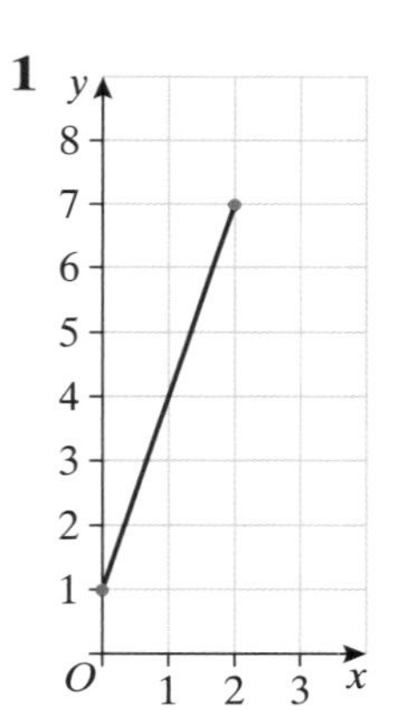

2

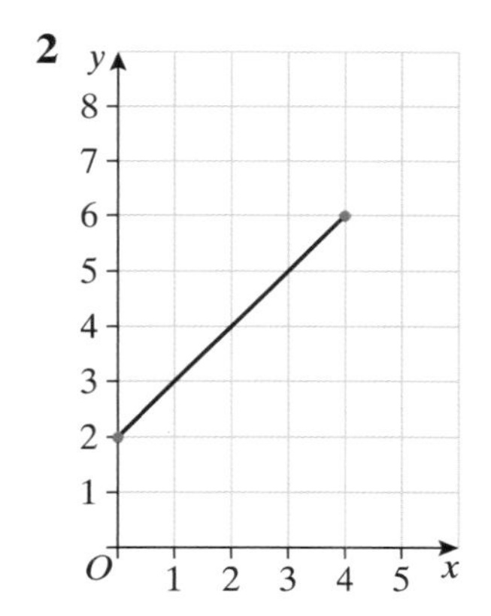

3

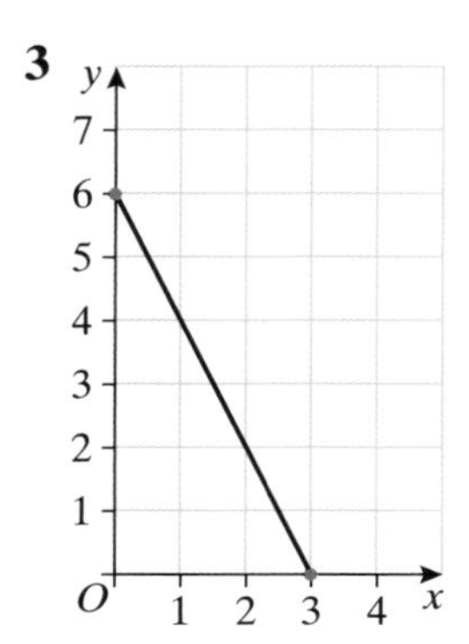

4

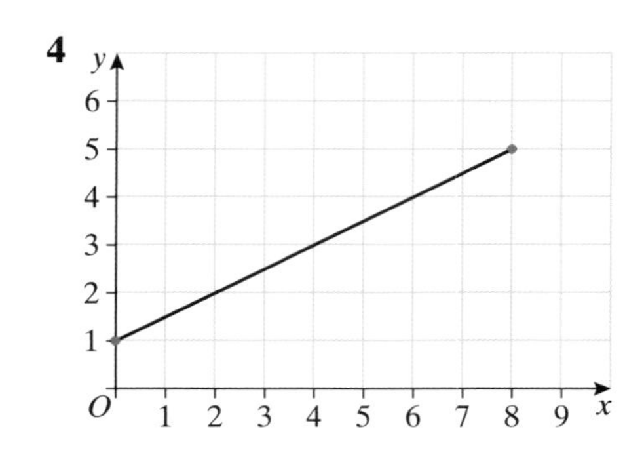

5

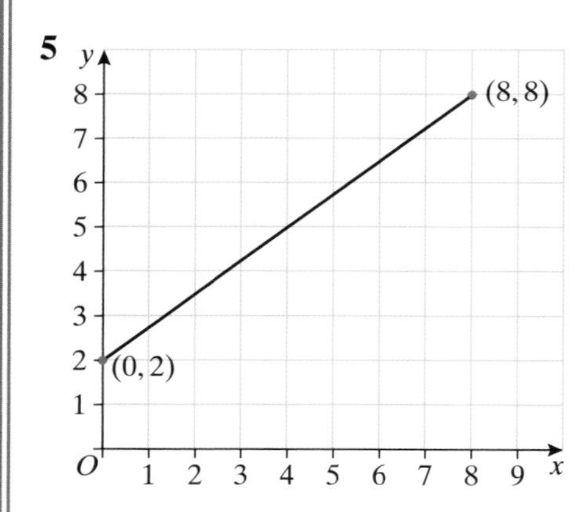

6

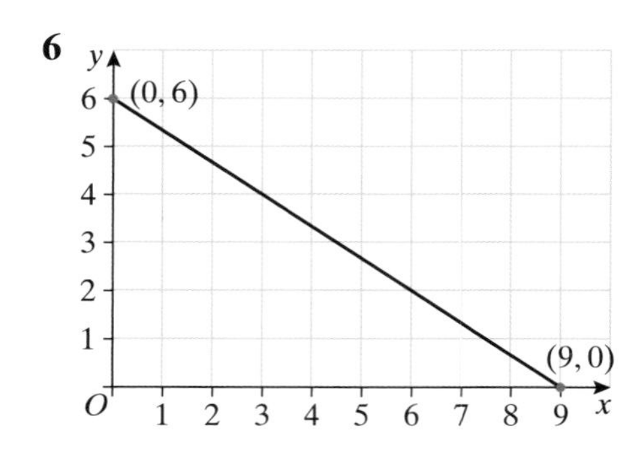

7 Find the gradient of the line joining the points (2, 2) and (14, 8).

8 Find the gradient of the line joining the points (2, 3) and (8, 0).

10.6 Equations and graphs

Here is the graph of $y = 2x + 3$

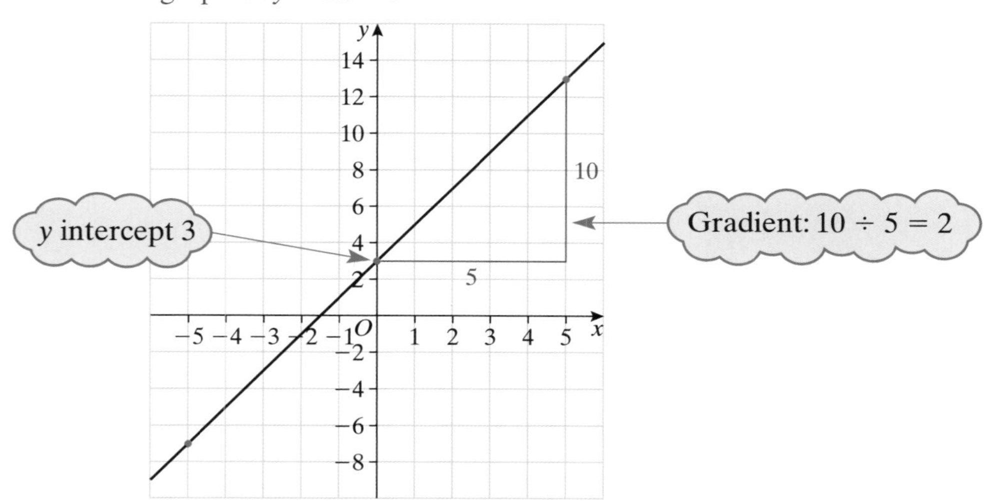

Notice that the line has gradient 2, and y intercept 3
These are the the values of the two coefficients that appear in the equation of the line.

numbers

This illustrates an important general result:

> **The graph of the function $y = mx + c$ has gradient m and y intercept c**

where the line cuts the y axis

EXAMPLE

Write down the gradient of the functions:

a) $y = 3x + 1$
b) $y = 5 - \frac{1}{2}x$

SOLUTION

a) The gradient is $\underline{3}$

$y = 3x + 1$ — 3: gradient; 1: y intercept
Note: The equation must be in the form $y = \ldots$

b) The gradient is $\underline{-\frac{1}{2}}$

$y = 5 - \frac{1}{2}x$ — 5: y intercept; $\frac{1}{2}$: gradient

You can use this idea the other way round, to find the equation of a given straight line graph.

EXAMPLE

Find the equation of this straight line:

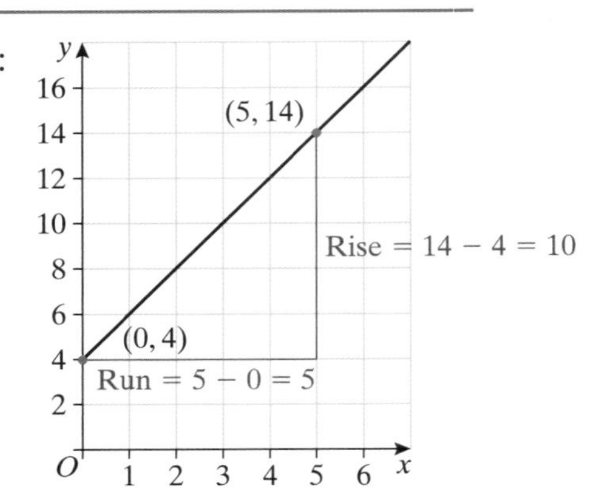

SOLUTION

The y intercept is $c = 4$

The gradient is $\frac{14-4}{5-0} = \frac{10}{5} = 2$

So the equation of the line is $y = 2x + 4$

EXERCISE 10.6

1 Write down the gradients of the lines:

a) $y = 3x$ **b)** $y = 4x$ **c)** $y = -5x$

d) $y = 2x - 1$ **e)** $y = x + 2$ **f)** $y = 4 - 3x$

2 a) Write down the coordinates of the points P and Q on the line.

b) Find the gradient and intercept of the line.

c) Hence write down the equation of the straight line.

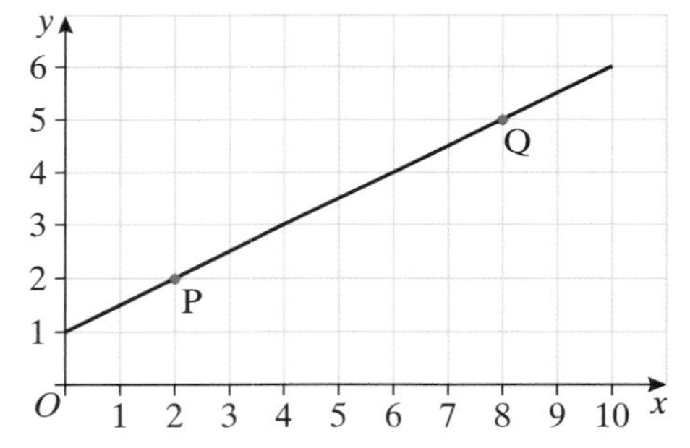

3 a) Find the gradient and intercept of the line.

b) Hence write down the equation of the straight line.

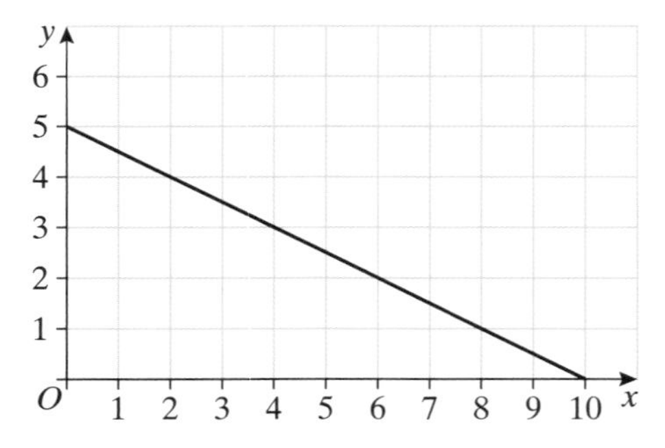

10.7 Simultaneous equations

Sometimes you will be asked to draw 2 lines on the same grid.
The point where the lines cross is called the **intersection**.
This gives the solution to a pair of simultaneous equations, as in the next example.

EXAMPLE

a) Draw the graphs of $x + y = 8$
and $y = 2x - 1$
on the same axes.

b) Write down the coordinates of the point of intersection.

c) Hence solve the simultaneous equations $x + y = 8$
$y = 2x - 1$

SOLUTION

a) $x + y = 8$

x	0	4	8
y	8	4	0

$y = 2x - 1$

x	0	4	8
y	-1	7	15

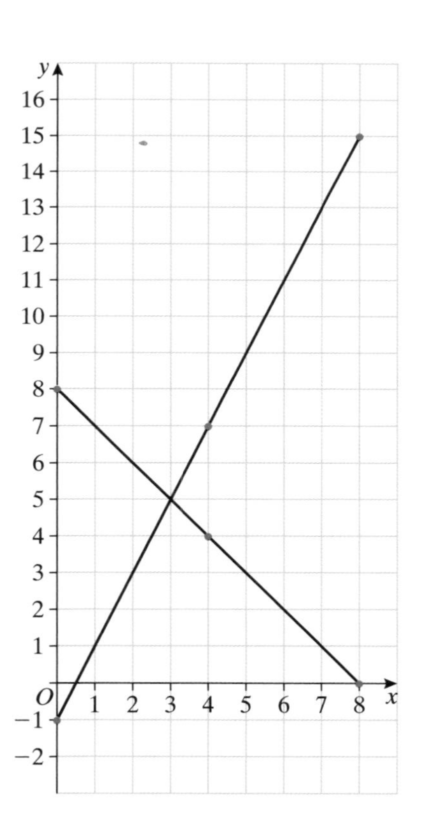

b) $(3, 5)$

This point lies on both lines.

c) $x = 3$
$y = 5$
Check: $3 + 5 = 8$
$5 = 2 \times 3 - 1$ ✓

EXERCISE 10.7

1 a) On the same axes, draw the graphs of:

$y = 2x + 1$ and $y = x - 2$

b) Write down the coordinates of the point of intersection.

c) Hence, solve the simultaneous equations:

$y = 2x + 1$
$y = x - 2$

2 a) On the same axes, draw the graphs of:

$y = x$ and $x = 3$

b) Write down the coordinates of the point of intersection.

c) Hence, solve the simultaneous equations:

$y = x$
$x = 3$

3 a) On the same axes, draw the graphs of:

$y = 2x$
and $y = x - 1$

b) Write down the coordinates of the point of intersection.

c) Hence, solve the simultaneous equations:

$y = 2x$
$y = x - 1$

4 a) On the same axes, draw the graphs of:

$y = x - 4$
and $y = \frac{1}{2}x$

b) Write down the coordinates of the point of intersection.

c) Hence, solve the simultaneous equations:

$y = x - 4$
$y = \frac{1}{2}x$

5 a) On the same axes, draw the graphs of:

$x + y = 9$
and $y = 2x$

b) Write down the coordinates of the point of intersection.

c) Hence, solve the simultaneous equations:

$x + y = 9$
$y = 2x$

10.8 Quadratic curves

Any equation with an 'x^2' term will be a curve.

The curve will be either $\smile$ or $\frown$ shaped depending on the sign in front of the 'x^2' term.

When the sign is positive, for example x^2, $2x^2$ or $3x^2$, the curve is $\smile$ shaped (**smiling**).

When the sign is negative, for example $-x^2$, $-2x^2$ or $-3x^2$, the curve is $\frown$ shape (**frowning**).

EXAMPLE

Draw the graph of $y = x^2 + 1$ for $-4 \leqslant x \leqslant 4$

This means use x values between -4 and 4

SOLUTION

Step 1 Make a table:

x	−4	−3	−2	−1	0	1	2	3	4
x^2	16	9	4	1	0	1	4	9	16
+1	+1	+1	+1	+1	+1	+1	+1	+1	+1
y	17	10	5	2	1	2	5	10	17

Remember when you multiply two negative numbers the answer is positive.

Step 2 Draw your axes so they extend to both the maximum and minimum values of x and y.

So the x axis should go from -4 to 4 and the y axis from 0 to 17

Step 3 Plot the points $(-4, 17), (-3, 10) \ldots (4, 17)$

Step 4 Join your points with a smooth curve *– never use a ruler to join points for graphs which curve!*

If one of your points doesn't lie on the curve, then check carefully for a mistake.

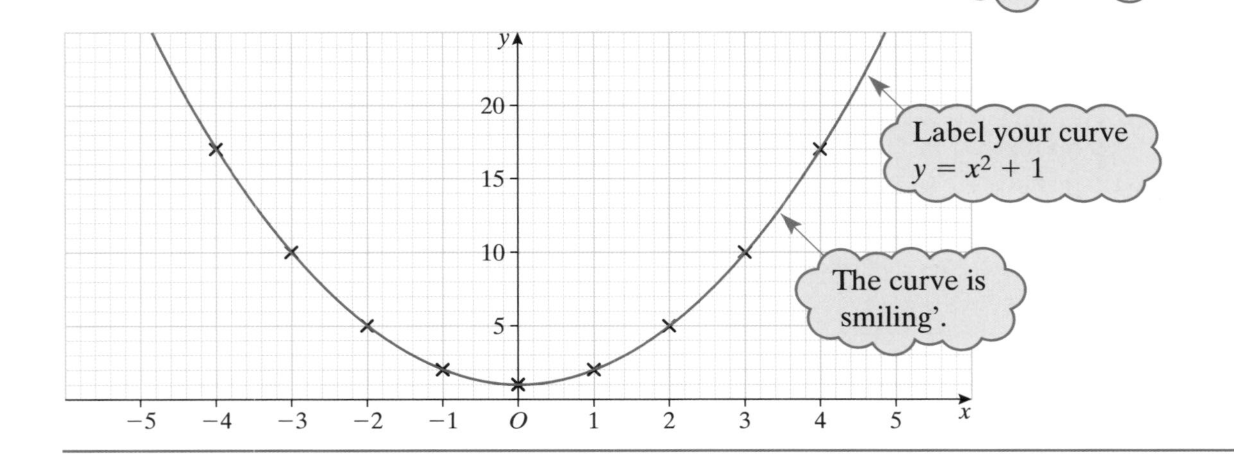

EXAMPLE

a) Draw the graph of $y = 8 - 2x^2$ for $-4 \leqslant x \leqslant 4$

b) Use your graph to solve the equation $8 - 2x^2 = 0$

Remember $-2x^2$ means -2 lots of x^2 and the graph will be 'frowning'.

SOLUTION

a) Step 1 Make a table:

x	−4	−3	−2	−1	0	1	2	3	4
x^2	16	9	4	1	0	1	4	9	16
8	8	8	8	8	8	8	8	8	8
$-2x^2$	−32	−18	−8	−2	0	−2	−8	−18	−32
y	−24	−10	0	6	8	6	0	−10	−24

Step 2 Draw your axes so they extend to both the maximum and minimum values of x and y

So the x axis should go from -4 to 4 and the y axis from -24 to 8

Step 3 Plot the points $(-4, -24), (-3, -10) \ldots (4, -24)$

Step 4 Join your points with a smooth curve
– never use a ruler to join points for graphs which curve!

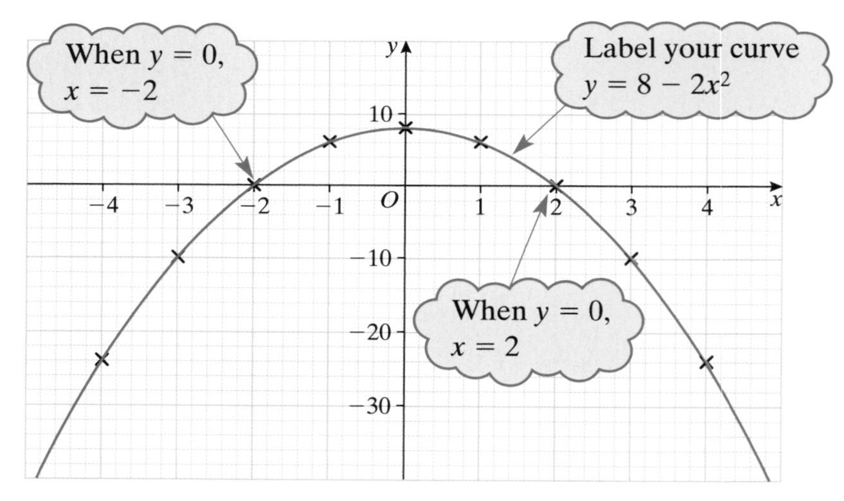

b) The solution to $8 - 2x^2 = 0$ is found when $y = 0$
This is where the curve crosses the x axis.
So $x = 2$ or $x = -2$

EXAMPLE

a) Draw the graph of $y = x^2 - 3x + 1$ for $-1 \leqslant x \leqslant 4$
b) Use your graph to solve the equation $x^2 - 3x + 1 = 0$

SOLUTION

a) Step 1 Make a table:

x	-1	0	1	2	3	4
x^2	1	0	1	4	9	16
$-3x$	3	0	-3	-6	-9	-12
$+1$	$+1$	$+1$	$+1$	$+1$	$+1$	$+1$
y	5	1	-1	-1	1	5

Step 2 Draw your axes so they extend to both the maximum and minimum values of x and y.

So the x axis should go from -1 to 4 and the y axis from -1 to 5

Step 3 Plot the points $(-1, 5)$, $(0, 1)$ … $(4, 5)$

Step 4 Join your points with a smooth curve – *never use a ruler to join points for graphs which curve!*

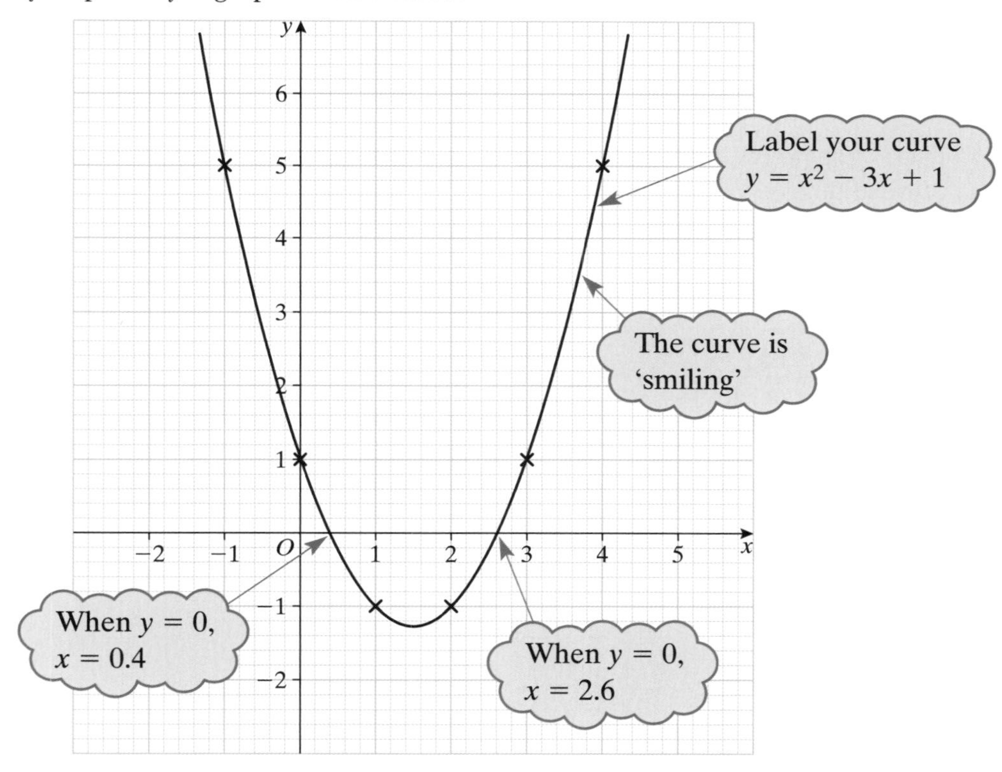

b) The solution to $x^2 - 3x + 1 = 0$ is found when $y = 0$
This is where the curve crosses the x axis.
So $x = 0.4$ or $x = 2.6$

EXERCISE 10.8

1 a) Complete the table of values for $y = x^2$

x	−3	−2	−1	0	1	2	3
x^2	9		1			4	
y	9		1			4	

b) Draw the graph of $y = x^2$

2 a) Complete the table of values for $y = x^2 + 3$

x	−4	−3	−2	−1	0	1	2	3	4
x^2	16			1	0	1		9	
+3	+3			+3		+3		+3	
y	19			4				12	

b) Draw the graph of $y = x^2 + 3$

3 a) Complete the table of values for $y = x^2 - 4$

x	−4	−3	−2	−1	0	1	2	3	4
x^2		9			0		4		
−4		−4		−4		−4	−4		
y		5					0		

b) Draw the graph of $y = x^2 - 4$

c) Use your graph to solve the equation $x^2 - 4 = 0$

4 a) Complete the table of values for $y = 2x^2 - 4$

x	−4	−3	−2	−1	0	1	2	3	4
x^2		9					4		
$2x^2$		18					8		
−4		−4		−4		−4	−4		
y		14					−4		

b) Draw the graph of $y = 2x^2 - 4$

c) Use your graph to solve the equation $2x^2 - 4 = 0$

5 a) Complete the table of values for $y = x^2 + 2x - 6$

x	-4	-3	-2	-1	0	1	2	3
x^2	16	9						9
$+2x$	-8	-6						6
-6	-6	-6						-6
y	2	-3						9

b) Draw the graph of $y = x^2 + 2x - 6$

c) Use your graph to solve the equation $x^2 + 2x - 6 = 0$

6 Draw the following graphs for $-4 \leqslant x \leqslant 4$:

a) $y = 2x^2$ **b)** $y = x^2 - 3$ **c)** $y = 2x^2 + 1$

d) $y = x^2 + x$ **e)** $y = 2x^2 - 3x$ **f)** $y = x^2 + 3x - 7$

REVIEW EXERCISE 10

Don't use your calculator for Questions 1–11.

1 a) Write down the coordinates of the points A to H.

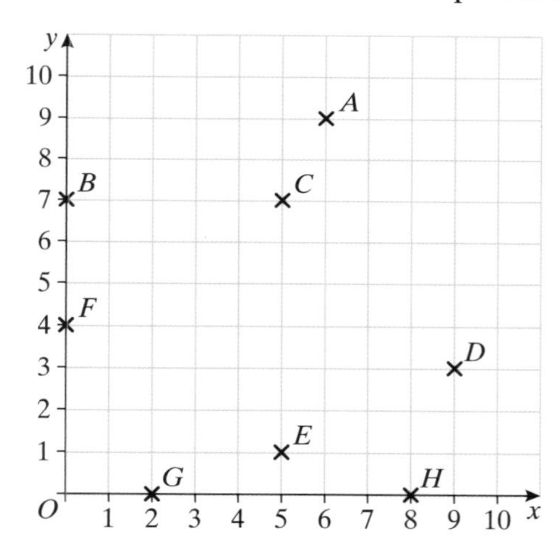

b) Write down the coordinates of the point that lies half way between:
(i) G and H **(ii)** B and F **(iii)** C and E **(iv)** B and D

2 Draw a pair of axes on a grid and label both axes from 0 to 6

a) Plot the points A (1, 2), B (1, 6) and C (6, 4)

A fourth point, D, is needed to form a parallelogram.

b) Write down the coordinates of D.

3 Draw a pair of axes on a grid and label both axes from -5 to 5
   a) Plot the points $A(-3, -2)$, $B(-3, 4)$ and $C(3, 4)$

   A fourth point, D, is needed to form a square.
   b) Write down the coordinates of D.
   c) Write down the coordinates of the centre of the square.

4 Write down the equation of each of the following lines:

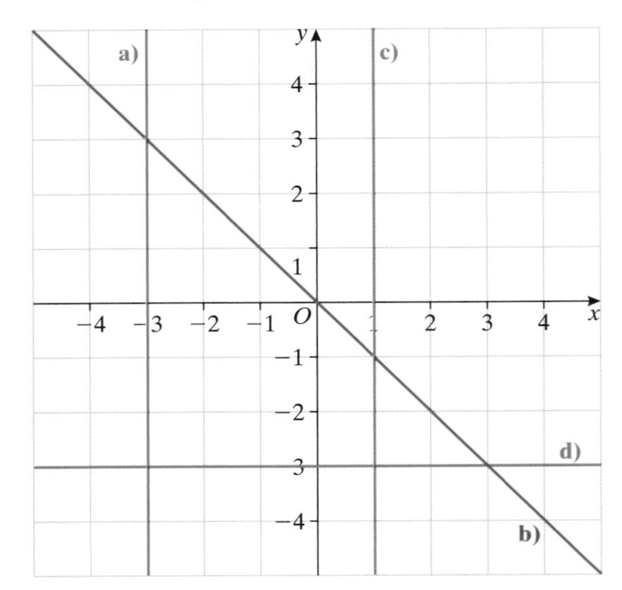

5 Write down the coordinates of the vertices of this cuboid.

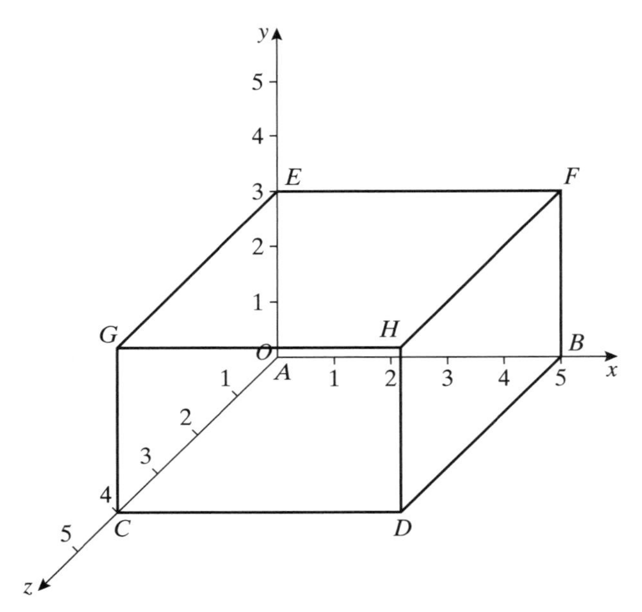

6 a) Complete the table of values for $y = 5x - 3$

x	-3	-2	-1	0	1	2	3
y				-3	2		

b) Draw the graph of $y = 5x - 3$

c) Use your graph to find:

(i) the value of y when $x = -2.3$ **(ii)** the value of x when $y = 9$

d) Find the gradient of the line $y = 5x - 3$

7 a) Complete the table of values for $y = 4x - 1$

x	-3	-2	-1	0	1	2	3
y	-13			-1			

b) Draw the graph of $y = 4x - 1$

c) Use your graph to find:

(i) the value of y when $x = 1.7$ **(ii)** the value of x when $y = -11$

d) Find the gradient of the line $y = 4x - 1$

8 a) Draw graphs of $y = x + 1$ and $y = 3x - 7$ on the same axes.

b) Write down the solution of the simultaneous equations $y = x + 1$ and $y = 3x - 7$

9

a) Write down the coordinates of the points:

(i) A **(ii)** B

b) Write down the coordinates of the midpoint of the line AB

c) Find the gradient of the line AB.

10

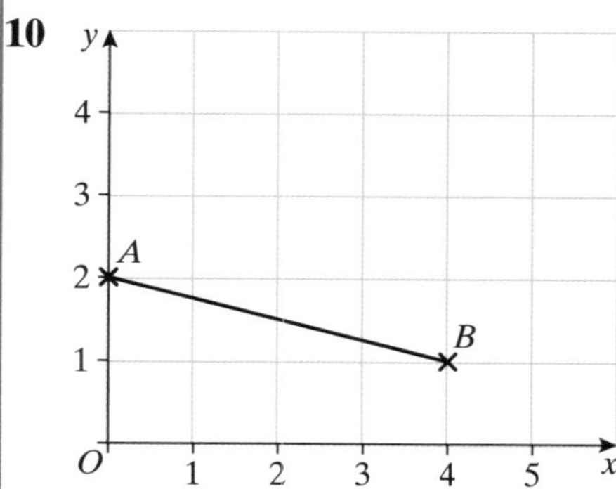

a) Write down the coordinates of the points:

(i) A **(ii)** B

b) Write down the coordinates of the midpoint of the line AB. **[Edexcel]**

11 a) Write down the coordinates of the point P.

b) (i) Copy the grid.
On your grid, plot the point $(0, 3)$
Label this point Q.

(ii) On your grid, plot the point $(-2, -3)$
Label this point R.

c) Write down the coordinates of the midpoint of the line QR.

[Edexcel]

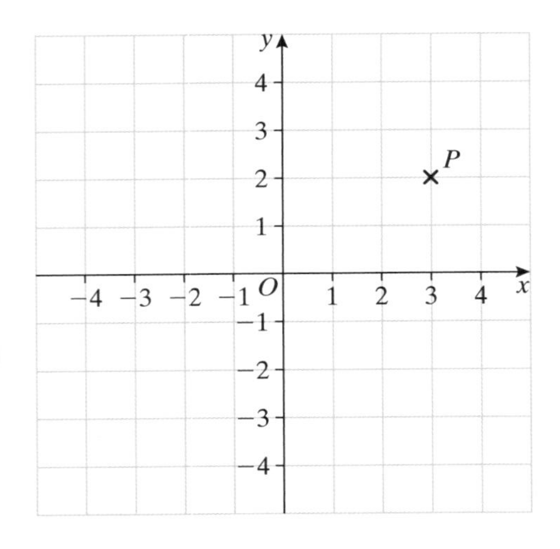

You can use your calculator for Questions 12–15.

12 a) Complete the table of values for $y = 3 - x^2$

x	−4	−3	−2	−1	0	1	2	3	4
3	3	3	3	3	3	3	3	3	3
$-x^2$		−9			0		−4		
y		−6					−1		

b) Draw the graph of $y = 3 - x^2$

c) Use your graph to solve the equation $3 - x^2 = 0$

13 a) Complete the table of values for $y = x^2 + 3x - 2$

x	−4	−3	−2	−1	0	1	2	3
x^2	16	9						9
$+3x$	−12	−9						9
−2	−2	−2						−2
y	2	−2						16

b) Draw the graph of $y = x^2 + 3x - 2$

c) Use your graph to solve the equation $x^2 + 3x - 2 = 0$

14 a) Complete the table of values for $y = 3x^2 - 10$

x	-4	-3	-2	-1	0	1	2	3	4
x^2		9					4		
$3x^2$		27					12		
-10		-10					-10		
y		17					2		

b) Draw the graph of $y = 3x^2 - 10$

c) Use your graph to solve the equation $3x^2 - 10 = 0$

15 a) Copy and complete the table of values for $y = 2x + 3$

x	-2	-1	0	1	2	3
y		1	3			

b) On a copy of the grid, draw the graph of $y = 2x + 3$

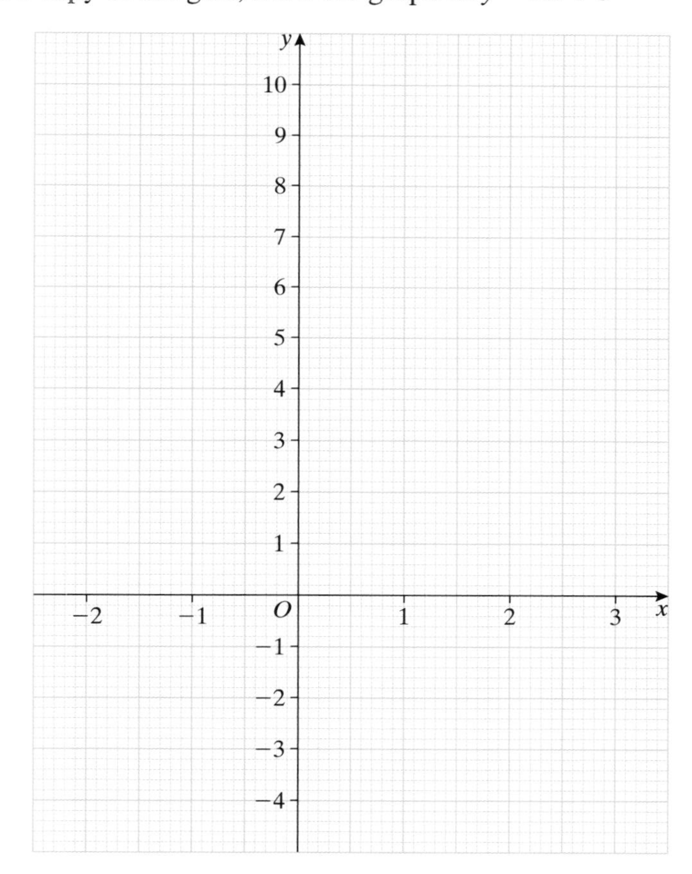

c) Use your graph to find:

(i) the value of y when $x = -1.3$

(ii) the value of x when $y = 5.4$

[Edexcel]

KEY POINTS

1. The position of a point is given by coordinates.
2. You can plot points on a pair of axes.
3. The horizontal axis is called the x axis.
4. The vertical axis is called the y axis.
5. The point (0, 0) or (0, 0, 0) is called the origin.
6. In two dimensions (2-D) the x coordinate (along) is given first and then the y coordinate (up or down).
7. In three dimensions (3-D) the order is always (x, y, z)
8. Equations like $y = 3x + 1$ can be plotted accurately by drawing a table of values.
9. Equations in the form $y = mx + c$, where m and c are numbers, give a straight line graph.
   m gives the gradient of the line.
10. The gradient tells you how steep a line is.
    An 'uphill' line has a positive gradient.
    A 'downhill' line has a negative gradient.
    $$\text{gradient} = \frac{\text{rise}}{\text{run}}$$

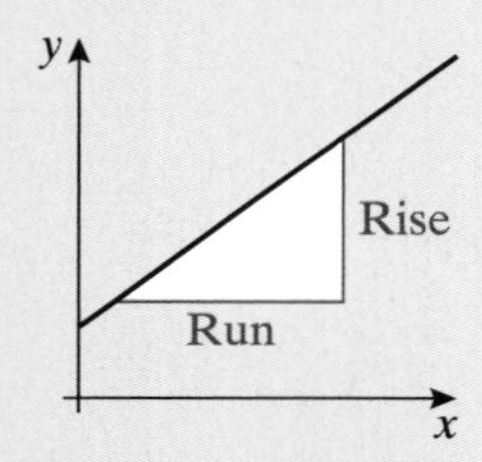

11. To solve a pair of simultaneous equations graphically:
    - draw the graph of each
    - write down the coordinates of the point where the lines cross.
12. A quadratic equation is an equation with an 'x^2' term and no other powers, for example: $y = x^2$ or $y = 2x^2 - 3$
    - These equations always give a curve.

Internet Challenge 10

Parallel lines

Use the internet to help you answer these questions about parallel things.

1 What name is given to a quadrilateral with two sets of parallel sides?

2 What name is given to a quadrilateral with exactly one set of parallel sides?

3 What is a parallelepiped?
How do you draw one?

4 Which iconic rock group recorded the album *Parallel Lines* in 1978?

5 What is the 49th parallel?

6 What are parallel universes?

7 What is the parallel postulate?

8 Where might you find a parallel port?

9 Where might you make a parallel turn?

10 'Parallel lines never meet' – true or false?

11 Is it possible for two curves to be parallel?

12 Who might choose to place things in parallel rather than in series?

13

a) How many of the lines running from left to right are parallel?
Now check your answer with a ruler or straight edge.

b) The picture is called *Café Wall*.
Find out the location of the café that inspired this picture, and the name of the mathematician who first described it.

CHAPTER 11

Measurements

In this chapter you will **revise earlier work on**:

- reading scales.

You will **learn how to**:

- use the 24-hour clock and read timetables
- use metric and imperial units
- convert between metric and imperial units
- use the formula speed $= \dfrac{\text{distance}}{\text{time}}$ to solve problems
- use the formula density $= \dfrac{\text{mass}}{\text{volume}}$ to solve problems.
- comment on the accuracy of measurements

You will also be **challenged to**:

- investigate speeds of man-made objects.

Starter: Scales

Read the scales and times shown below and opposite.

1 a)

b)

c)

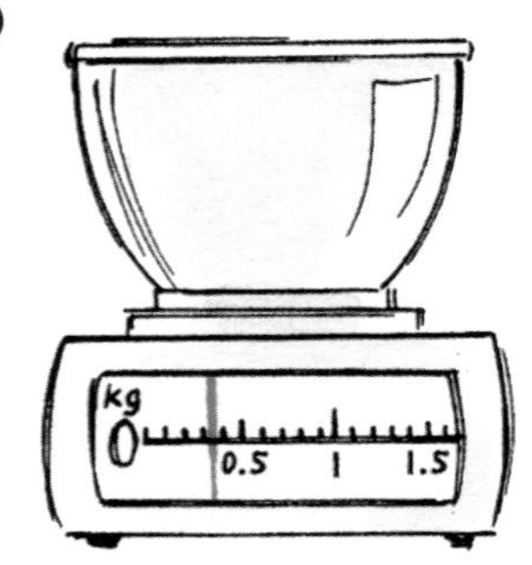

2 a)

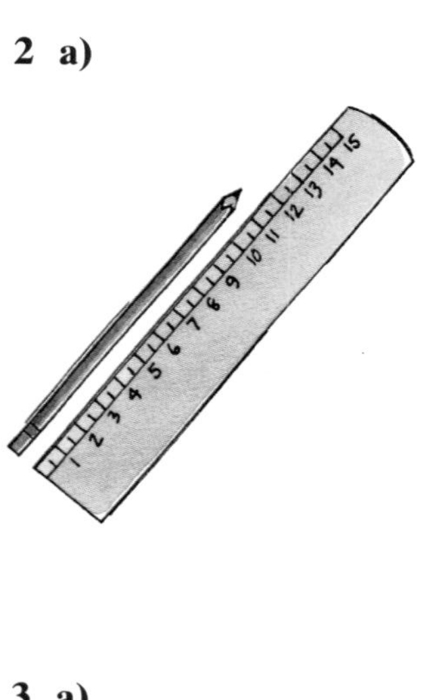

b)

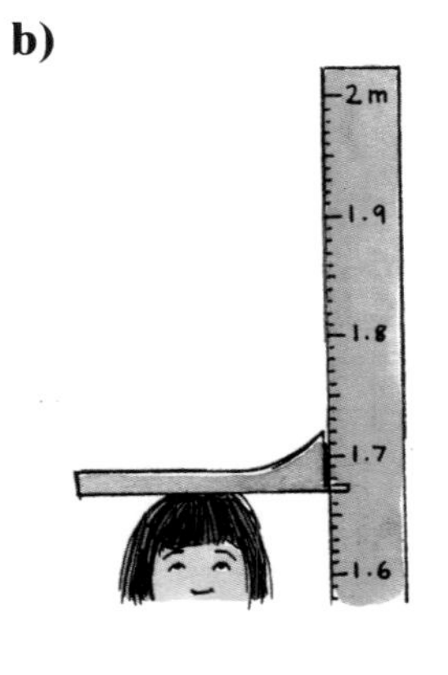

c)

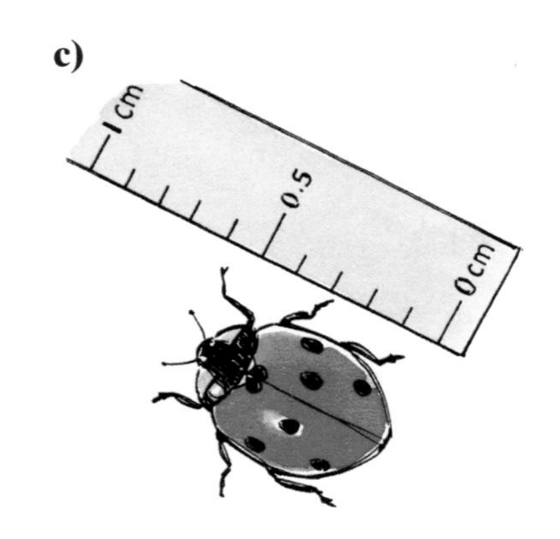

3 a)

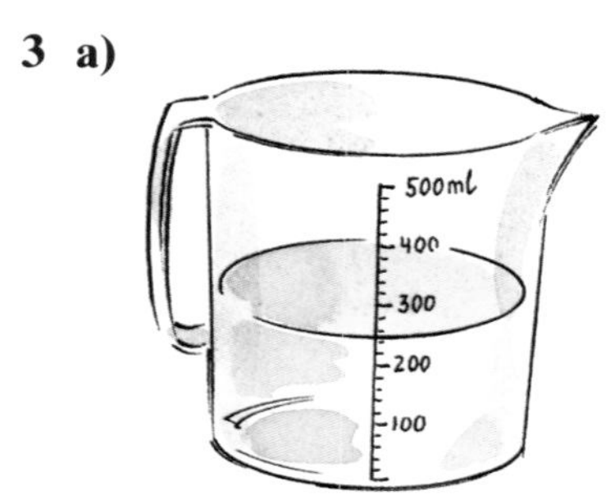

b)

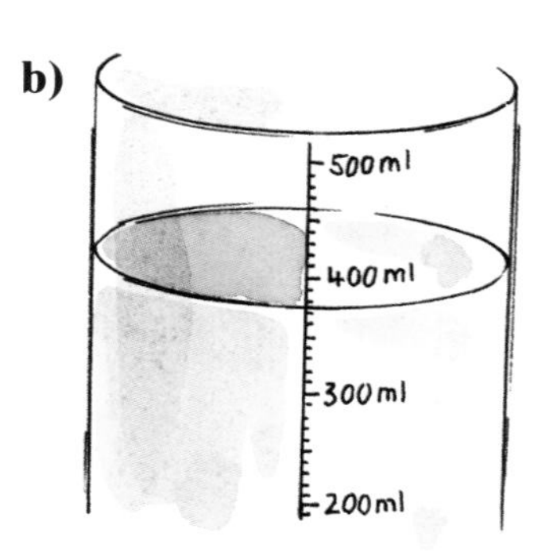

c)

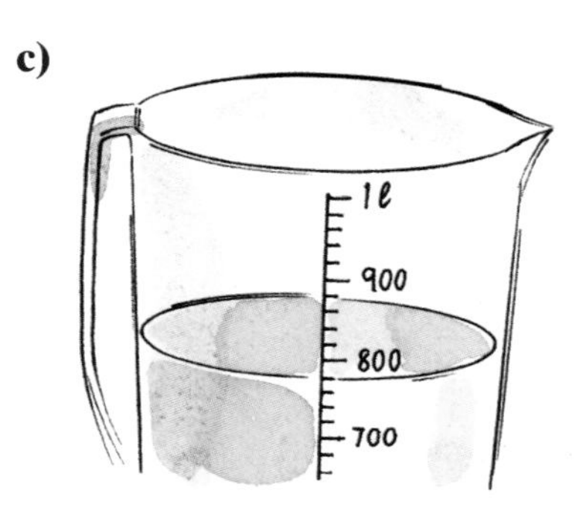

4 a)

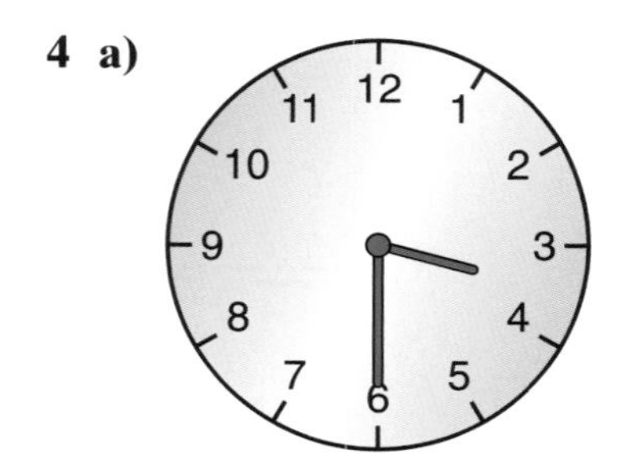

b)

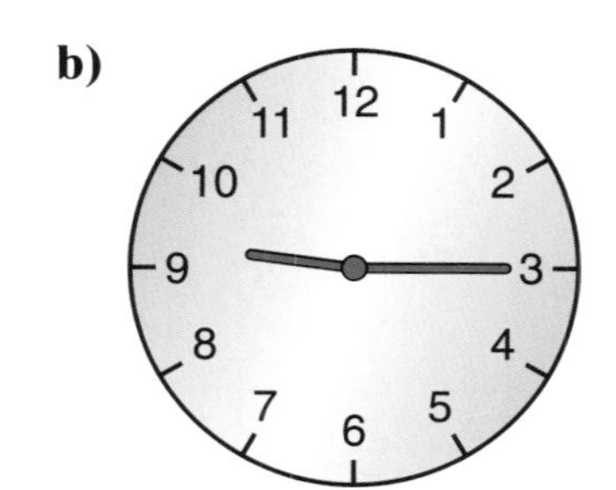

c)

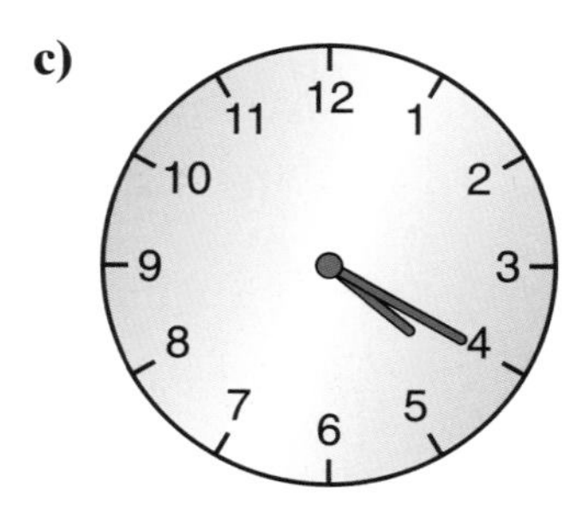

d)

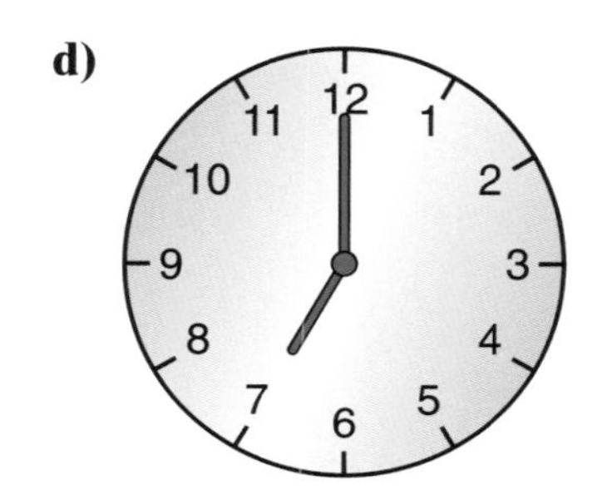

e)

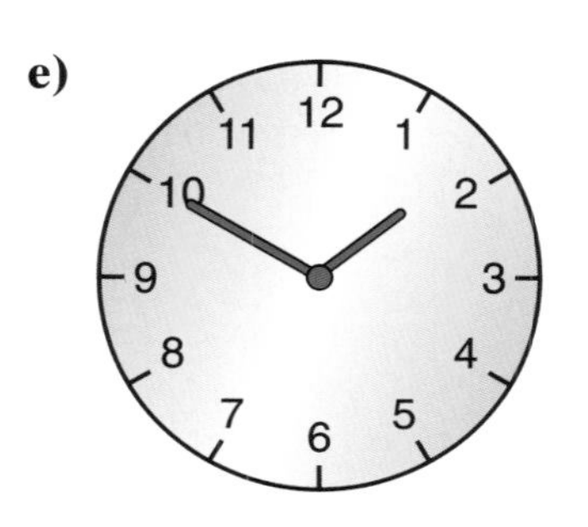

f)

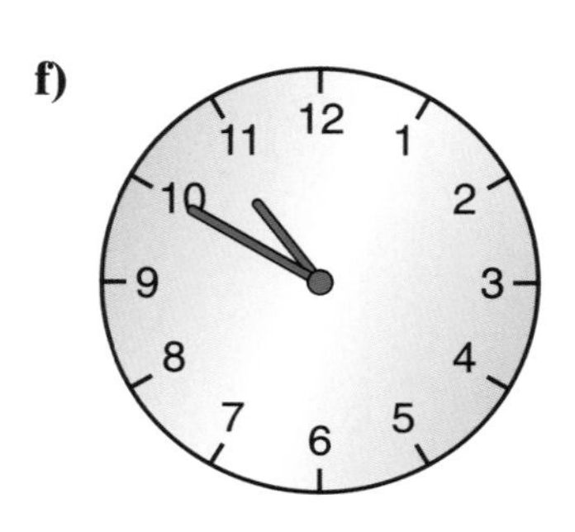

11.1 Time

Digital clocks and timetables often use the 24-hour clock.

A 24-hour clock tells you how many hours past midnight the time is.
For example:

11 am is 11 hours past midnight, so this is 11 00 hours.

2 pm is 14 hours past midnight, so this is 14 00 hours.

12.15 am is 15 minutes after midnight, so this is 00 15 hours.

12 midnight is written as 00 00

> To convert from the 12-hour clock to the 24-hour clock:
> - the 24-hour clock *always* has four digits
> - the 12-hour clock *always* has am or pm after the digits.
>
> For times before 12 noon:
> put a '0' in front of any single-digit 'hours' number.
>
> For times after 12 noon:
> add 12 to the hours number.

So 9.35 am is 09 35 and 11.14 am is 11 14

So 6.15 pm is 18 15

You need to be able to change between hours and minutes.

There are 60 minutes in an hour.

> To change: hours to minutes $\times$ by 60
> minutes to hours $\div$ by 60

EXAMPLE

Change:

a) $3\frac{3}{4}$ hours to minutes

b) 135 minutes to hours.

$\frac{1}{4}$ of 60 = 60 ÷ 4 = 15
So $\frac{3}{4}$ of 60 = 3 × 15 = 45

SOLUTION

a) 3 hours $= 3 \times 60 = 180$ minutes

$\frac{3}{4}$ of an hour $= \frac{3}{4} \times 60 = 45$ minutes

So $3\frac{3}{4}$ hours $= 180 + 45$ minutes

$=$ 225 minutes

With a calculator you can just do 3.75 × 60

Remember: 0.25 hours is not 25 minutes; 0.25 hours = 0.25 × 60 = 15 minutes

b) $135 \div 60 = 2.25$ hours $= 2\frac{1}{4}$ hours

EXAMPLE

Ben catches a train at 20 43
The journey lasts 1 hour and 39 minutes.

a) Write down the time that Ben arrives at his destination using the 24-hour clock.

b) Convert this time to the 12-hour clock.

SOLUTION

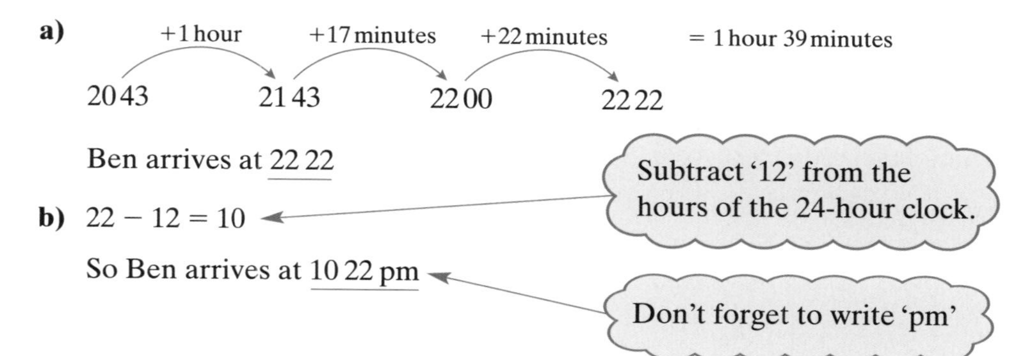

a) Ben arrives at <u>22 22</u>

b) 22 − 12 = 10

So Ben arrives at <u>10 22 pm</u>

EXAMPLE

The time is 10.37 am
Millie's maths lesson starts at 11.20 am
How long has Millie got until her lesson?

SOLUTION

+23 minutes +20 minutes

10.37 am → 11.00 am → 11.20 am

23 minutes + 20 minutes = <u>43 minutes</u>

EXAMPLE

Here is part of a bus timetable:

High Street	09 35	10 15	–	10 35	10 57
Bow Street	–	10 22	10 30	10 42	–
Kings Road	09 50	–	–	10 55	–
Berry Lane	10 07	10 50	10 55	–	–
Rose Road	10 22	11 05	–	11 25	11 35
Old Church Lane	10 30	11 13	11 15	–	11 43

Billy catches the 10 22 bus from Bow Street.

a) At what time should the bus arrive at Rose Road?

Anna arrives at the High Street at 10 30

b) How long should Anna expect to wait for a bus to take her to Old Church Lane?

c) How long should Anna expect the bus journey to take?

SOLUTION

High Street	09 35	10 15	–	10 35	10 57
Bow Street	–	10 22	10 30	10 42	–
Kings Road	09 50	–	–	10 55	–
Berry Lane	10 07	10 50	10 55	–	–
Rose Road	10 22	11 05	–	11 25	11 35
Old Church Lane	10 30	11 13	11 15	–	11 43

a) 11 05

b) The 10 35 bus doesn't go to Old Church Lane.
Anna has to wait for the 10 57 bus.
Anna expects to wait 10 57 − 10 30 = 27 minutes

c)

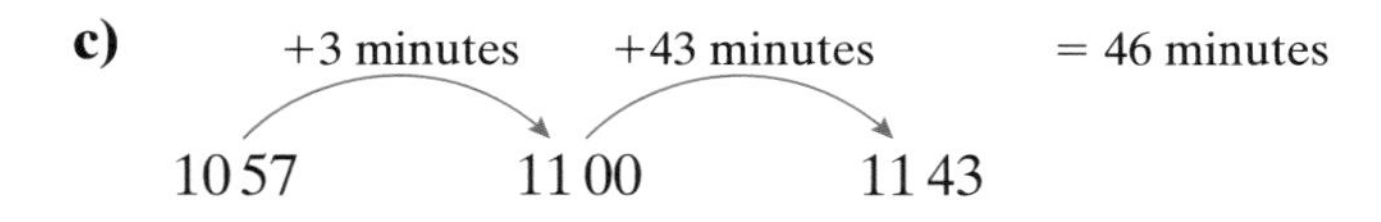

The journey should take 46 minutes

EXERCISE 11.1

Don't use your calculator for Questions 1–4.

1 Change these times to the 24-hour clock times.

a) **am**

b)  **pm**

c)  **pm**

d) 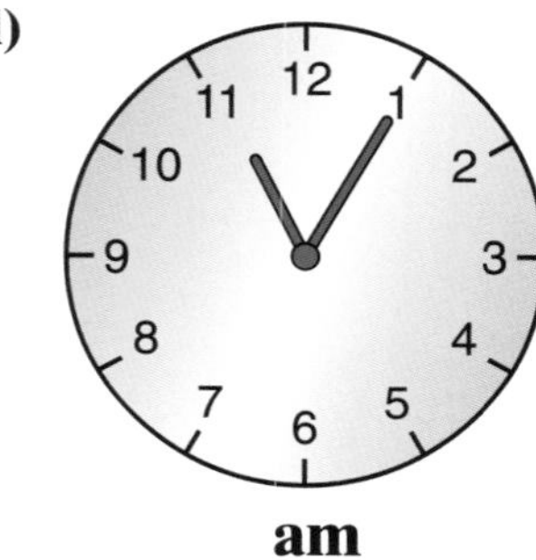 **am**

e)  **pm**

f)  **am**

2 These digital clocks are showing 24-hour clock times.
Change them to 12-hour clock times.
Don't forget to write 'am' or 'pm'

a) 18:36 **b)** 22:43 **c)** 00:13

d) 11:10 **e)** 13:56 **f)** 02:27

3 The time is 9.27 am
Martha has 2 hours and 45 minutes before she has to catch a bus.
What time is Martha's bus due?

4 The time is 15 46
Ben finishes work at 17 30
How much longer has Ben got to work for?

You can use your calculator for Questions 5–9.

5 Change these hours to minutes:

a) 4 hours **b)** 0.5 hours **c)** $\frac{1}{3}$ of an hour

d) 0.25 hours **e)** 3.6 hours **f)** 4.2 hours

6 Change these minutes to hours:

a) 120 minutes **b)** 300 minutes **c)** 150 minutes

d) 45 minutes **e)** 15 minutes **f)** 75 minutes

7 Look at this digital clock.
Write down what the display should show in:

a) $\frac{3}{4}$ of an hour's time **b)** $1\frac{1}{2}$ hours' time **c)** $2\frac{1}{4}$ hours' time

8 Here is part of a bus timetable.

High Street	14 26	14 56	15 17	15 34
Library Road	14 30	15 00	15 21	–
Mulberry Walk	14 37	–	15 28	15 48
Leafy Lane	14 42	–	15 33	15 53
Regent Crescent	–	–	15 37	–
Old Oak Lane	14 51	15 12	15 44	16 02

Maria catches the 15 21 bus from Library Road.

a) At what time should the bus arrive at Regent Crescent?

Ahmed arrives at the High street at 2.50 pm

b) **(i)** How long should Ahmed expect to wait for a bus to take him to Leafy Lane?

(ii) How long should Ahmed expect the bus journey to take?

9 Here is part of a train timetable.

Appleton	17 20	17 28	17 39	17 53
Fishville	17 32	17 40	–	–
Newtown	17 45	17 53	–	–
Kingsford	18 01	18 09	18 12	18 26
Queenswood	18 17	–	18 28	18 41
Littleton	18 28	18 35	18 39	–

Claire plans to catch the 17 20 from Appleton to go to Queenswood.

a) How long should Claire expect the journey to take?

Claire doesn't arrive at Appleton station until 17 24

b) What time should Claire expect to arrive in Queenswood?

11.2 Metric units

You need to be able to use the following **metric** units.

	Metric units	Conversions
Length	kilometres (km) metres (m) centimetres (cm) millimetres (mm)	1 km = 1000 m 1 m = 100 cm 1 m = 1000 mm 1 cm = 10 mm
Mass	tonne (t) kilograms (kg) grams (g)	1 tonne = 1000 kg 1 kg = 1000 g
Capacity (volume)	litres (ℓ) centilitres (cl) millilitres (ml) cubic centimetres (cm^3)	1 ℓ = 100 cl 1 ℓ = 1000 ml 1 cl = 10 ml 1 ml = 1 cm^3

kilo' means 'thousand' – so 1 kilogram = 1000 grams

'centi' means 'hundredth' – so 1 centilitre = $\frac{1}{100}$ of a litre

'milli' means 'thousandth' – so 1 millimetre = $\frac{1}{1000}$ of a metre

EXAMPLE

Write down a sensible **metric** unit that should be used to measure:

a) the weight of a person
b) the length of a pen
c) the volume of petrol in a car's petrol tank.

SOLUTION

a) kilograms **b)** centimetres **c)** litres

You also need to be able to convert between metric units.

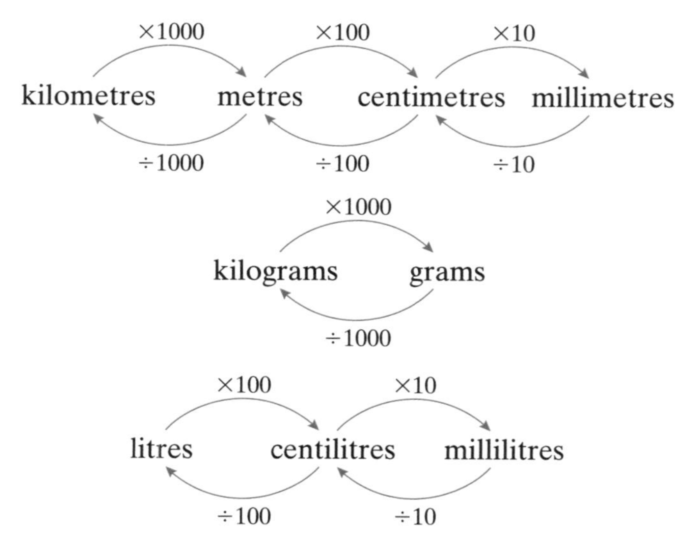

See pages 9 and 10 for a reminder on how to multiply and divide by 10, 100 and 1000

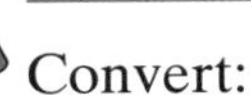

EXAMPLE

Convert:

a) 3.3 litres to millilitres
b) 127 centimetres to metres
c) 47 grams to kilograms.

> To change from a larger unit to a smaller unit you multiply.

SOLUTION

a) Remember: $1\,\ell = 1000$ ml
To change litres to millilitres, multiply by 1000
3.3 litres = 3300 ml

> To change from a smaller unit to a larger unit you divide.

b) Remember: 1 m = 100 cm
To change centimetres to metres, divide by 100
127 cm = 1.27 m

c) Remember: 1 kg = 1000 g
To change grams to kilograms, divide by 1000
47 g = 0.047 kg

Sometimes you may want to convert an area or a volume from one set of units to another.
This needs to be done carefully!

There are, for example, 10 mm in 1 cm, but there are **more than** 10 mm^2 in 1 cm^2
The diagram shows why:

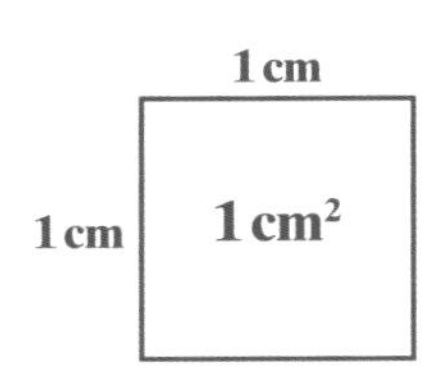

Area = 1 cm × 1 cm = 1 cm^2

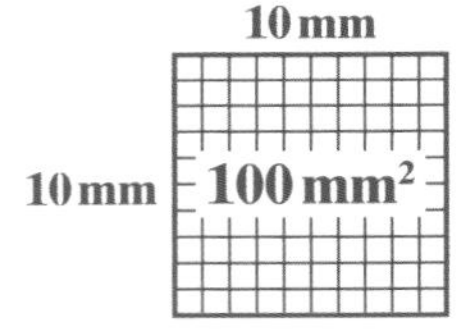

Area = 10 mm × 10 mm = 100 mm^2

Now look at these cubes:

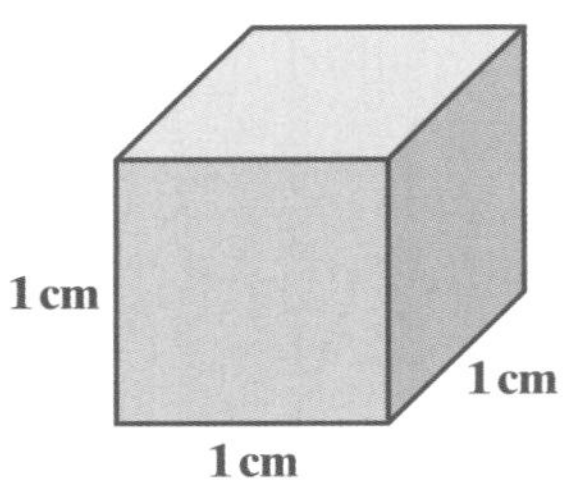

Volume = 1 cm × 1 cm × 1 cm
= 1 cm^3

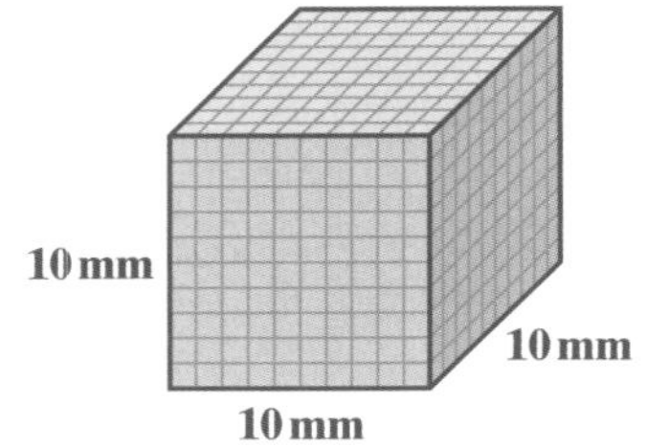

Volume = 10 mm × 10 mm × 10 mm
= 1000 mm^3

> The power on the unit tells you how many times to multiply or divide by the conversion factor.

So we have:

For a **length**	1 cm = 10 mm	(multiply by 10 **once**) × 10
For an **area**	1 cm^2 = 100 mm^2	(multiply by 10 **twice**) × 10 × 10
For a **volume**	1 cm^3 = 1000 mm^3	(multiply by 10 **three times**) × 10 × 10 × 10

EXAMPLE

Change:

a) $2\,m^2$ to cm^2 **b)** $5000\,mm^3$ to cm^3

SOLUTION

a) Remember: 1 m = 100 cm
To change m to cm multiply by 100
To change m^2 to cm^2 multiply by 100 twice

$2 \times 100 \times 100 = 20\,000$

So $2\,m^2 = \underline{20\,000\,m^2}$

b) Remember: 1 cm = 10 mm
To change mm to cm divide by 10
To change mm^3 to cm^3 divide by 10 three times

$5000 \div 10 \div 10 \div 10 = 5$ ←

So $5000\,mm^3 = \underline{5\,cm^3}$

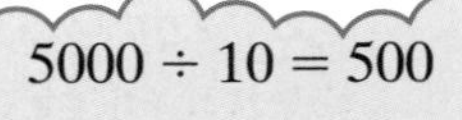

$5000 \div 10 = 500$

$500 \div 10 = 50$

$50 \div 10 = 5$

EXERCISE 11.2

Don't use your calculator for Questions 1–7.

1 Write down a sensible *metric* unit that should be used to measure:

a) the height of a two storey house
b) the length of a garden
c) the capacity of fizzy drink can
d) the weight of a baby sparrow
e) the distance between Glasgow and London
f) the capacity of a bottle of wine
g) the length of a pencil case
h) the thickness of ten sheets of paper
i) the volume of milk in a large carton
j) the weight of a car
k) the length of a car
l) the weight of a sack of potatoes

2 Convert the following measurements to metres:

a) 123 cm **b)** 3 km **c)** 8 cm
d) 0.5 km **e)** 1 cm **f)** 3.7 km
g) 1234 mm **h)** 10 km **i)** 57 cm

3 Change the following measurements to centimetres:

a) 12 mm **b)** 4 m **c)** 0.5 m
d) 6 mm **e)** 57 mm **f)** 2.8 m
g) 3.67 m **h)** 28 mm **i)** 138 mm

4 Change the following measurements:

a) 1.2 kilograms to grams
b) 450 kilograms to tonnes
c) 500 grams to kilograms
d) 3.24 kilograms to grams
e) 3.5 tonnes to kilograms
f) 0.4 kilograms to grams

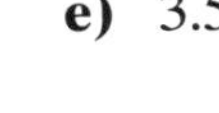

5 Change the following measurements:

a) 2 litres to millilitres
b) 70 centilitres to litres
c) 4.2 litres to centilitres
d) 0.6 litres to millilitres
e) 450 millilitres to litres
f) 300 millilitres to centilitres

6 Convert the following measurements:

a) 20 centimetres to millimetres
b) 4000 metres to kilometres
c) 2 metres to millimetres
d) 1250 metres to kilometres
e) 0.4 centimetres to millimetres
f) 600 metres to kilometres

7 Write down sensible *estimates* using appropriate *metric* units for:

a) the height of a person
b) the weight of a person
c) the length and width of your classroom
d) the height of the door into your classroom
e) the length of your pen
f) the weight of a textbook
g) the capacity of a drink carton with straw
h) the weight of an apple

You can use your calculator for Question 8.

8 Change the following measurements:

a) $3\ m^2$ to cm^2
b) $400\ mm^2$ to cm^2
c) $2\ m^3$ to cm^3
d) $2.5\ cm^3$ to mm^3
e) $500\ mm^3$ to cm^3
f) $0.03\ m^3$ to cm^3
g) $7.5\ cm^2$ to mm^2
h) $3.2\ cm^3$ to mm^3
i) $15\,000\ mm^3$ to cm^3
j) $0.25\ m^3$ to cm^3
k) $300\ mm^2$ to cm^2
l) $4.2\ m^3$ to cm^3
m) $3.5\ cm^3$ to mm^3
n) $2000\ cm^2$ to m^2

11.3 Imperial units

You need know about these **imperial units**:

	Imperial unit
Length	miles (m) yards (yd) feet (′) inches (″)
Mass	stones pounds (lb) ounces (oz)
Capacity (volume)	gallons pints

EXAMPLE

Write down a sensible *imperial* unit that should be used to measure

a) the weight of a person
b) the length of a pen
c) the volume of petrol in a car's petrol tank

SOLUTION

a) stones or pounds **b)** inches **c)** gallons

You need to be able to convert between some metric and imperial units.
You need to learn these conversions:

2.2 pounds	$\approx$	1 kg
1 inch	$\approx$	2.5 cm
1 foot	$\approx$	30 cm
5 miles	$\approx$	8 km
$1\frac{3}{4}$ pints	$\approx$	1 litre
1 gallon	$\approx$	4.5 litres

'$\approx$' means 'is approximately equal to'

EXAMPLE

Change: **a)** 40 miles to kilometres **b)** 6 gallons to litres **c)** 11 pounds to kilograms

SOLUTION

a)

$5\text{ miles} \approx 8\text{ km}$

$\times 8$ $\times 8$

$40\text{ miles} \approx 64\text{ km}$

$40 \div 5 = 8$

b)

$1\text{ gallon} \approx 4.5\text{ litres}$

$\times 6$ $\times 6$

$6\text{ gallons} \approx 27\text{ litres}$

$4 \times 6 = 24$
$0.5 \times 6 = 3$
$24 + 3 = 27$

c)

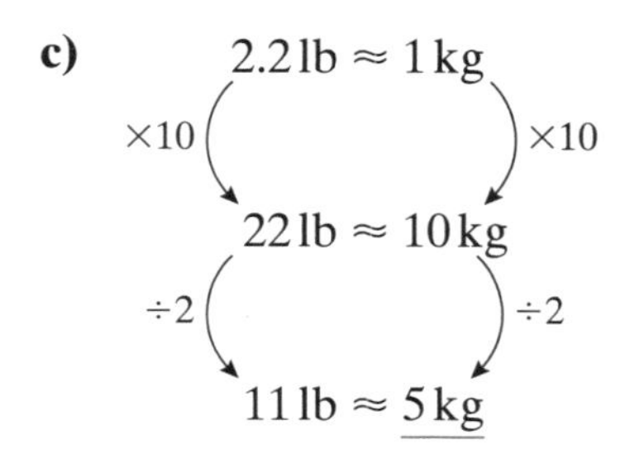

EXAMPLE

Change 42 kilometres to miles.

SOLUTION

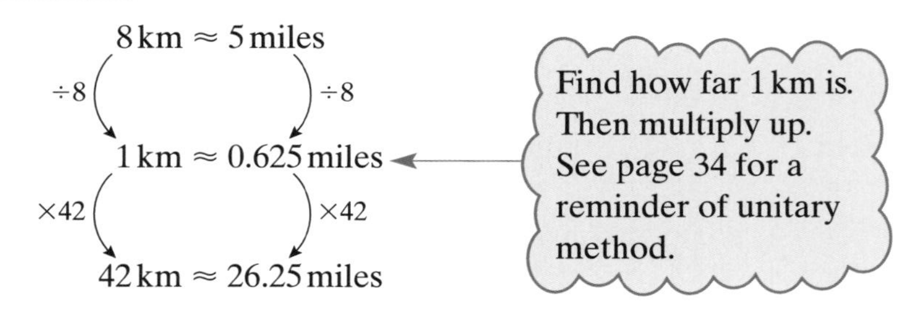

8 km ≈ 5 miles
÷8
÷8
1 km ≈ 0.625 miles
×42
×42
42 km ≈ 26.25 miles
Find how far 1 km is. Then multiply up. See page 34 for a reminder of unitary method.

So 42 km ≈ 26 miles

EXERCISE 11.3

Don't use your calculator for Questions 1–3.

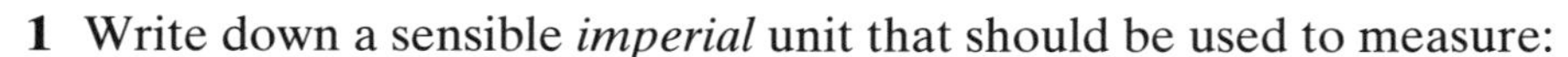

1 Write down a sensible *imperial* unit that should be used to measure:

a) the height of a two-storey house
b) the length of a garden
c) the weight of a baby sparrow
d) the distance between Glasgow and London
e) the length of a pen
f) the volume of milk in a large carton
g) the weight of a sack of potatoes

2 Change:

a) 50 miles to kilometres
b) 30 miles to kilometres
c) 16 kilometres to miles
d) 25 miles to kilometres
e) 160 kilometres to miles
f) 24 kilometres to miles

3 Change:

a) 20 gallons to litres
b) 110 pounds to kilograms
c) 9 litres to gallons
d) 90 centimetres to feet
e) 10 litres to pints
f) 10 centimetres to inches

You can use your calculator for Questions 4–7.

4 Change:

a) 66 pounds to kilograms
b) 18 miles to kilometres
c) 12 gallons to litres
d) 15 kilograms to pounds
e) 20 inches to centimetres
f) 15 pints to litres

5 There are 12 inches to 1 foot.
Hiromi is 5 feet 6 inches tall.
a) What is Hiromi's height in metres?
Hiromi weighs 140 pounds.
b) What is Hiromi's weight in kilograms?
Give your answer correct to 2 significant figures.

6 Liam buys 10 gallons of petrol.
Approximately how many litres of petrol does Liam buy?

7 Kathy drives 220 miles.
What is 220 miles in kilometres?

11.4 Speed

Speed is a measure of how fast an object is travelling.

Speed usually has units of miles per hour (mph) or kilometres per hour (km/h or kph) or metres per second (m/s).

The speed of an object tells you how far an object travels in a given time.
For example, an object travelling at:

- **60 mph travels 60 miles in 1 hour**
- **20 m/s travels 20 metres in 1 second**

You need to learn the formula:

$$\text{Average speed} = \frac{\text{total distance}}{\text{total time}}$$

or

$$s = \frac{d}{t}$$

You might find this 'cover-up' triangle helps:

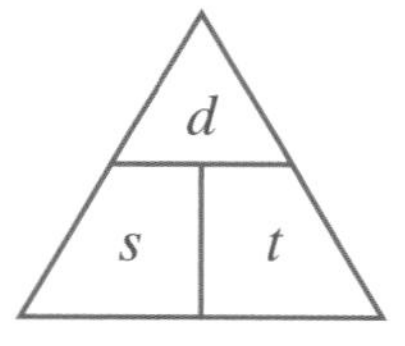

EXAMPLE

A man drives for 3 hours at an average speed of 60 km/h
How far does the man travel?

SOLUTION

Cover up 'd' in the triangle.

$d = s \times t$

$s = 60$ km and $t = 3$ hours

$d = 60 \times 3$

$d = \underline{180 \text{ km}}$

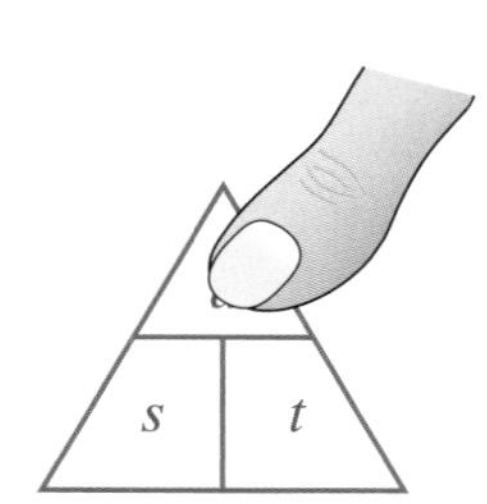

EXAMPLE

A girl cycles 15 miles in 90 minutes.
What is her average speed in miles per hour?

SOLUTION

You have been asked for her speed in miles per hour.
So you need to change 90 minutes to hours.

90 minutes = 1.5 hours

Divide by 60 to change minutes to hours.

Cover up 's' in the triangle.

$d = 15$ miles and $t = 1.5$ hours

$$s = \frac{d}{t}$$

$$s = \frac{15}{1.5}$$

To make this easier to work out, you can multiply 'top' and 'bottom' by 10.
This gives $s = \frac{150}{15} = 10$ mph

$s = \underline{10\text{ mph}}$

EXAMPLE

A boy runs 100 m at an average speed of 6.25 m/s

a) How long does he take?

b) What is his speed in kilometres per hour?

SOLUTION

a) Cover up 't' in the triangle.

$d = 100$ m and $s = 6.25$ m/s

$$t = \frac{100}{6.25} = \underline{16\text{ s}}$$

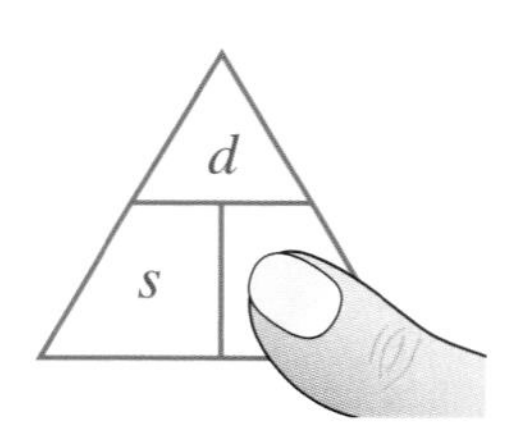

b) 6.25 m/s means in 1 second the boy runs 6.25 m

×60 ×60

so in 60 seconds (1 min) he runs 375 m

×60 ×60

and in 60 minutes (1 hour) he runs 22 500 m

22 500 m = 22.5 km

So his speed is $\underline{22.5\text{ km/h}}$

EXERCISE 11.4

Don't use your calculator for Questions 1–7.

1 A car travels for 5 hours at an average speed of 70 mph
How far does the car travel?

2 A car travels for $1\frac{1}{2}$ hours at an average speed of 100 km/h
How far does the car travel?

3 Karin jogs for half an hour at an average speed of 6 km/h
How far does Karin jog?

4 Hilmar cycles for 90 minutes at an average speed of 12 km/h
How far does Hilmar cycle?

5 A car travels 75 miles in an hour and a half.
What is the average speed of the car?

6 Michelle cycles 30 miles in 2 hours.
What is her average speed?

7 Phil drives 300 km at an average speed of 60 km/h
How long does the journey take?

You can use your calculator for Questions 8–11.

8 Simon sprints for 20 seconds at an average speed of 6 m/s
How far does Simon run?

9 Hans drives 40 km at an average speed of 60 km/h
How long, in minutes, does the journey take?

10 Siobhan cycles 200 metres in 50 seconds.
a) What is her average speed in metres per second?
b) What is Siobhan's average speed in kilometres per hour?

11 An intercity train travels at 120 km/h
a) How far does it travel in:
(i) 15 minutes **(ii)** 10 minutes **(iii)** 45 minutes?
b) How long in minutes does it take to travel:
(i) 60 km **(ii)** 12 km **(iii)** 200 km?
c) What is the speed of the train in metres per second?
Give your answer correct to 1 decimal place.

11.5 Density

This old chestnut tries to trick you into answering 'lead'.

Of course a kilogram of feathers weighs the same as a kilogram of lead.

However, the feathers would take up a lot more space!

Lead has a greater density than feathers.

In everyday English we often use the word **'weight'** to mean **'mass'**.

Density is a measure of how much mass a certain volume of a material has.

Density usually has units of kilograms per cubic metre (kg/m^3) or grams per cubic centimetre (g/cm^3)

For example, a material with a density of:

- **20 kg/m^3 means 1 m^3 of the material has a mass of 20 kg**
- **15 g/cm^3 means 1 cm^3 of the material has a mass of 15 g**

You need to learn the formula:

$$\text{density} = \frac{\text{mass}}{\text{volume}} \quad \text{or} \quad D = \frac{M}{V}$$

You might find this 'cover-up' triangle helps.

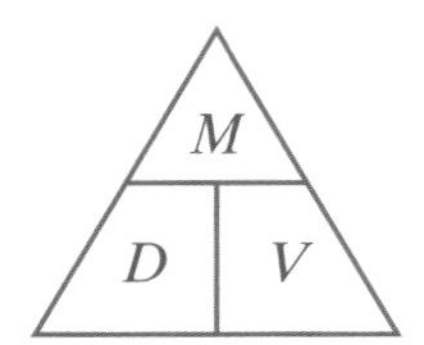

EXAMPLE

$0.25\ m^3$ of copper has a mass of 2230 kg
Work out:

a) the density of copper in kg/m^3
b) the mass, in kilograms, of $5\ m^3$ of copper
c) the volume, in m^3, of 1000 kg of copper – give your answer to 3 significant figures.

SOLUTION

a) Cover up 'D' in the triangle.

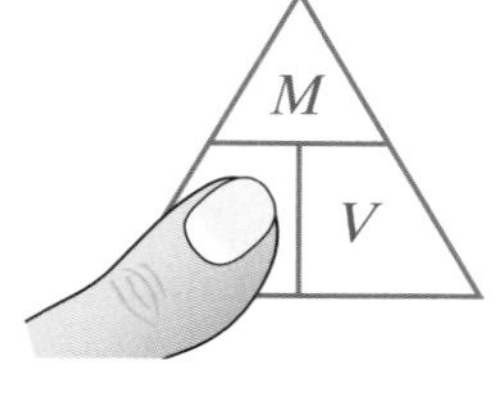

$M = 2230$ and $V = 0.25$

$$D = \frac{M}{V}$$

$$D = \frac{2230}{0.25}$$

$$D = \underline{8920\ kg/m^3}$$

b) Cover up 'M' in the triangle.

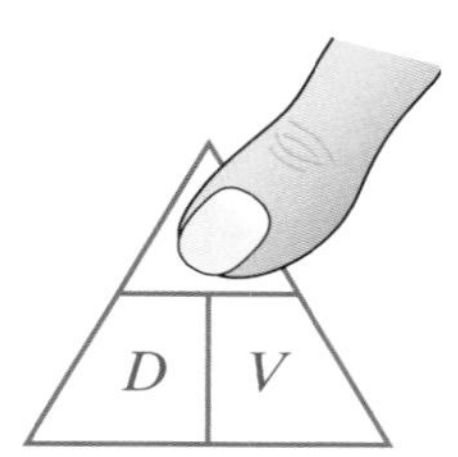

$D = 8920$ and $V = 5$

$$M = 8920 \times 5$$

$$M = \underline{44\,600\ kg}$$

c) Cover up 'V' in the triangle.

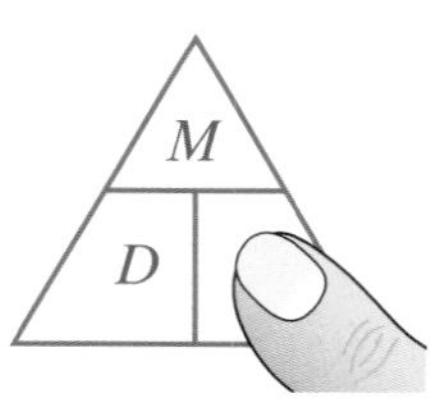

$M = 1000$ and $D = 8920$

$$V = \frac{M}{D}$$

$$V = \frac{1000}{8920}$$

$$V = 0.1121\ \ldots\ m^3$$

$$V = \underline{0.112\ m^3} \text{ to 3 s.f.}$$

EXERCISE 11.5

1 The density of water is $1000\ kg/m^3$

a) Find the mass of:
(i) $5\ m^3$ of water **(ii)** $20\ m^3$ of water **(iii)** $0.5\ m^3$ of water **(iv)** $0.75\ m^3$ of water

b) Find the volume of:
(i) 200 kg of water **(ii)** 6000 kg of water **(iii)** 850 kg of water **(iv)** 1500 kg of water

You can use your calculator for Questions 2 and 3.

2 The density of silver is 10.49 g/cm^3
Give your answers to 3 significant figures.
a) Find the volume of silver used to make a silver ring of mass 25 g
b) Find the volume of silver used to make a silver spoon of mass 125 g

A silver coin has a volume of 2.5 cm^3
c) What is the mass of the coin?

3 An aluminium hub-cap has a volume of 2510 cm^3
The mass of the hub-cap is 680 g
Give your answers to 2 significant figures.
a) Work out the density of aluminium in g/cm^3
b) What is the mass in grams of 1 cm^3 of aluminium?
c) What is the mass in grams of 1 m^3 of aluminium?
d) What is the mass in kilograms of 1 m^3 of aluminium?
e) What is the density of aluminium in kg/m^3?

11.6 Accuracy

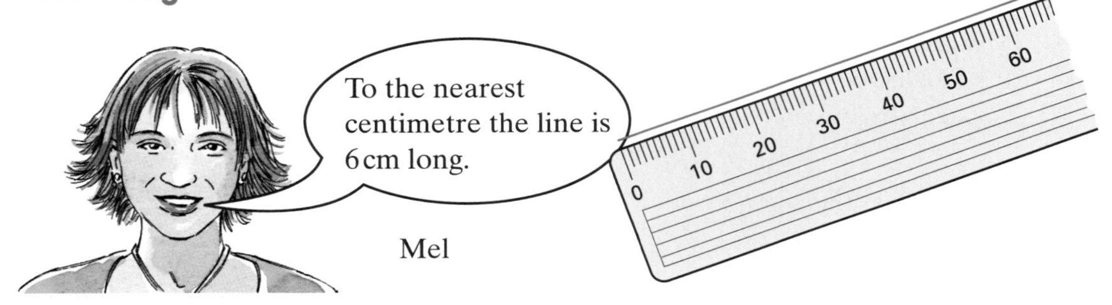

When a measurement is taken, it is important to say how accurate the measurement is.

Mel has measured the line to the nearest centimetre.
This means she has rounded the actual measurement up or down to 6 cm

A line 5.5 cm long has a length of 6 cm to the nearest centimetre.
But a line 5.4 cm long has a length of 5 cm to the nearest centimetre.

The line can't be 6.5 cm (as this rounds up to 7 cm) but it could be so close (e.g. 6.499999) that 6.5 is accepted as the upper bound.

So a line measuring 6 cm to the nearest cm could be as *short* as 5.5 cm
We say this is the **lower bound** for the length of the line.

Likewise, a line measuring 6 cm to the nearest cm could be as *long* as 6.5 cm
We say this is the **upper bound** for the length of the line.

The measurement could be 'out' by half a centimetre either way.

You can write this as an inequality:

$$5.5 \text{ cm} \leqslant \text{length of line} < 6.5 \text{ cm}$$

or

$$5.5 \leqslant l < 6.5$$

where l is the length of the line in centimetres.

Any measurement to the nearest whole unit can be as much as half a unit out in either direction.

EXERCISE 11.6

1 Write down the **(i)** lower and **(ii)** upper bound of these numbers, which have all been written to the nearest whole number:

a) 26 **b)** 793 **c)** 80 **d)** 5 **e)** 490

2 Write down the **(i)** lower and **(ii)** upper bound of these measurements:

a) an apple weighing 128 g to the nearest gram
b) a woman weighing 63 kg to the nearest kilogram
c) a child who is 120 cm tall to the nearest centimetre
d) a line which is 8 cm long to the nearest centimetre
e) the distance between two towns is 23 km to the nearest kilometre
f) a room which is 4 m long to the nearest metre.

REVIEW EXERCISE 11

Don't use your calculator for Questions 1–3.

1 **a)** Write down a sensible *metric* unit that should be used to measure:
(i) the length of a text book
(ii) the weight of a CD.
b) Write down a sensible *imperial* unit that should be used to measure:
(i) the distance between London and Brighton
(ii) the weight of a box of chocolates.

2 **a)** Copy and complete the table by writing a sensible metric unit on each dotted line. The first one has been done for you.

The distance from London to Manchester	222 kilometres
The volume of coffee in a mug	310 ....
The height of a door	215 ....
The weight of a one pound coin	12 ....

b) Change 8 kilometres to metres. **[Edexcel]**

3

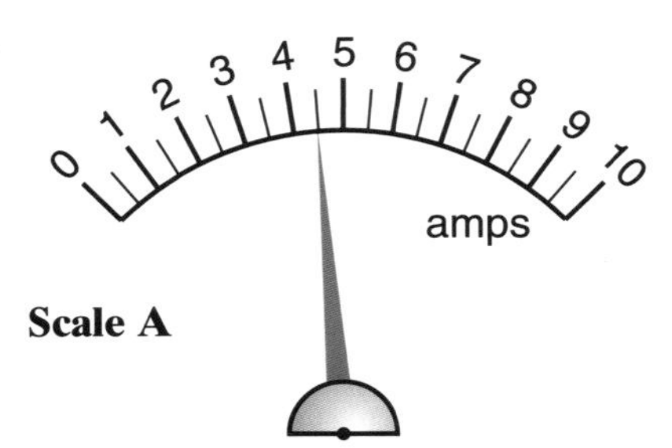

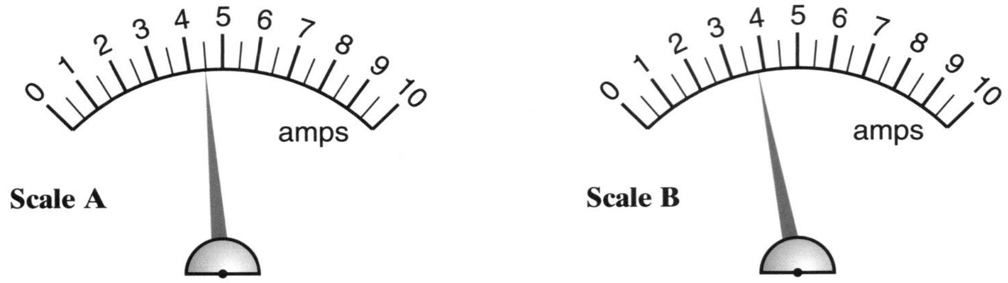

Scale A

Scale B

a) What is the reading on scale A?

b) What is the reading on scale B?

c) On a copy of the scale, draw an arrow to show a reading of 6.2 amps. Be as accurate as you can.

d) The diagram can be used to convert between temperatures in °C and temperatures in °F.

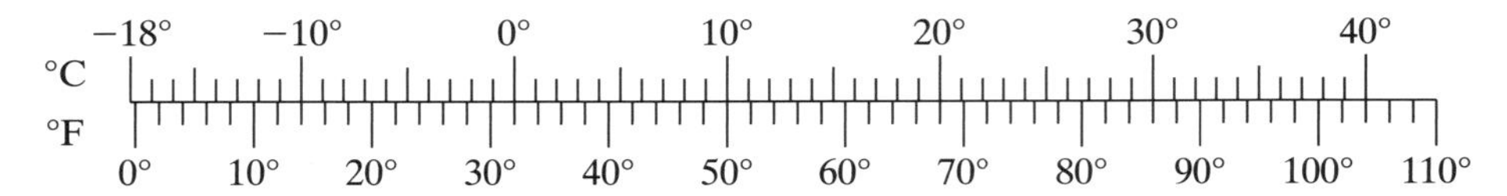

Use the diagram to convert:

(i) 10°C to °F **(ii)** 14°F to °C **[Edexcel]**

You can use your calculator for Questions 4–11.

4 Work out:

a) the speed of a plane flying 600 km in $1\frac{1}{2}$ hours

b) the distance travelled by a cyclist cycling at an average speed of 12 km/h for 30 minutes

c) the time taken, in minutes, to walk 600 m at an average speed of 2 m/s

5 Change:

a) 10 m^2 to cm^2 **b)** 4.5 m^2 to cm^2 **c)** 330 mm^3 to cm^3 **d)** 2 cm^3 to mm^3

6 Nickel has a density of 8900 kg/m^3

a) Work out the mass of 0.4 m^3 of nickel.

b) Work out the volume of 2225 kg of nickel.

7 Here is part of a railway timetable.

Manchester	07 53	09 17	10 35	11 17	13 30	14 36	16 26
Stockport	08 01	09 26	10 43	11 25	13 38	14 46	16 39
Macclesfield	08 23	09 38	10 58	11 38	13 52	14 58	17 03
Congleton	08 31	–	–	11 49	–	15 07	17 10
Kidsgrove	08 37	–	–	–	–	–	17 16
Stoke-on-Trent	08 49	10 00	11 23	12 03	14 12	15 19	17 33

A train leaves Manchester at 10 35

a) At what time should it arrive in Stoke-on-Trent?

Doris has to go to a meeting in Stoke-on-Trent.
She will catch the train in Stockport.
She needs to arrive in Stoke-on-Trent before 2 pm for her meeting.

b) Write down the time of the latest train she can catch in Stockport.

c) Work out how many minutes it should take the 14 36 train from Manchester to get to Stoke-on-Trent.

The 14 36 train from Manchester to Stoke-on-Trent takes less time than the 16 26 train from Manchester to Stoke-on-Trent.

d) How many minutes less? **[Edexcel]**

8 The picture shows a man standing next to a giraffe.
The man and the giraffe are drawn on the same scale.

a) Write down an estimate for the height, in metres, of the man.

b) Estimate the height, in metres, of the giraffe. **[Edexcel]**

9 Here is part of a train timetable from Crewe to London.

Station	Time of leaving
Crewe	08 00
Wolverhampton	08 40
Birmingham	09 00
Coventry	09 30
Rugby	09 40
Milton Keynes	10 10

a) At what time should the train leave Coventry?

The train should arrive in London at 10 45

b) How long should the train take to travel from Crewe to London?

Verity arrived in Milton Keynes station at 09 53

c) How many minutes should she have to wait before the 10 10 train leaves? **[Edexcel]**

10 a) Complete the following sentences by writing a sensible metric unit on the dotted line.

(i) There are 330 .... of cola in a can.

(ii) A television set weighs 21 ....

(iii) A brick is 215 .... long.

(iv) There are 425 .... of baked beans in a tin.

b) Change 10 kilograms to pounds. **c)** Change 7 pints to litres. **[Edexcel]**

11 James and Sam went on holiday by plane.
The pilot said the speed of the plane was 285 kilometres per hour.

James told Sam that 285 kilometres per hour was about the same as 80 metres per second.

Was James correct?
Show working to justify your answer. **[Edexcel]**

KEY POINTS

1 To convert between hours and minutes, remember there are 60 minutes in an hour:

- so to change hours to minutes you multiply by 60
- and to change minutes to hours you divide by 60.

2 To convert from the 12-hour clock to the 24-hour clock:

- times before 12 noon – the hours number stays the same
- times after 12 noon – add 12 to the hours number.

3

	Metric units	Conversions
Length	kilometres (km) metres (m) centimetres (cm) millimetres (mm)	1 km = 1000 m 1 m = 100 cm 1 m = 1000 mm 1 cm = 10 mm
Mass	tonne (t) kilograms (kg) grams (g)	1 tonne = 1000 kg 1 kg = 1000 g
Capacity (volume)	litres (ℓ) centilitres (cl) millilitres (ml) cubic centimetres (cm^3)	1 ℓ = 100 cl 1 ℓ = 1000 ml 1 cl = 10 ml 1 ml = 1 cm^3

4

	Imperial unit	Metric equivalent
Length	miles (m) feet (′) inches (″)	5 miles ≈ 8 km 1 foot ≈ 30 cm 1 inch ≈ 2.5 cm
Mass	pounds (lb)	2.2 lb ≈ 1 kg
Capacity (volume)	gallons pints	1 gallon ≈ 4.5 litres 1.75 pints ≈ 1 litre

5 Average speed $= \dfrac{\text{total distance}}{\text{total time}}$

d
s t

6 Density $= \dfrac{\text{mass}}{\text{volume}}$

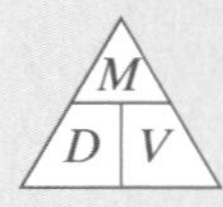

7 Any measurement to the nearest whole unit can be as much as half a unit out in either direction.

Internet Challenge 11

Faster and faster

Some man-made objects are capable of travelling at very high speeds.

Use your judgment to arrange these in order of speed, slowest to fastest.

Then use the internet to check if your order was correct.

Intercity 225 train

Porsche 911 GT3 RS car

Apollo 11 spacecraft

Speed of sound (in air)

Eurofighter *Typhoon* jet aircraft

Challenger 2 tank

Disney's *Space Mountain* roller coaster (Paris)

Orbiting space shuttle

Boeing 747-400 passenger jet aircraft

The tea clipper *Cutty Sark*

CHAPTER 12

Interpreting graphs

In this chapter you will **learn how to:**

- interpret graphical information
- use a conversion graph
- interpret a travel graph
- draw a travel graph.

You will also be **challenged to:**

- name famous graph curves.

Starter: Lower and lower

All the containers are filled with water.
There is a hole in the base of each container.
The graphs show depth of water against time.

Match each container with its graph.
The first container and graph have been matched for you.

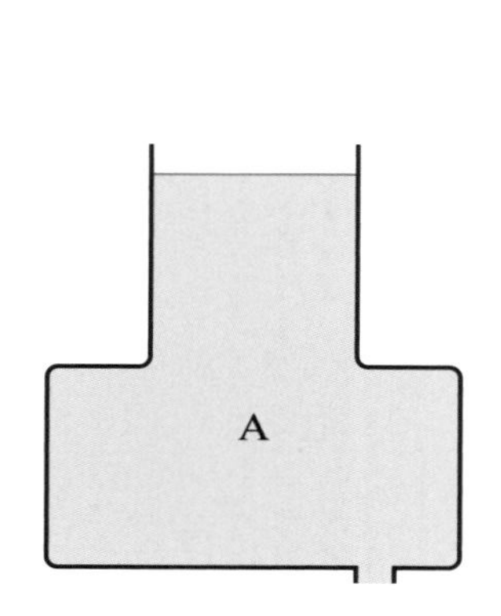

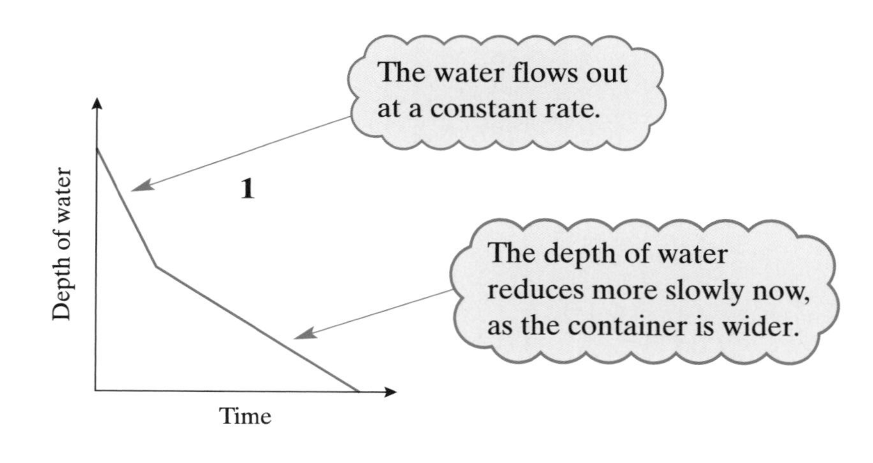

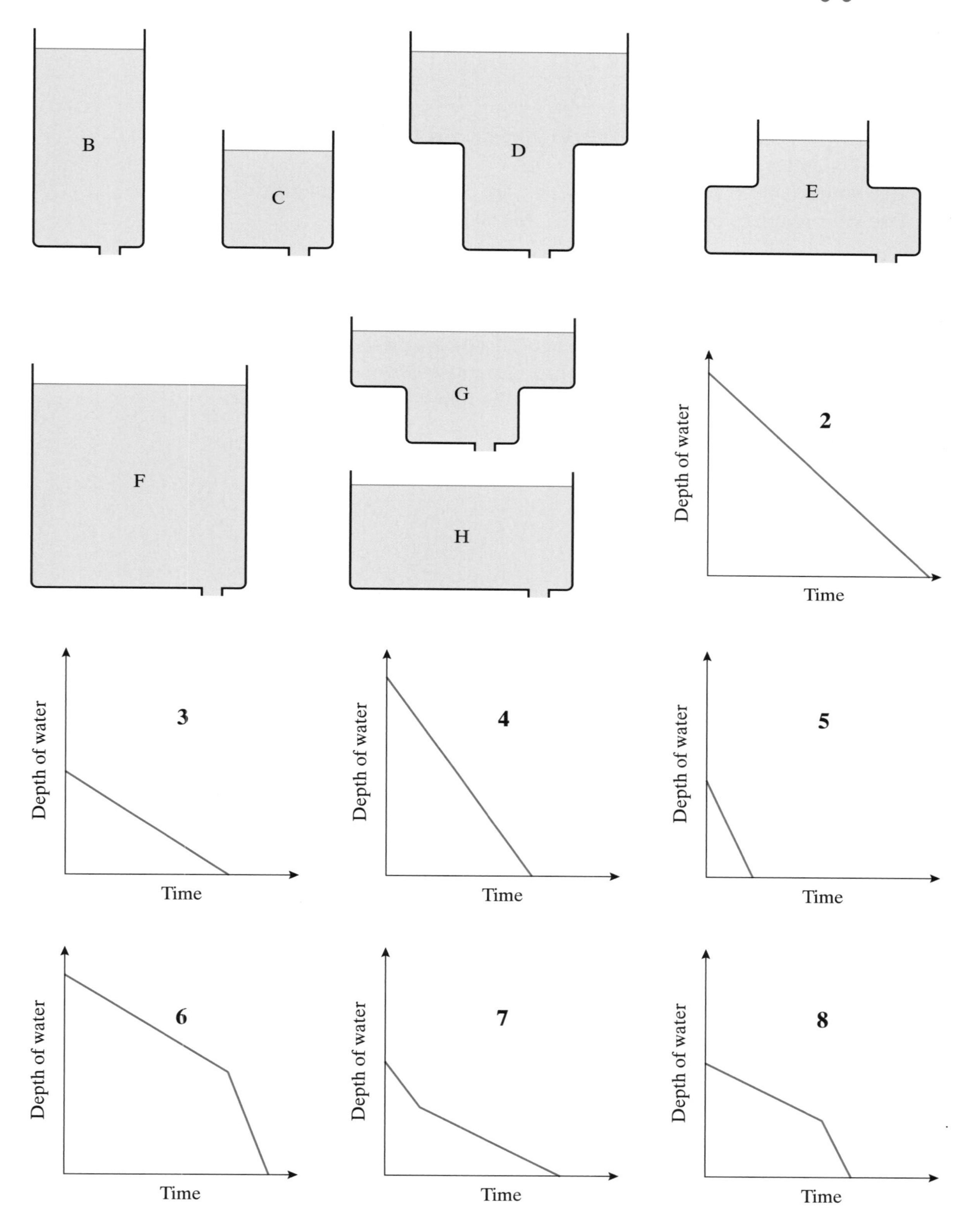
B
C
D
E
F
G
H
2
3
4
5
6
7
8
Depth of water
Time

12.1 Real-life graphs

EXAMPLE

a) Use the fact that 50 miles ≈ 80 km to produce a conversion graph between miles and kilometres.
b) Use your graph to convert **(i)** 20 miles **(ii)** 30 miles to kilometres.
c) Use your graph to convert **(i)** 40 km **(ii)** 60 km to miles.

SOLUTION

a) To draw a straight line you need only two points. One is given in the question; 50 miles ≈ 80 km means that the point (50, 80) lies on the line.
Also 0 miles = 0 kilometres so the line also passes through (0, 0)
So plot (0, 0) and (50, 80) on a grid. Join the points with a straight line.

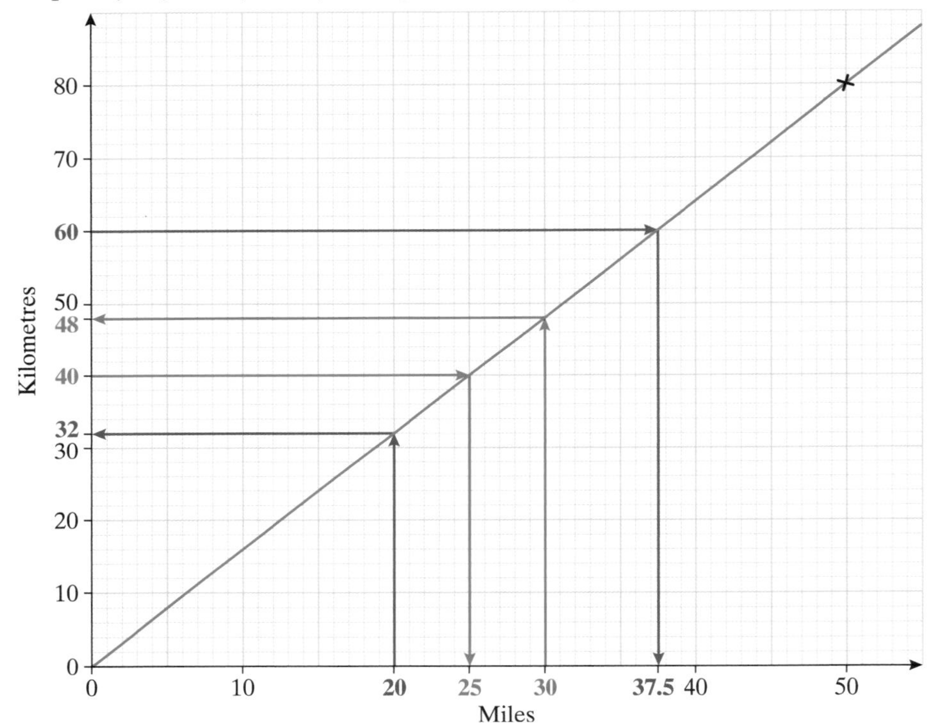

a) **(i)** Read off the graph from where the **number of miles = 20**
So 20 miles = 32 km

(ii) Read off the graph from where the **number of miles = 30**
So 30 miles = 48 km

c) **(i)** Read off the graph from where the **number of kilometres = 40**
So 40 km = 25 miles

(ii) Read off the graph from where the **number of kilometres = 60**
So 60 km = 37.5 miles

EXERCISE 12.1

1 This conversion graph converts between stones and kilograms.

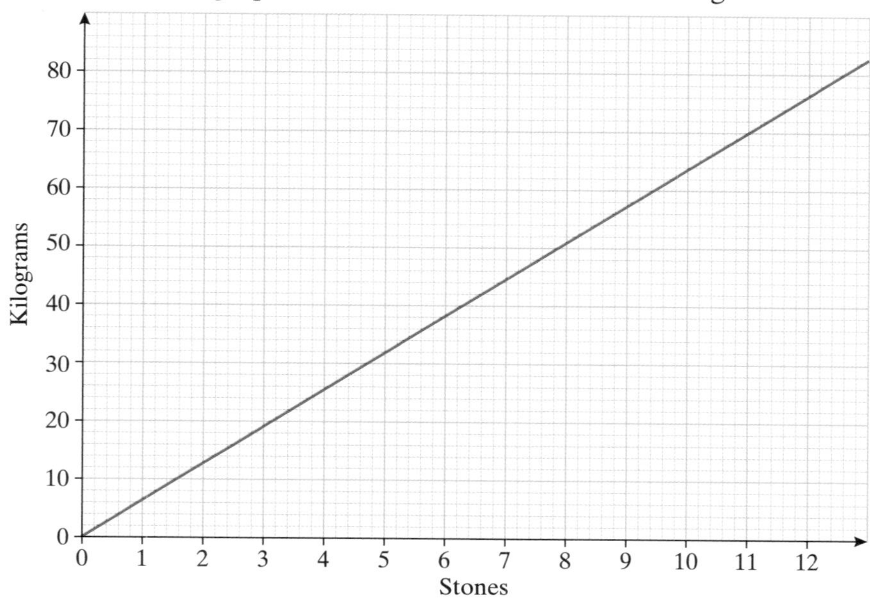

a) Henry weighs 11 stones. What is his weight in kilograms?
b) Catherine weighs 9 stones. What is her weight in kilograms?
c) Baby Isobel weighs 1 stone. What is her weight in kilograms?
d) Rachel weighs 50 kg What is her weight in stones?
e) Phil has been on a diet and lost 15 kg What is 15 kg in stones?
f) Mark weighs 25 kg more than Rosie. What is 25 kg in stones?

2 This conversion graph converts between pounds (lb) and ounces (oz).

a) A bag of sugar weighs 2 lb. How many ounces are in 2 lb?

b) A newborn baby weighs $7\frac{1}{2}$ lb. What is the weight of the baby in ounces?

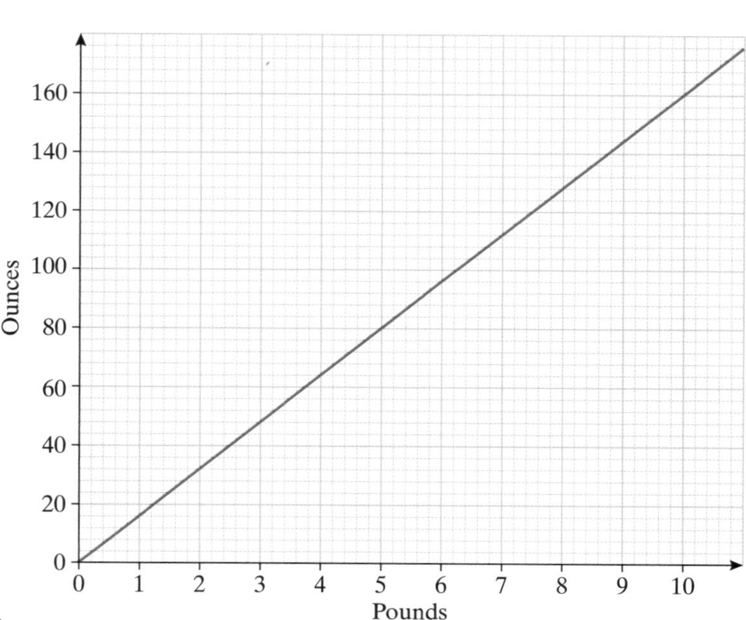

A large birthday cake needs 56 ounces of flour and 36 ounces of dried fruit.

c) (i) How many pounds of flour are needed?

(ii) How many pounds of dried fruit are needed?

3 This conversion graph converts the temperature in Celsius to the temperature in Fahrenheit.

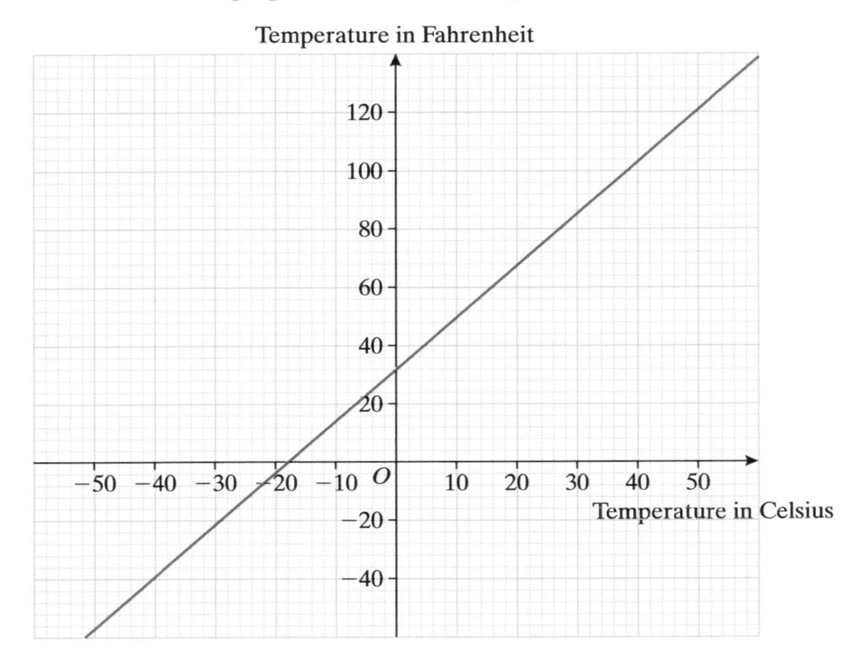

a) Martin has a high temperature of 40 °C
What is his temperature in Fahrenheit?

b) At a village fete the temperature soared to 100 °F
What is this in degrees Celsius?

The Met office issued a severe weather warning and stated that temperatures could drop as low as −10 °C.

c) What is −10 °C in degrees Fahrenheit?

d) Simon measured the temperature of a substance.

Did you measure the temperature in Fahrenheit or Celsius?
It doesn't matter; they are both the same at this temperature.

What is the temperature of the substance?

4 The graph shows the temperature of a liquid as it cools down.

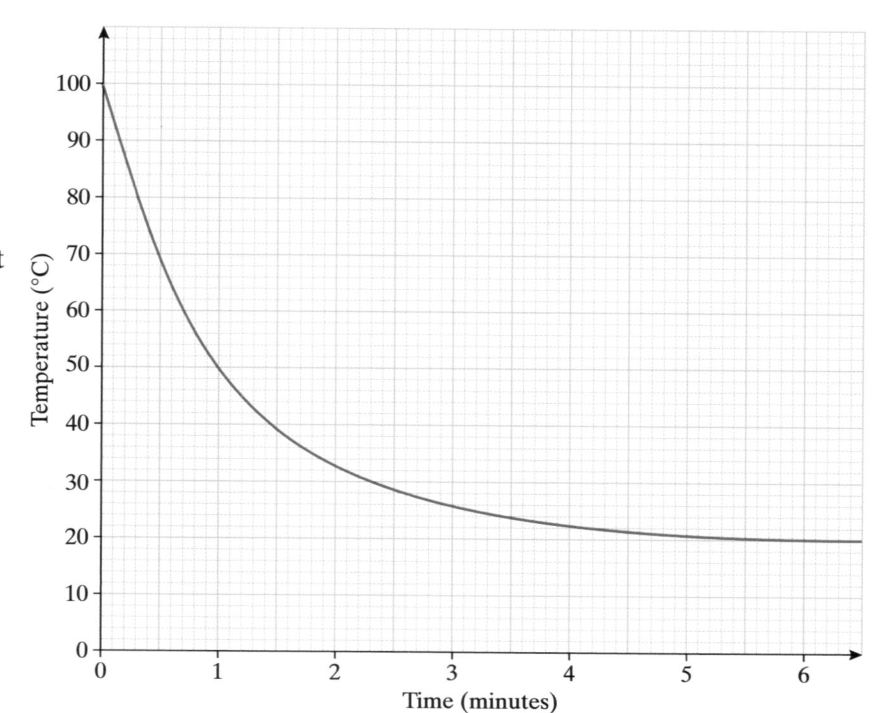

a) What is the temperature of the liquid when it starts cooling?

b) What is the temperature of the liquid after:

(i) 2 minutes

(ii) 6 minutes?

c) How long does it take for the liquid to cool down to 50 °C?

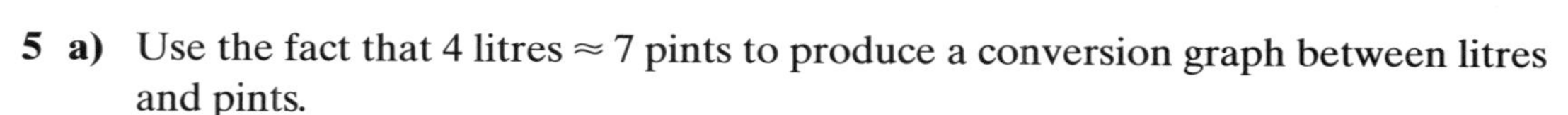

5 a) Use the fact that 4 litres ≈ 7 pints to produce a conversion graph between litres and pints.

b) Use your graph to convert:

(i) 2 pints **(ii)** 10 pints to litres.

c) Use your graph to convert:

(i) 8 litres **(ii)** 3 litres to pints.

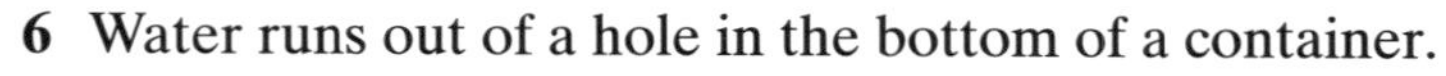

6 Water runs out of a hole in the bottom of a container.
The water runs out at a steady rate.
The diagram below shows how the depth of water in the container varies over time.

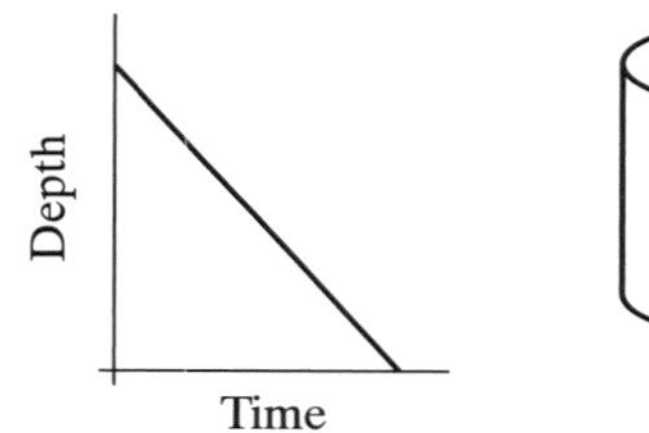

Say which one of containers A, B or C, best matches this graph.
Explain your reasoning.

12.2 Travel graphs

A travel graph is used to show the different stages of a journey.

- The horizontal axis shows the time of the journey.
- The vertical axis shows the distance travelled.

You will need to be able to draw and interpret travel graphs.

Look at this travel graph.
It shows Ben going on a cycle ride and returning home.

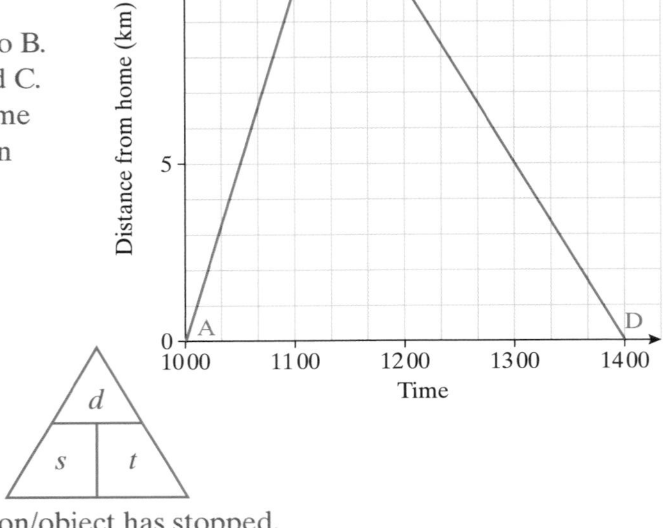

So this travel graph shows that:

- Ben is moving with the greatest speed between A and B.
  He is moving at a steady rate from A to B.
- Ben is at rest (stopped) between B and C.
- Between C and D Ben is returning home at a steady rate but at more slowly than before (it takes him twice the time to travel the same distance).
- At D Ben is back at his starting point.
- Altogether Ben has cycled 20 km in 4 hours.

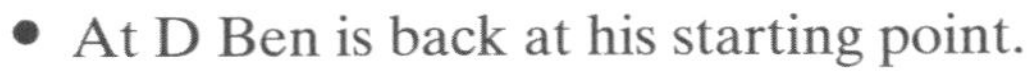

Ben's **average speed** is:

$$s = \frac{d}{t} = \frac{20}{4} = 5 \text{ km/h}$$

Some important points to remember:

- When the graph is horizontal, the person/object has stopped.
- The steeper the line, the greater the speed of the person/object.

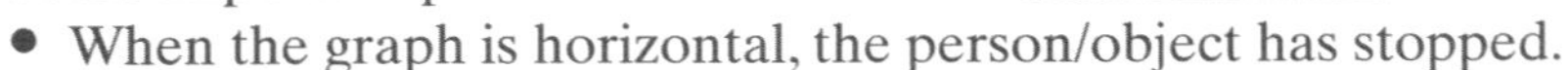

EXAMPLE

Isobel goes to visit a friend.
She leaves home at 10 am
The travel graph shows Isobel's journey.

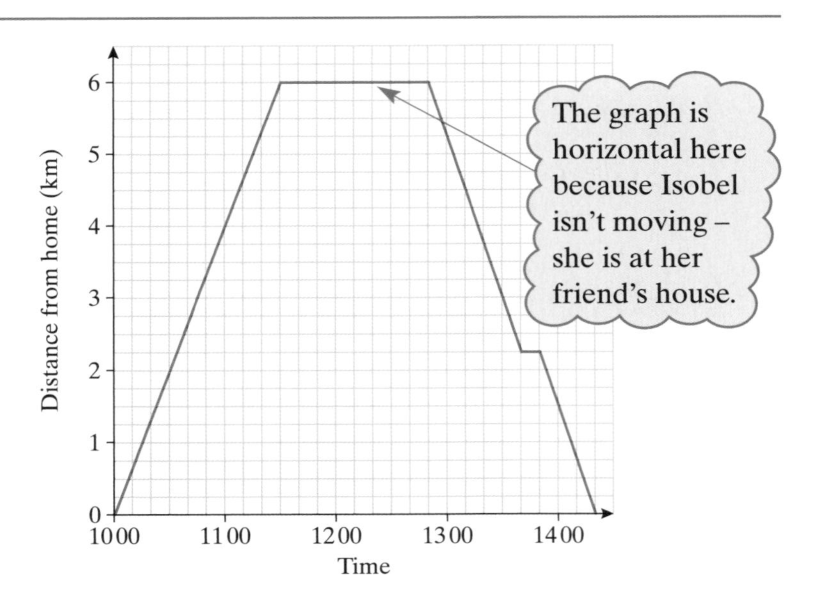

a) At what time does Isobel arrive at her friend's house?
b) How long does Isobel stay at her friend's house?
c) How far does Isobel travel altogether?

On the way home Isobel stops for a rest.
d) How long does she stop for?
e) How far from home is Isobel when she stops?
f) What is Isobel's average speed, in km/h, on her journey to her friend's house?

SOLUTION

a) First work out what time one small square represents:
6 small squares = 1 hour

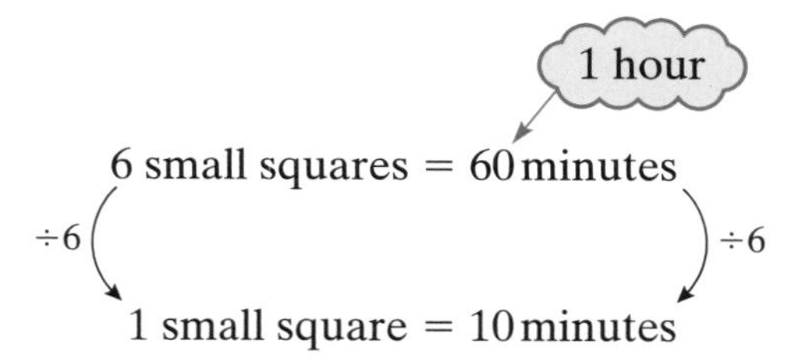

She arrives at 11 30

b) 80 minutes = 1 hour 20 minutes

c) 12 km ← 6 km there and 6 km back

d) 10 minutes

e) Work out what distance one small square represents.

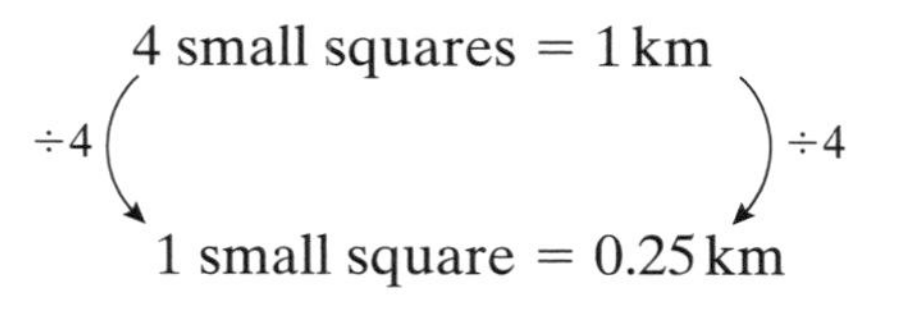

She is 2.25 km from home.

Make sure you use hours. Divide the number of minutes by 60

f) It takes Isobel 90 minutes = $\frac{90}{60}$ hours = 1.5 hours to travel 6 km

Speed = $\frac{\text{distance}}{\text{time}} = \frac{6}{1.5}$ = 4 km/h

To find the average speed without a calculator you can say:

It takes Isobel 90 minutes to walk 6 km
÷3 ÷3
So it takes 30 minutes to walk 2 km
×2 ×2
And 60 minutes (1 hour) to walk 4 km

So her average speed is 4 km/h

EXERCISE 12.2

Do not use your calculator for Questions 1–3.

1 Olivia walks to the cinema.
Here is a description of her journey:
Olivia leaves her house at 6 pm
She walks 1 km and arrives at her friend's house at 6.15 pm
Olivia waits 15 minutes for her friend.
Olivia then walks $1\frac{1}{2}$ km and arrives at the cinema at 7 pm
The film finishes at 9 pm
Olivia walks straight home and arrives at 9.30 pm

a) Draw a travel graph of Olivia's journey.
b) What is Olivia's average speed, in km per hour, on her walk home?

2 Every Sunday Phil drives to visit his girlfriend.
The travel graph shows part of his journey one Sunday.
Phil stops for petrol on the way to his girlfriend's house.

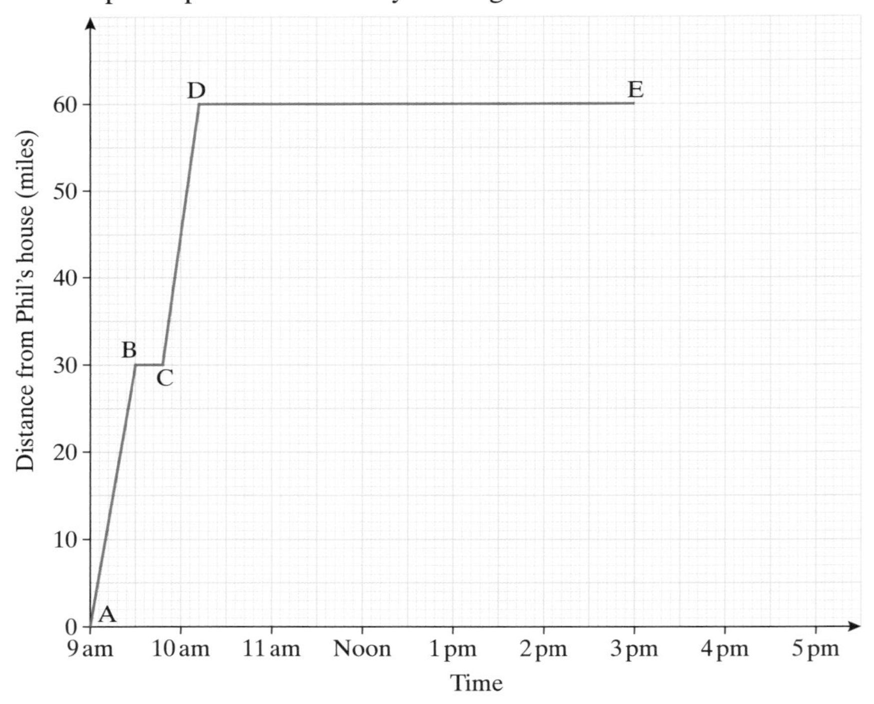

a) What time does Phil arrive at his girlfriend's house?

Phil stops for petrol on the way.
b) How long does he stop for?
Phil stays at his girlfriend's house until 3 pm and drives straight home at a steady speed.

He arrives home at 5 pm
c) Copy and complete the travel graph for Phil's journey.
d) What is Phil's average speed, in miles per hour, on the journey home?

3 One Saturday Jamie walks into town, does some shopping and returns home on the bus.
The travel graph shows his journey.

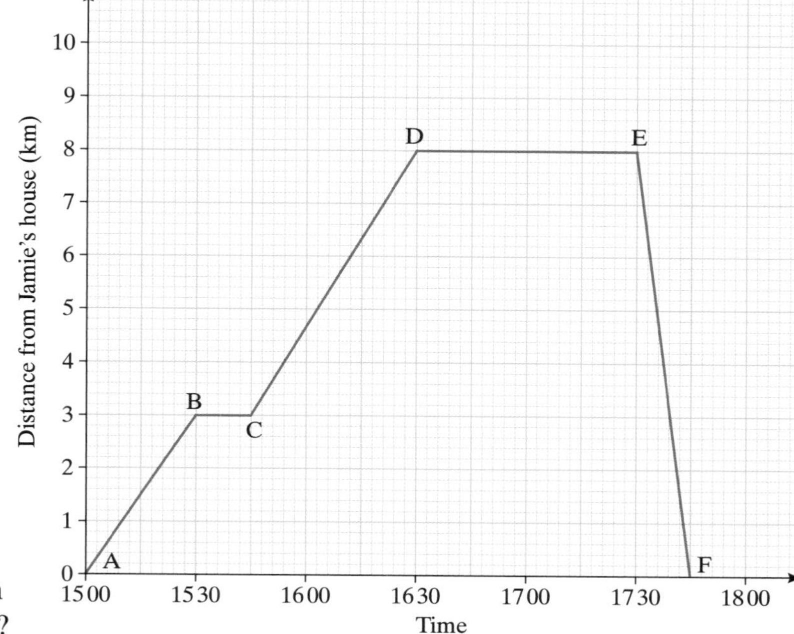

Jamie stops to talk to a friend on his way to town.

a) How long does Jamie stop for on his way to town?

b) What time does Jamie get to town?

c) How far is the town from Jamie's house?

d) Is Jamie's speed greater between A and B or between C and D? Explain how you can tell.

e) What time does Jamie arrive home?

You can use your calculator for Questions 4–9.

4 Chloe leaves her house for school at 8 am each morning.
She walks to the bus stop and waits for her bus.
Here is a distance–time graph for Chloe's journey to school.

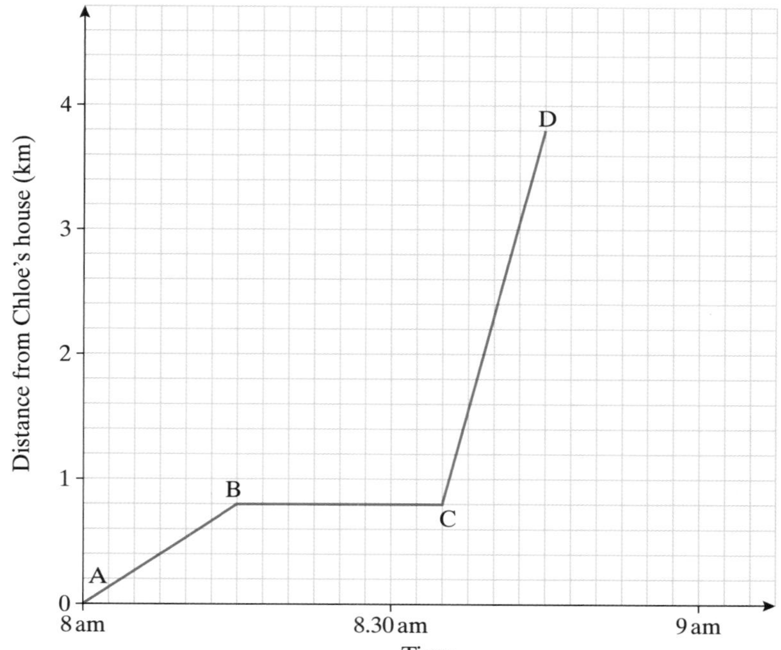

a) What time does Chloe reach the bus stop?

b) How long does she have to wait for the bus?

c) How far away from Chloe's house is the:

(i) bus stop

(ii) school?

d) What is Chloe's average speed, in km per hour, on her journey from:

(i) A to B **(ii)** C to D?

5 One morning Myra leaves Oldsville to visit her grandma in Little Chipping.
That same day, Petra leaves Little Chipping to visit her uncle in Oldsville.
Here is a travel graph of their journeys.

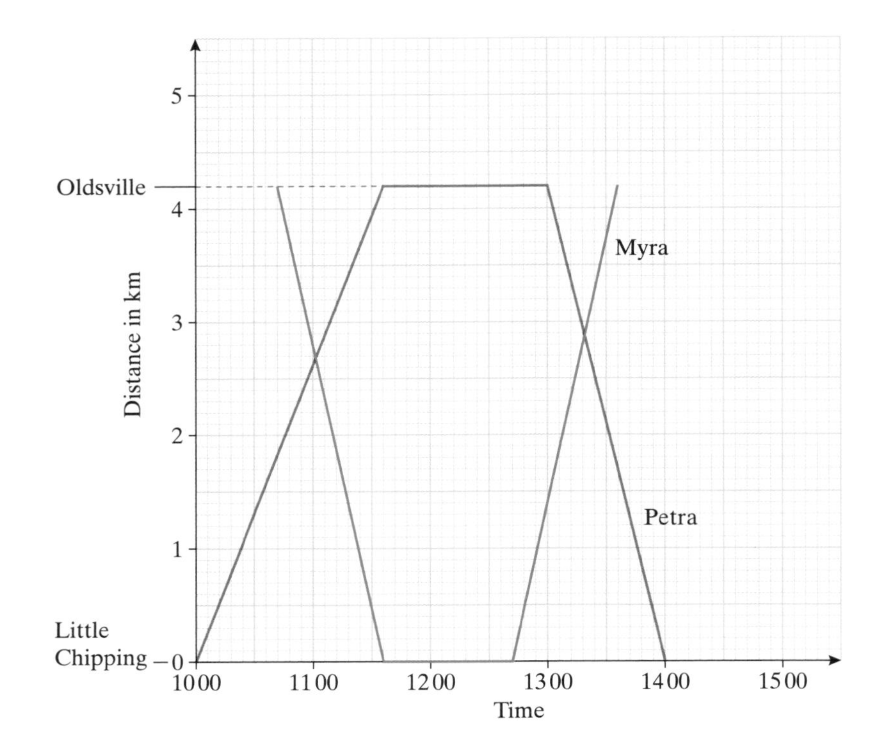

a) Give the distance between Oldsville and Little Chipping.
b) What time does Petra leave Little Chipping?
c) What time does Myra leave Oldsville?
d) How far away from Little Chipping are they when they meet for the first time?
e) How long does Petra spend in Oldsville?
f) What time do they meet on their journey home?

6 Tim cycles from home to his friend's house for tea.
He has a puncture on the way.
He fixes the puncture, and is able to complete his journey.
After tea, he cycles back home again.
The travel graph shows his journey.

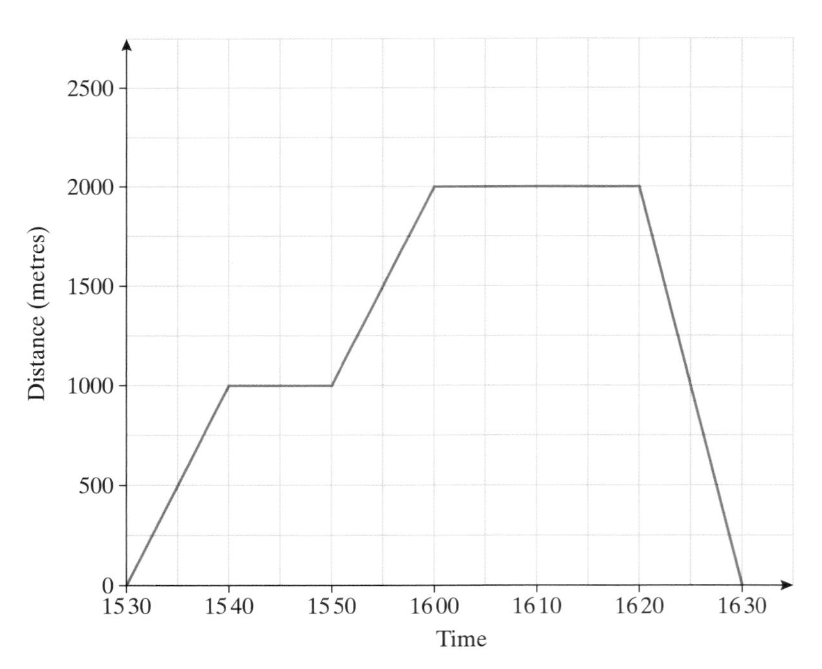

a) How long did Tim spend at his friend's house?
b) Work out his speed, in km per hour, for the journey home from his friend's house.
c) Did he cycle at the same speed as this on the outward journey?

7 The diagram shows a distance–time graph for a train travelling between Ayton and Beesville. The train leaves Ayton at 12 00 for its outward journey to Beesville.

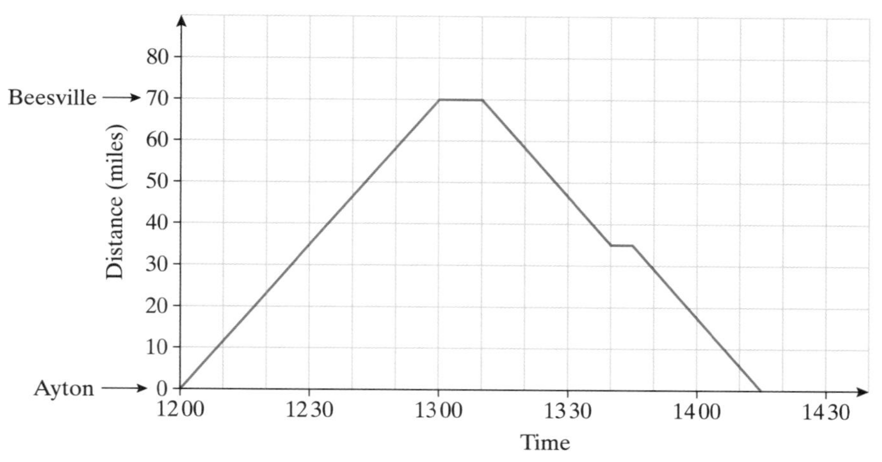

a) Work out the speed of the train on its journey from Ayton to Beesville.

The train leaves Beesville at 13 10 for its return journey to Ayton.

b) State one difference between the outward journey and the return journey.

c) State one thing that is the same on the outward journey and the return journey.

At 12 10 a second train leaves Beesville.
It travels towards Ayton at a constant speed of 60 miles per hour.

d) Draw the journey of the second train on a copy of the graph.

e) At what time do the two trains pass each other?

8 A group of teenagers are doing an outdoor walk.
They set off from their base at 09 00
They walk for 2 hours at 4 km/h
Then they rest for 1 hour.
After their rest, they walk on for a further 2 hours at 5 km/h

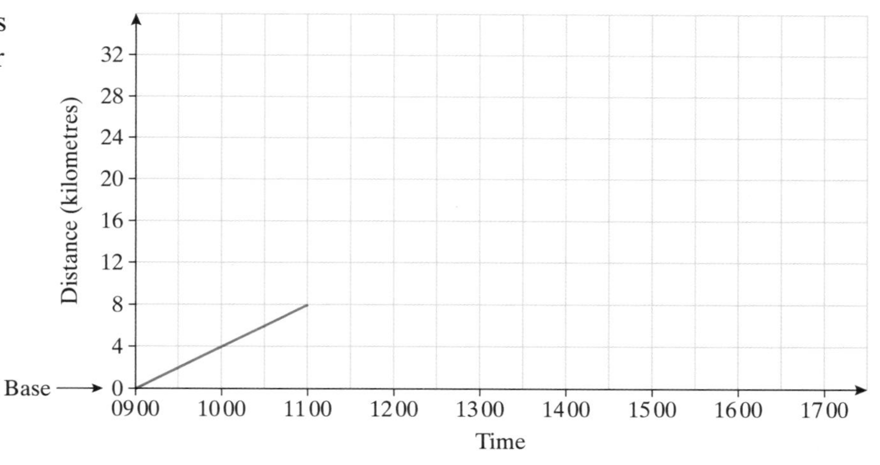

Then they rest for 1 hour again.
Finally, they walk for another 2 hours at 5 km/h

a) On a copy of the grid, complete the travel graph.

A teacher is camped 20 km from base.
At 12 00 he starts walking towards the group at 4 km/h
He keeps walking until he meets the group.

b) Add a line on your graph to show the teacher's journey.

c) At what time does the teacher meet the group?

9 John climbs a mountain.
He gains height at a rate of 10 metres per minute.
It takes him 45 minutes to reach the top.
He then stops for 20 minutes to have lunch.
Then he descends, at 15 metres per minute.

a) How high is the mountain?

b) How long does John's descent take?

c) Copy and complete the graph, numbering the scales on both axes.

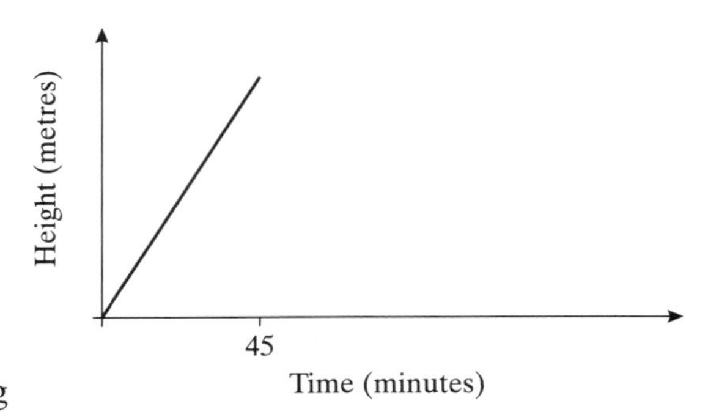

REVIEW EXERCISE 12

Do not use your calculator for Questions 1–5.

1 The diagram below shows a bowl.
It is in the shape of a hemisphere (half a sphere).
Water is poured into the bowl at a steady rate.

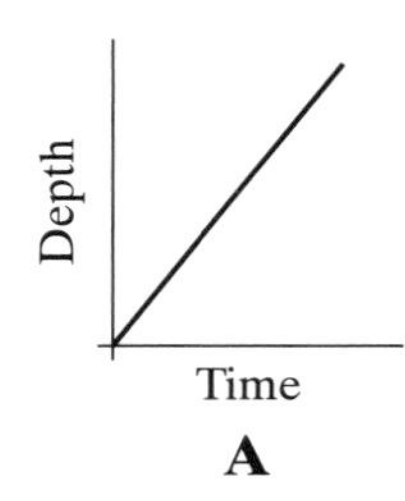

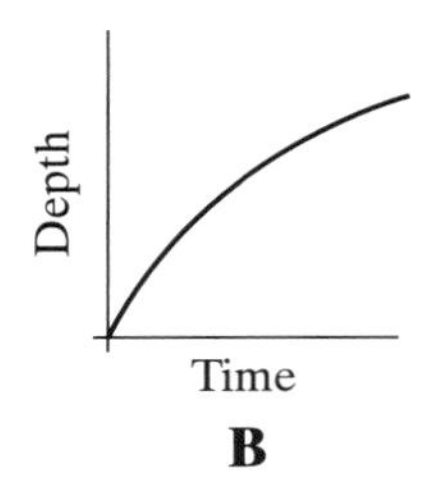

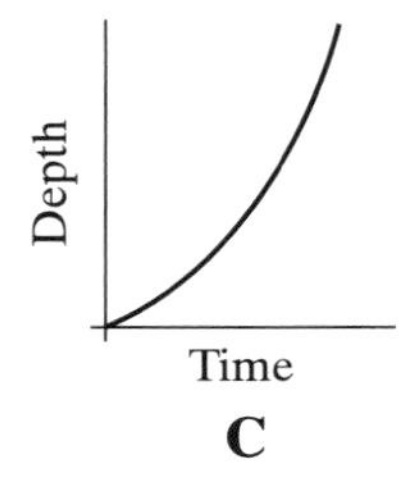

Say which graph, A, B or C, best describes how the depth of water in the bowl varies over time.

2 Water is poured at a steady rate into four different containers A, B, C and D.
The graphs P, Q, R and S show how the depth of water in each container changes over time.
Match the shapes to their corresponding graphs.

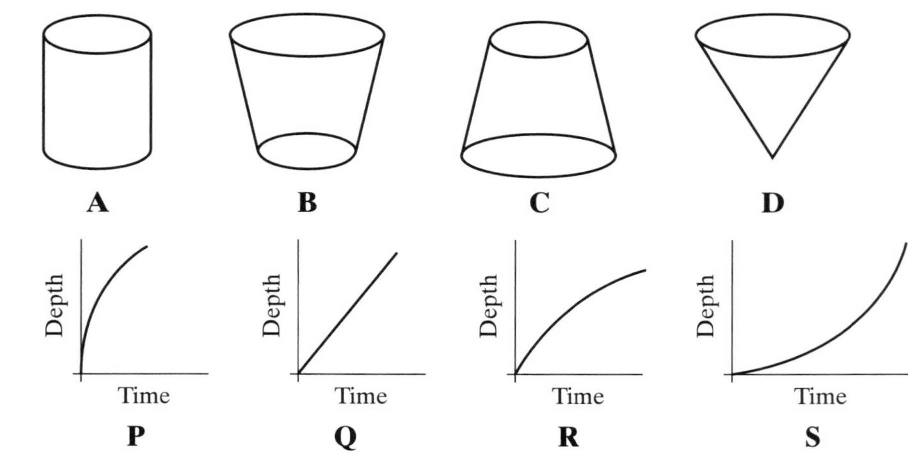

3 Anil cycled from his home to the park. Anil waited in the park.
Then he cycled back home. Here is a distance–time graph for Anil's complete journey.

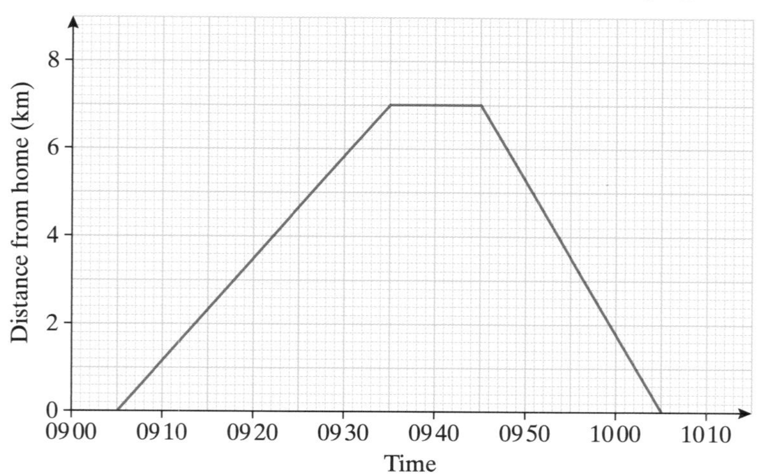

a) At what time did Anil leave home?
b) What is the distance from Anil's home to the park?
c) How many minutes did Anil wait in the park?
d) Work out Anil's average speed on his journey home.
Give your answer in kilometres per hour. **[Edexcel]**

4 Here is part of a travel graph of Sian's journey from her house to the shops and back.

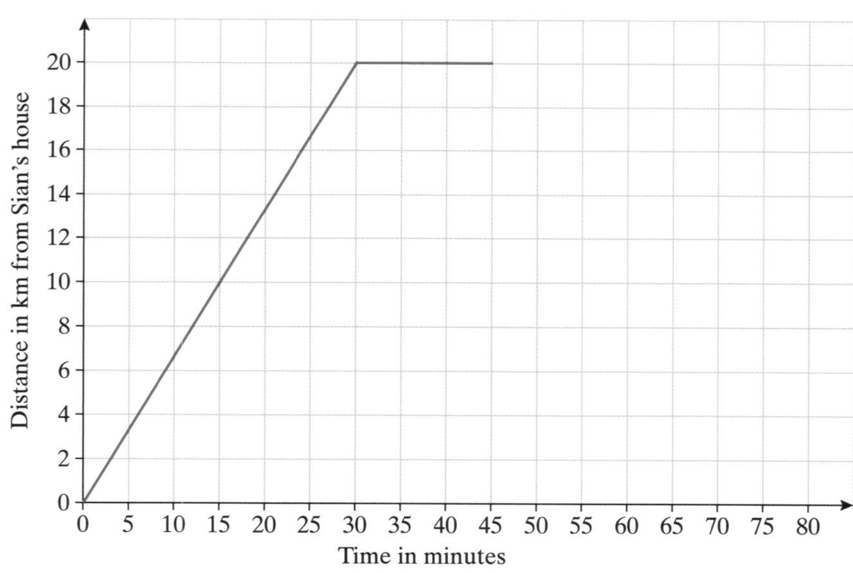

a) Work out Sian's speed for the first 30 minutes of her journey. Give your answer in km/h

Sian spends 15 minutes at the shops.
She then travels back to her house at 60 km/h
b) Copy and complete the travel graph. **[Edexcel]**

5 The diagram shows four empty containers.

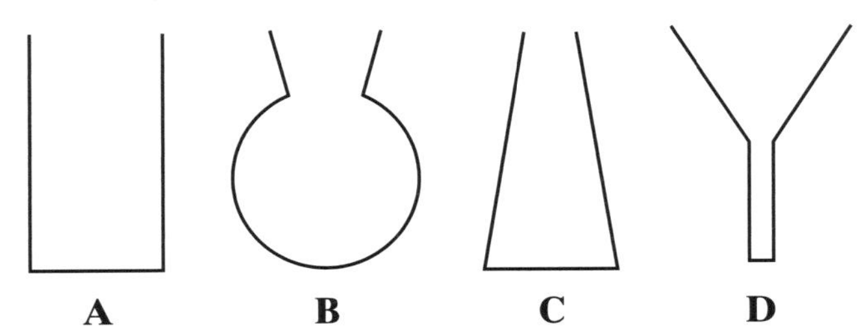

Water is poured at a constant rate into each of these containers.
Each sketch graph shows the relationship between the height of water in a container and the time as the water is poured in.

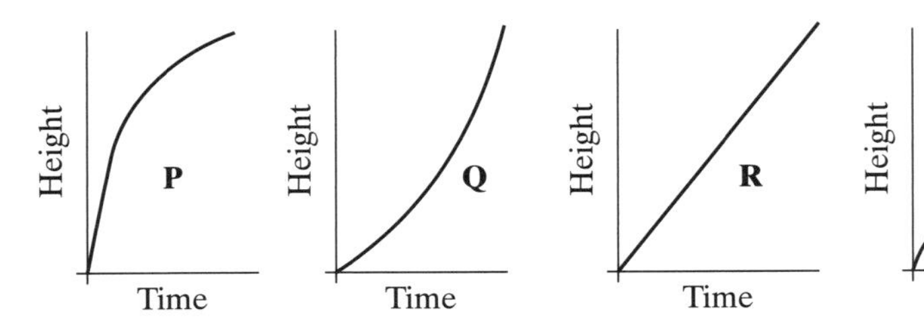

Copy this table, and write the letter of each graph in the correct place.

Container	Graph
A	
B	
C	
D	

[Edexcel]

You can use your calculator for Questions 6–10.

6 Helen writes down the reading on her gas meter on the first day of each month:

Reading on 1st January: 3580 units
Reading on 1st February: 3742 units

Gas is charged at 56p for each unit used.

a) Work out how much Helen is charged for the gas used in January.

In February Helen used 165 units of gas.
$\frac{1}{5}$ of these units were used in the first week.

b) How many units did she use in the rest of February?

The gas company increases its charges for units of gas used.
Helen works out the amount she will now be charged for gas used.

She uses the graph opposite.

c) Use the graph to write down:

(i) the amount Helen will be charged for using 100 units of gas

(ii) the number of units of gas used when Helen is charged £90

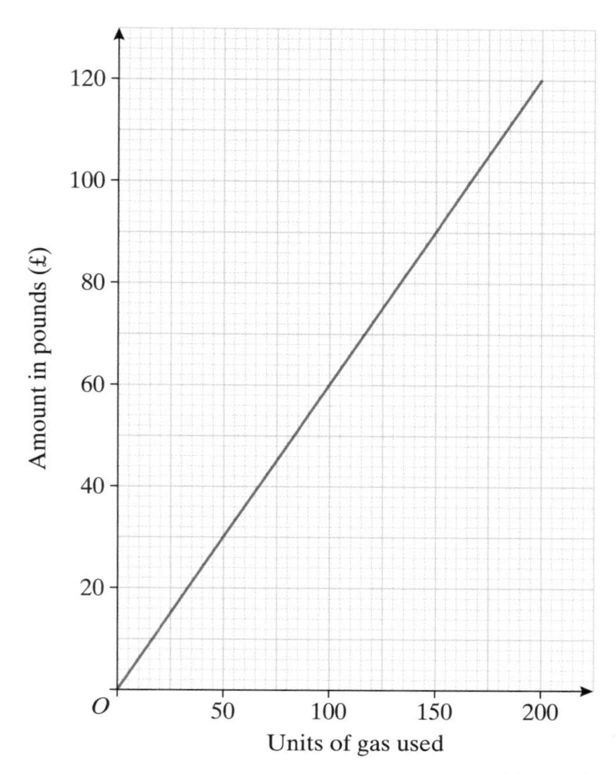

[Edexcel]

7 The conversion graph opposite can be used for changing between kilograms and pounds.

a) Use the graph to change 22 pounds to kilograms.

b) Use the graph to change 2.5 kilograms to pounds.

Firoza weighs 110 pounds.

c) Change 110 pounds to kilograms.

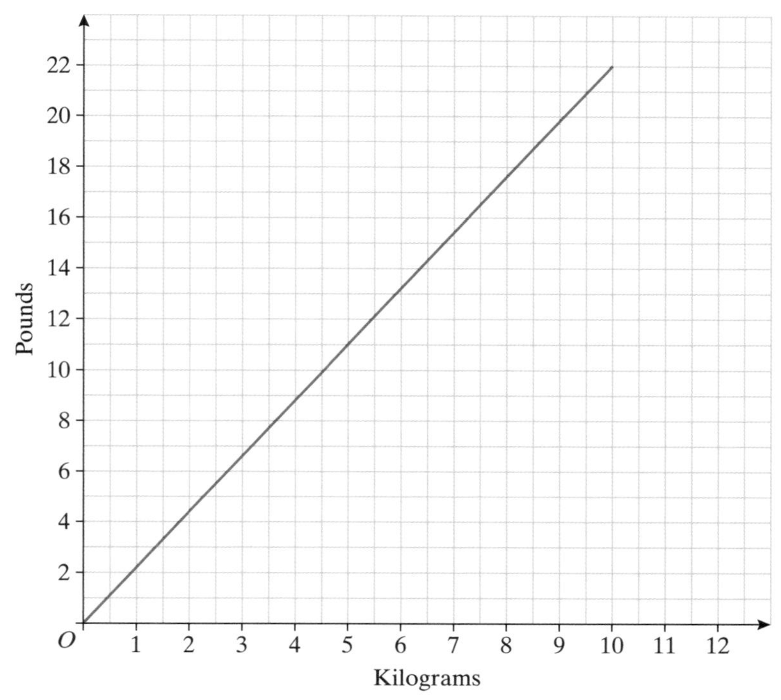

[Edexcel]

8 A man left home at 12 noon to go for a cycle ride.
The travel graph represents part of the man's journey.

At 12.45 pm the man stopped for a rest.

a) For how many minutes did he rest?

b) Find his distance from home at 1.30 pm.

The man stopped for another rest at 2 pm
He rested for 1 hour.
Then he cycled home at a steady speed.
It took him 2 hours.

c) Copy and complete the travel graph.

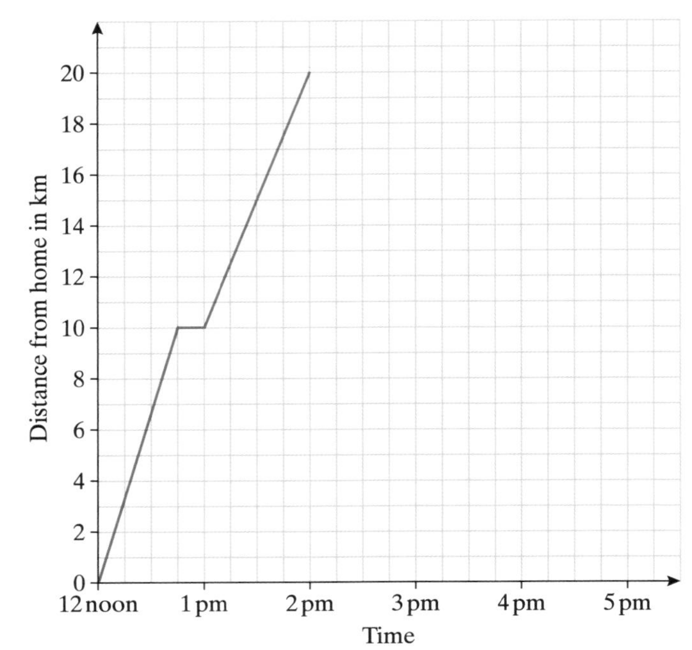

[Edexcel]

9 Ken and Wendy go from home to their caravan site.
The caravan site is 50 km from their home.
Ken goes on his bike.
Wendy drives in her car.
The diagram shows information about the journeys they made.

a) At what time did Wendy pass Ken?

b) Between which two times was Ken cycling at his greatest speed?

c) Work out Wendy's average speed for her journey.

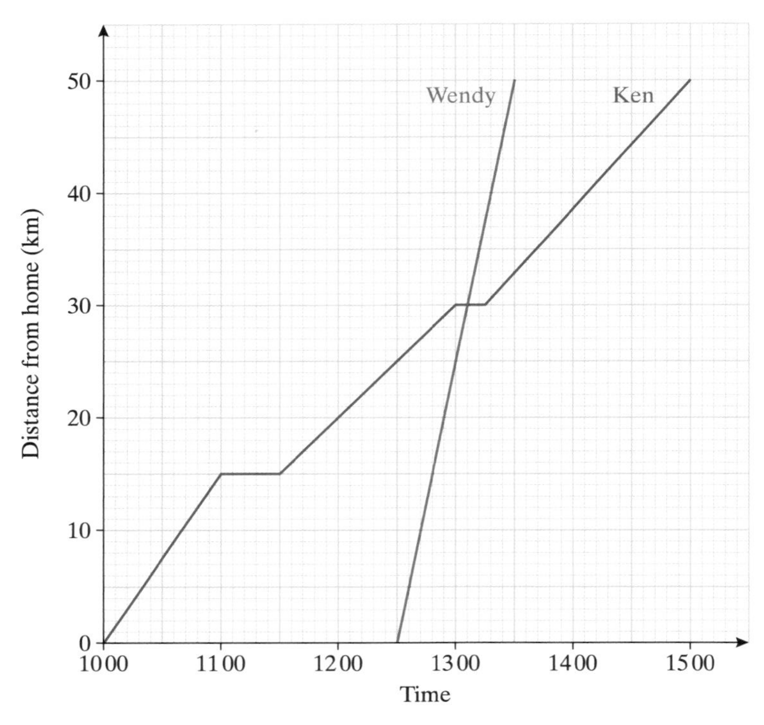

[Edexcel]

10 Linford runs in a 100 metres race.
The graph shows his speed, in metres per second, during the race.

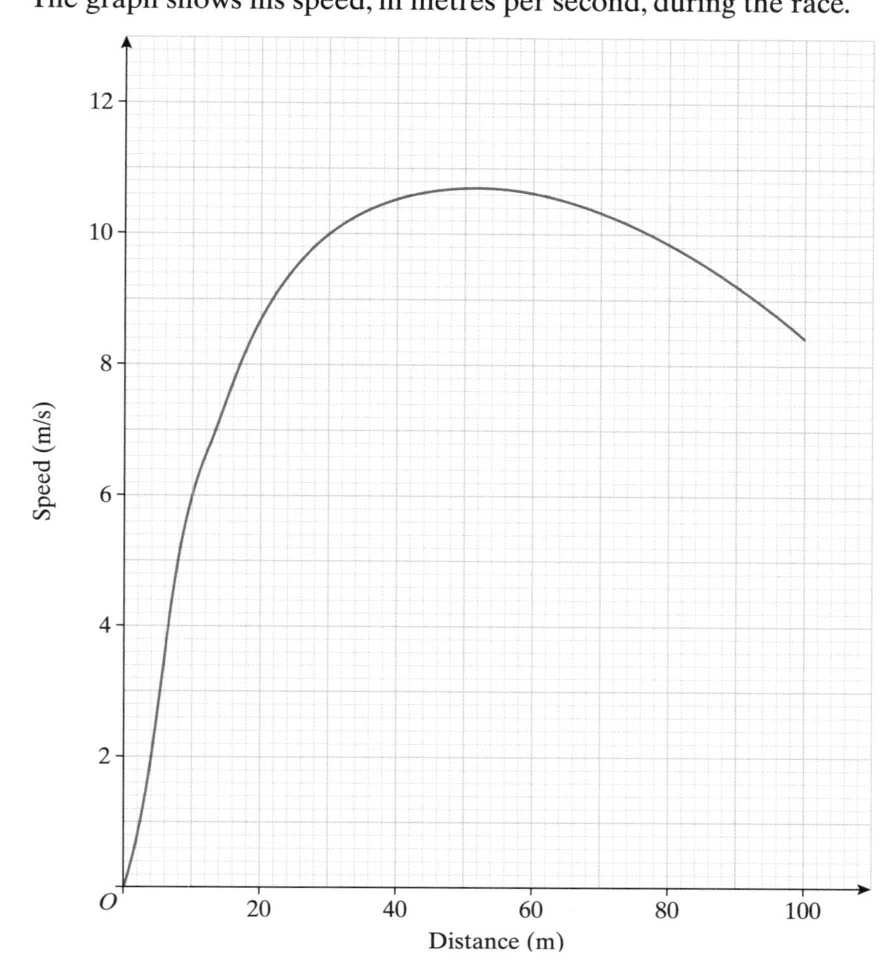

a) Write down Linford's speed, after he has covered a distance of 10 m
b) Write down Linford's greatest speed.
c) Write down the distance Linford has covered when his speed is 7.4 m/s **[Edexcel]**

KEY POINTS

1. When containers like cylinders or cuboids are filled at a steady rate, then the graph of depth against time is a straight line.
2. Other shaped containers, such as cones or spheres, generate curved graphs.
3. A conversion graph can be use to convert between one quantity and another. Always check the scale carefully when reading from a graph.
4. On a distance–time (travel) graph:
   - a horizontal line shows the object is at rest
   - the steeper the line the greater the speed.
5. Divide by 60 to convert minutes to hours when calculating the speed of an object.
6. Average speed $= \dfrac{\text{total distance}}{\text{time}}$

Internet Challenge 12

Famous curves

Below are some descriptions of famous curves based on mathematical graphs.

Use the internet to find the correct name for each one.

Try to find an image, such as a photograph, of an object for each curve.

You could make a poster of your images.

1 A cylinder is a solid object with a □□□□□□ for its cross section.

2 When a ball is thrown through the air, its path traces out a □□□□□□□□.

3 A cable hanging under its own weight forms a □□□□□□□□.
The support cables for suspension bridges take up this shape.

4 Many galaxies have the shape of a □□□□□□.
So does a sea creature called the chambered nautilus.

5 When a circle is tilted at an angle, it appears oval.
The correct name for this shape is an □□□□□□□.

6 A curve with infinitely many smaller curves inside it is known as a □□□□□□□.
A famous one takes its name from the mathematician Mandelbrot.

7 The □□□□□□□□ is a heart-shaped curve.
It can be made by rotating a circle around a second, fixed, circle, and tracing a point on the circumference of the moving circle.

8 The □□□□□□□ was studied extensively by mathematicians in the 17th century.
They thought it both beautiful and controversial, and so nicknamed it 'the Helen of geometry' after Helen of Troy.

CHAPTER 13

Angles

In this chapter you will **revise earlier work on**:

- how to measure angles.

You will **learn how to**:

- use angles along a straight line and around a point
- use corresponding and alternate angles
- work with angles in triangles and quadrilaterals
- calculate interior and exterior angles of polygons
- use 3-figure bearings.

You will be **challenged to**:

- investigate the four-colour theorem.

Starter: Measuring angles

1 Write down an estimate for each of the angles **a)** to **g)**.

2 Measure each angle to the nearest degree.

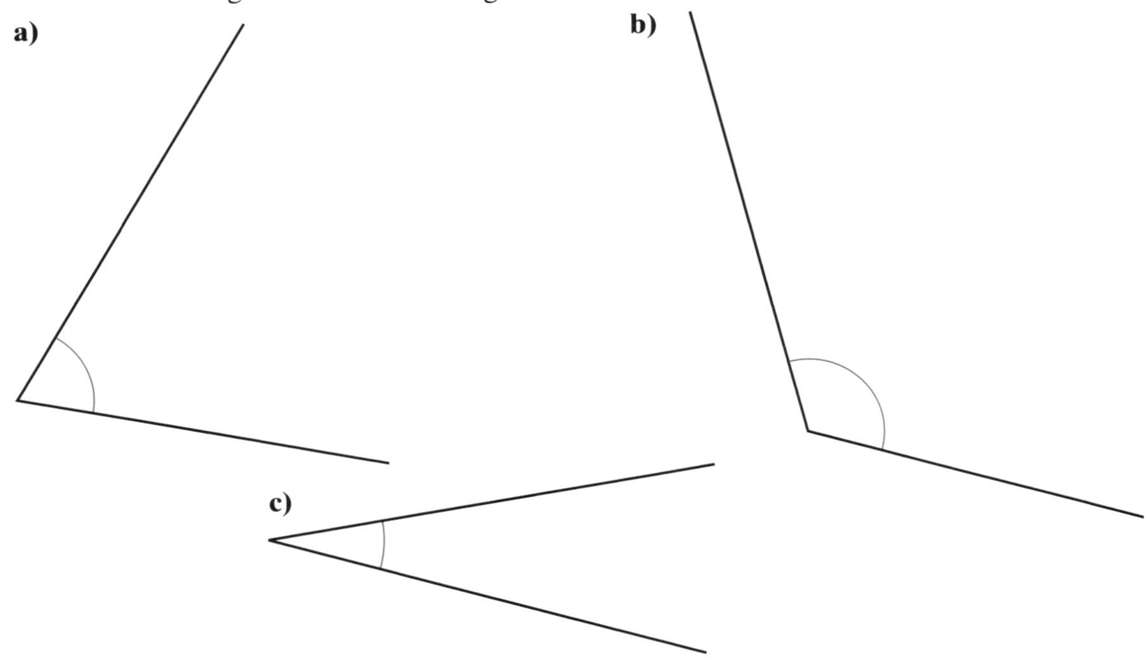

d)

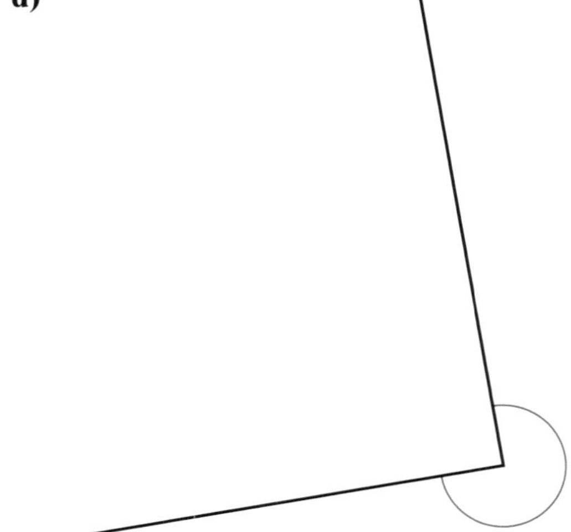

e)

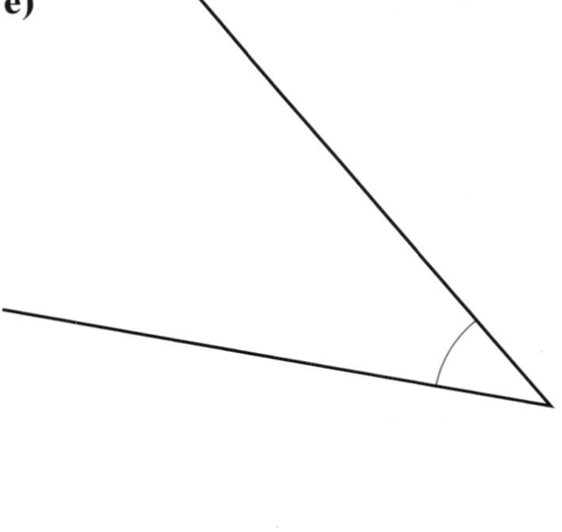

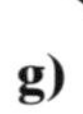

f)

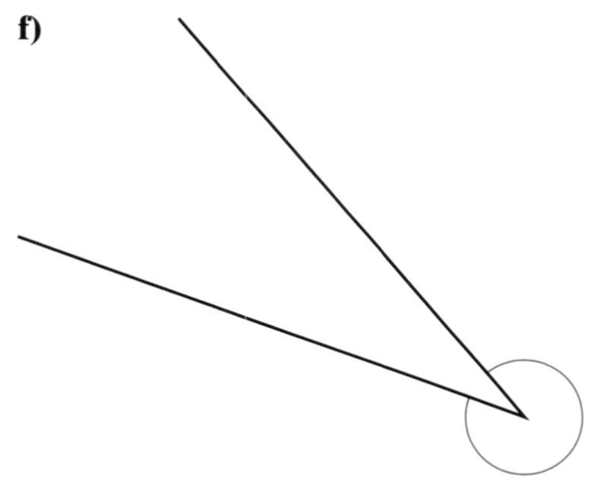

g)

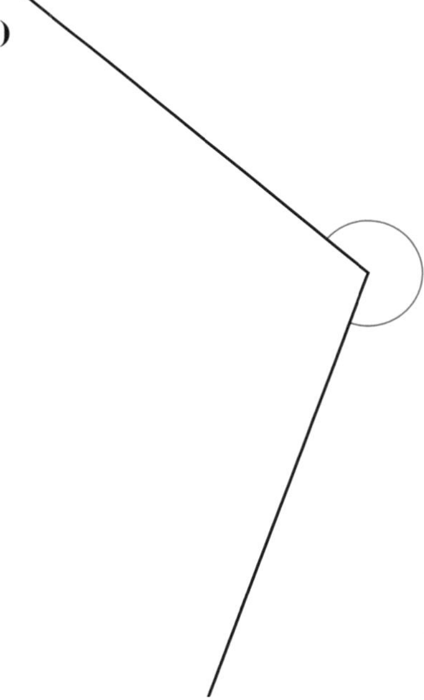

13.1 Angle facts

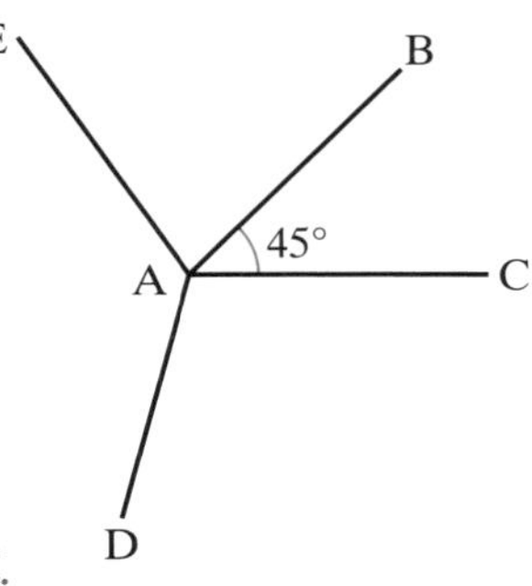

An **angle** is made whenever two lines meet.
Angles are measured in **degrees**, °

There are four different angles at the point A
The one marked as 45° is called angle BAC or ∠BAC:

- the middle letter, A, tells you the point (or **vertex**) where the angle is
- the other letters tell you the 'arms' of the angle – they are BA and AC.

You need to learn the following:

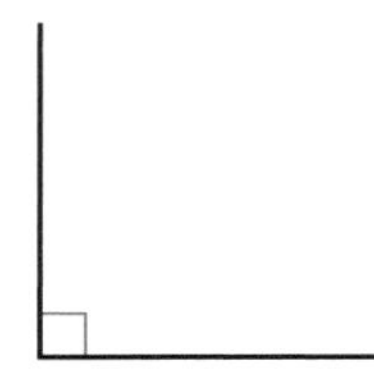

A **right angle** is 90°

The angle along a straight line is 180°

An **acute angle** is smaller than 90°

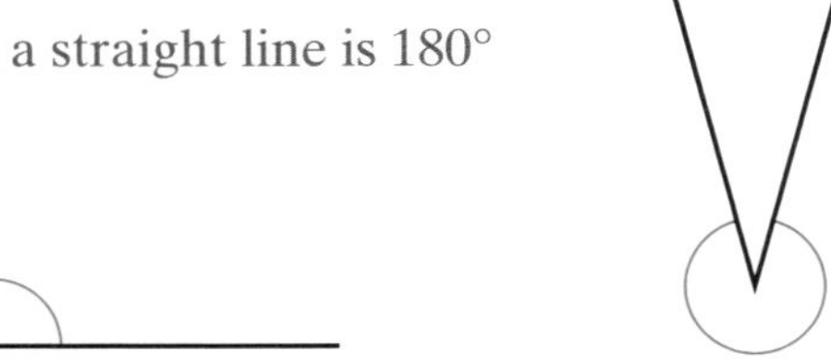

An **obtuse angle** is larger than 90° and less than 180°

A **reflex angle** is larger than 180°

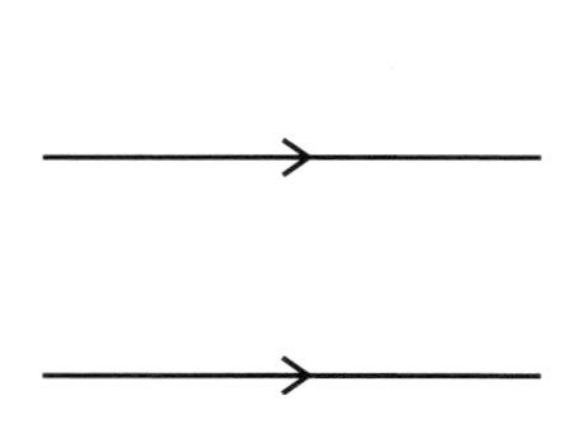

Parallel lines will never meet – even if they are continued forever.

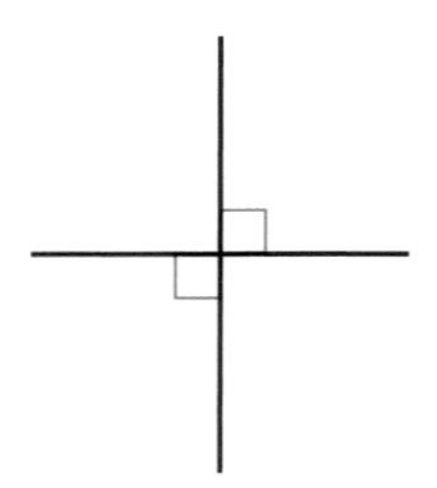

Perpendicular lines meet at right angles

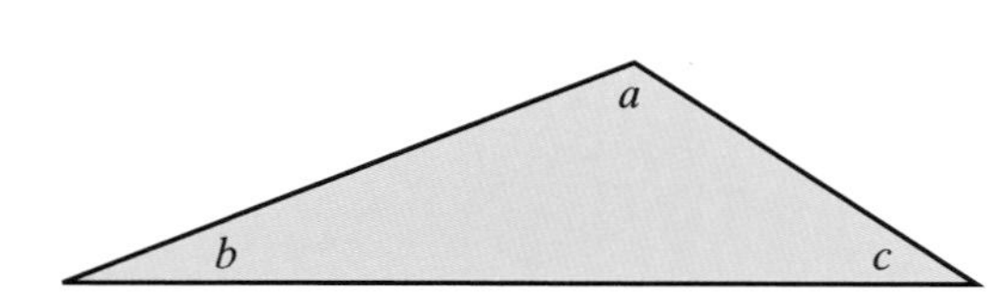

The angles in any triangle add up to 180°
A **scalene triangle** has three unequal angles and sides

$a + b + c = 180°$

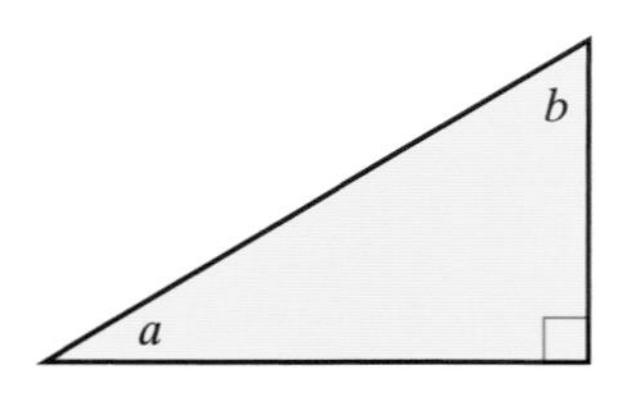

A **right-angled triangle** has one 90° angle

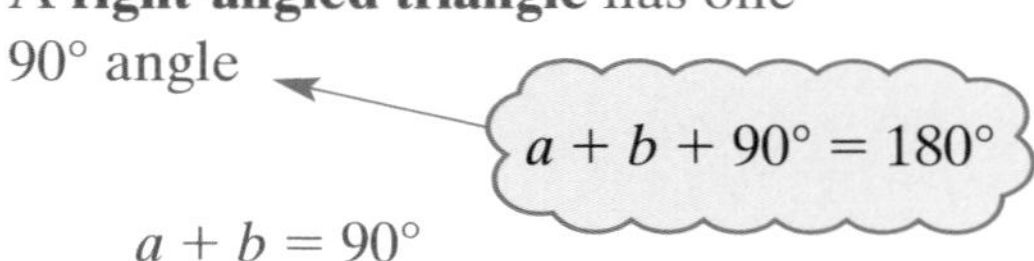

$a + b = 90°$

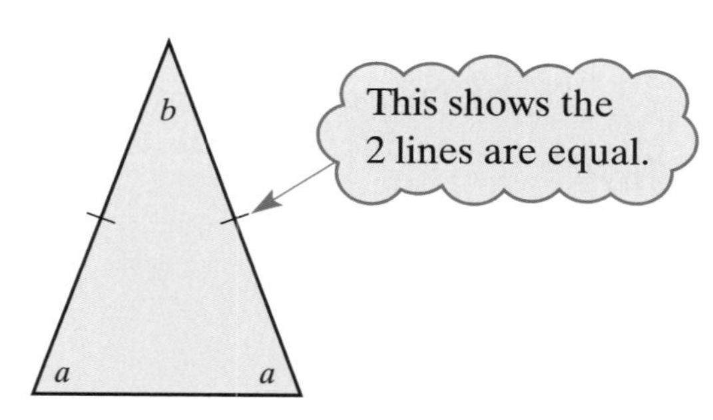

An **isosceles triangle** has two equal angles and 2 equal sides

$$2a + b = 180°$$

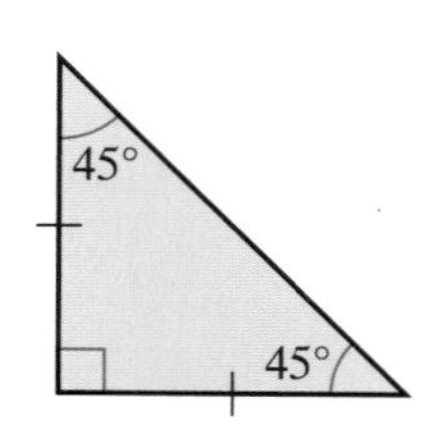

This is a **right-angled isosceles** triangle

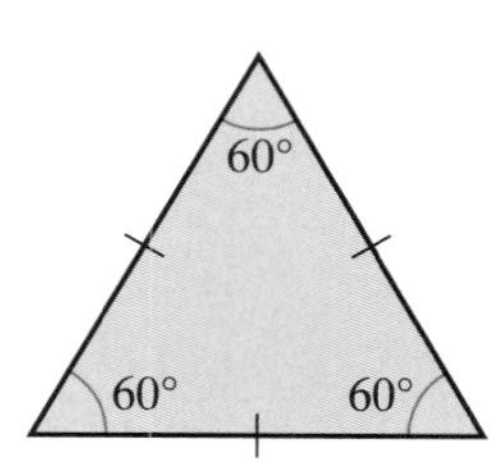

An **equilateral triangle** has three equal sides and three angles of 60°

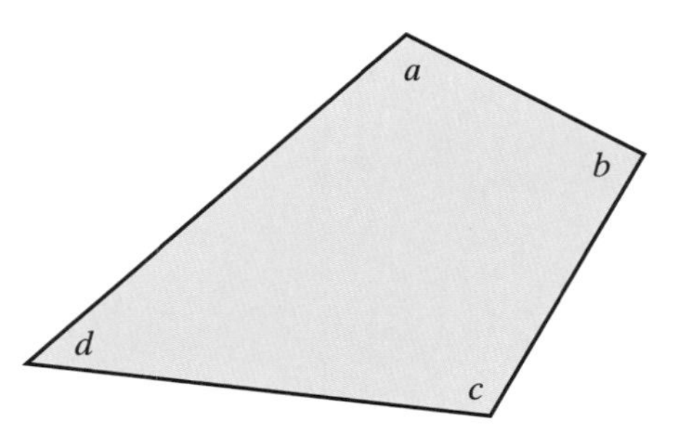

A **quadrilateral** has four sides and the angles add up to 360°

$$a + b + c + d = 360°$$

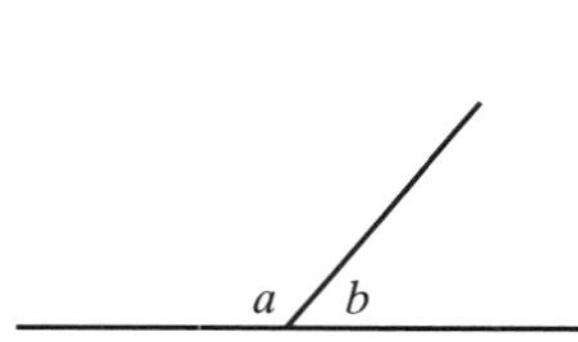

Angles along a straight line add up to 180°

$$a + b = 180°$$

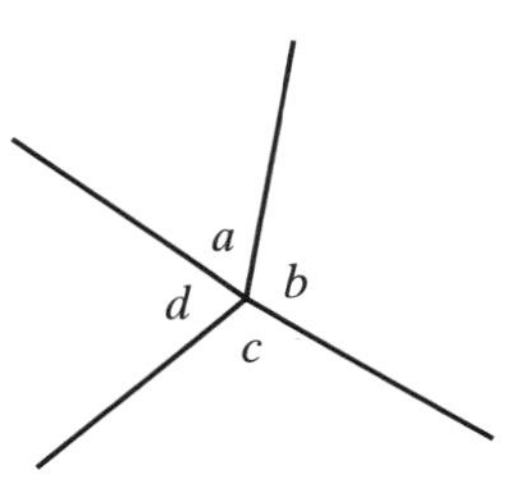

Angles around a point add up to 360°

$$a + b + c + d = 360°$$

Here is a proof that the angles inside any quadrilateral must add up to 360°

A theorem is a statement that can be proved to be true.

Theorem

The angles in a quadrilateral add up to 360°

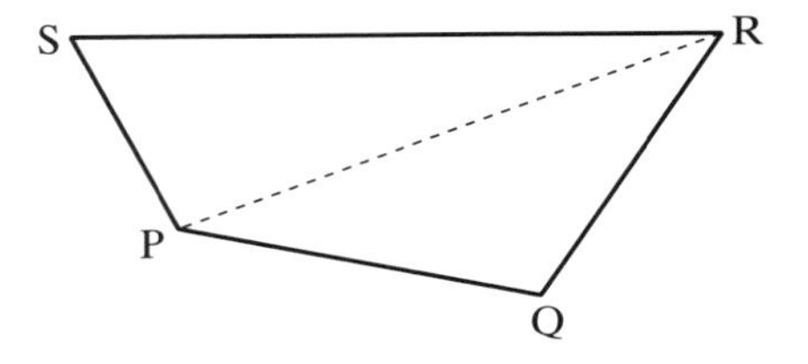

Proof

Consider any quadrilateral PQRS, and draw the diagonal PR, so as to divide it into two triangles.

Now consider the angles inside each of the two triangles.

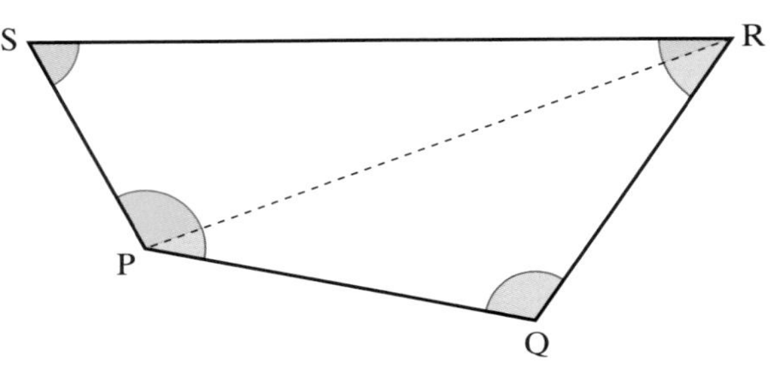

The total of the angles inside quadrilateral PQRS is equal to the sum of the angles in triangle PSR plus the sum of the angles in triangle PQR.
But the sum of the angles in a triangle is 180°

So the sum of the angles in quadrilateral PQRS $= 180° + 180°$
$= \underline{360°}$

EXAMPLE

Find the value of the angles marked by the letters a and b

a)

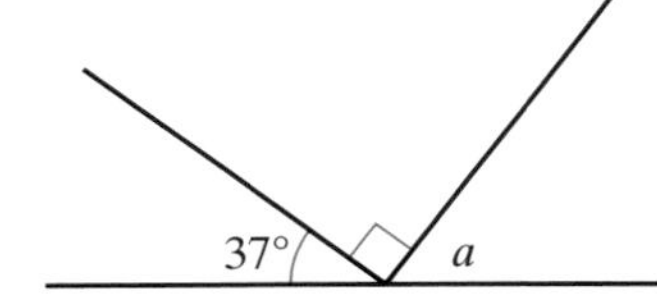

b)

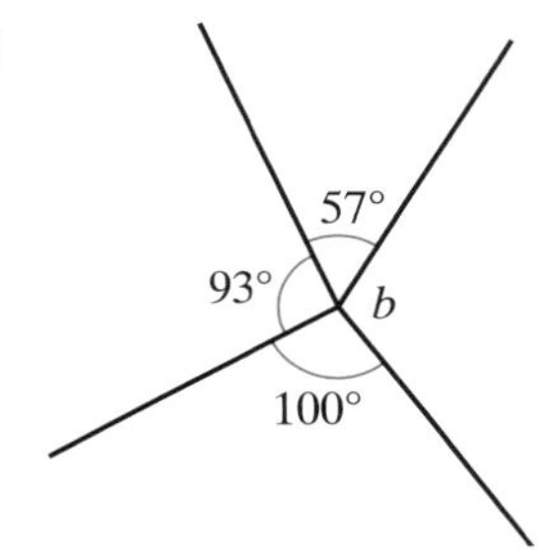

SOLUTION

a) The angles on a straight line add up to 180°

So $37° + 90° + a = 180°$

$127° + a = 180°$

$-127°$ $\quad$ $-127°$

$\underline{a = 53°}$

b) The angles around a point add up to 360°

So $100° + 93° + 57° + b = 360°$

$250° + b = 360°$

$-250°$ $\quad$ $-250°$

$\underline{b = 110°}$

EXAMPLE

Find the size of the angles marked by the letters a and b.

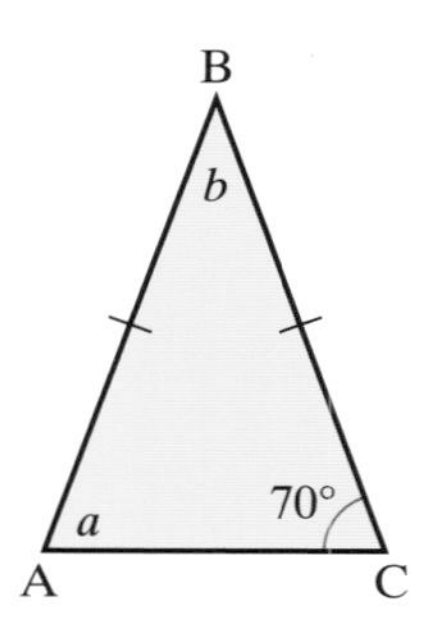

SOLUTION

This is an isosceles triangle so the sides AB and BC are the same length.

So angle BAC and angle BCA are the same.
So $\underline{a = 70°}$

The angles in a triangle add up to 180°
So $70° + 70° + b = 180°$

$$140° + b = 180°$$

$-140°$ (both sides)

$$\underline{b = 40°}$$

EXAMPLE

The diagram shows a quadrilateral PQRS.
Find the size of the angle x.

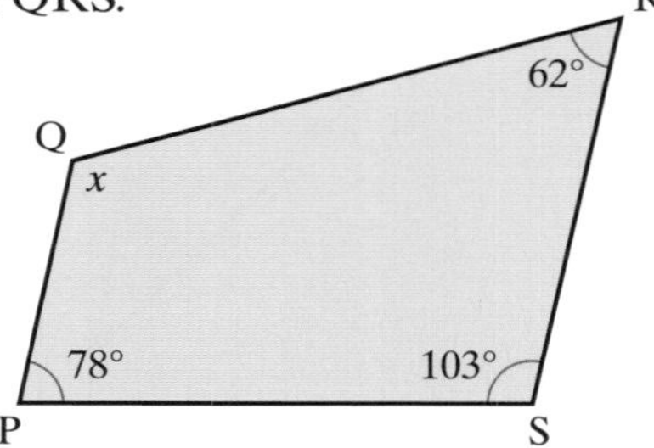

SOLUTION

The angles in a quadrilateral add up to 360°
So $x + 78° + 103° + 62° = 360°$

$$x + 243° = 360°$$

$-243°$ (both sides)

$$\underline{x = 117°}$$

EXERCISE 13.1

1 Write down what type of angle (acute, obtuse or reflex) each angle **a)** to **g)** is in the Starter exercise of this chapter.

2 For each of the following:

(i) work out the angles marked with letters

(ii) give a reason for your answer.

a)

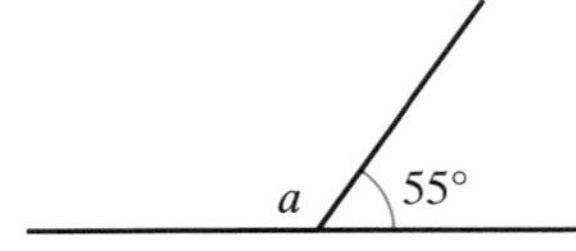

b)

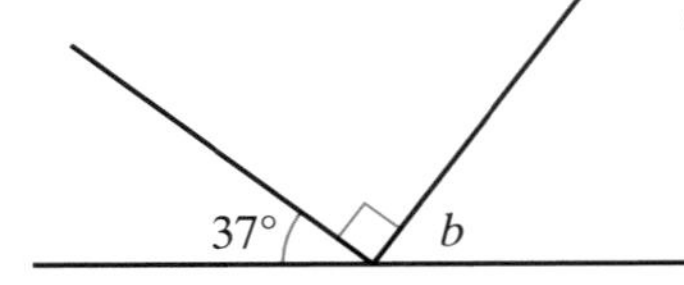

c)

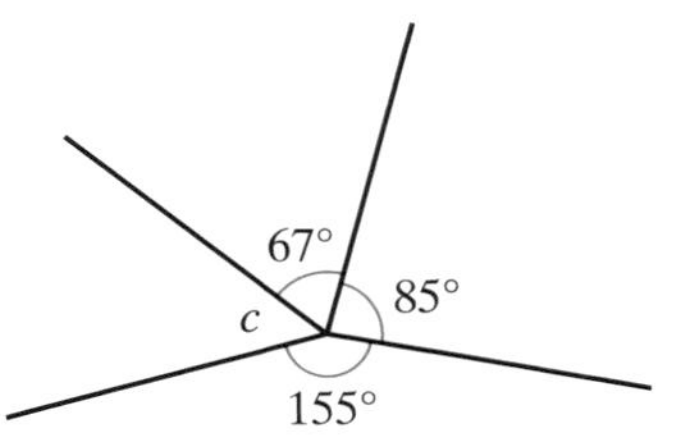

d)

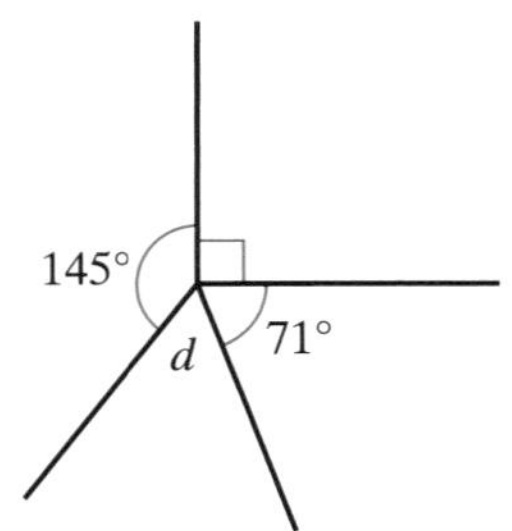

e)

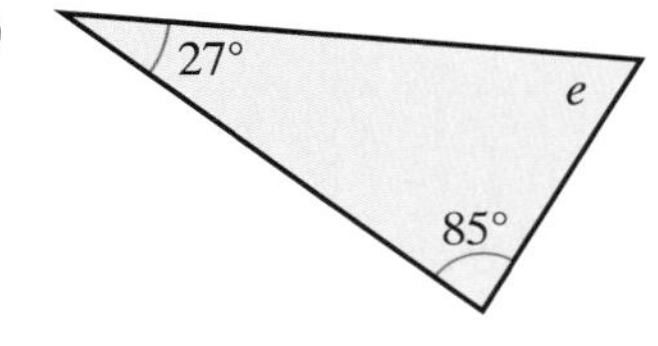

f)

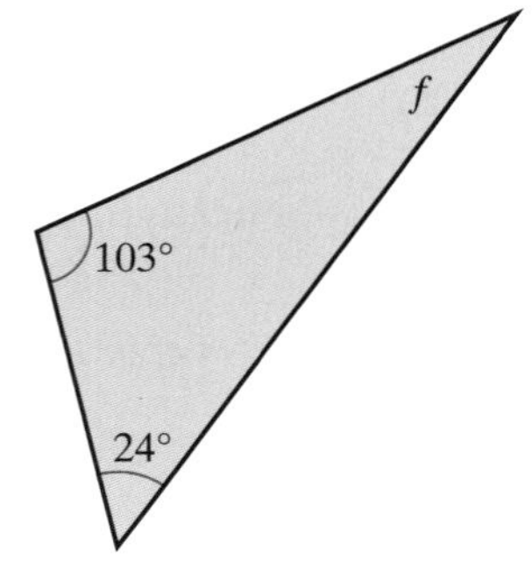

g)

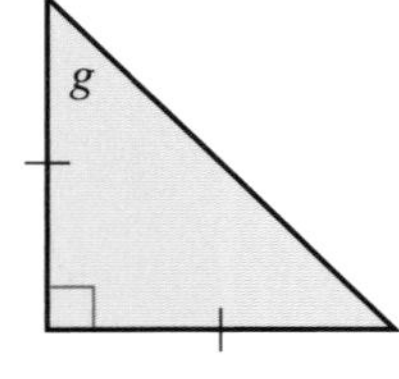

h)

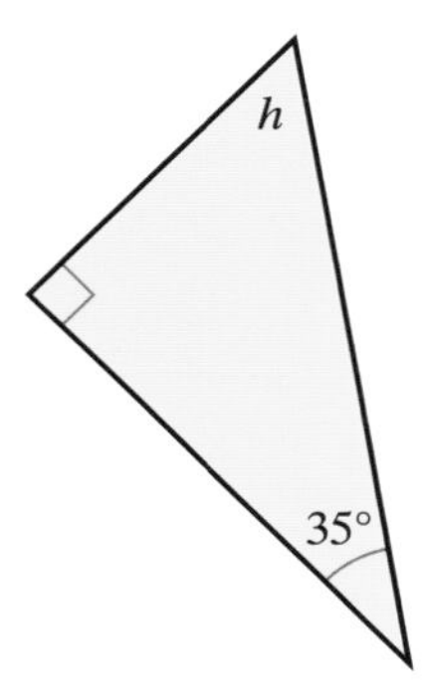

i)

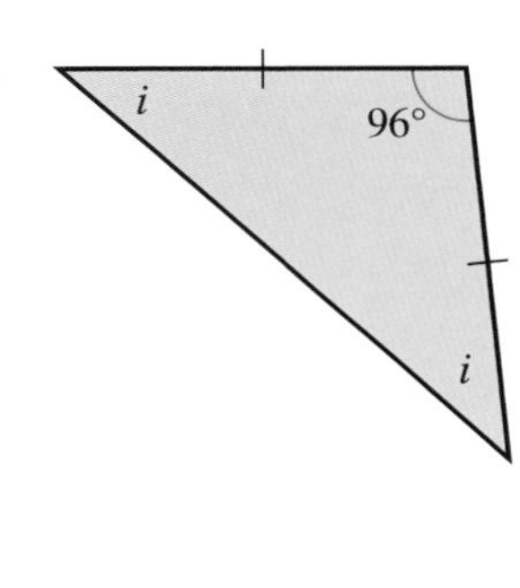

j)

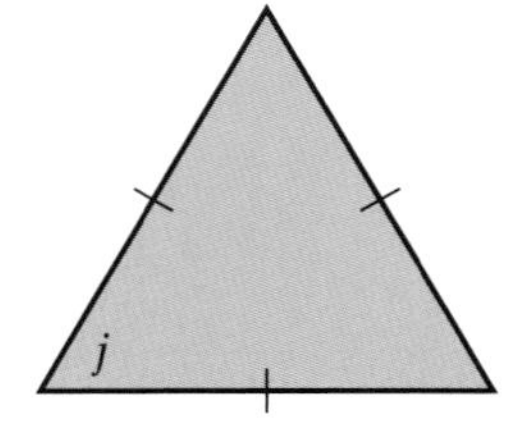

k)

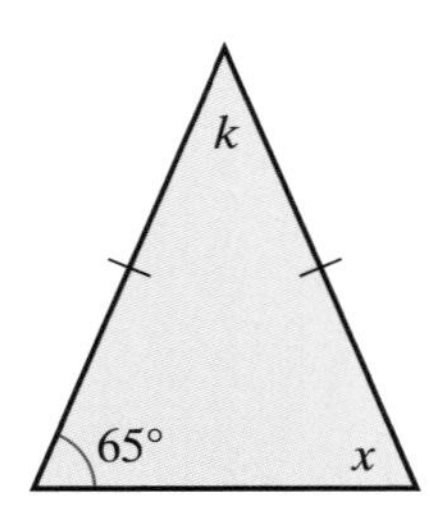

l)

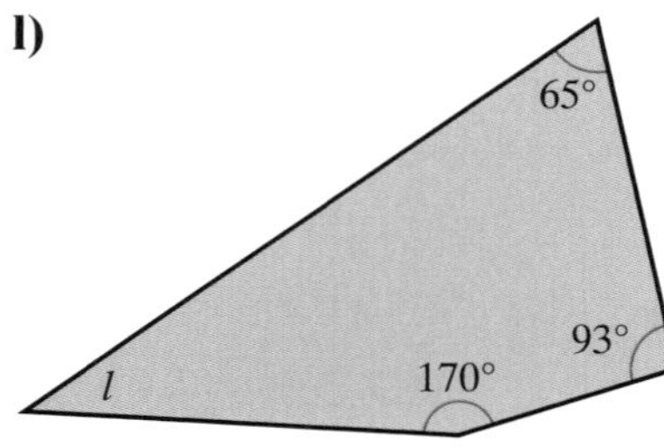

2 m)

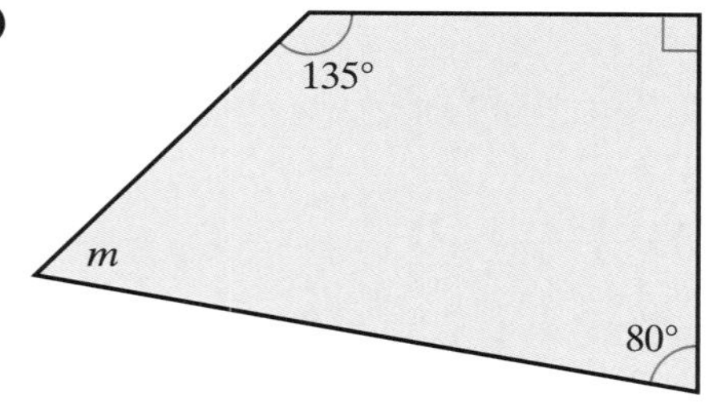

n)

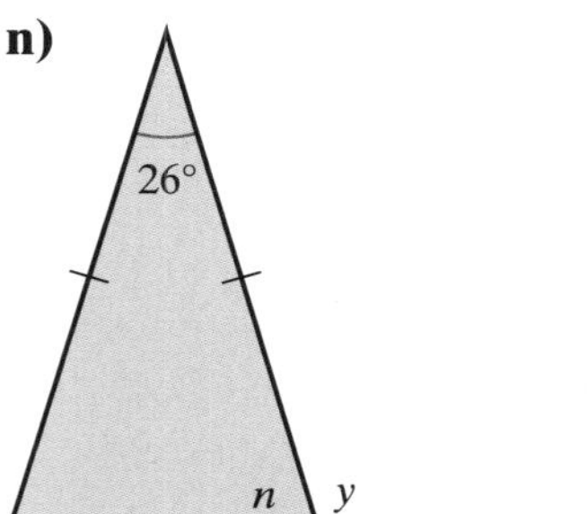

o)

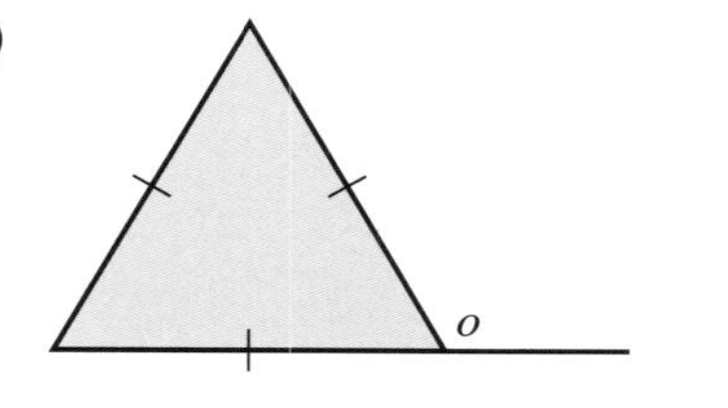

p)

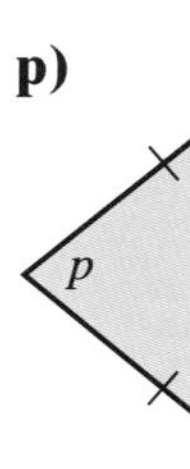

13.2 Using equations to solve problems

Some examination questions set problems on angles that lead to equations.

EXAMPLE

The diagram shows a quadrilateral ABCD.

a) Find the value of x

b) Hence find the sizes of each of the angles.

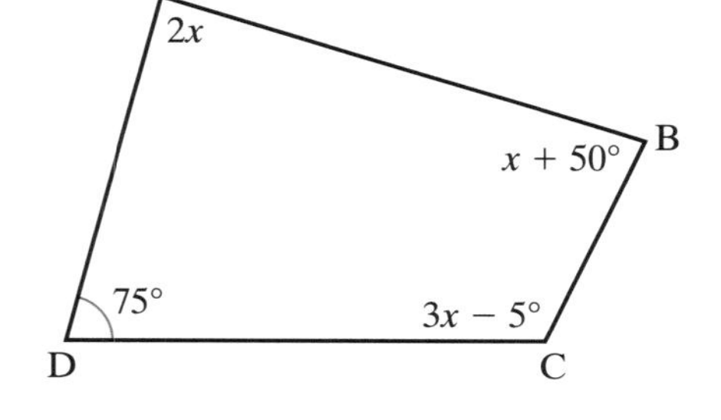

SOLUTION

a) Since the angles in a quadrilateral add up to 360°

$$2x + x + 50° + 3x - 5° + 75° = 360°$$

Simplifying: $6x + 120° = 360°$

$-120°$ (both sides)

$$6x = 240°$$

$\div 6$ (both sides)

$$x = 40°$$

$2x + x + 3x = 6x$

b) Then angle DAB $= 2x = 2 \times 40° = \underline{80°}$

angle ABC $= x + 50° = 40° + 50° = \underline{90°}$

angle BCD $= 3x - 5° = 3 \times 40° - 5° = \underline{115°}$

Check:
$80° + 90° + 115° + 75° = 360°$ ✔

EXERCISE 13.2

1 Find the value of a

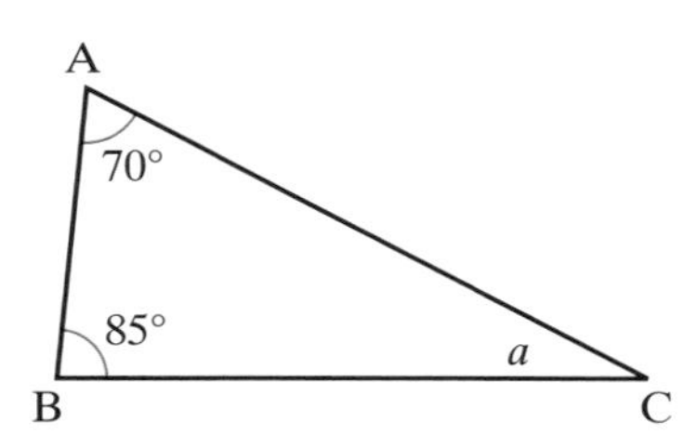

2 Find the value of x

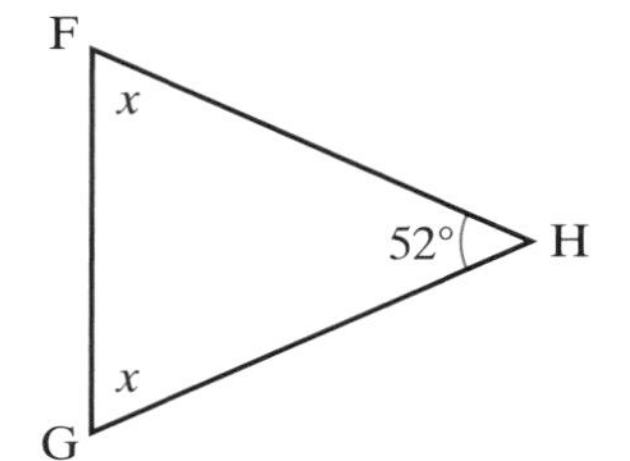

3 The diagram shows a quadrilateral.
Work out the value of m

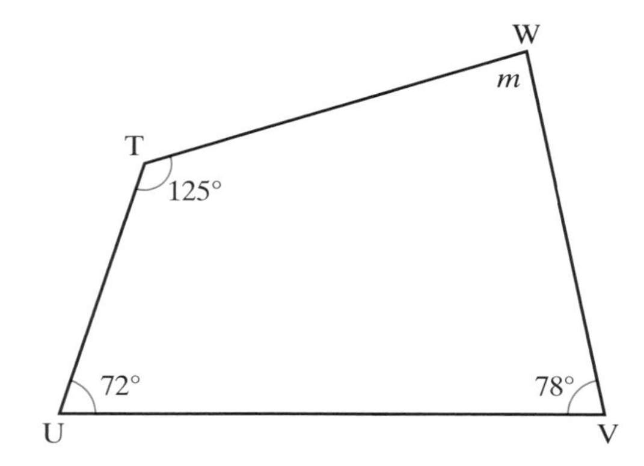

4 The diagram shows a quadrilateral.
Work out the value of y

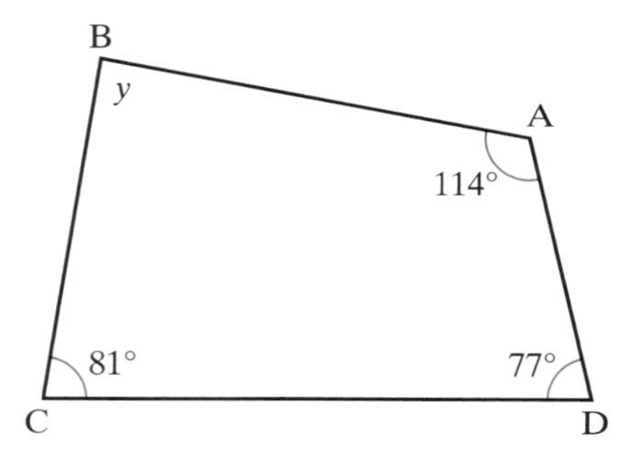

5 Find the size of the angles marked z

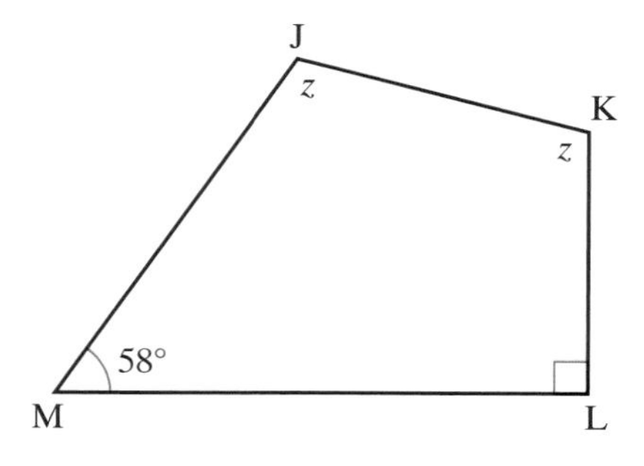

6 a) Write an expression for the sum of the angles in the triangle. Simplify your expression.
b) Write down an equation for the sum of the angles in the triangle.
c) Solve your equation to find the value of x
d) Hence work out the value of each angle in triangle MLN.

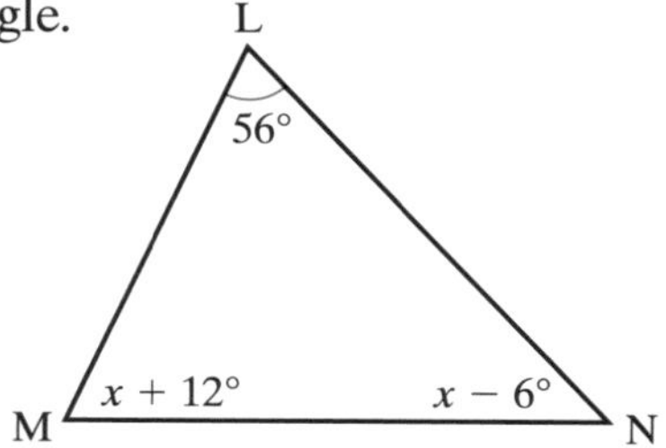

7 a) Write an expression for the sum of the angles in the triangle. Simplify your expression.
b) Write down an equation for the sum of the angles in the triangle.
c) Solve your equation to find the value of y
d) Hence work out the value of each angle.

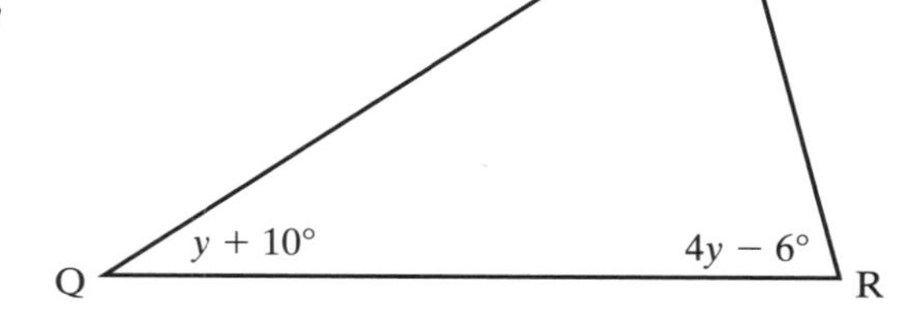

8 A triangle has angles $x + 8°$, $2x - 8°$ and $90°$
a) Set up an equation in x
b) Solve your equation to find the value of x
c) Work out the size of the angles in the triangle

9 The angles in a triangle are $4c + 4°$, $5c - 7°$ and $7c + 7°$
a) Set up an equation in c
b) Solve your equation to find the value of c
c) Work out the size of the angles in the triangle.
d) What kind of triangle is this?

10 a) Form an equation in y
b) Solve your equation to find y
c) Hence find the size of the angles in the quadrilateral.

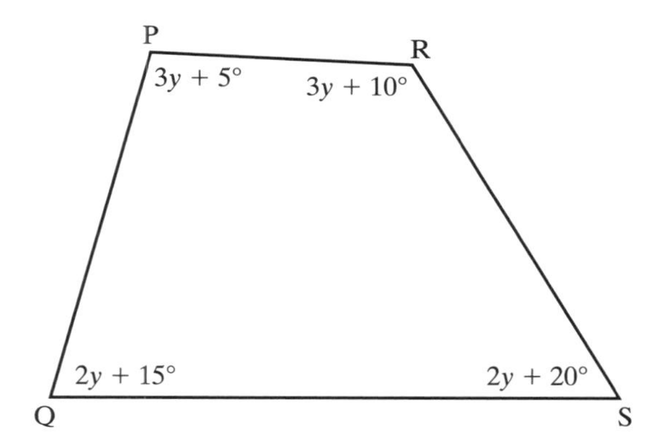

11 a) Form an equation in k

b) Solve your equation to find k

c) Hence find the size of the angles in the quadrilateral.

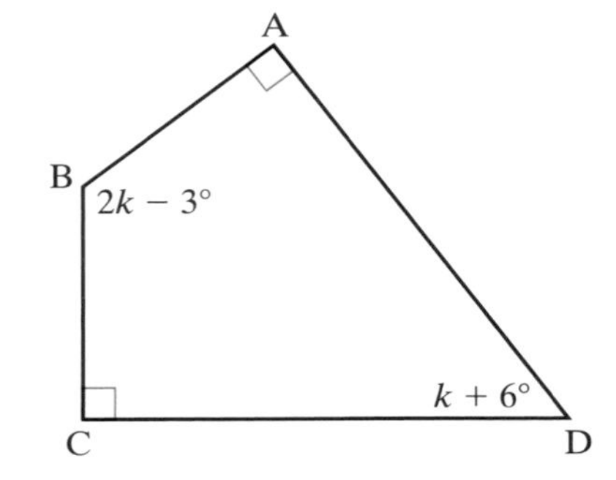

12 The angles in a quadrilateral are $x + 16°, 2x - 2°, 3x + 9°$ and $5x - 15°$

a) Set up an equation in x

b) Solve your equation to find the value of x

c) Work out the size of the angles in the quadrilateral.

d) Check that your four answers add up to 360°

13.3 Lines and angles

Imagine an infinitely long railway track, made up of two rails and a set of sleepers.

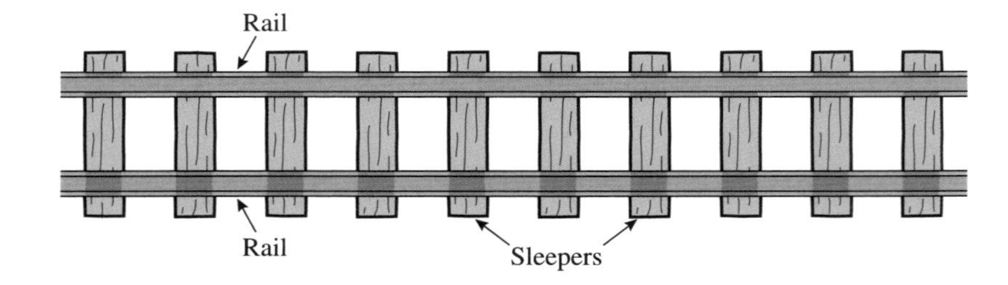

The rails and the sleepers are made up of straight lines, but there is a subtle mathematical distinction between them:

The rails are infinitely long straight lines.

The rails are **lines**.

The sleepers are pieces of straight line with definite start and finish points, so they are finite in length:

The sleepers are **line segments**.

Corresponding and alternate angles

The diagram shows two parallel lines, and two line segments that cross the parallel lines at an angle.

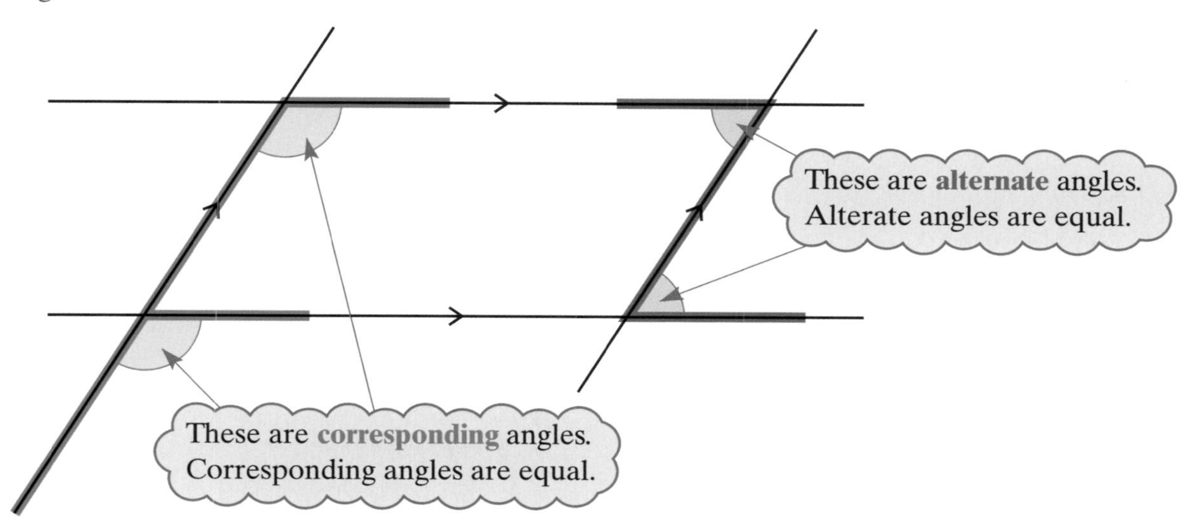

These are **alternate** angles.
Alterate angles are equal.
These are **corresponding** angles.
Corresponding angles are equal.

Sometimes corresponding angles are called *F*-angles and alternate angles are called *Z*-angles, because of the resemblance to those letters.

A GCSE examiner will know what you mean by these terms, but it is better to get into the habit of using the correct mathematical names for them.

EXAMPLE

Find the angles represented by letters a, b and c in the diagram. Give a reason in each case.

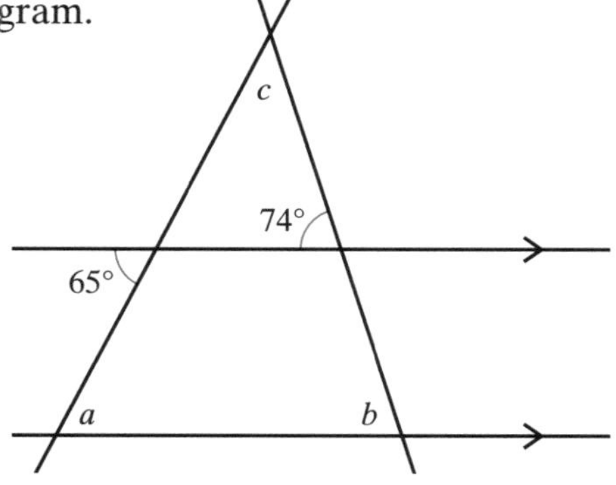

SOLUTION

$\underline{a = 65°}$ (alternate to marked 65° angle)

$\underline{b = 74°}$ (corresponding to marked 74° angle)

Angles a, b and c are at the three **vertices** of a triangle, so they add up to 180°

Always explain your reasons.

So $c + 65° + 74° = 180°$

$c + 139° = 180°$

$-139°$ $-139°$

$\underline{c = 41°}$

Theorem

The angles in a triangle add up to 180°

Proof

Construct a line through one vertex, parallel to the opposite side.

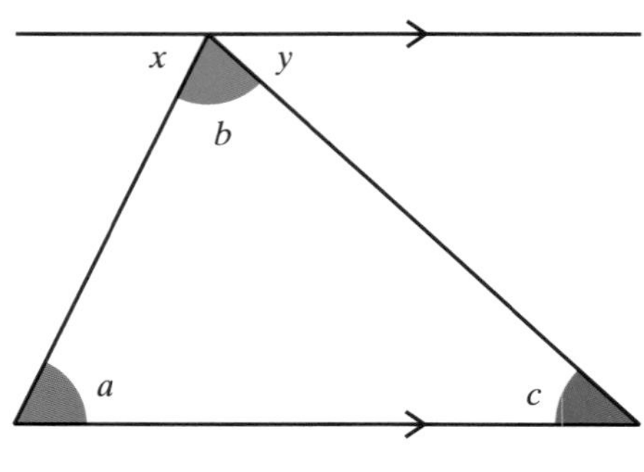

Angles x and a are alternate, so $x = a$
Likewise, angles y and c are alternate, so $y = c$

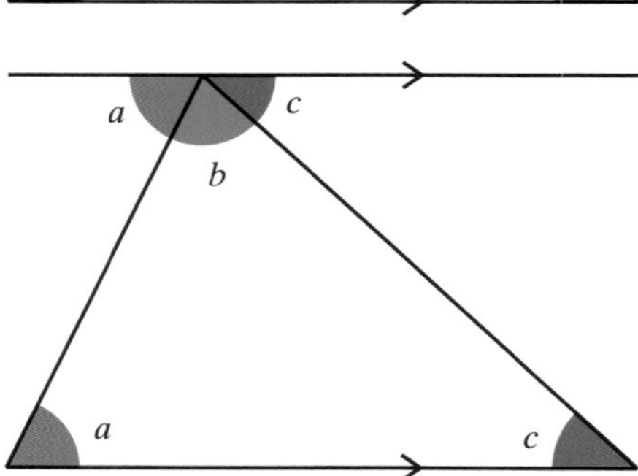

Now angles a, b and c form a straight line at the top of the diagram.

Therefore $a + b + c = 180°$

So the original angles in the triangle add up to 180°

Theorem

The exterior angle at the vertex of a triangle is equal to the sum of the interior opposite angles.

Proof

Construct a line through one vertex, parallel to the opposite side.

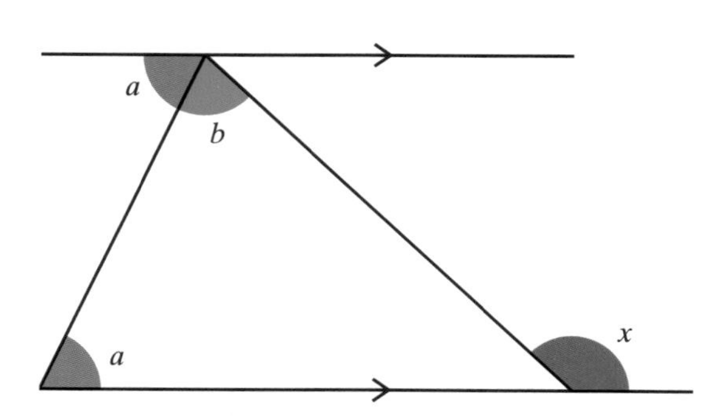

The angle alongside b is equal to a (alternate angles).

The angle x is alternate to the combined angle $a + b$

Therefore $x = a + b$

EXAMPLE

a) Form an equation in x
b) Solve your equation to find x

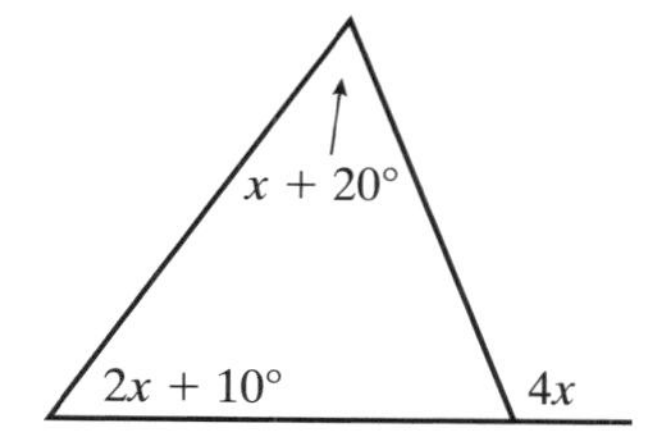

SOLUTION

a) Using the fact that the exterior angle at the vertex of a triangle is equal to the sum of the interior opposite angles:

$$4x = (2x + 10°) + (x + 20°)$$

Simplifying, $\underline{4x = 3x + 30°}$

b) Solve $4x = 3x + 30°$

$$\text{Solve } 4x = 3x + 30°$$

$-3x$ (both sides)

$$\underline{x = 30°}$$

Here are two more facts about equal angles.

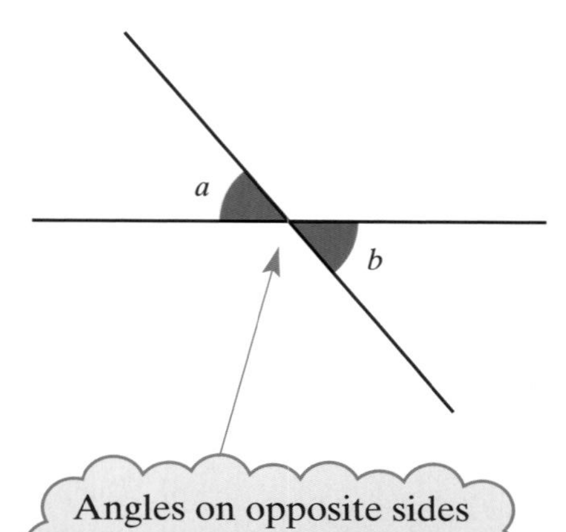

a
b
Angles on opposite sides of a vertex are equal. They are called **vertically opposite**.

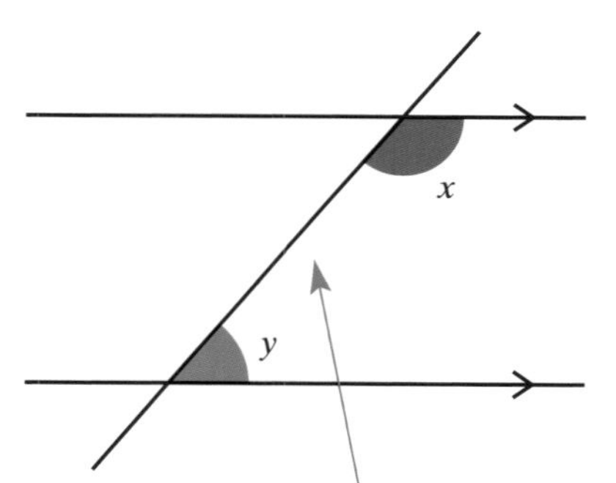

Angles inside two parallels add up to 180°
They are sometimes called **allied** angles or **co-interior** angles.

In this next exercise you may use any angle properties you know, including those about vertically opposite angles, allied angles, alternate angles and corresponding angles.

EXERCISE 13.3

Find the sizes of each of the angles represented by the letters in each diagram.

1

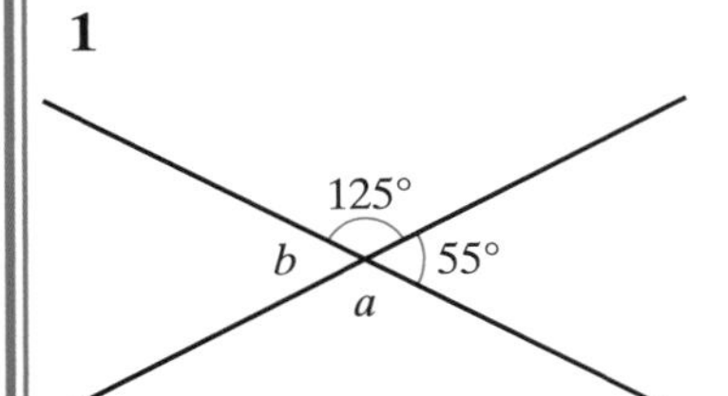

2

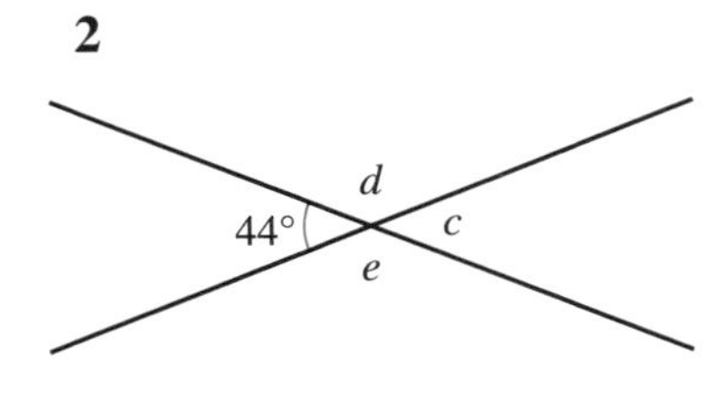

3

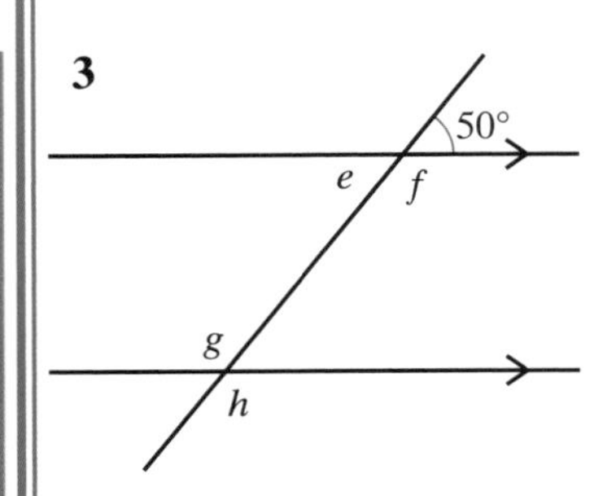

4

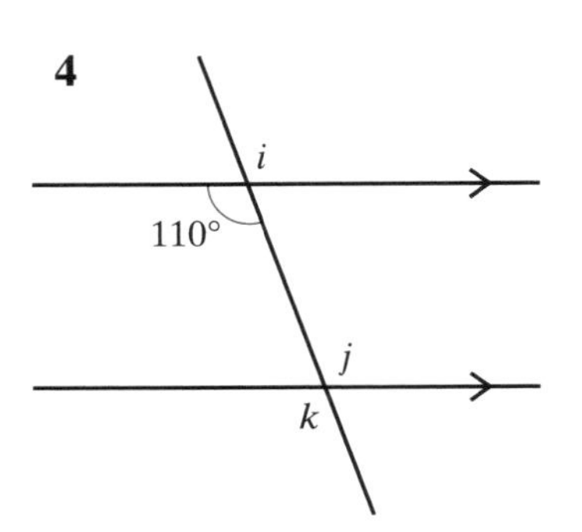

5

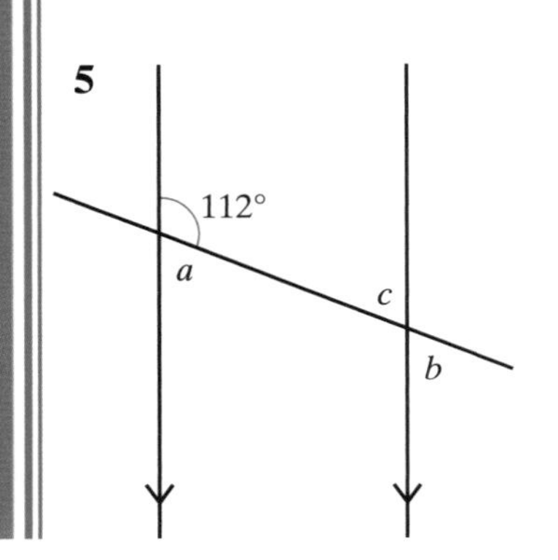

6

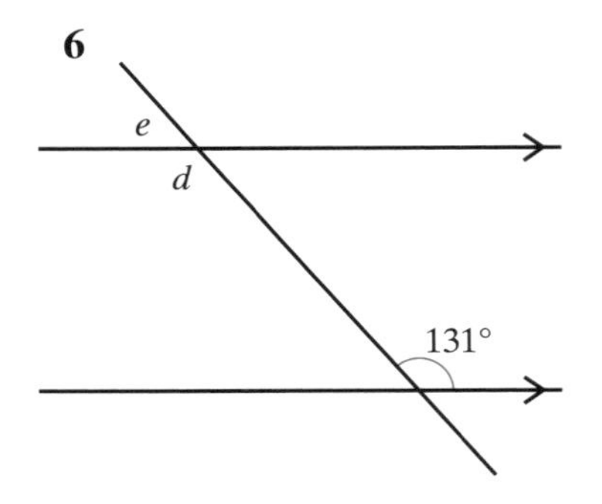

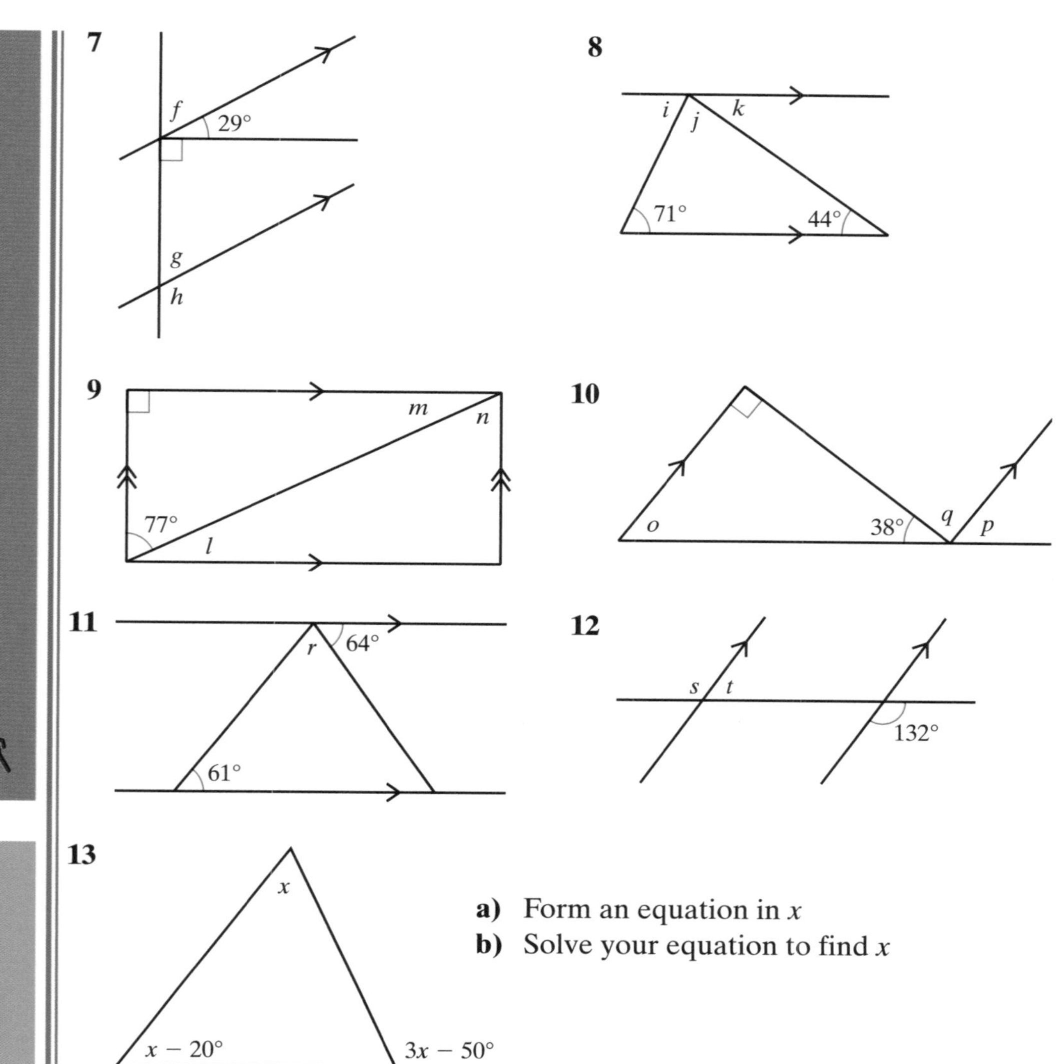

a) Form an equation in x

b) Solve your equation to find x

13.4 Angles in polygons

A polygon is a many-sided shape.

Here are the names of some polygons:

Name of polygon	**Number of sides**
Triangle	3
Quadrilateral	4
Pentagon	5
Hexagon	6
Heptagon	7
Octagon	8
Nonagon	9
Decagon	10

The angles in a triangle add up to 180° and those in a quadrilateral add up to 360°

For polygons with more sides, you just add on another 180° for each extra side.

The angles in a pentagon, for example, must add up to 360° + 180° = 540°

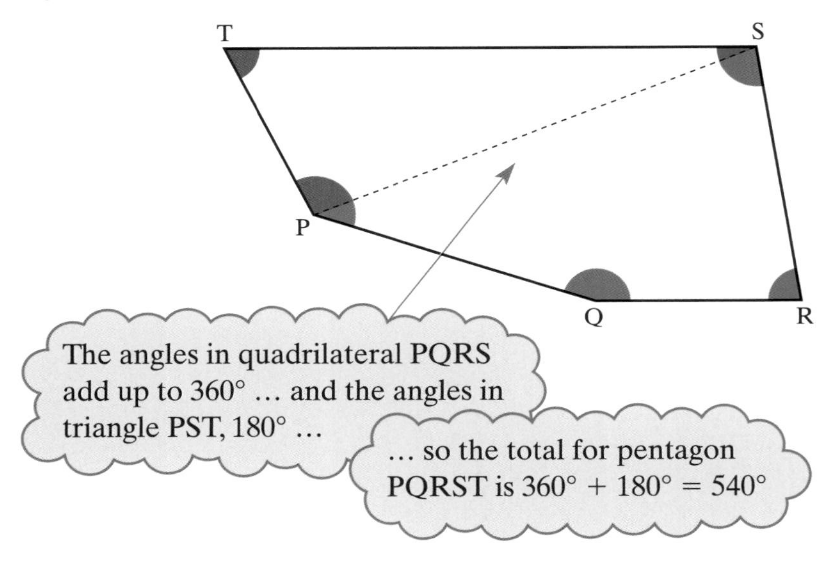

T
S
P
Q
R
The angles in quadrilateral PQRS add up to 360° … and the angles in triangle PST, 180° …
… so the total for pentagon PQRST is 360° + 180° = 540°

This can also be expressed as a formula:

The sum of the interior angles of a polygon with n sides = $180° \times (n - 2)$

EXAMPLE

Find the sum of all the interior angles of a hexagon.

SOLUTION

A hexagon has six sides, so use $n = 6$ in the formula:

$$\text{angle sum} = 180° \times (n - 2)$$

So $180° \times (6 - 2) = 180° \times 4$

$= \underline{720°}$

EXAMPLE

Six of the angles in a seven-sided polygon are:

$100°, 110°, 130°, 145°, 145°$ and $150°$

Find the size of the seventh angle.

SOLUTION

The angle sum for a seven-sided polygon is $180° \times (7 - 2) = 900°$
The given angles have a sum of:

$$100° + 110° + 130° + 145° + 145° + 150° = 780°$$

So the remaining angle is $900° - 780° = \underline{120°}$

If you were to travel all the way around the perimeter of a polygon, you would need to change direction at each corner, or **vertex**.

The angle by which you change direction is called the **exterior angle** at that vertex.

To show an exterior angle on a diagram, you need to extend each of the sides slightly, in the same sense (clockwise or anticlockwise) each time.

The diagram shows the exterior angles of a pentagon, with each side extended in a clockwise direction.

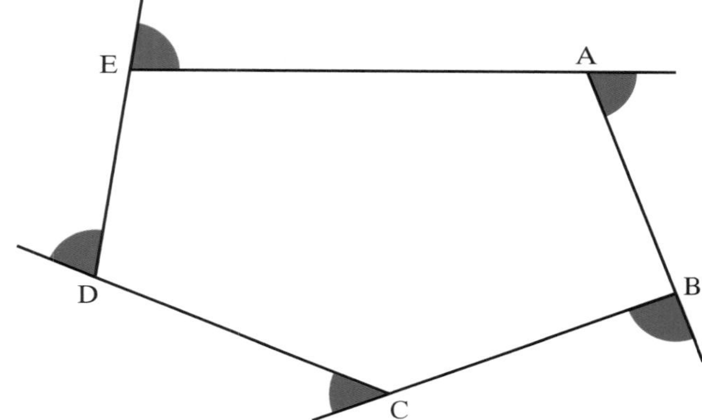

The sum of the exterior angles is simply the total angle you would turn through by travelling all the way around the perimeter, and this must be a complete turn – or 360°

> The sum of the exterior angles of any polygon = 360°

This only works for a regular polygon – when all the interior (or exterior) angles are the same size.

Regular polygons have all their sides the same length and all their angles equal.

> For a regular polygon with n sides, each exterior angle is $360° \div n$

The interior angle and exterior angle lie along a straight line, so:

> Interior angle = 180° – exterior angle

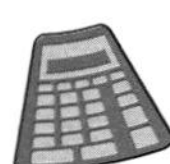

EXAMPLE

A regular polygon has 12 sides.

a) Calculate the size of each exterior angle.

b) Hence find the size of each interior angle.

SOLUTION

a) Exterior angle = $360° \div 12 = \underline{30°}$

b) Interior angle = $180° - 30° = \underline{150°}$

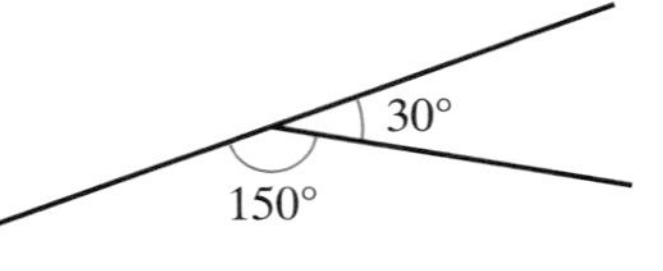

EXAMPLE

The diagram shows one vertex of a regular polygon with n sides.

Calculate the value of n

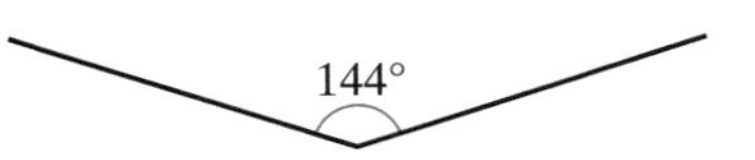

SOLUTION

Each exterior angle is $180° - 144° = 36°$

Number of sides is $360° \div 36° = 10$ sides

So $\underline{n = 10}$

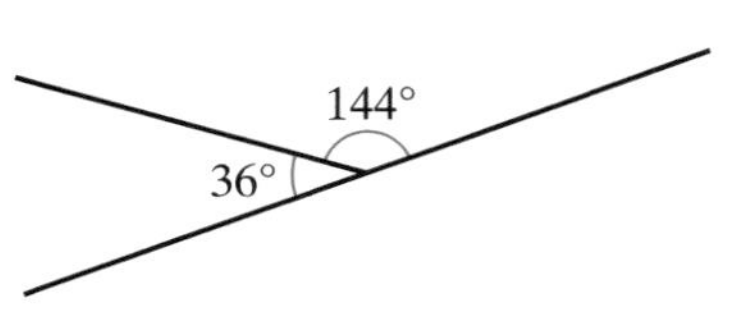

Some regular polygons fit together exactly and could be used to make tiles for a floor or a wall.

This is called **tessellating**.
A regular polygon will **tessellate** if its interior angle is a factor of 360

Regular hexagons tessellate because their interior angle is 120°

$$360° \div 120° = 3$$

So three hexagons meet around a point.

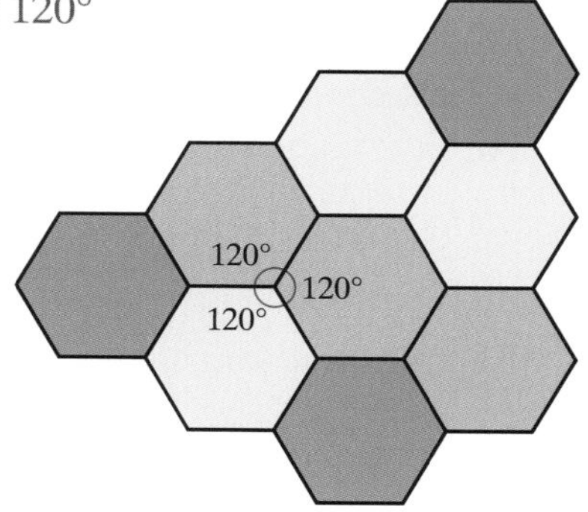

Regular pentagons don't tessellate because their interior angle is 108°

$$360° \div 108° = 3 \text{ remainder } 36$$

So when three pentagons meet around a point there is a gap.

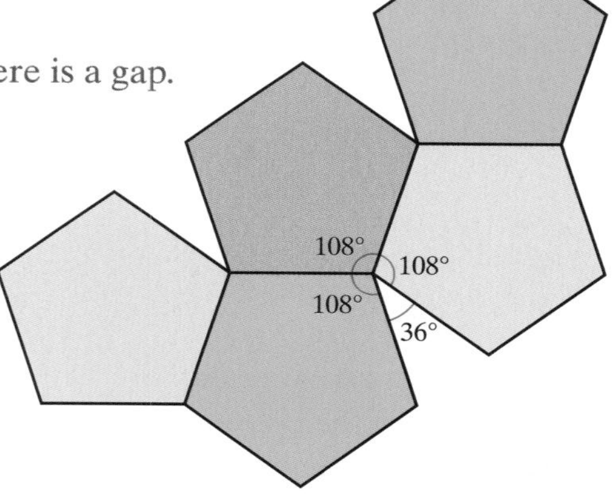

EXERCISE 13.4

1 a) Find the sum of the interior angle at each vertex of:
(i) a regular octagon **(ii)** a regular 20-sided polygon.

b) Find the size of each interior angle of:
(i) a regular octagon **(ii)** a regular 20-sided polygon.

c) Does: **(i)** a regular octagon **(ii)** a regular 20-sided polygon tessellate?
Give a reason for your answers.

2 a) Work out the size of the exterior angle at each vertex of:
(i) a regular hexagon **(ii)** a regular 15-sided polygon.

b) Work out the size of the interior angle at each vertex of:
(i) a regular hexagon **(ii)** a regular 15-sided polygon.

c) Does: **(i)** a regular hexagon **(ii)** a regular 15-sided polygon tessellate?
Give a reason for your answers.

3 Five of the angles in a hexagon are 102°, 103°, 118°, 125° and 130°
Find the sixth angle.

4 **a)** The diagram shows part of a regular polygon.

Work out how many sides the polygon has.

b) Albert draws this diagram.
He says it shows part of a regular polygon.

Explain how you can tell that Albert must have made a mistake.

5 The diagram shows an irregular pentagon.

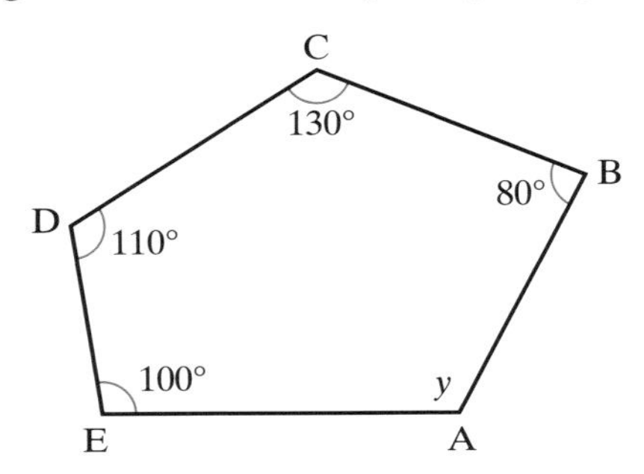

Work out the value of y

6 **a)** Write down an expression for the sum of the angles in the pentagon.
Simplify your expression.

b) Write down an equation for the sum of the angles in the pentagon.

c) Solve your equation to find the value of x

d) Hence work out the value of each angle.

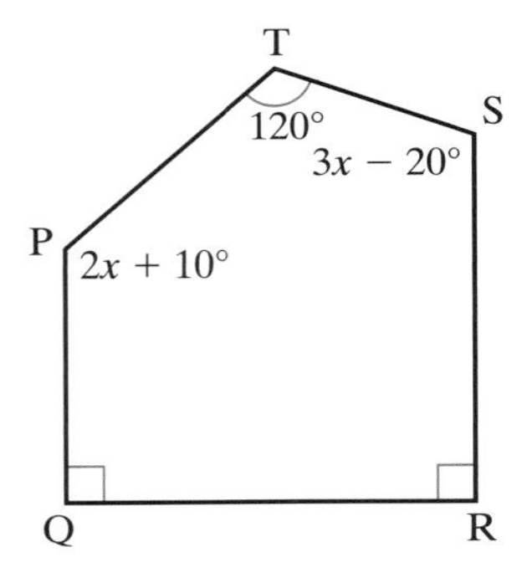

7 The diagram shows a hexagon.
All the angles marked a are equal.
Calculate the value of a.

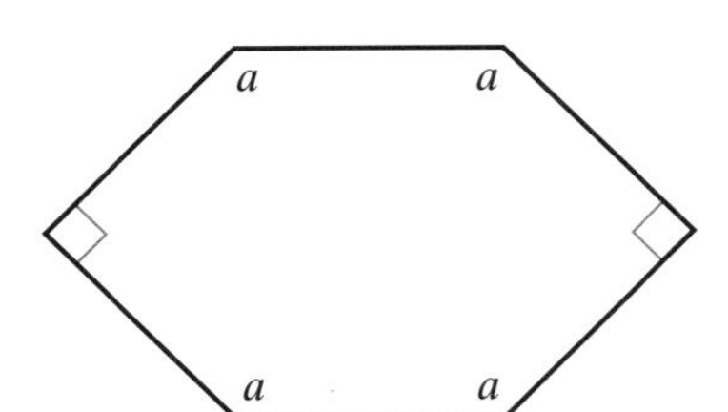

8 Follow these instructions to make an accurate drawing of a regular hexagon.

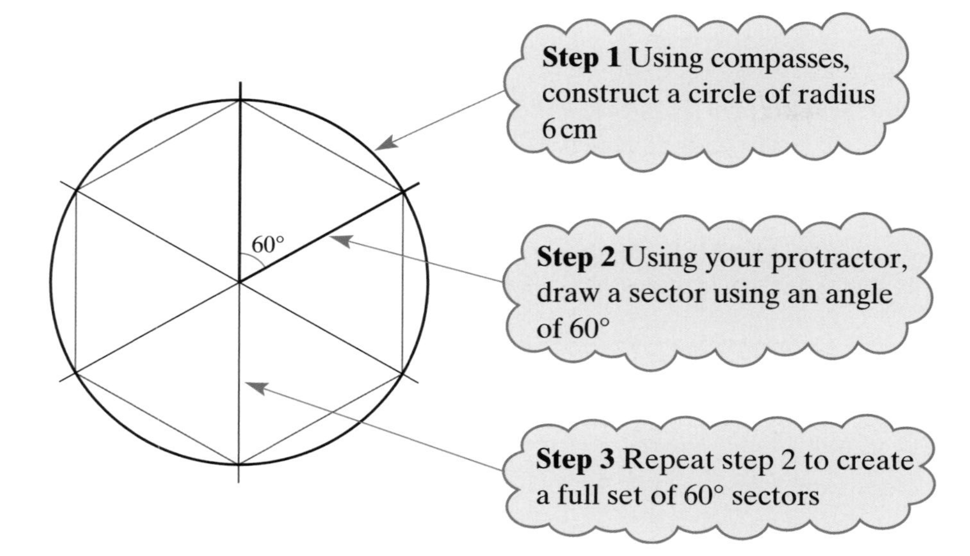

Step 1 Using compasses, construct a circle of radius 6 cm
60°
Step 2 Using your protractor, draw a sector using an angle of 60°
Step 3 Repeat step 2 to create a full set of 60° sectors

Complete the construction by joining the six points around the circumference of the circle.

9 Adapt the instructions from Question **8** to make:

a) a regular octagon

b) a regular 9-sided polygon

c) a 9-pointed star.

13.5 Bearings

In a straight line.

To get to Newtown from Oldtown 'as the crow flies' you need to know the distance between the towns and the direction in which to travel.

Newtown •

• Oldtown

This direction is given as a **three-figure bearing,** such as 125° or 062°

A bearing is an angle measured in a *clockwise* direction from *North*.

EXAMPLE

a) Measure the bearing of Newtown from Oldtown.
b) Use your answer to calculate the bearing of Oldtown from Newtown.

SOLUTION

a) **Step 1** Draw a **North line** at the place after the word 'from', that is, Oldtown.

Step 2 Draw a **line** linking the two places.

Step 3 Measure the angle made in a clockwise direction from the North line.

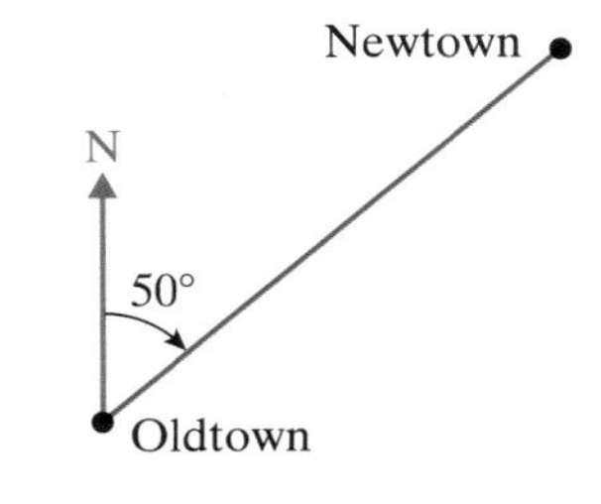
Newtown
N
50°
Oldtown

So the bearing of Newtown from Oldtown is <u>050°</u>

When the angle is less than 100° write a '0' in front so the bearing has three figures.

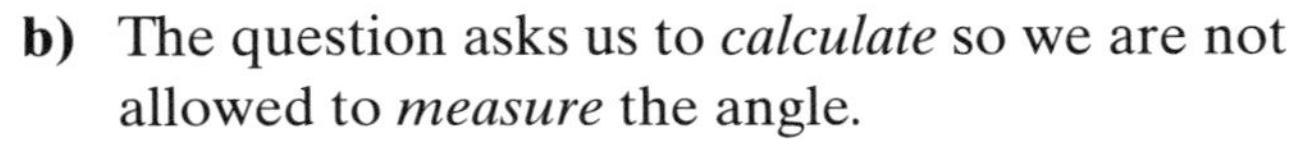

b) The question asks us to *calculate* so we are not allowed to *measure* the angle.

Step 1 Draw a **North line** at the place after the word 'from', that is, Newtown.

Step 2 Continue the **line** linking the two towns beyond Newtown.

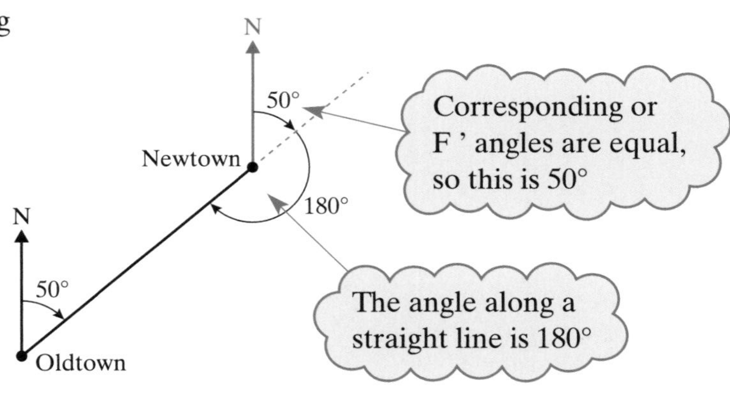
N
50°
Newtown
180°
N
50°
Oldtown
Corresponding or F' angles are equal, so this is 50°
The angle along a straight line is 180°

The total angle is 50° + 180° = 230°

So the bearing of Oldtown from Newtown is <u>230°</u>

You will find it helpful to know the compass points:

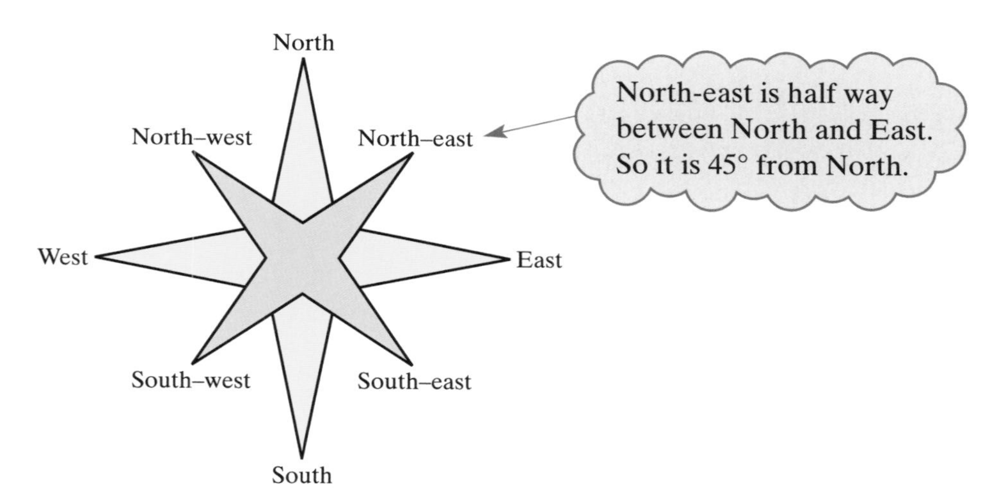

EXERCISE 13.5

1 Measure the bearings of each of the following towns from Newtown.

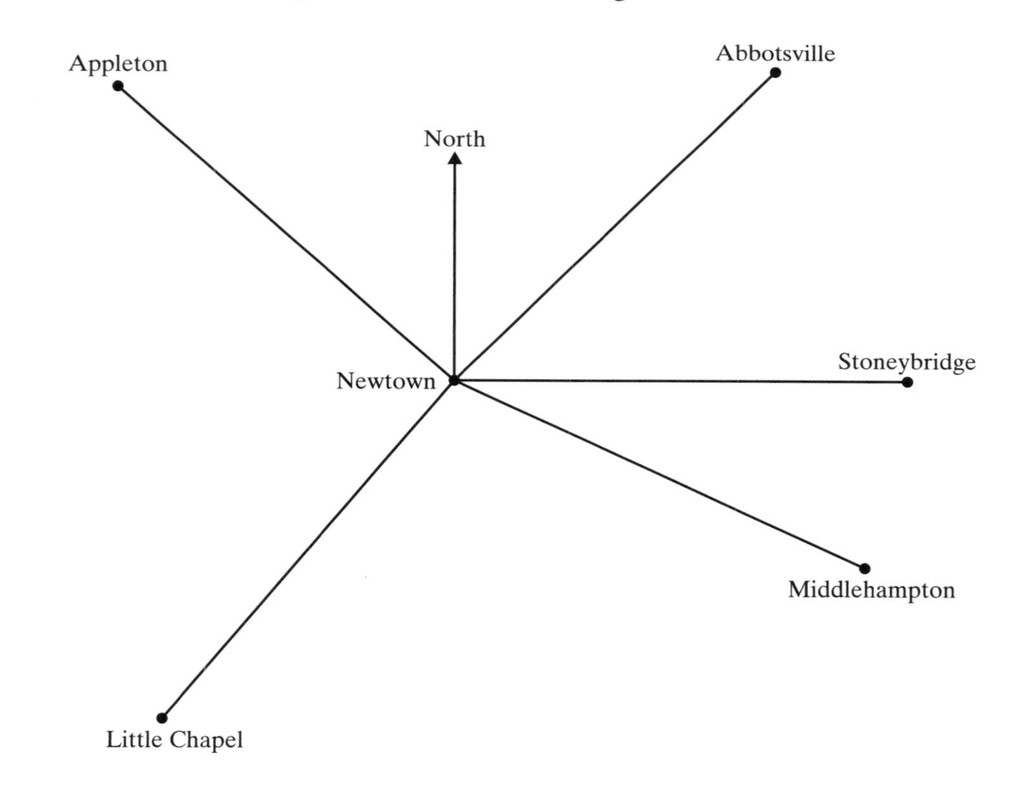

2 Write down the bearing on which a plane flies when it is flying:
a) due East **b)** due West **c)** South-east **d)** North-west.

3 Eastfarthing is 8 km away from Hampton on a bearing of 080°
Make a scale drawing of the two towns.
Use a scale of 1 cm to 1 km

4 A ship is 14.5 km away from a lighthouse on a bearing of 230°

a) Make a scale drawing of the ship and the lighthouse.
Use a scale of 1 cm to 1 km

A ferry is 10 km away from the ship on a bearing of 035°

b) Add the position of the ferry to your scale drawing.

c) Measure the bearing of the ferry from the lighthouse.

5 Look at this sketch – it is not drawn to scale.

a) Write down the bearing of B from A.

b) Work out the bearing of A from B.

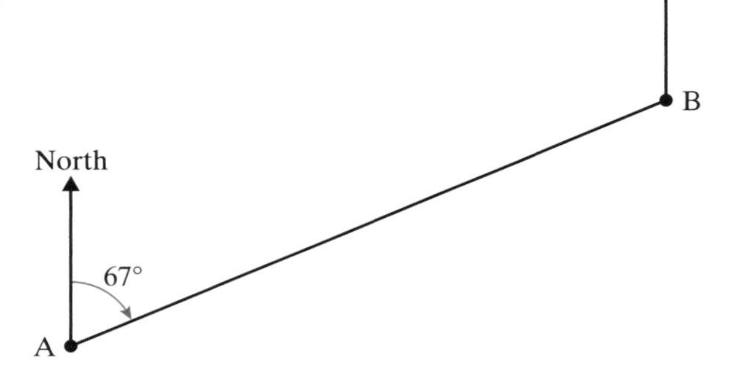

6 Look at this sketch – it is not drawn to scale.

a) Work out the bearing of R from P.

b) Work out the bearing of P from R.

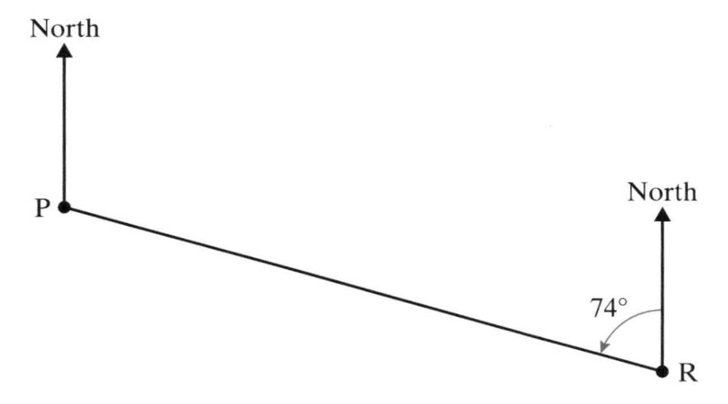

7 Otterbourne is South-west of Queensmead.

a) What is the bearing of Otterbourne from Queensmead?

b) What is the bearing of Queensmead from Otterbourne?

8 A helicopter and a plane are flying at the same altitude. ← height
The bearing of the helicopter from the plane is 147°
What is the bearing of the plane from the helicopter?

9 Queensworthy is on a bearing of 032° from Abbotsworthy.
What is the bearing of Abbotsworthy from Queensworthy?

10 This sketch shows the position of three boats A, B and C.
It is not a scale drawing.

a) Write down the bearing of B from A.

b) Work out the bearing of:

(i) A from B
(ii) C from B
(iii) B from C
(iv) A from C
(v) C from A.

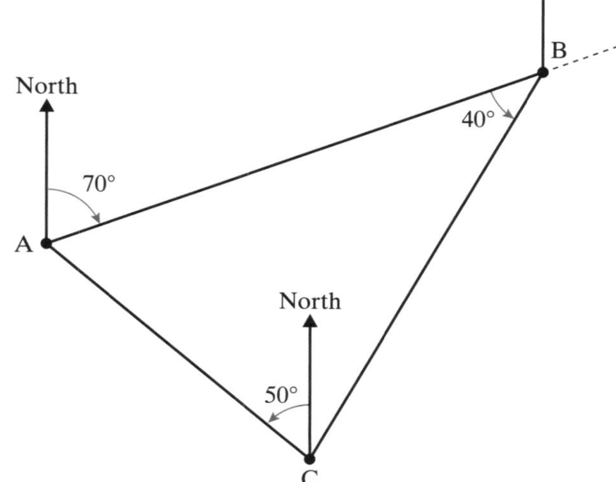

REVIEW EXERCISE 13

Don't use your calculator for Questions 1–14.

1 Work out the values of the missing letters in each of these angle diagrams.

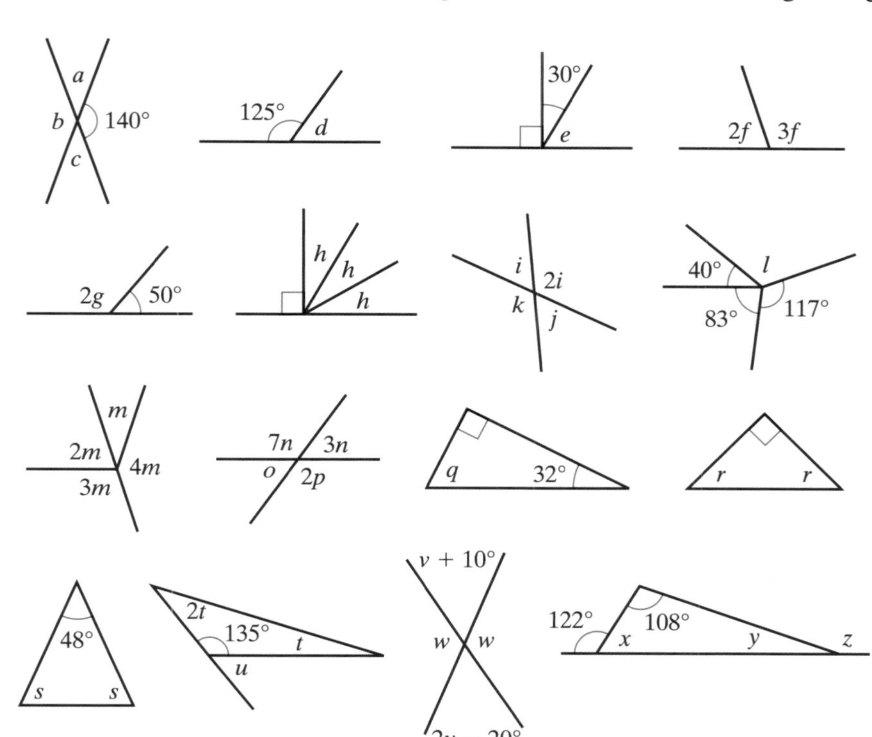

2 Find the values of angles a and b, indicated in the diagram.

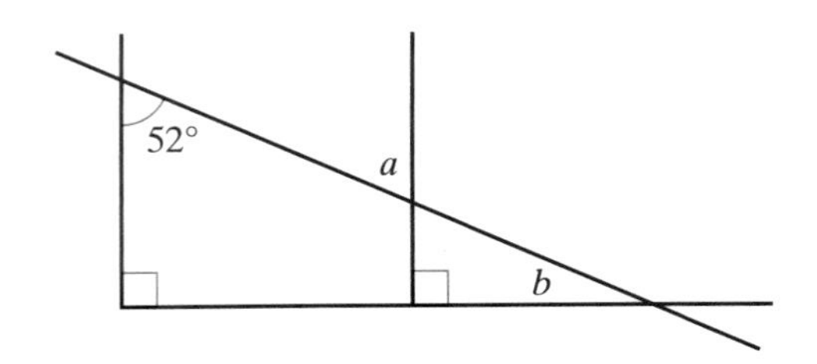

3 The bearing of town A from town B is 118°
What is the bearing of town B from town A?

4 Some rocks, R, are 500 m away from a lighthouse, L, on a bearing of 065°
a) What is the bearing of the lighthouse from the rocks?
b) Make a scale drawing of the rocks and lighthouse.
Use a scale of 1 cm to 50 m

A ship, S, is 600 m away from the lighthouse on a bearing of 120°
c) Add the ship to your scale drawing.
d) Measure the bearing of the ship from the rocks.

5 The angles in a quadrilateral are $4y - 10°$, $y + 40°$, $3y + 20°$ and $2y + 10°$ in order as you go around the quadrilateral.
a) Write down an expression for the sum of the angles in the quadrilateral.
Simplify your expression.
b) Set up and solve an equation to find the value of y
c) Hence work out the value of each angle.
d) What type of quadrilateral is this?

6 **a)** Work out the size of each exterior angle of a regular 12-sided polygon.
b) Find the size of each interior angle.
c) Will a regular 12-sided polygon tessellate?

7 The angles inside a certain polygon add up to 1980°
How many sides has the polygon?

8 A regular polygon has interior angles of size 176°
How many sides has the polygon?

9 The lines in the diagram opposite are straight.
Copy the diagram.
a) Mark with arrows (>) a pair of parallel lines.
b) Mark with a letter R, a right angle.
c) What type of angle is shown by the letter:
(i) x **(ii)** y? **[Edexcel]**

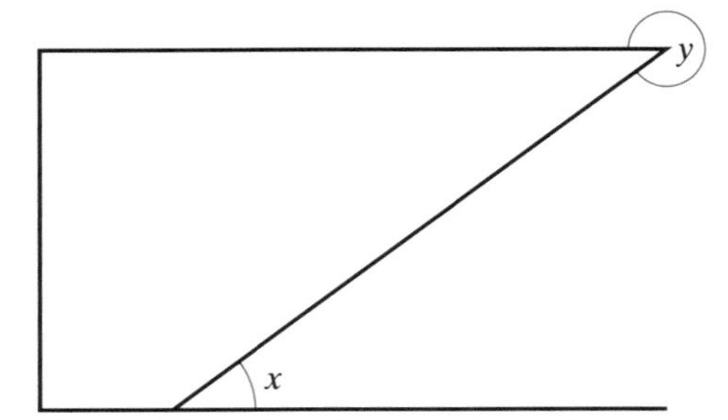

10 The diagram shows a 5-sided shape.
All the sides of the shape are equal in length.

a) Find the value of x
Give a reason for your answer.

b) Work out the value of y

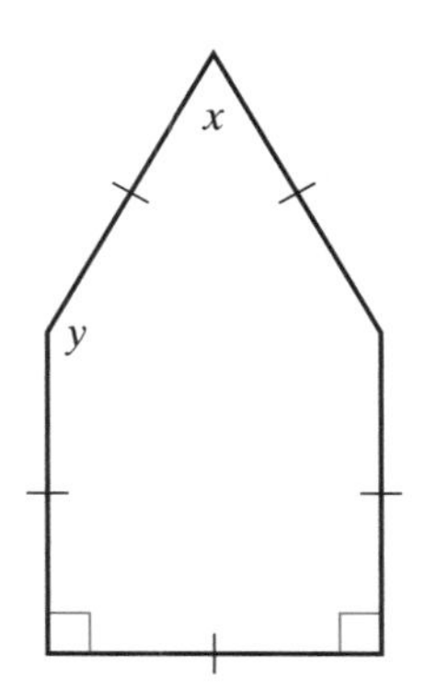

Diagram *not* accurately drawn

[Edexcel]

11 The diagram shows a shape.
The shape is a 6-sided polygon.

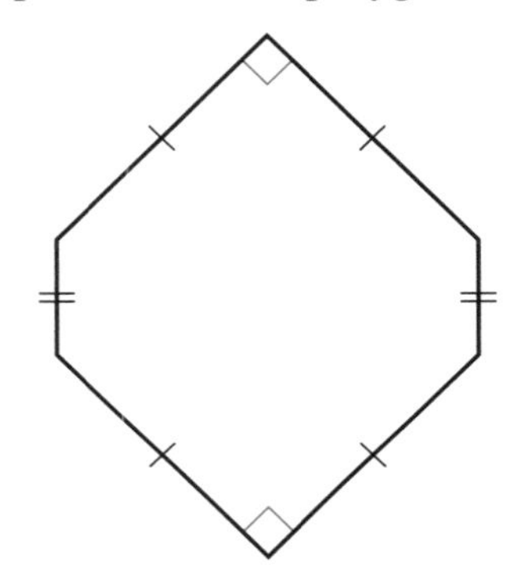

Diagram *not* accurately drawn

a) Write down the name for a 6-sided polygon.

The diagram below shows how the shape tessellates.

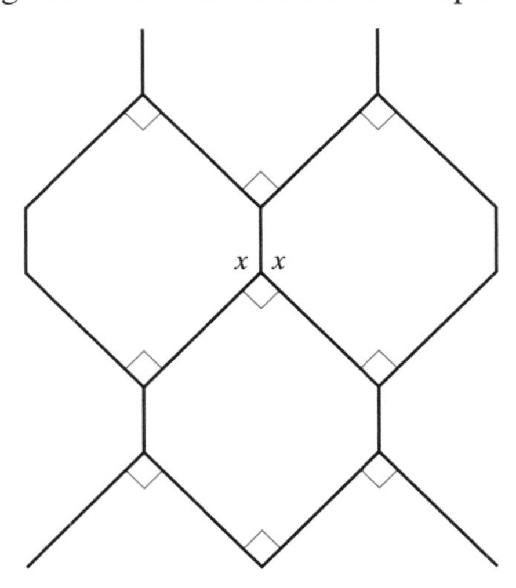

Diagram *not* accurately drawn

The size of each of the angles marked x is 135°

b) Give reasons why.

[Edexcel]

12

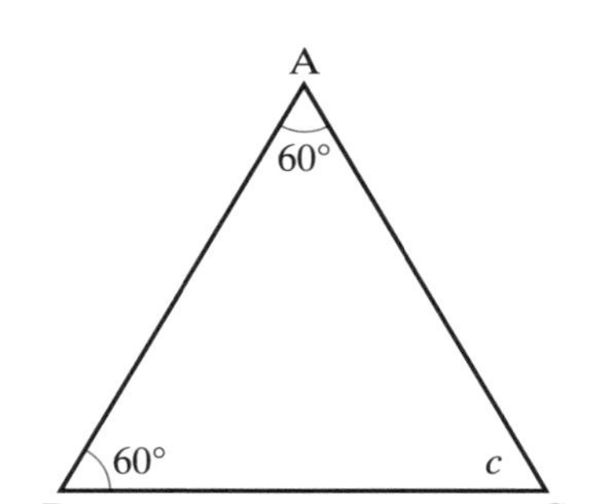

Diagram *not* accurately drawn

a) **(i)** Find the size of angle c

(ii) Triangle ABC is equilateral. Explain why.

b)

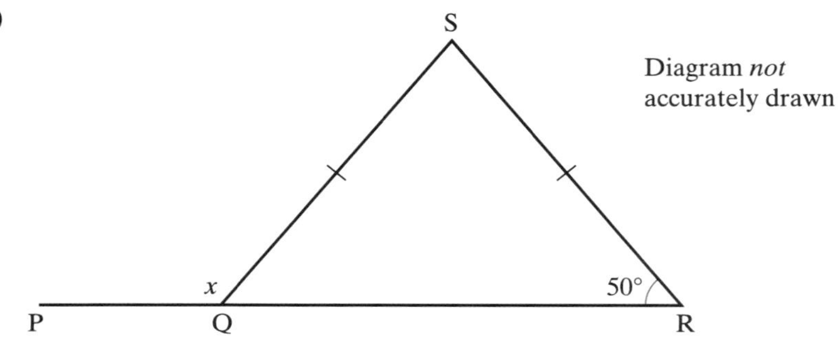

Diagram *not* accurately drawn

PQR is a straight line, and SQ = QR

(i) Work out the size of the angle marked x

(ii) Give reasons for your answer.

c)

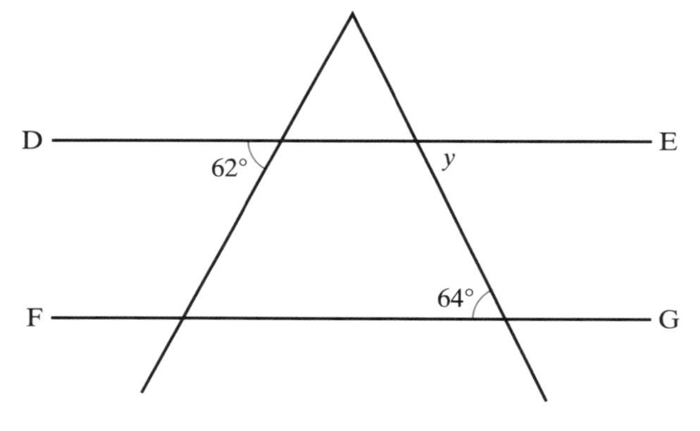

Diagram *not* accurately drawn

DE is parallel to FG.
Find the size of the angle marked y

[Edexcel]

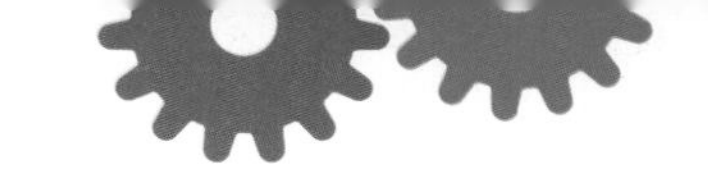

13 The crosses on the diagram show the positions of three places A, B and C.
The scale of the diagram is 1 cm to 10 km
Tariq cycled in a straight line from A to C.
He left A at 1.30 pm
He cycled at an average speed of 10 kilometres per hour.

a) Find the time he arrived at C.

b) Find the bearing of:
(i) B from A; **(ii)** A from C. [Edexcel]

14 The grid represents part of a map.

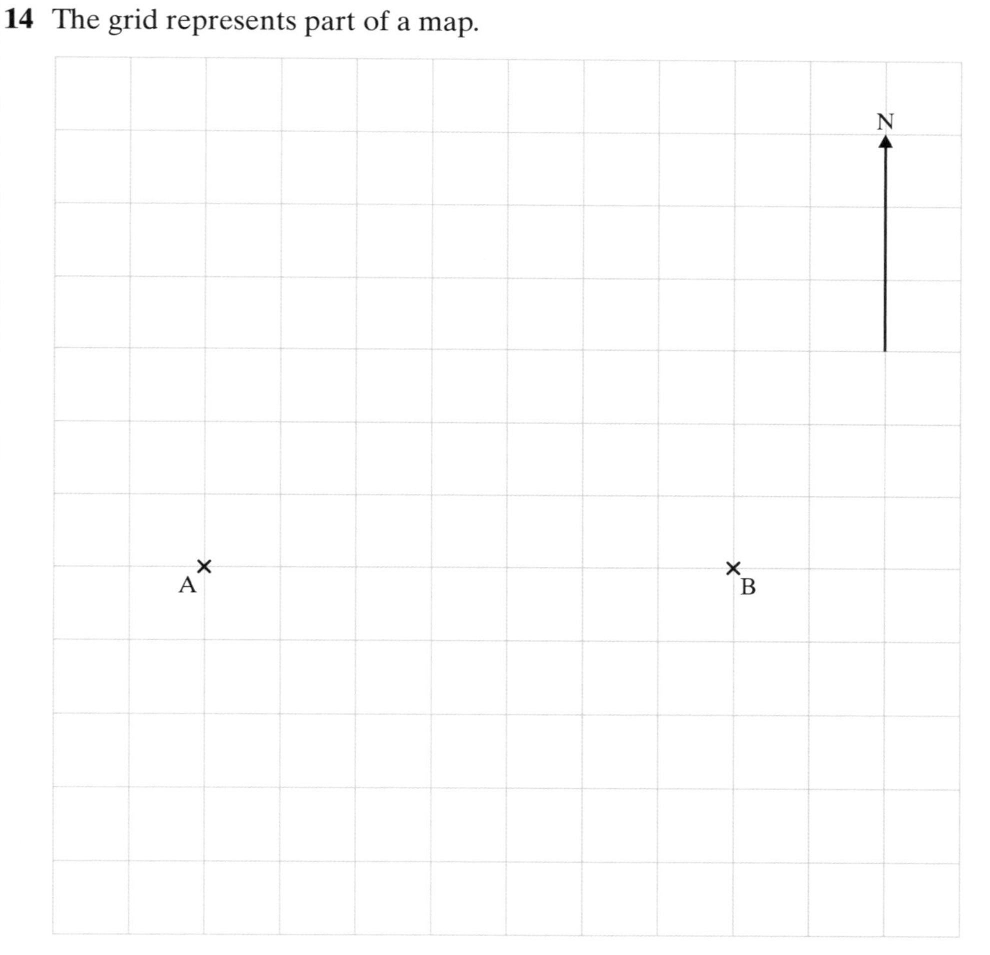

a) On a copy of the grid, draw a line on a bearing of 037° from the point marked A.

The point C is on a bearing of 300° from the point marked B.
C is also 3 cm from B.

b) Mark the position of the point C on the grid and label it with the letter C. [Edexcel]

You can use your calculator for Questions 15–25.

15 a) Write down the special name for this type of angle:

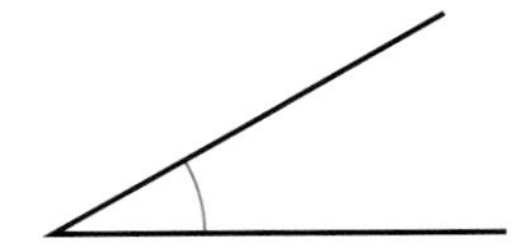

b) Write down the special name for this type of angle:

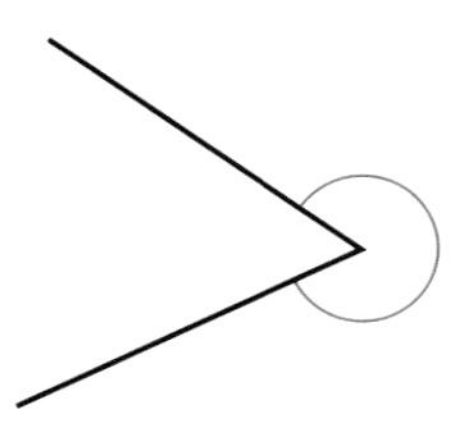

c) This diagram is wrong.
Explain why.

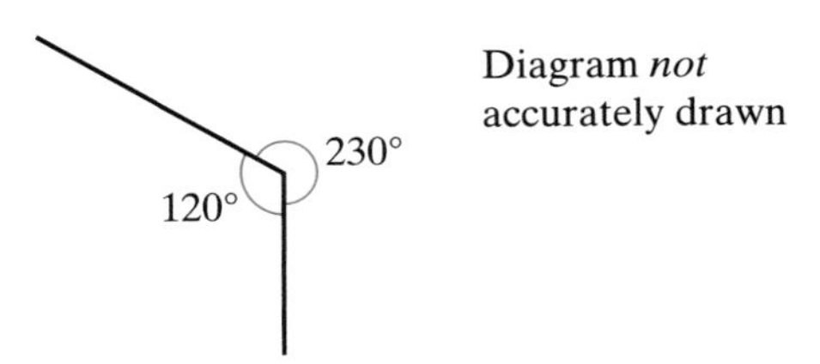

[Edexcel]

16 The diagram shows a triangle drawn on a grid of centimetre squares.

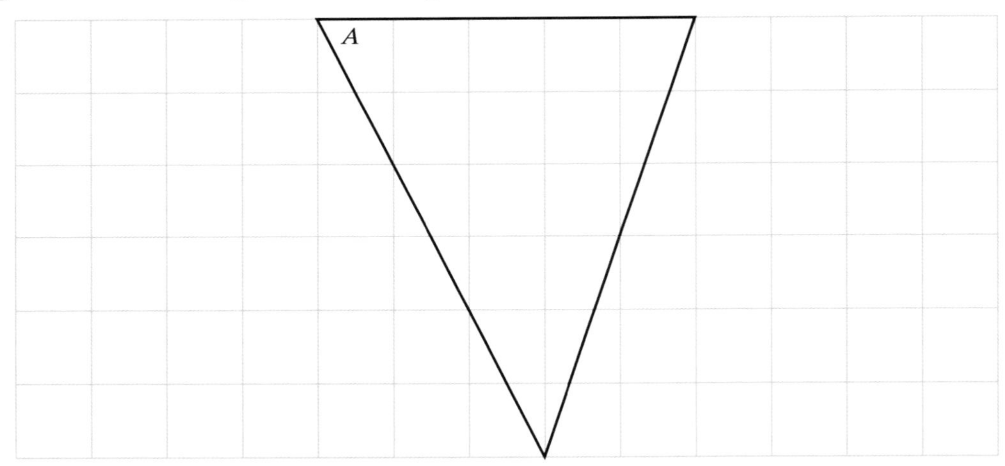

a) What type of triangle is this?
b) Measure the size of the angle marked with the letter *A*.
c) What type of angle have you measured?

[Edexcel]

17 PQ is a straight line.

a) Work out the size of the angle marked x

b) Work out the size of the angle marked y

Give reasons for your answers.

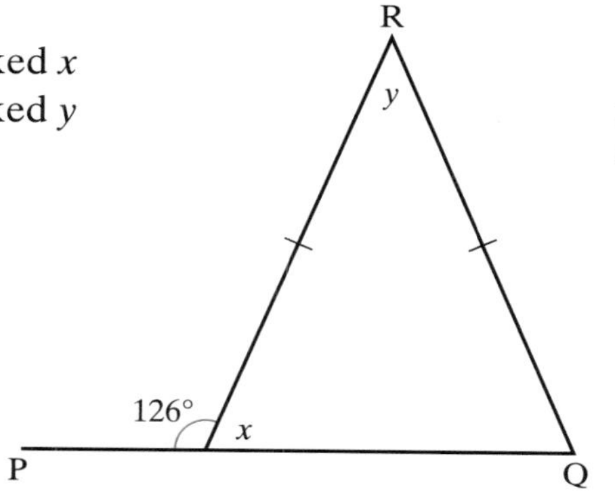

[Edexcel]

18 PQRS is a parallelogram.

Angle QSP = 47°

Angle QSR = 24°

PST is a straight line.

a) **(i)** Find the size of the angle marked x

(i) Give a reason for your answer.

b) **(i)** Work out the size of angle PQS.

(ii) Give a reason for your answer.

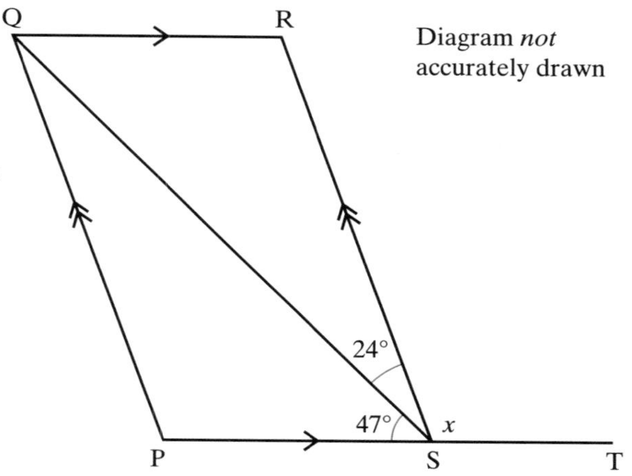

[Edexcel]

19 ABC and EBD are straight lines.

BD = BC.

Angle CBD = 42°

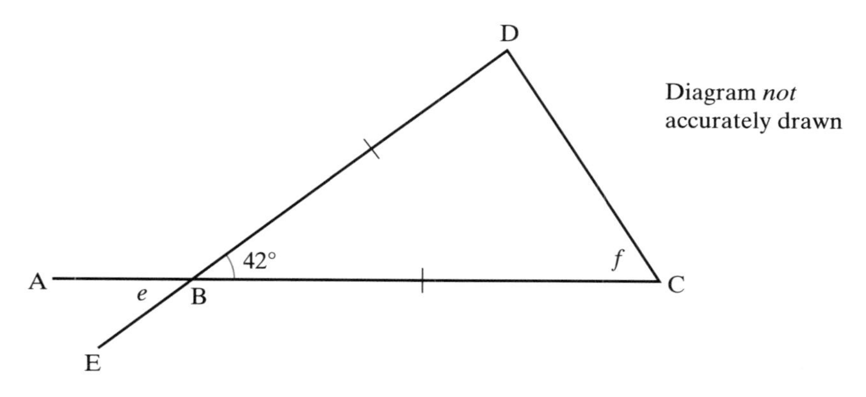

a) Write down the size of the angle marked e

b) Write down the size of the angle marked f

[Edexcel]

20 In this diagram, the lines AB and CD are parallel.
CRQ is a straight line.
Angle CRS = 94°
Angle QRB = 56°
Angle CSR = $x°$
Find the value of x

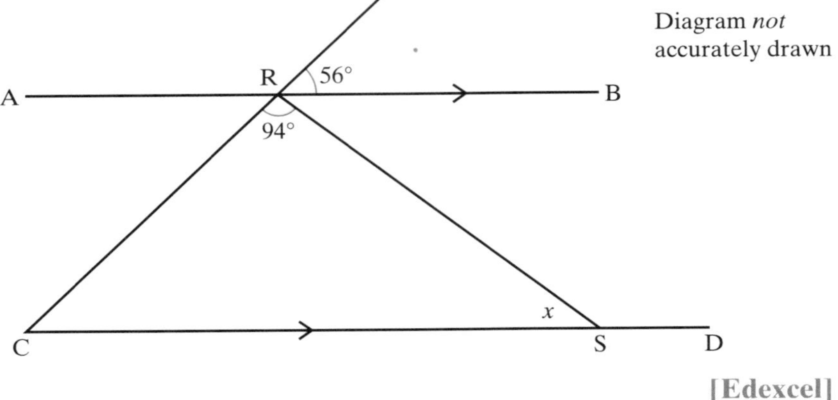

[Edexcel]

21 This is part of the design of a pattern found at the theatre of Diana at Alexandria.
It is made up of a regular hexagon, squares and equilateral triangles.

a) Write down the size of the angle marked x

b) Work out the size of the angle marked y

The area of each equilateral triangle is 2 cm^2

c) Work out the area of the regular hexagon.

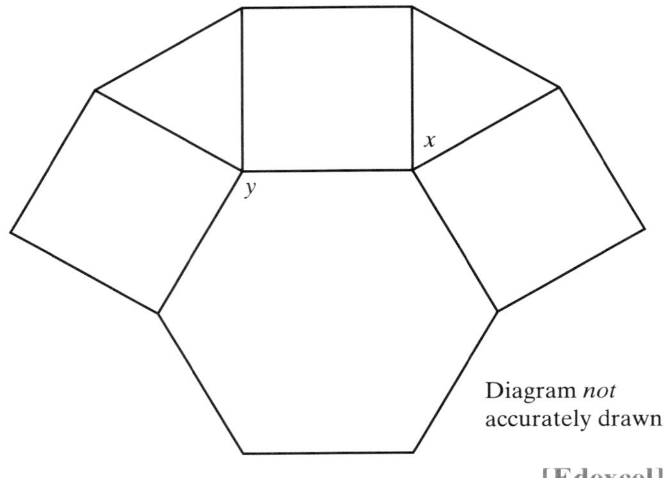

[Edexcel]

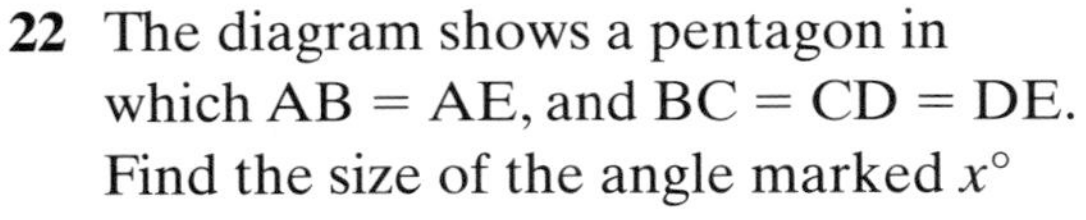

22 The diagram shows a pentagon in which AB = AE, and BC = CD = DE.
Find the size of the angle marked $x°$

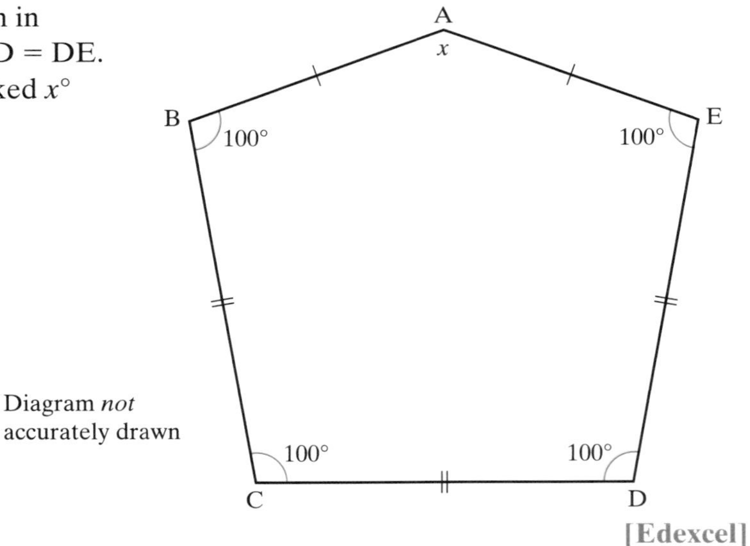

[Edexcel]

23 Triangle ABC is isosceles, with AC = BC.
Angle ACD = 62°
BCD is a straight line.

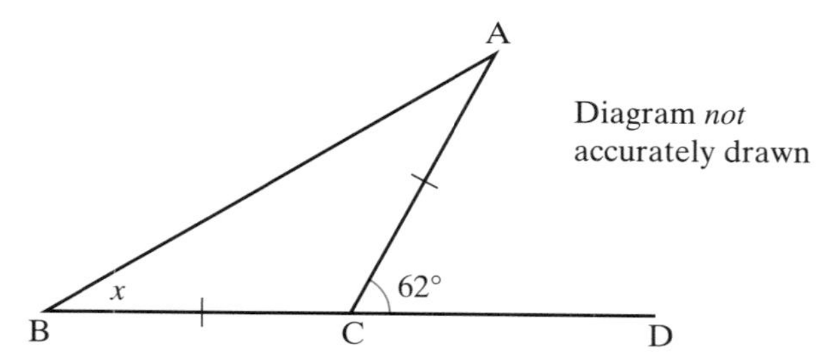

a) Work out the size of angle x

b) The diagram below shows part of a regular octagon.

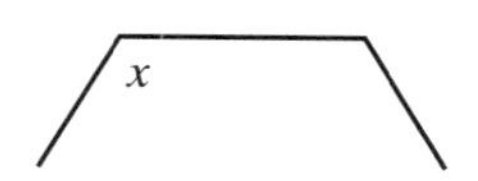

Diagram *not* accurately drawn

Work out the size of angle x

[Edexcel]

24 The diagram shows a regular hexagon.

a) Work out the value of x

b) Work out the value of y

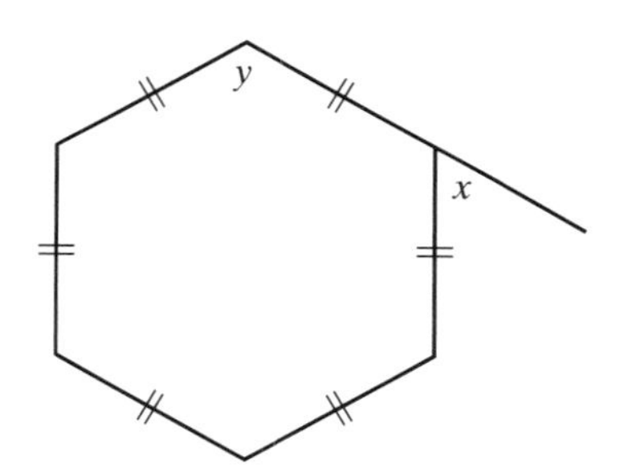

Diagram *not* accurately drawn

[Edexcel]

25 ABCD is a quadrilateral.
Work out the size of the largest angle in the quadrilateral.

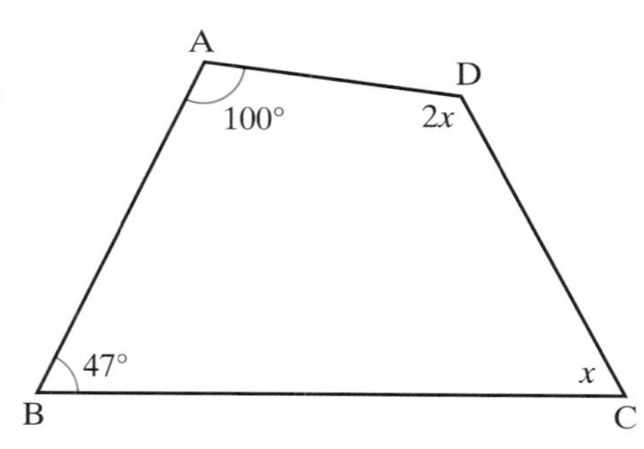

Diagram *not* accurately drawn

[Edexcel]

KEY POINTS

1. An acute angle is smaller than 90°
2. An obtuse angle is larger than 90° and smaller than 180°
3. A reflex angle is more than 180°
4. A right angle is 90°
5. Angles on a straight line add up to 180°
6. Angles around a point add up to 360°
7. Angles in a triangle add up to 180°
8. Angles in a quadrilateral add up to 360°
9. Vertically opposite angles are equal:

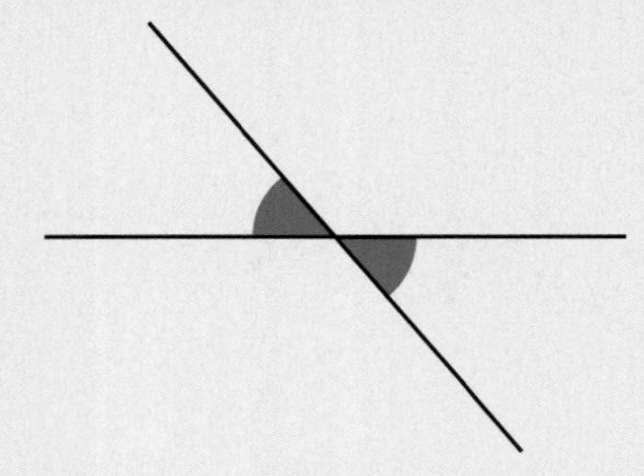

10. Angles inside two parallels add up to 180°:

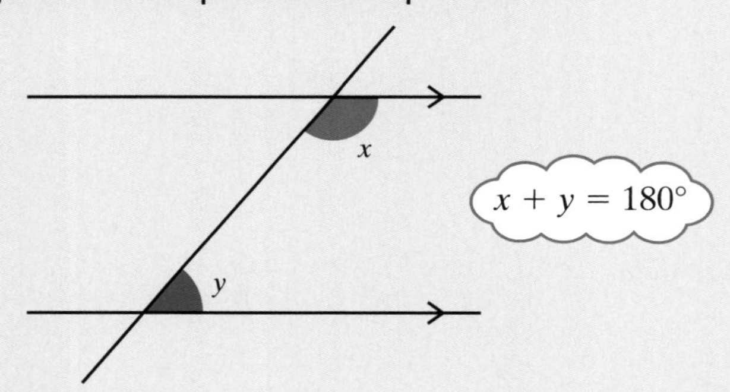

11. Alternate angles are equal.

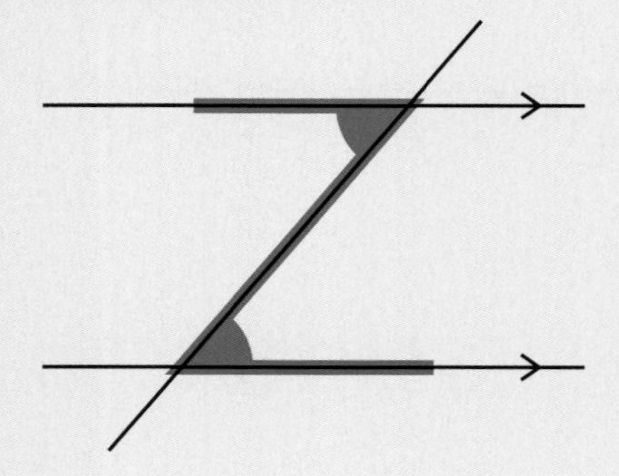

12 Corresponding angles are equal.

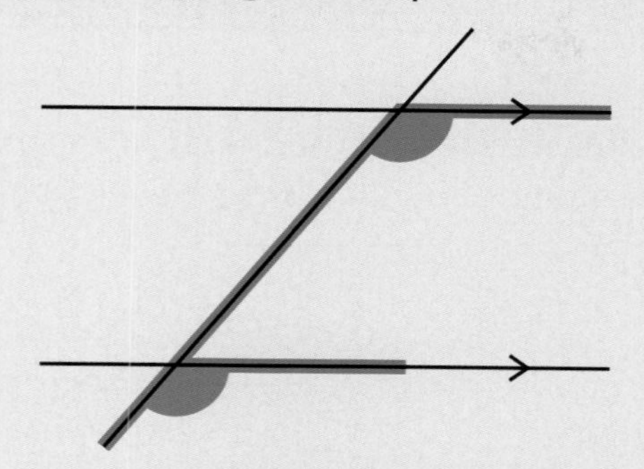

13 For a polygon with n sides, the interior angles will sum to $180° \times (n - 2)°$

14 The exterior angles of any polygon add up to 360°

15 A regular polygon has equal sides and equal angles.

16 Each exterior angle of an n-sided regular polygon is $360° \div n$

17 An interior angle in a polygon $= 180° -$ the exterior angle

18 Bearings are measured clockwise from North.
Bearings are always three figures, for example 043° or 196°

Internet Challenge 13

The four-colour theorem

The four-colour theorem claims that four colours are sufficient to colour in a map, in such a way that no two regions share the same colour along a boundary (except at a point).

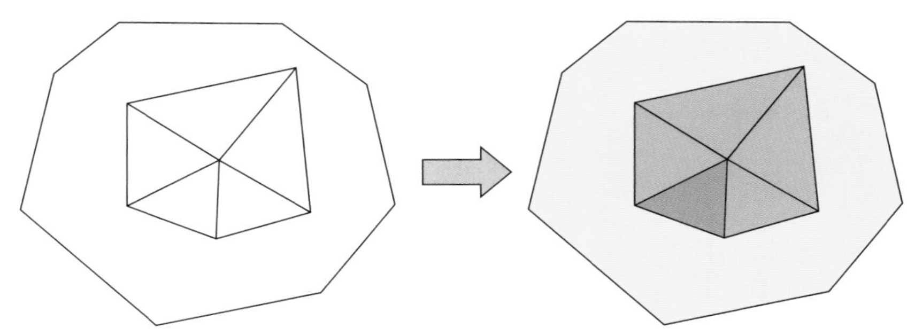

Try drawing some maps of your own, and colouring them in.

Does the four-colour theorem seem to be true?

Now use the internet to help answer these questions.

1 Who first proposed this theorem, in 1852?

2 Who presented a flawed proof in 1879?

3 Who presented another flawed proof in 1880?

4 When was the four-colour theorem first successfully proved?

5 Who achieved this first successful proof?

6 What major innovation was used to support the proof?

7 Suppose a map is drawn on a sphere instead of a plane.
How many colours are sufficient now?

8 Mathematicians refer to a 3-D ring doughnut shape as a *torus*.
How many colours are sufficient to colour in any map on a torus?

9 What April Fool's joke concerning the four-colour theorem was perpetrated by the mathematician and puzzler Martin Gardner in 1975?

10 Why might a real map-maker need to use more than four colours?
(Clue: What is unusual about Alaska?)

CHAPTER 14

2-D and 3-D shapes

In this chapter you will **revise earlier work on**:

- perimeter and area.

You will **learn how to**:

- find the perimeter of a shape
- find areas of triangles and quadrilaterals
- calculate surface areas and volumes of solids.

You will be **challenged to**:

- investigate triangles.

Starter: Inside and all round

You can find the area of these shapes by counting the number of squares in each one.

1 Find the area and perimeter of the following shapes.

Each shape is drawn on 1 cm^2 paper.

The perimeter is the distance all the way around the outside edge.

A

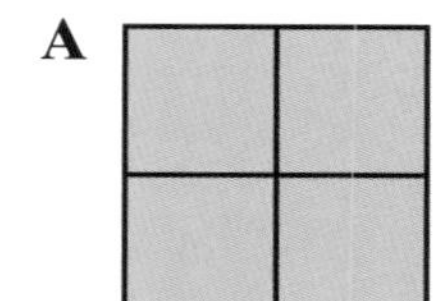

B

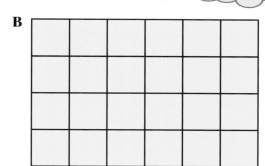

C

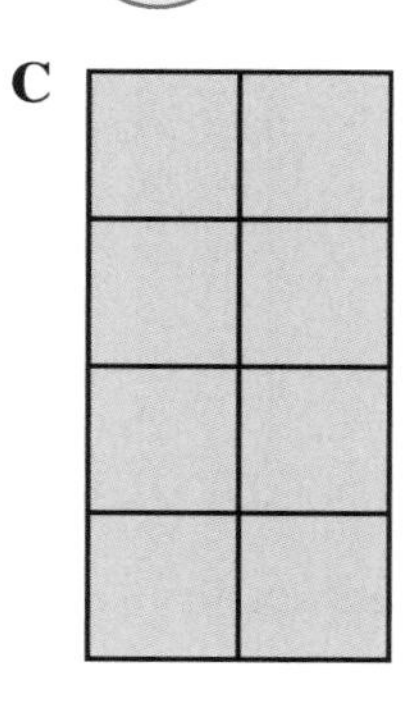

D

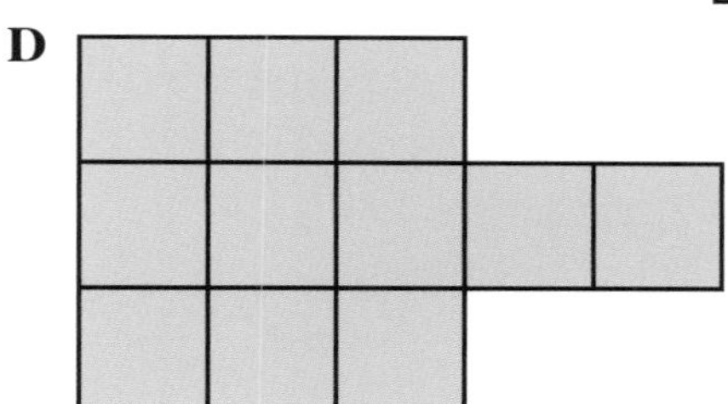

E

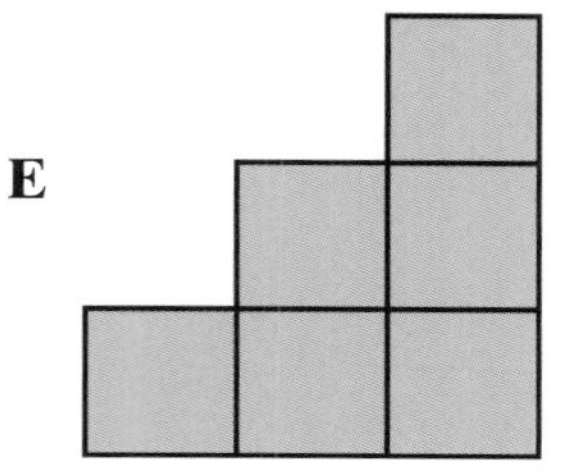

2 Find the area of these shapes.

A **B** **C**

3 Use a 3 by 3 pin board and an elastic band to make as many different triangles as you can.

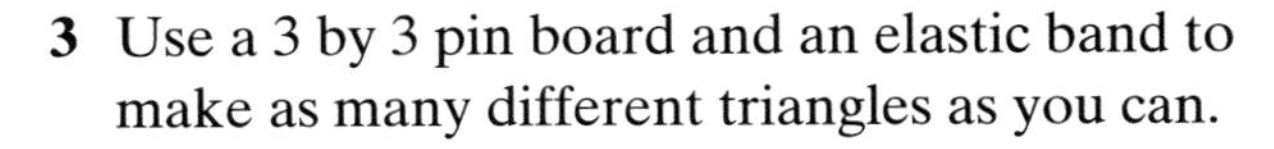

Draw a sketch of each triangle and write down what type of triangle it is.

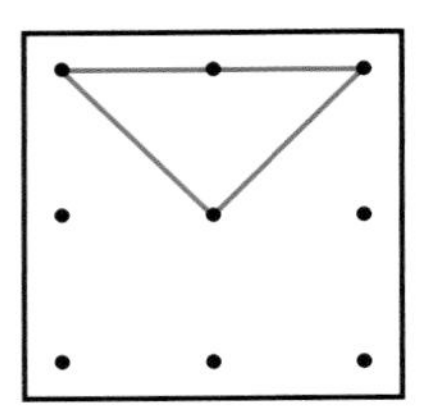

4 Use a 3 by 3 pin board and an elastic band to make as many different quadrilaterals as you can.
Draw a sketch of each quadrilateral.

14.1 Special quadrilaterals

Here is a reminder of the special quadrilaterals, and their geometric properties.

Square

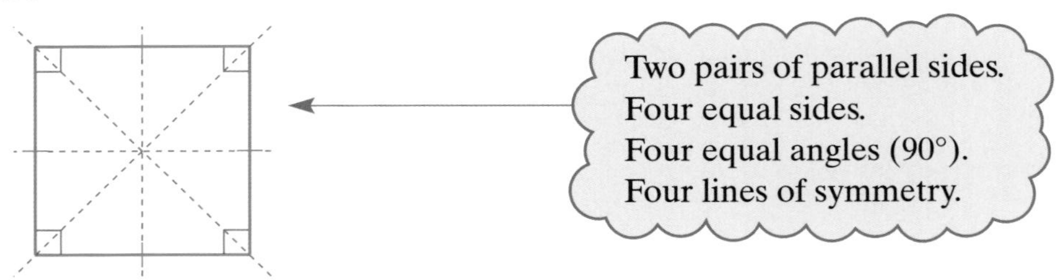
Two pairs of parallel sides.
Four equal sides.
Four equal angles (90°).
Four lines of symmetry.

Rectangle

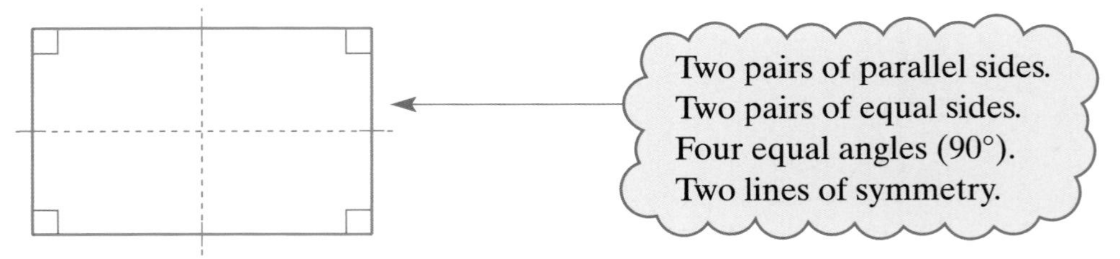
Two pairs of parallel sides.
Two pairs of equal sides.
Four equal angles (90°).
Two lines of symmetry.

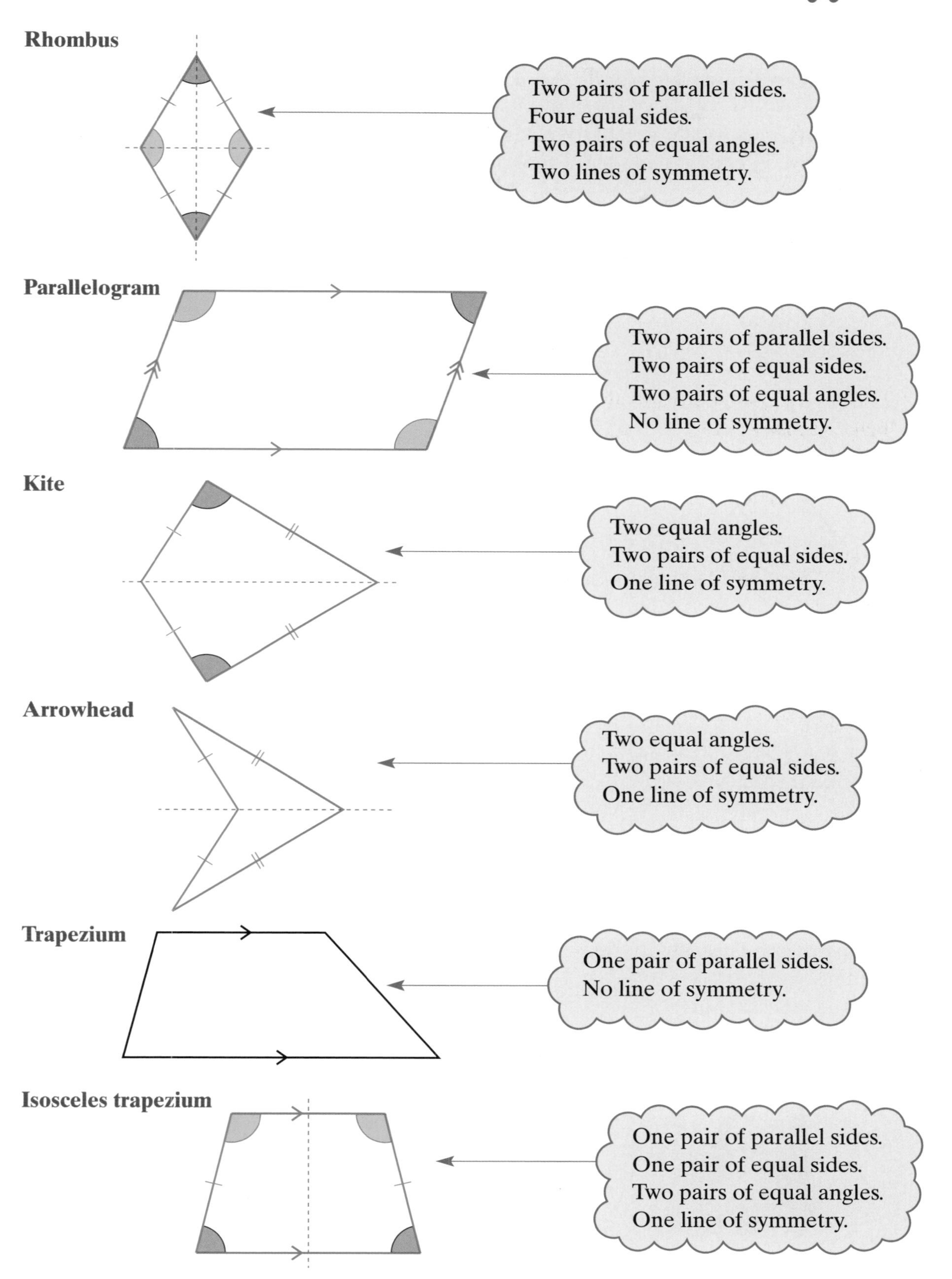

Rhombus
Two pairs of parallel sides.
Four equal sides.
Two pairs of equal angles.
Two lines of symmetry.
Parallelogram
Two pairs of parallel sides.
Two pairs of equal sides.
Two pairs of equal angles.
No line of symmetry.
Kite
Two equal angles.
Two pairs of equal sides.
One line of symmetry.
Arrowhead
Two equal angles.
Two pairs of equal sides.
One line of symmetry.
Trapezium
One pair of parallel sides.
No line of symmetry.
Isosceles trapezium
One pair of parallel sides.
One pair of equal sides.
Two pairs of equal angles.
One line of symmetry.

EXERCISE 14.1

1 Classify all the quadrilaterals you found in Question 4 on page 322.

2 Name these quadrilaterals and triangles.
See Section 13.1 for a reminder of special triangles.

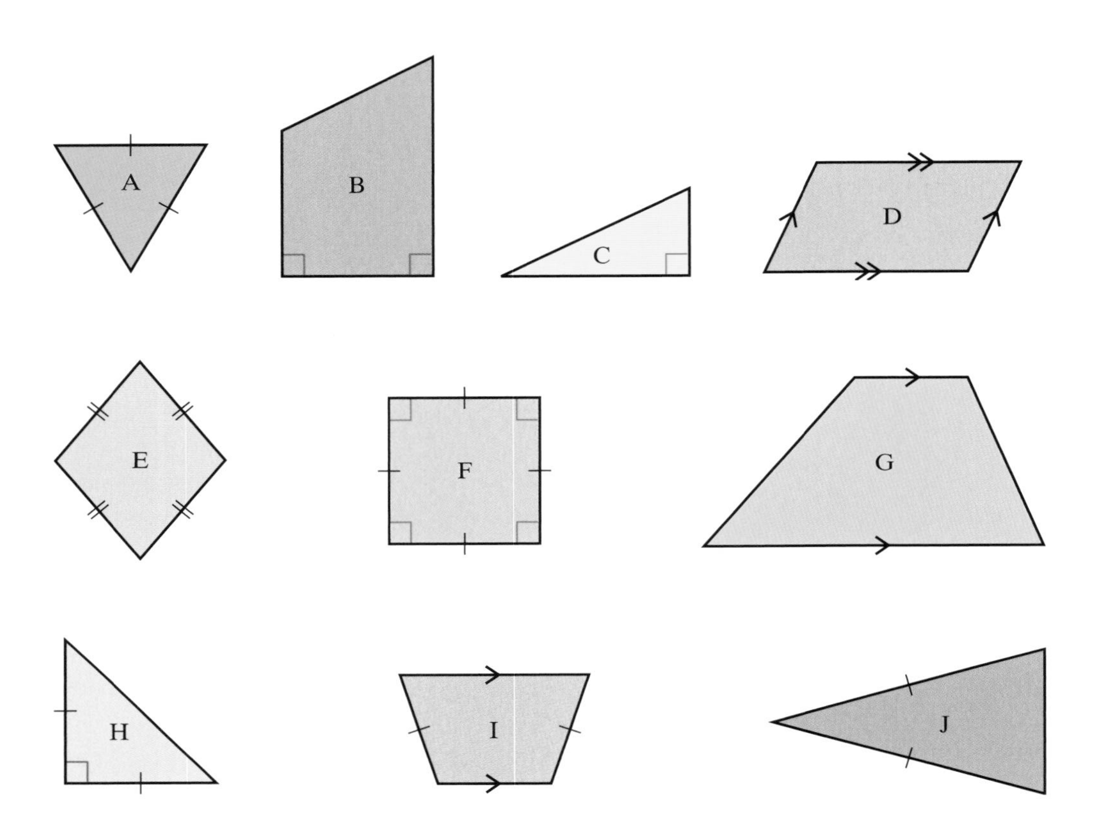

3 A quadrilateral has exactly one set of parallel sides.
What type of quadrilateral is it?

4 A quadrilateral has all four sides the same length.
Is Emily right?
Explain your answer.

14.2 Perimeters and areas of rectangles and triangles

The **area** of a shape is a measure of how much space there is inside the shape.

Area is measured in mm^2, cm^2, m^2 or km^2

The **perimeter** of a shape is the distance around the outside edge.

Perimeter is measured in mm, cm, m or km

EXAMPLE

This shape is drawn on 1 cm^2 paper.
Work out:

a) the perimeter
b) the area of this shape.

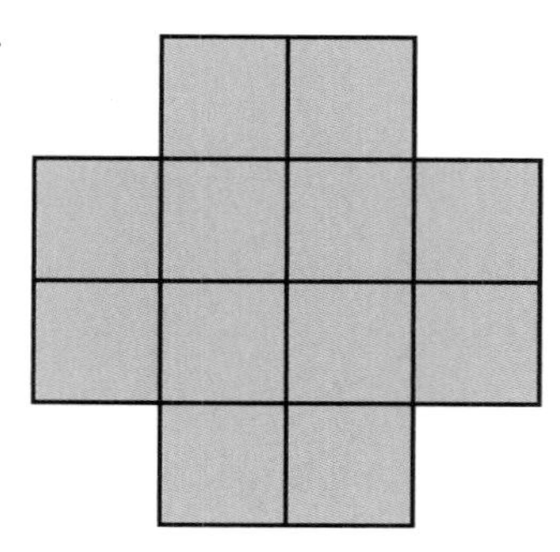

SOLUTION

It is a good idea to tick off the sides and squares as you count them.

a) Perimeter = distance around the outside edge.
So starting at the top of the shape and working round:

$$2 + 1 + 1 + 2 + 1 + 1 + 2 + 1 + 1 + 2 + 1 + 1 = \underline{16\text{ cm}}$$

b) Area = amount of space contained in the shape.
So counting squares: area = $\underline{12\text{ cm}^2}$

You need to be able to work out the area and perimeters of rectangles and squares.

Rectangle

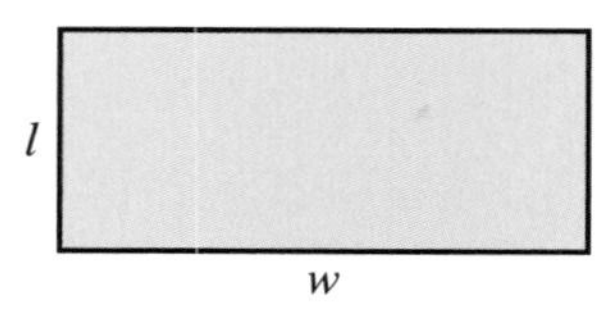

Area of a rectangle = length × width
$A = l \times w$

Perimeter $= l + w + l + w$
$= 2l + 2w$
$= 2(l + w)$

Square

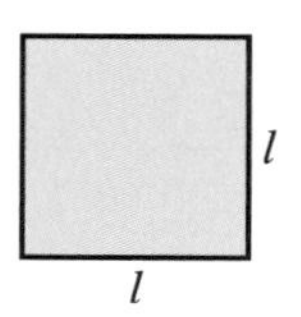

Area of a square = length × length
$A = l \times l$
$A = l^2$

Perimeter $= l + l + l + l$
$= 4l$

EXAMPLE

Work out: **a)** the area
b) the perimeter of this rectangle.

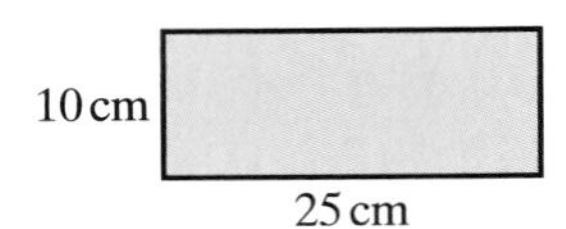

SOLUTION

a) Area of a rectangle = length × width

$= 25 \times 10$

$= \underline{250\text{ cm}^2}$

You must write down the units (cm^2)

b) Perimeter $= 10 + 25 + 10 + 25$

$= \underline{70\text{ cm}}$

EXAMPLE

The area of this rectangle is 35 cm^2

What is the perimeter of the rectangle?

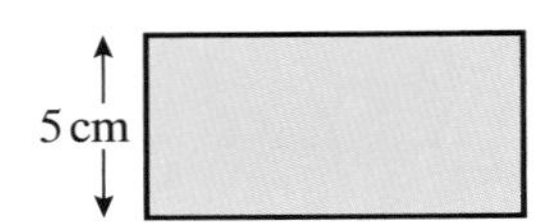

SOLUTION

We need to find the length of the rectangle.

We know that area = length × width = 35 cm^2

Length × 5 = 35

÷5 ÷5

Length = 7 cm

Width = 5 cm

So the perimeter $= 2 \times 5 + 2 \times 7 = \underline{24\text{ cm}}$

You can cut up any triangle and make it into half a rectangle.

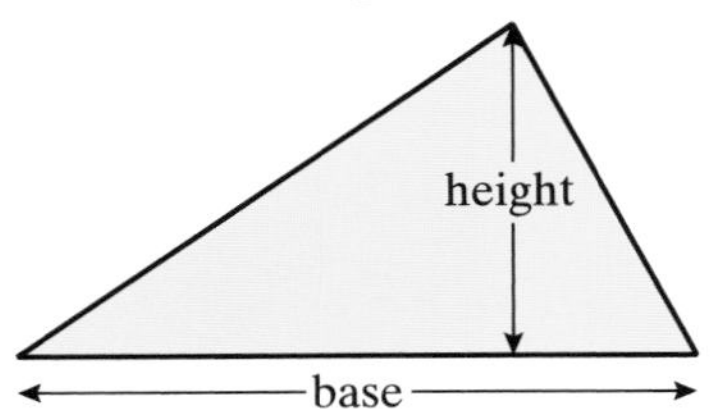

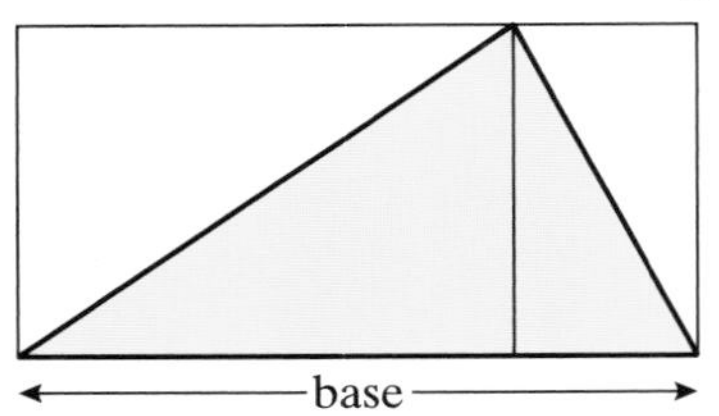

So the area of a triangle is half that of a rectangle.

Area = half × base × height

$A = \frac{1}{2}bh$

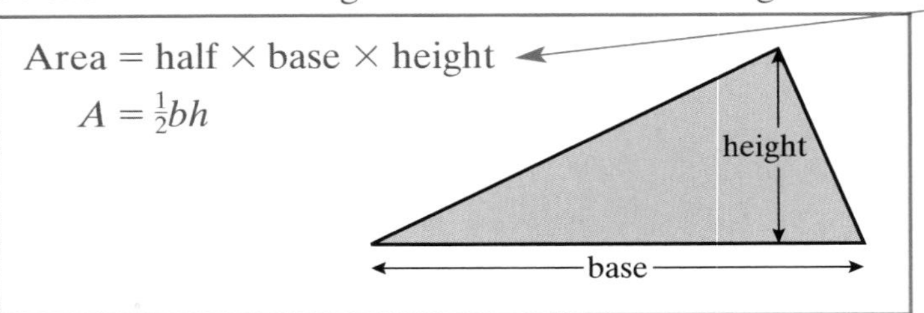

You may have learnt this as: 'half base times height' ($\frac{1}{2}b \times h$), or 'base times height divided by 2' $\left(\frac{b \times h}{2}\right)$
These are all different ways of writing the same formula.

EXAMPLE

Work out:

a) the area

b) the perimeter of this triangle.

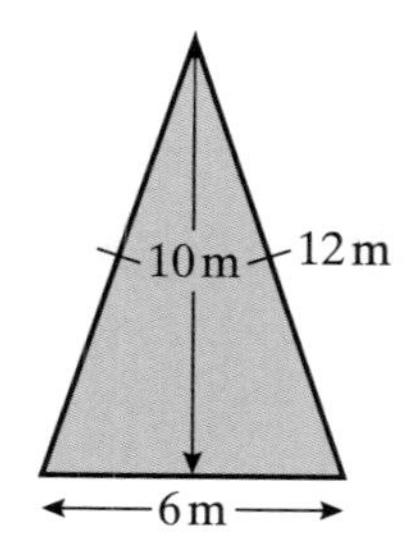

SOLUTION

a) Area of a triangle $= \frac{1}{2}bh$

$= \frac{1}{2} \times 6 \times 10$

$= 3 \times 10$

$= 30\ \text{m}^2$

The area is $\underline{30\ \text{m}^2}$

b) The triangle is an isosceles triangle, so the 3rd side is 12 m

So the perimeter $= 6 + 12 + 12 = \underline{30\ \text{m}}$

Sometimes you need to turn the triangle around to find the base. Of course, rotating a shape doesn't change its area.

Remember the height of a triangle is how 'tall' the triangle is. This is not always the same as one of the sides!

EXAMPLE

Work out:

a) the area

b) the perimeter of this triangle.

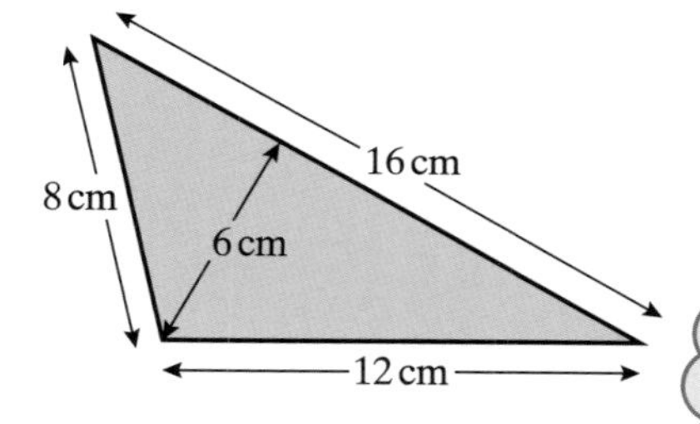

You can turn your textbook or exam paper around!

SOLUTION

a) Turn the triangle around so that the 16 cm side is the base. Now it is easy to see that the height of the triangle is 6 cm.

Area of a triangle $= \frac{1}{2}bh$

$= \frac{1}{2} \times 16 \times 6$

$= 8 \times 6$

$= 48\ \text{cm}^2$

So the area of the triangle is $\underline{48\ \text{cm}^2}$

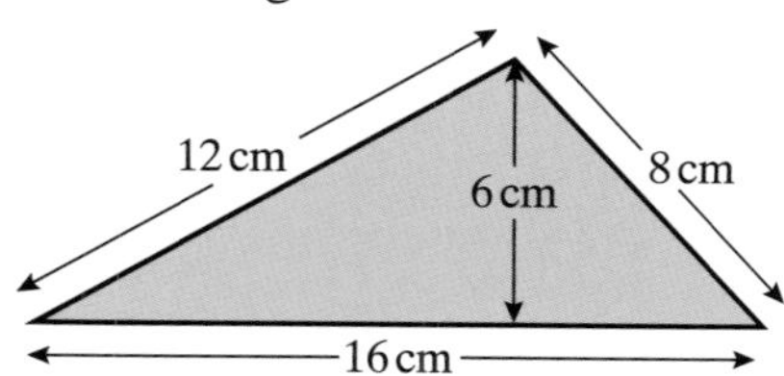

b) To find the perimeter, simply add up the three sides of the triangle:

$16 + 12 + 8 = 36$

So the perimeter is $\underline{36\ \text{cm}}$

EXERCISE 14.2

1 Work out: **(i)** the perimeter, **(ii)** the area of each of these rectangles.

a)

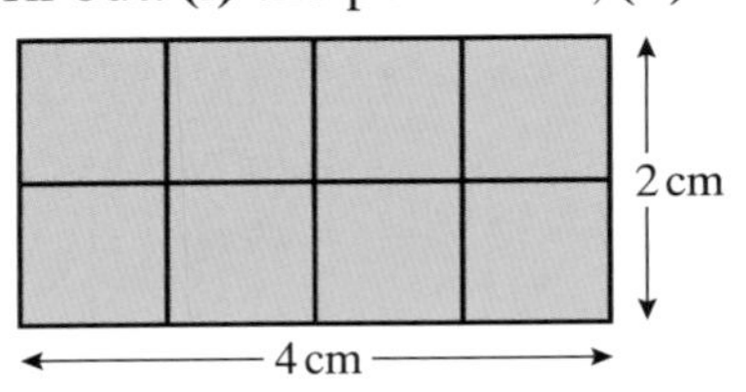

b)

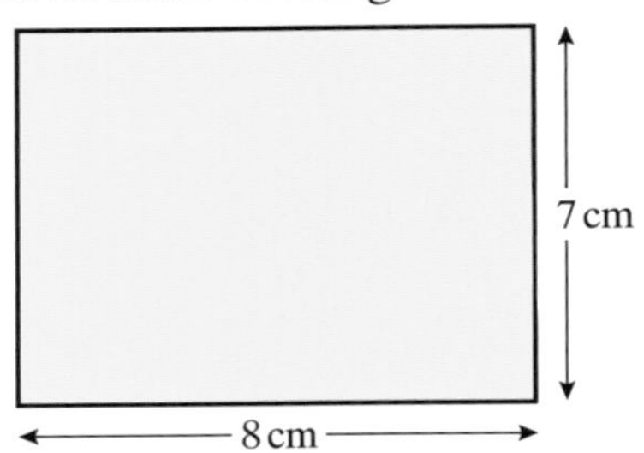

c)

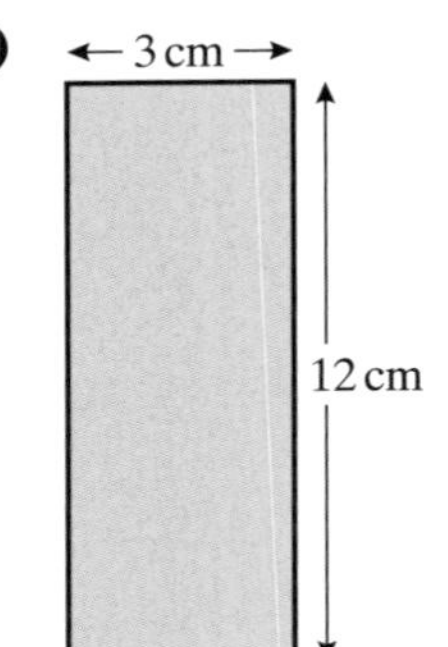

d)

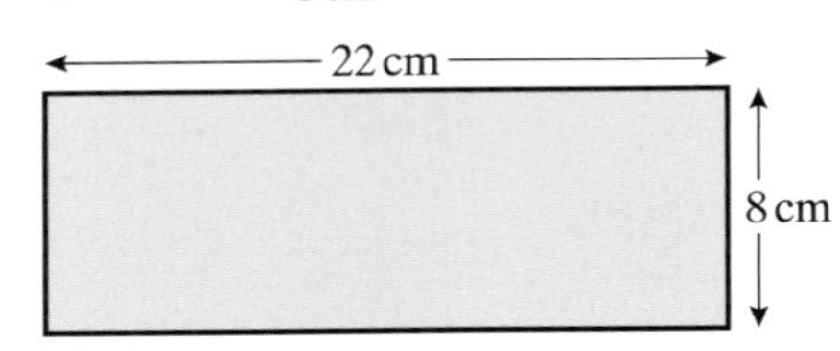

You can use your calculator for Questions 2–10.

2 Work out: **(i)** the perimeter, **(ii)** the area of each of these squares.

a)

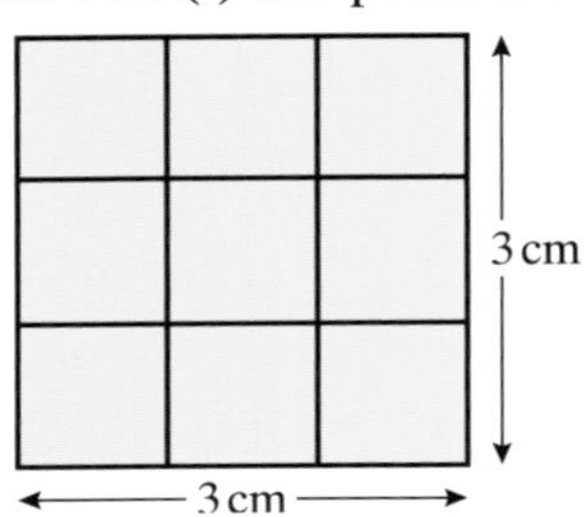

b)

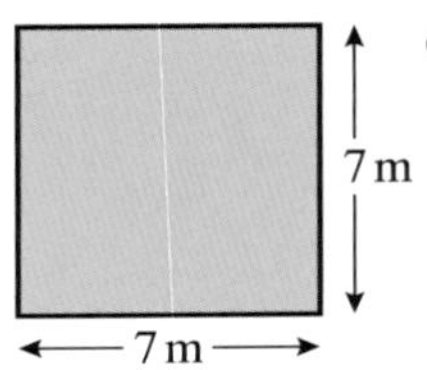

c)

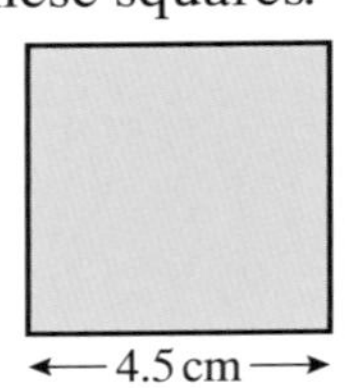

d)

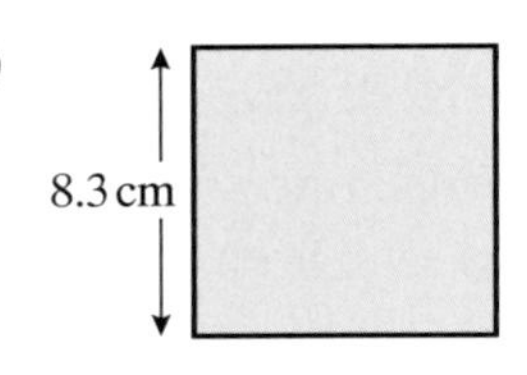

3 Work out: **(i)** the perimeter, **(ii)** the area of each of these triangles.

a)

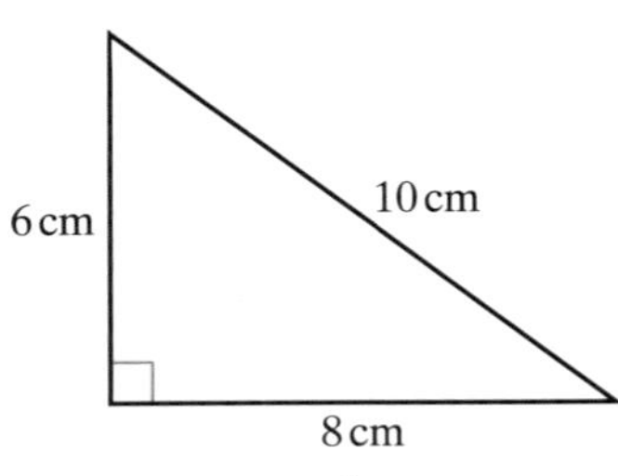

b)

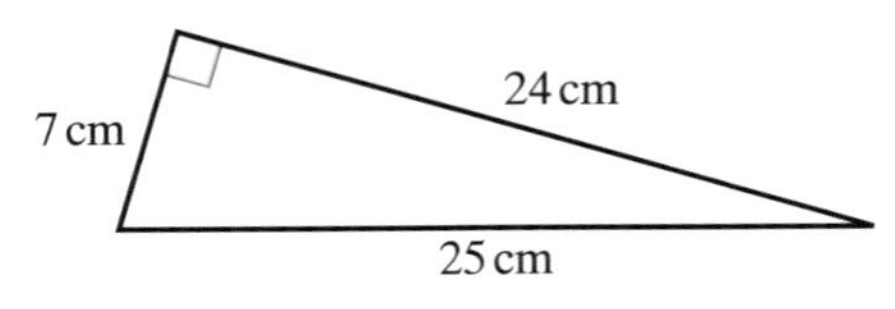

c)

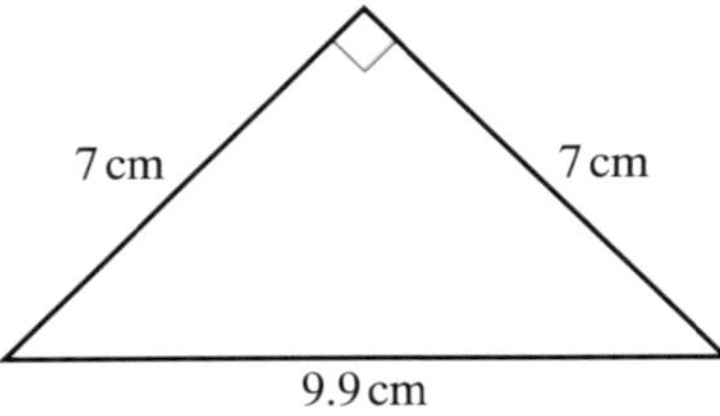

d)

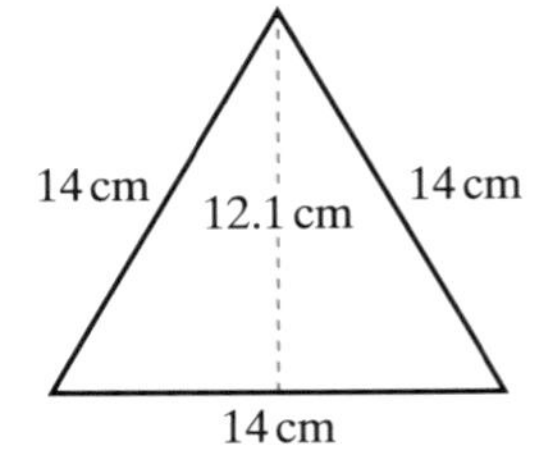

4 Work out the area of these triangles.

a)

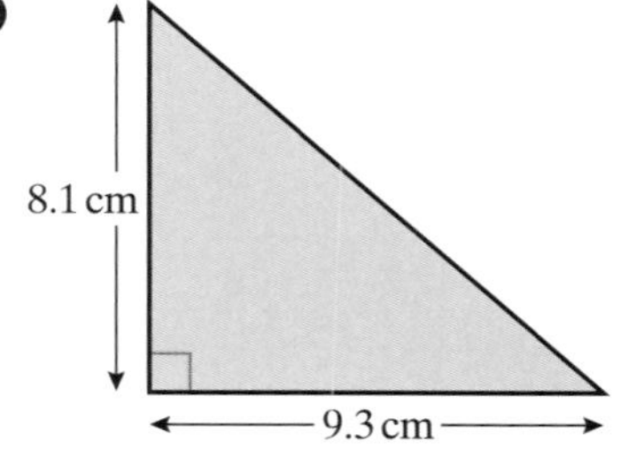

b)

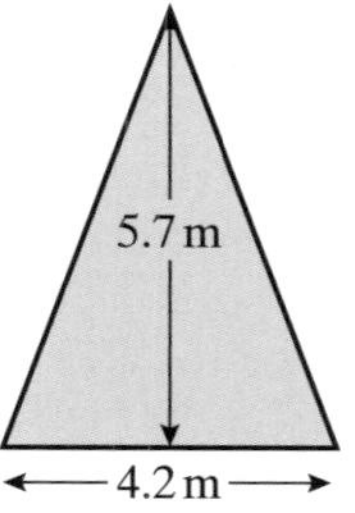

c)

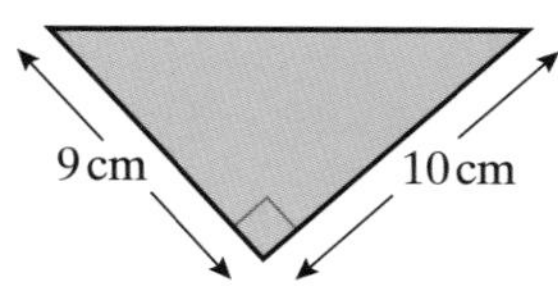

d)

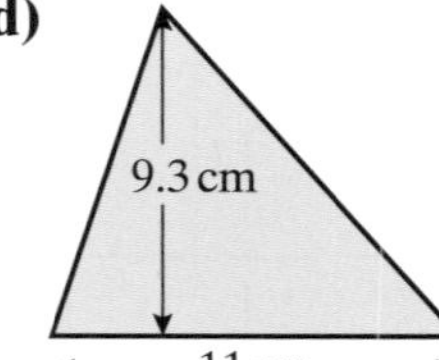

e)

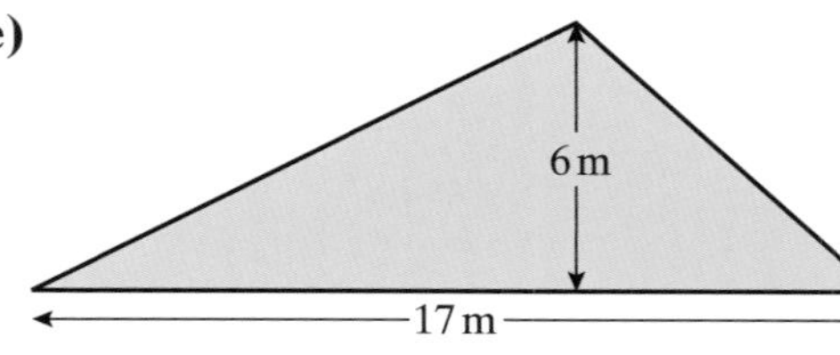

f)

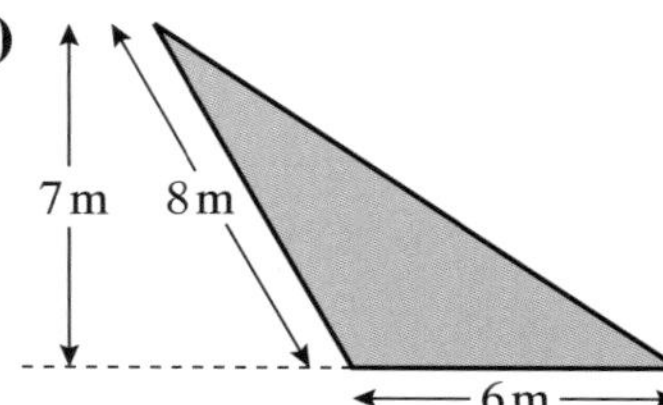

5 Find:

a) the area

b) the perimeter of this triangle.

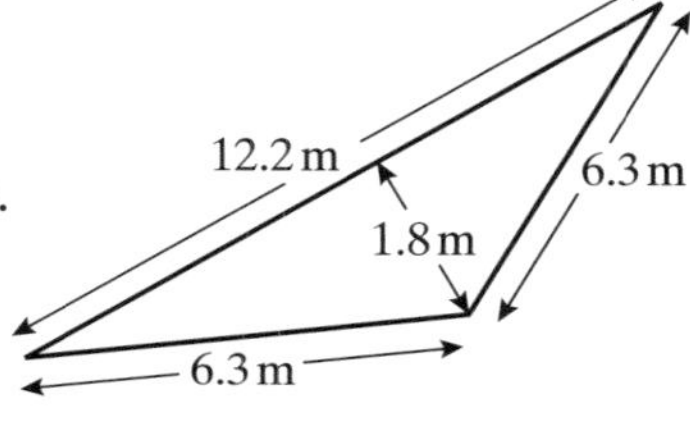

6 A rectangle has an area of 20 cm^2
Its sides are a whole number of centimetres.

a) On squared paper, draw three different rectangles with an area of 20 cm^2

20 cm^2

The perimeter of the rectangle is 18 cm

b) What are the dimensions of the rectangle?

The length and width.

7 A square has a perimeter of 32 cm
What is the area of the square?

8 A square has an area of 36 cm^2
What is the perimeter of the square?

9 This garden has an area of 54 m^2

a) What is the length of the garden?

b) What is the perimeter of the garden?

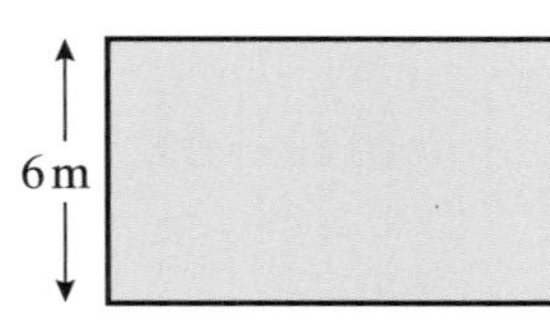

10 A farmer puts 120 m of fencing around rectangular field.

a) What is the length of the field?

b) What is the area of the field?

14.3 Using algebra

Some questions on perimeter may be suitable for solving with algebra.

EXAMPLE

This rectangle has a perimeter of 58 cm
Work out the value of x

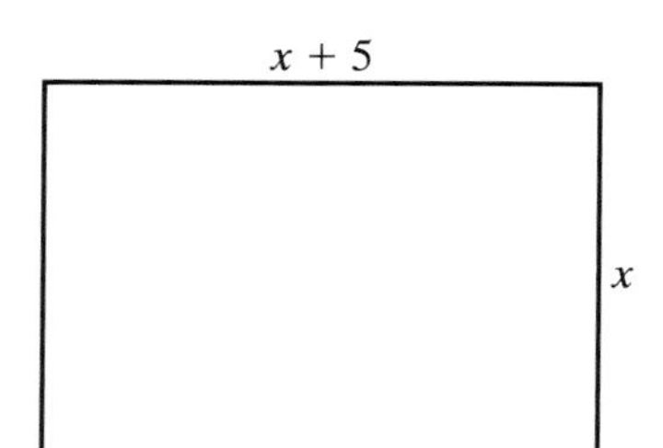

SOLUTION

The rectangle has sides of length $x, x + 5, x$ and $x + 5$

Its perimeter is $x + (x + 5) + x + (x + 5) = 4x + 10$

So $4x + 10 = 58$

$-10 \quad -10$

$4x = 48$

$\div 4 \quad \div 4$

$x = 12$

EXERCISE 14.3

1 The diagram shows a rectangle.
The perimeter of the rectangle is 32 cm

a) Write down an expression for the perimeter of the rectangle.
Simplify your expression.

b) Write down an equation in x for the perimeter of the rectangle.

c) Solve your equation to find the value of x

d) Find the length and width of the rectangle.

$x + 6$

x

2 The diagram shows an isosceles triangle.
AB = CB

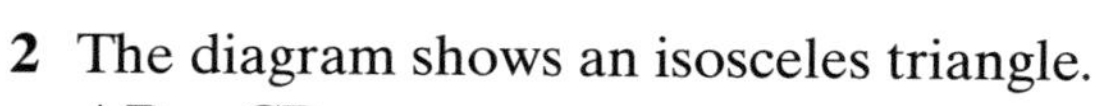

a) Use the information that AB = CB to set up an equation in x

b) Solve your equation to find the value of x

3 The diagram shows a triangle.

a) What type of triangle is this?

b) Set up, and solve, an equation in x

c) Hence work out the perimeter of the triangle.

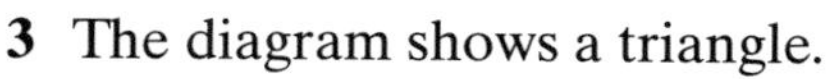

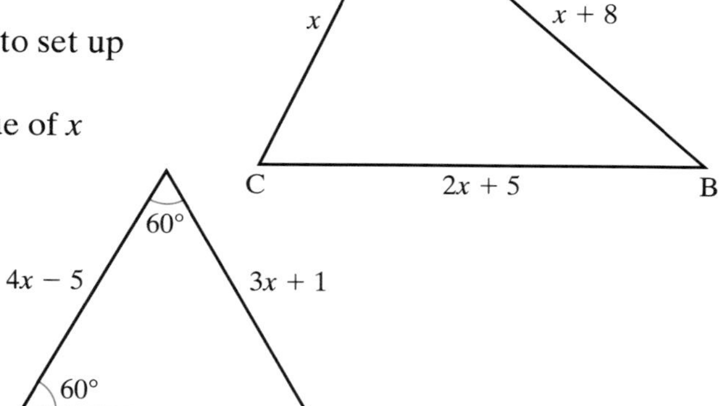

14.4 Area of a parallelogram and a trapezium

The area of a parallelogram can be found by cutting it up into two identical triangles.

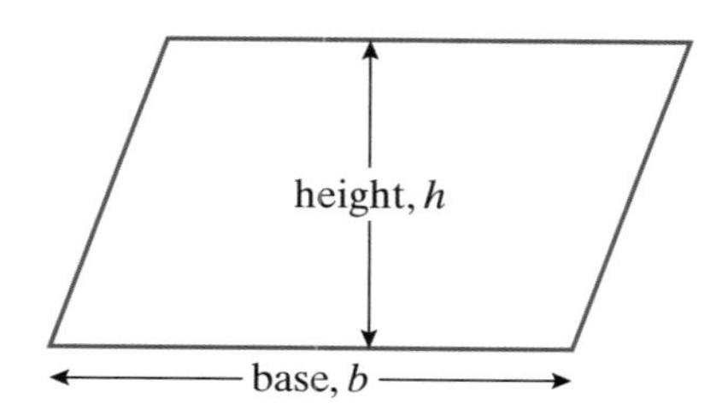
height, h
base, b

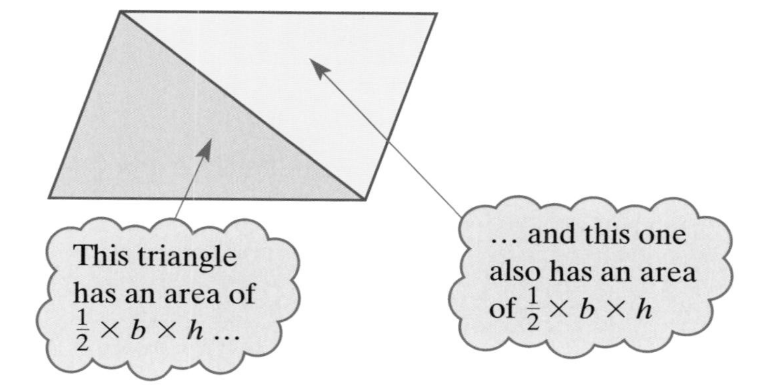
This triangle has an area of $\frac{1}{2} \times b \times h$ ...
... and this one also has an area of $\frac{1}{2} \times b \times h$

$$\text{Area of parallelogram} = \tfrac{1}{2} \times b \times h + \tfrac{1}{2} \times b \times h$$
$$A = b \times h$$

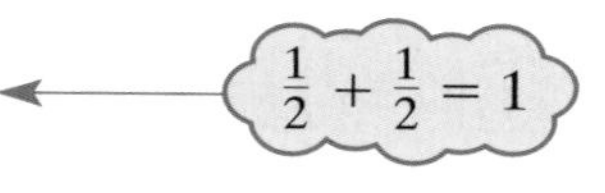
$\frac{1}{2} + \frac{1}{2} = 1$

EXAMPLE

Find the area of this parallelogram.

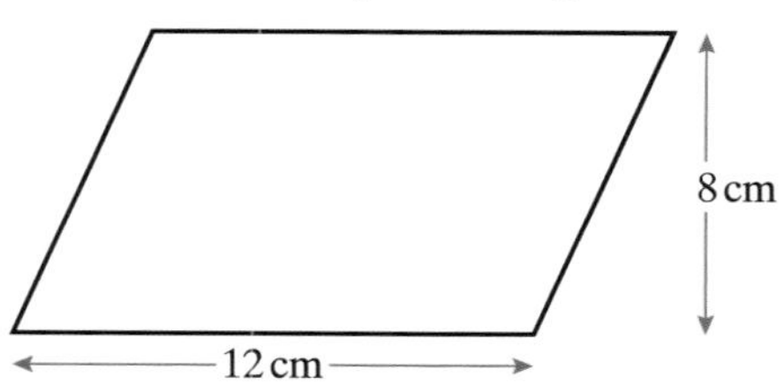
8 cm
12 cm

SOLUTION

The parallelogram has a base $b = 12$ and a height $h = 8$

$$\text{Area} = b \times h$$
$$= 12 \times 8$$
$$= \underline{96\text{ cm}^2}$$

Finally, there is a formula for finding the area of a trapezium. It requires some 'surgery'.

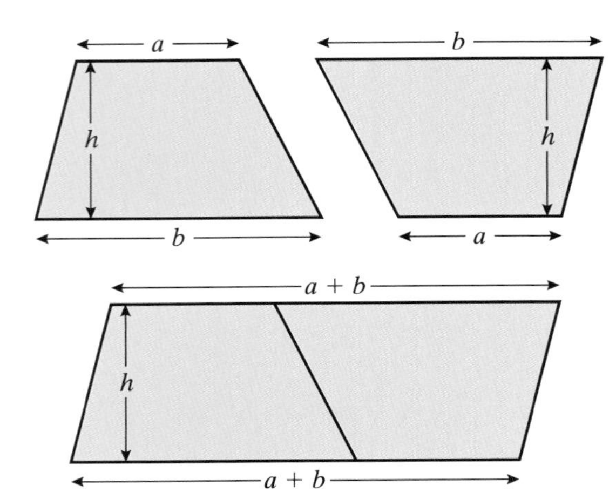

By putting two identical trapeziums together you can make a parallelogram with twice the area.

The area of the parallelogram is $(a + b) \times h$

The area of one trapezium is half of the parallelogram $= \frac{1}{2}(a + b) \times h$

This formula is often written as:

$$\text{Area of trapezium} = \tfrac{1}{2}(a + b)h$$

EXAMPLE

Find the area of this trapezium.

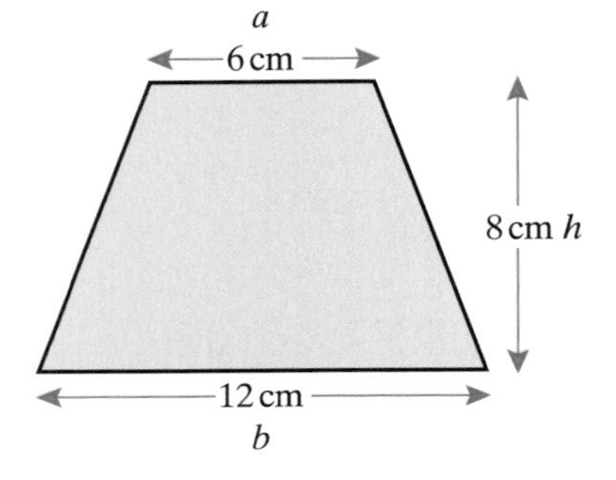

SOLUTION

The trapezium has parallel sides $a = 6$ and $b = 12$, and height $h = 8$

$$\begin{aligned} \text{Area} &= \tfrac{1}{2}(a + b)h \\ &= \tfrac{1}{2} \times (6 + 12) \times 8 \\ &= \tfrac{1}{2} \times 18 \times 8 \\ &= 9 \times 8 \\ &= 72\ \text{cm}^2 \end{aligned}$$

The area of the trapezium is $\underline{72\ \text{cm}^2}$

Areas of shapes like triangles, rectangles, parallelograms and trapeziums can be found directly – by using the appropriate formulae. Make sure you learn them!

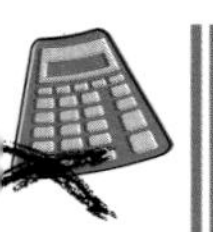

EXERCISE 14.4

Don't use your calculator for Questions 1 and 2.

1 Find the area of each of these parallelograms

a)

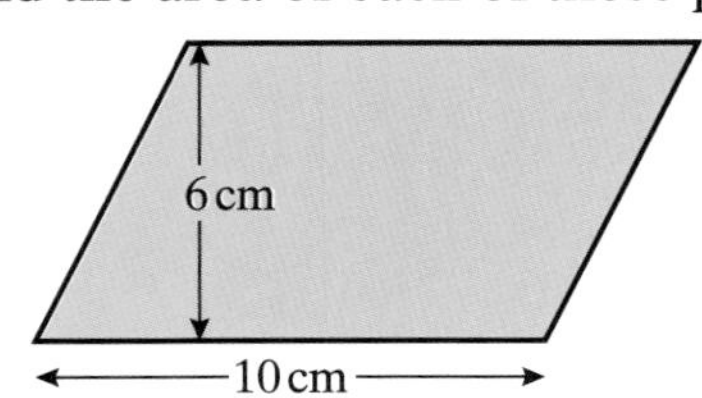

b)

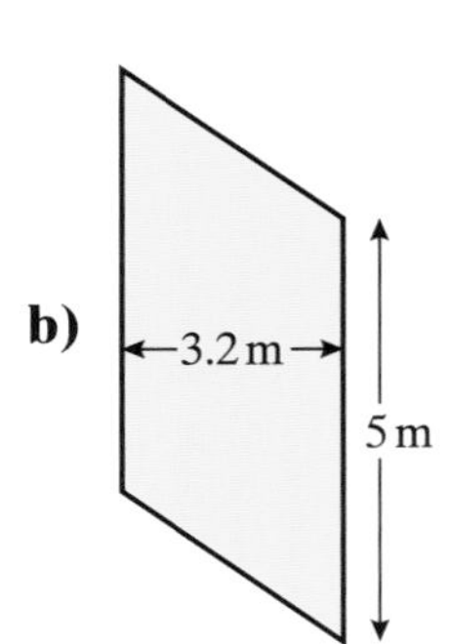

2 Find the area of each of these trapeziums

a)

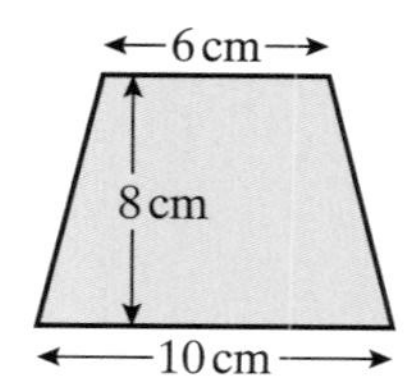

b)

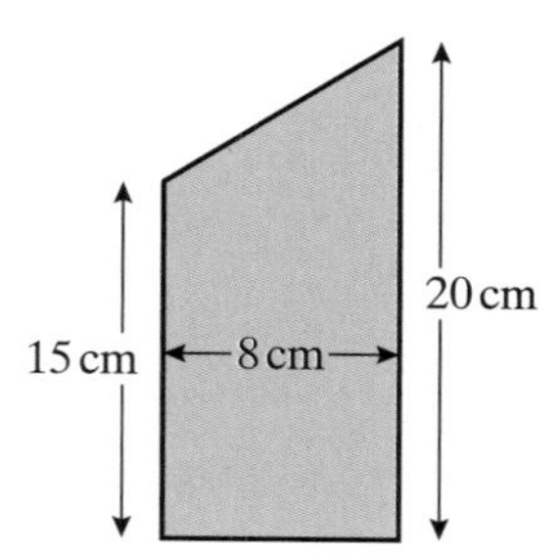

You can use your calculator for Questions 3 and 4.

3 Find the area of each shape.

a)

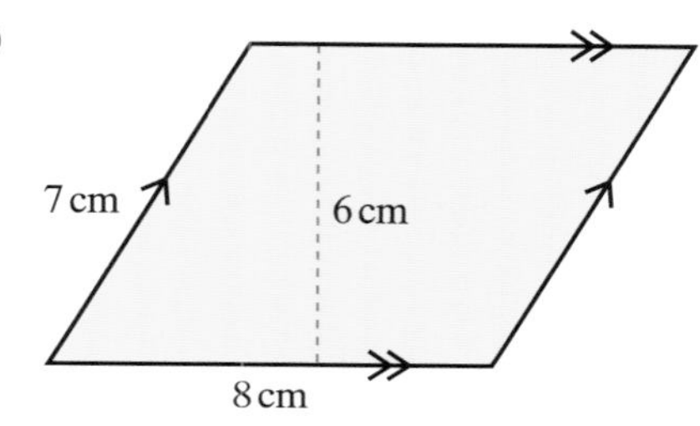

b)

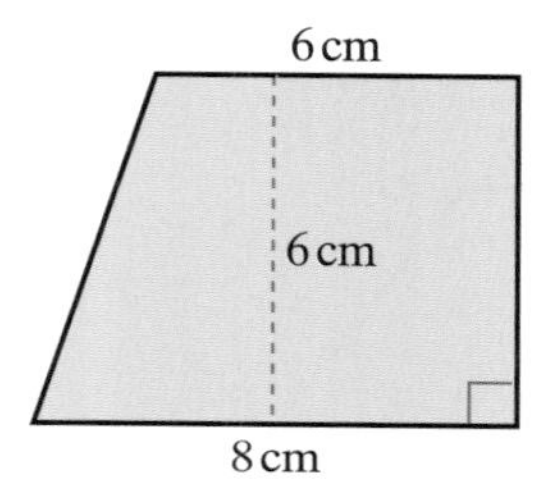

c)

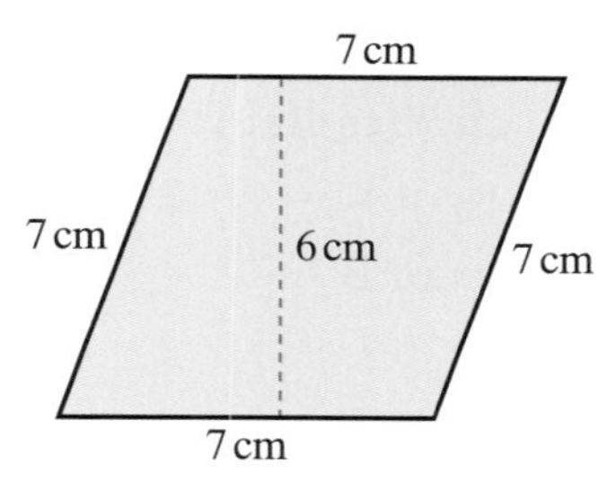

d)

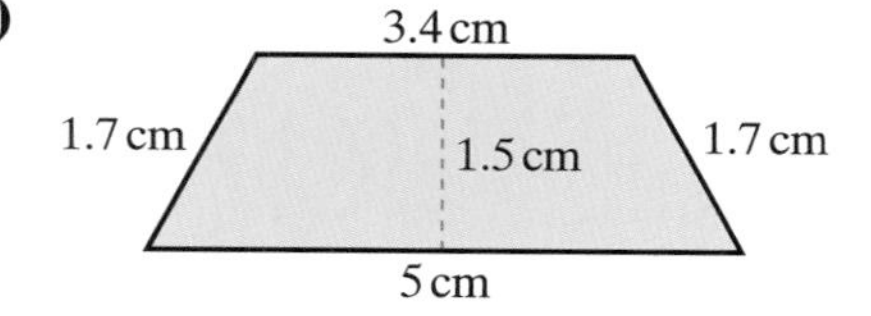

4 The diagram shows a parallelogram.

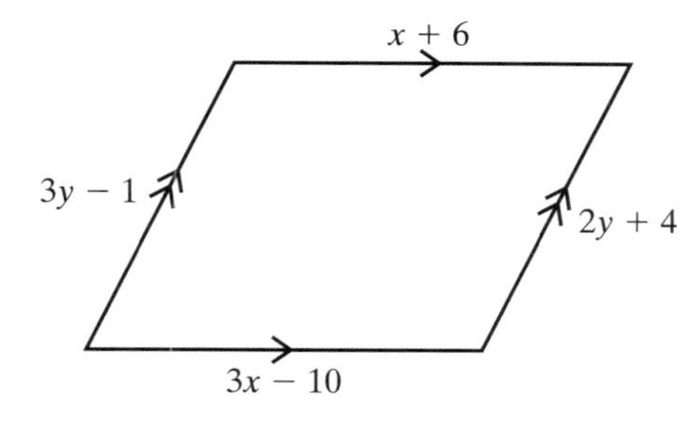

a) Set up, and solve, an equation in x

b) Set up, and solve, an equation in y

c) Hence work out the lengths of the sides of the parallelogram.

d) Suggest a better name for this shape.

14.5 Area of compound shapes

A **compound shape** is made when two or more shapes are put together.

You may need to work out the areas of compound shapes by breaking them down into two or more simpler pieces.

EXAMPLE

The diagram shows a shape made up from rectangles and a triangle.

a) Calculate the perimeter of the shape.

b) Calculate the area of the shape.

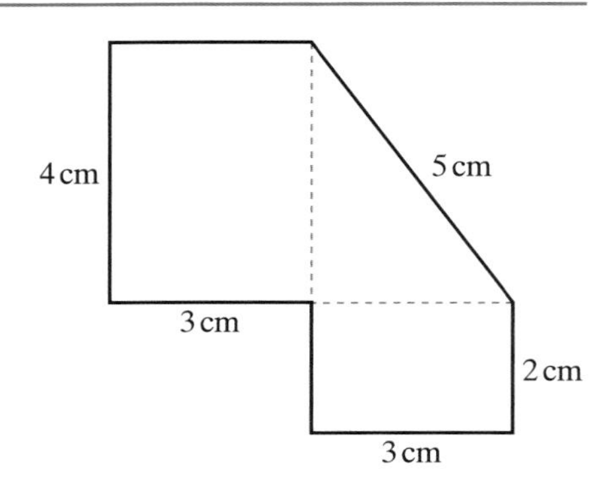

SOLUTION

a) Marking the missing lengths:

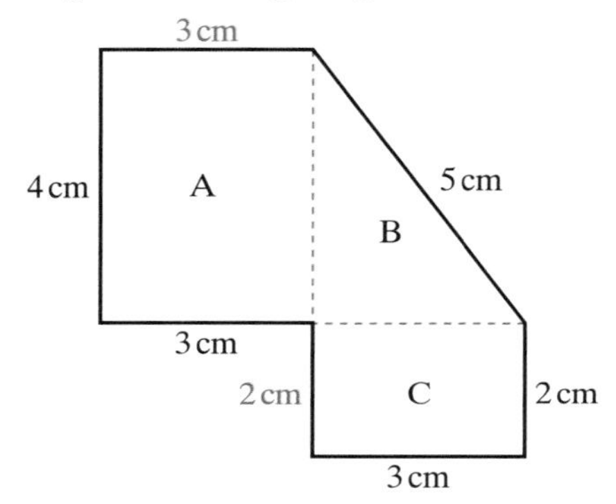

Perimeter $= 3 + 4 + 3 + 2 + 3 + 2 + 5$
$= \underline{22 \text{ cm}}$

b) Denoting the three parts as A, B and C (see diagram left), then:

area of A $= 4 \times 3 = 12 \text{ cm}^2$

area of B $= \frac{1}{2} \times 4 \times 3 = 6 \text{ cm}^2$

area of C $= 3 \times 2 = 6 \text{ cm}^2$

Total area $= 12 + 6 + 6 = \underline{24 \text{ cm}^2}$

Sometimes a shape has a hole inside it.

When this is the case, work out the area of the whole shape and then subtract the area of the hole.

EXAMPLE

Work out the shaded area in this square which has a triangular hole cut out of it.

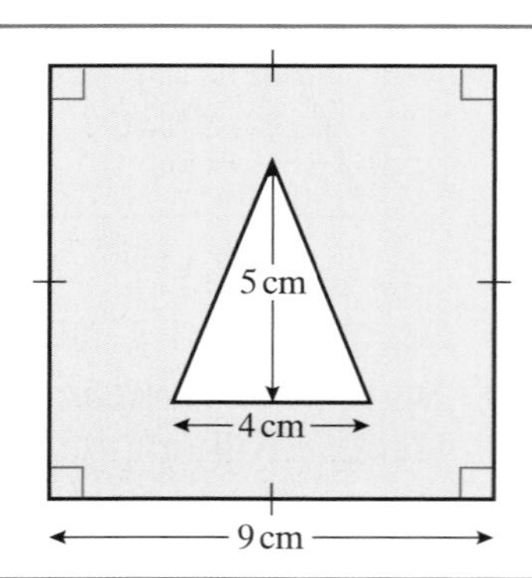

SOLUTION

Area of square $= 9 \times 9 = 81 \text{ cm}^2$

Area of triangle $= \frac{1}{2}bh = \frac{1}{2} \times 4 \times 5 = 10 \text{ cm}^2$

Shaded area = area of square − area of triangle
$= 81 - 10 = \underline{71 \text{ cm}^2}$

EXERCISE 14.5

1 Find the shaded area of each shape.

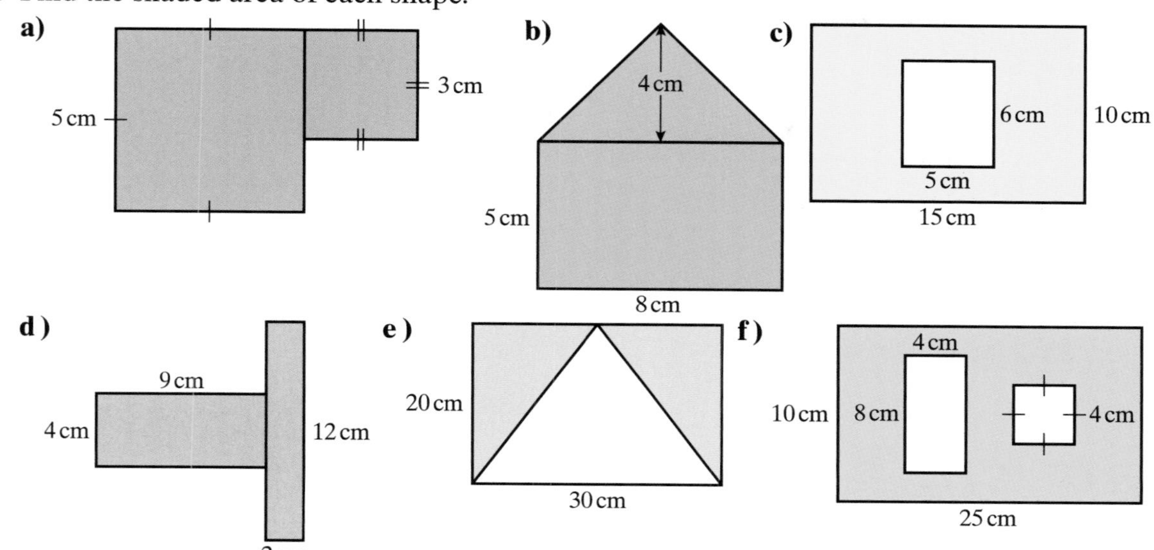

You can use your calculator for Questions 2 and 3.

2 Calculate: **(i)** the perimeter, **(ii)** the area of each shape.
State the units in each case.

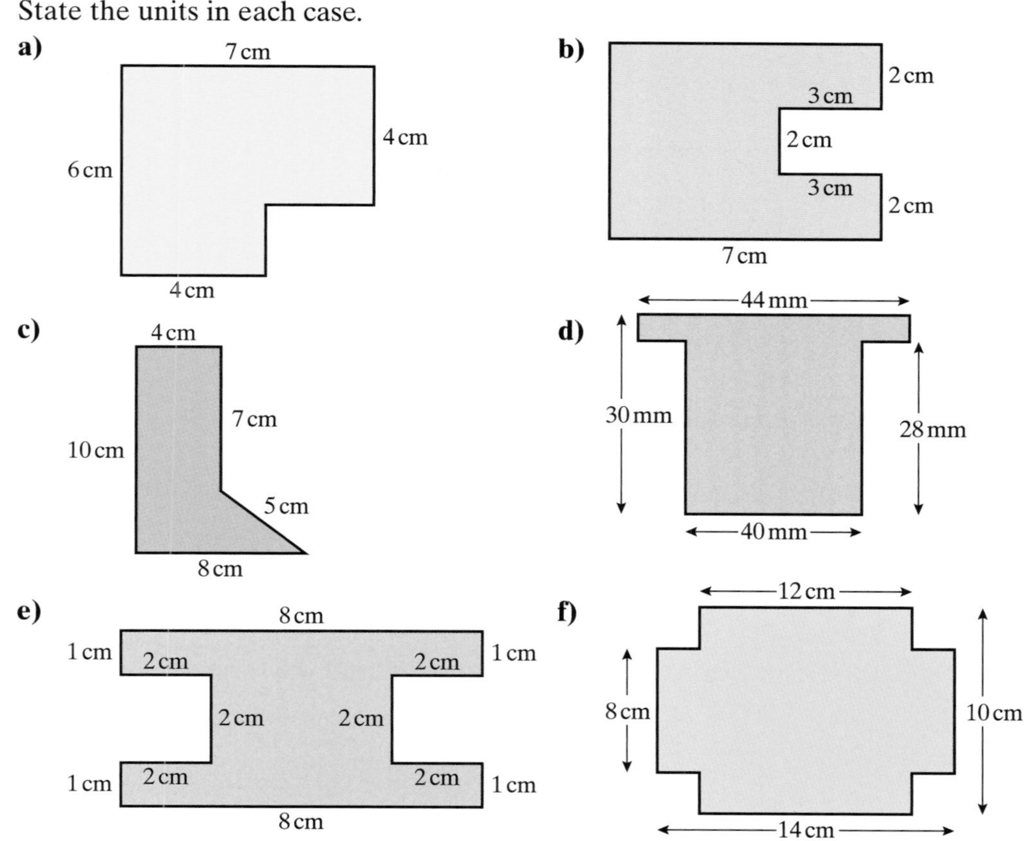

3 Find the shaded area of each shape.

a)

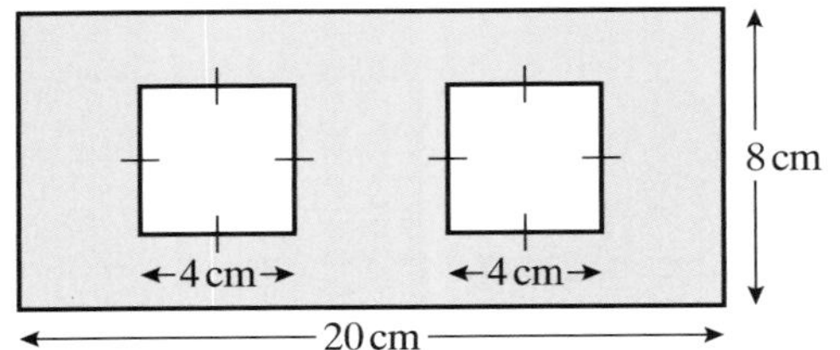

b)

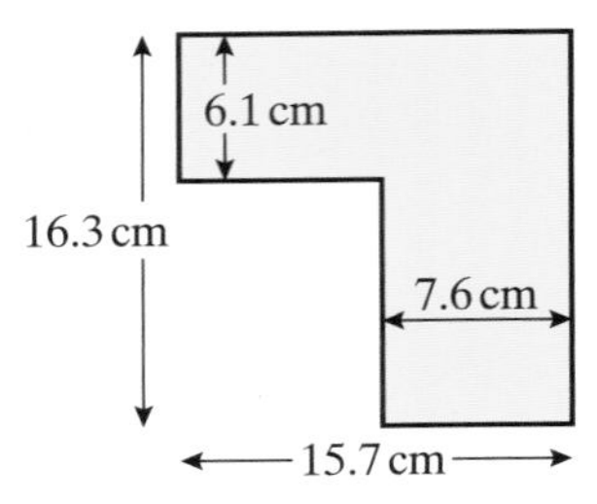

c)

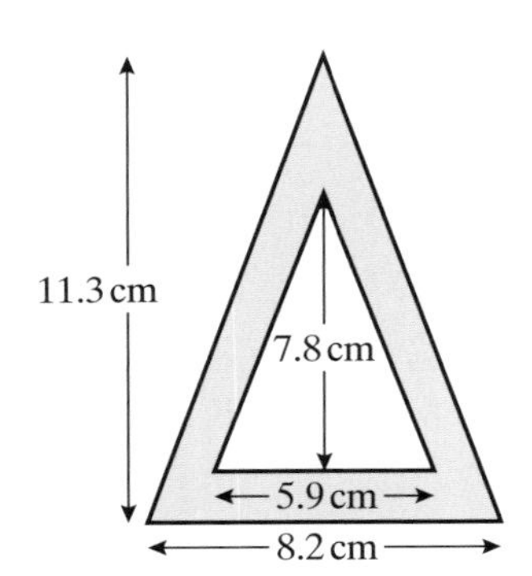

d)

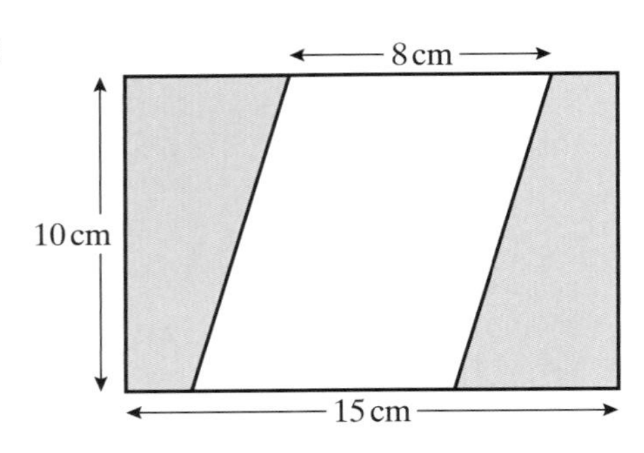

14.6 Surface area and volume

Here are some common 3-D shapes you will need to recognise.

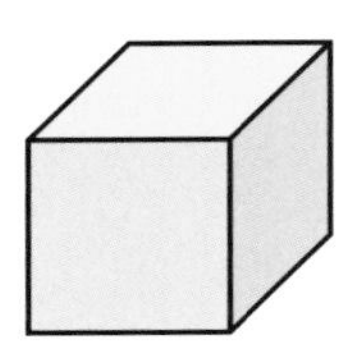

Cube
Has six faces.
Each face is a square.

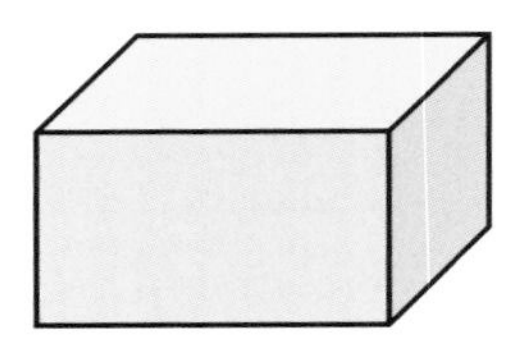

Cuboid
Has six faces.
Each face is a rectangle.

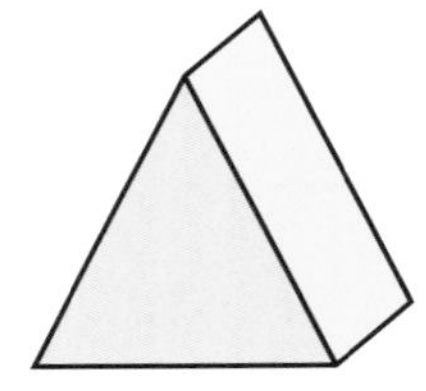

Triangular prism
Has five faces.
The two end faces (the cross section) are both triangles.
The other three faces are rectangles.

The **surface area** of a cuboid is found by calculating the areas of its six separate faces and then adding them together.

The volume of a cuboid is found by multiplying the three sides of the cuboid together:

> volume of a cuboid = length × width × height

The units of volume are mm^3, cm^3 or m^3

EXAMPLE

Find: **a)** the surface area, **b)** the volume of this solid cuboid.

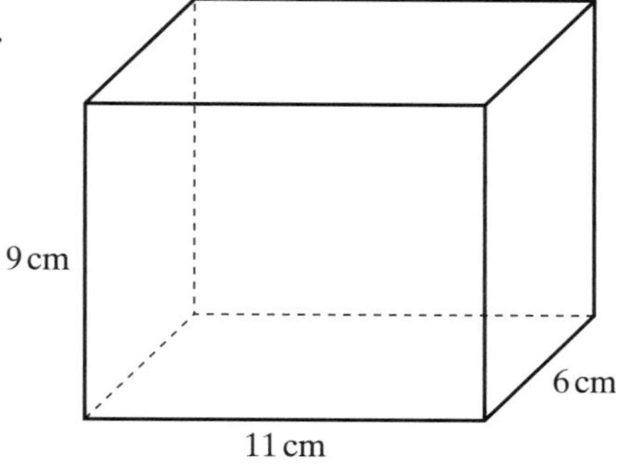
9 cm
6 cm
11 cm

SOLUTION

a) Consider the left and right ends:

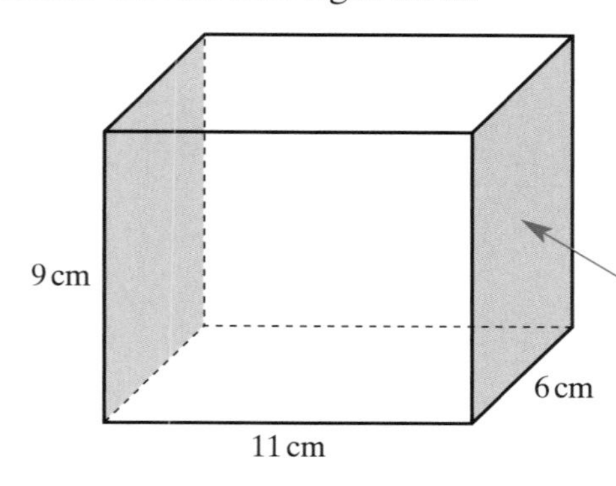
9 cm
6 cm
11 cm
Each end has an area of $9 \times 6 = 54\ \text{cm}^2$
There are two ends, so the total is $54 \times 2 = 108\ \text{cm}^2$

Similarly for the top and bottom:

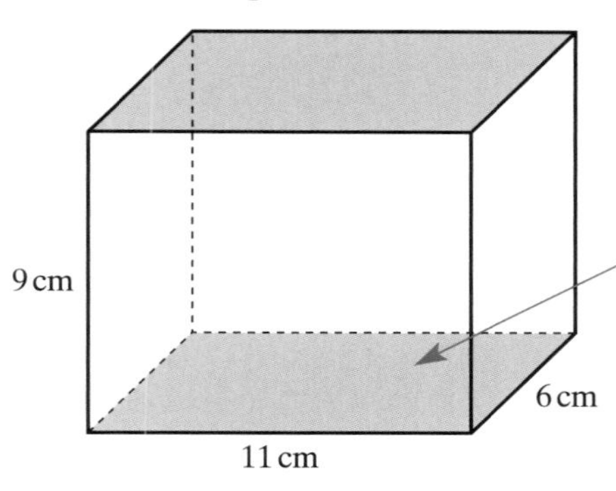
9 cm
6 cm
11 cm
Each shaded rectangle has an area of $11 \times 6 = 66\ \text{cm}^2$
There are two rectangles, so the total is $66 \times 2 = 132\ \text{cm}^2$

Finally, look at the front and back:

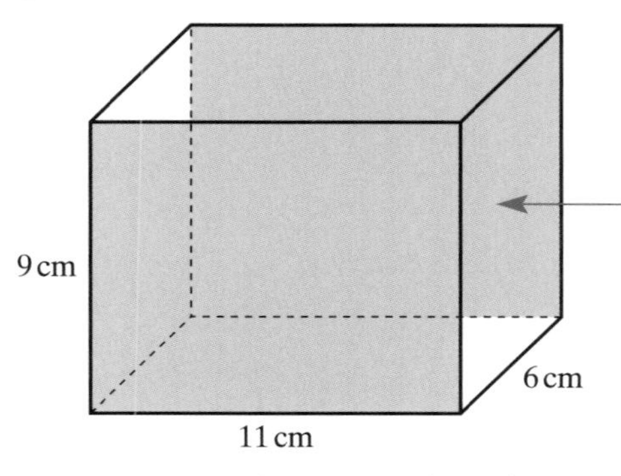
9 cm
6 cm
11 cm
Each shaded rectangle has an area of $11 \times 9 = 99\ \text{cm}^2$
There are two rectangles, so the total is $99 \times 2 = 198\ \text{cm}^2$

So the total area is $(9 \times 6 \times 2) + (11 \times 6 \times 2) + (11 \times 9 \times 2)$

$= 108 + 132 + 198$

$= \underline{438\ \text{cm}^2}$

b) The volume is $11 \times 9 \times 6 = \underline{594\ \text{cm}^3}$

A cuboid is a simple example of a **prism**.

Prisms are 3-D solids with a constant cross-section.
To find the volume of a prism, multiply its cross-sectional area by its length.

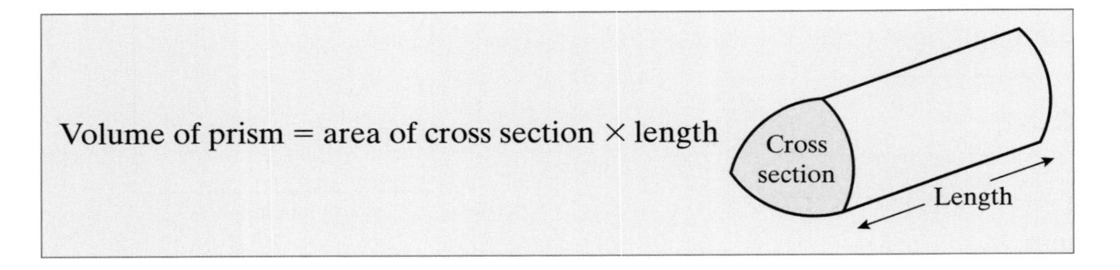

Volume of prism = area of cross section × length
Cross section
Length

Sometimes you will be told the cross-sectional area, and you can then simply multiply it by the length.

EXAMPLE

The diagram shows a prism of length 10 centimetres and cross-sectional area 8 cm^2

Calculate its volume.

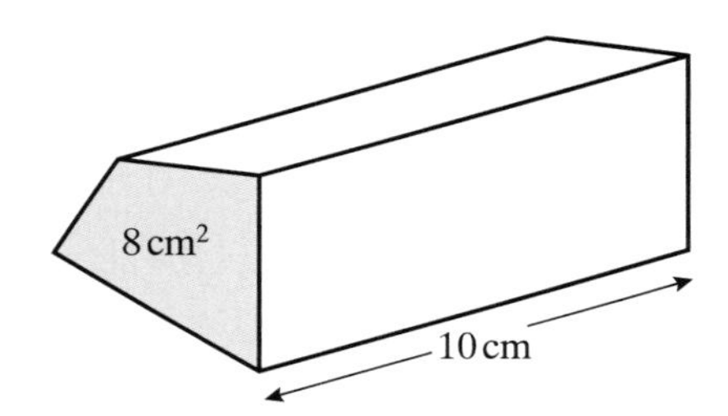

SOLUTION

$$\begin{aligned} \text{Volume} &= \text{area of cross section} \times \text{length} \\ &= 8 \times 10 \\ &= \underline{80\ \text{cm}^3} \end{aligned}$$

If the cross section is a simple shape, such as a triangle, then you might be asked to work its area out first.

EXAMPLE

The diagram shows a prism.
The cross section of the prism is a right-angled triangle.

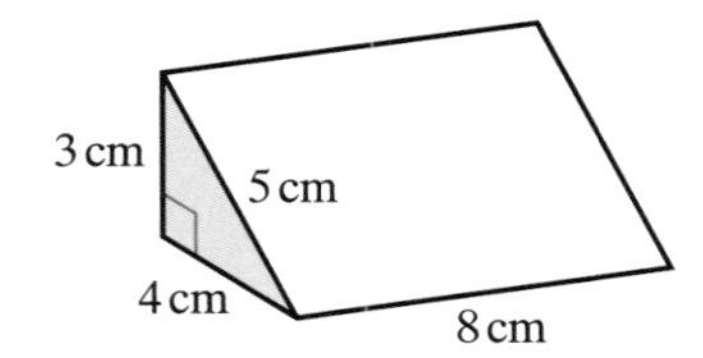

a) Calculate the area of the cross section.

b) Find the volume of the prism.

c) Work out the surface area of the prism.

SOLUTION

a) Area of cross section $= \frac{1}{2} \times 4 \times 3$

$= \underline{6\text{ cm}^2}$

b) Volume of prism $= 6 \times 8$

$= \underline{48\text{ cm}^3}$

c) The two triangular faces have areas of 6 cm^2 each.
The three rectangular faces have areas of:

$5 \times 8 = 40\text{ cm}^2$

$4 \times 8 = 32\text{ cm}^2$

$3 \times 8 = 24\text{ cm}^2$

Total surface area $= 6 + 6 + 40 + 32 + 24 = \underline{108\text{ cm}^2}$

EXERCISE 14.6

Don't use your calculator for Questions 1–3.

1 Find:

(i) the volume

(ii) the surface area of these shapes.

represents 1 cm^3

a)

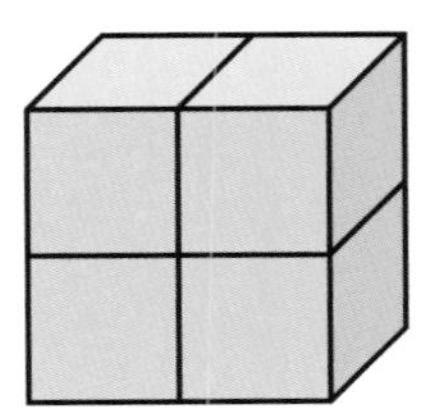

b)

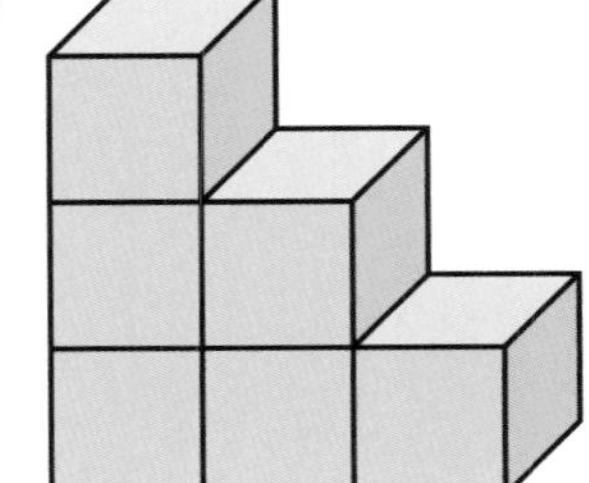

c)

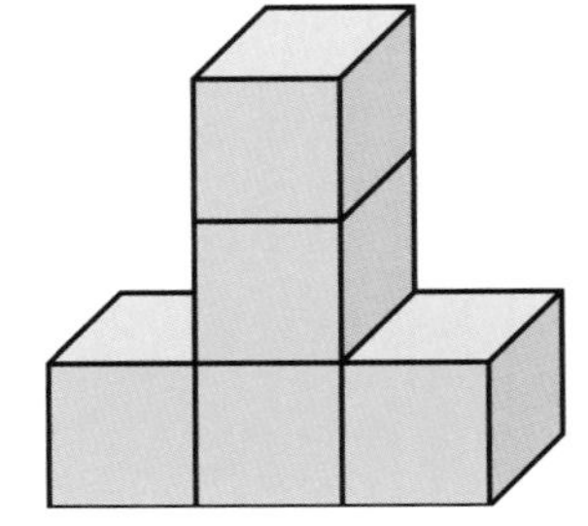

2 The diagram shows a cube of side 10 cm
Calculate its:
a) surface area **b)** volume.
State the units in your answers.

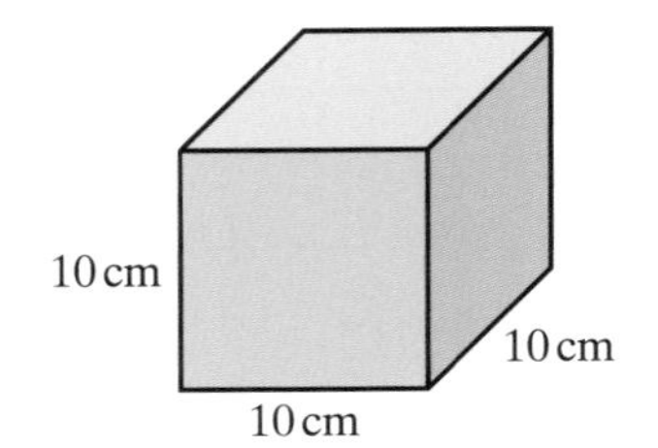

3 The diagram shows a prism.
Its cross section is formed by a right-angled triangle of sides 5 cm, 12 cm and 13 cm
The prism has a length of 6 cm

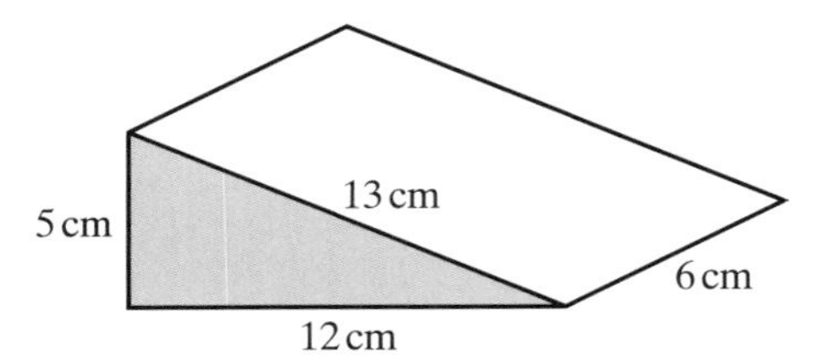

a) Calculate the area of the cross section shaded in the diagram.
b) Work out the volume of the prism.
c) Calculate the surface area of the prism.

You can use your calculator for Questions 4–11.

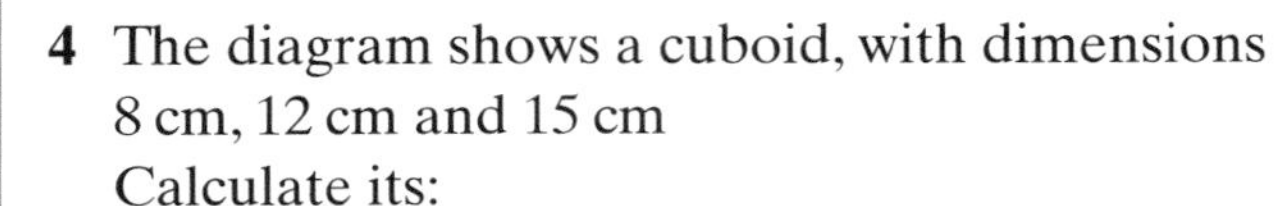

4 The diagram shows a cuboid, with dimensions 8 cm, 12 cm and 15 cm
Calculate its:
a) surface area
b) volume.
State the units in your answers.

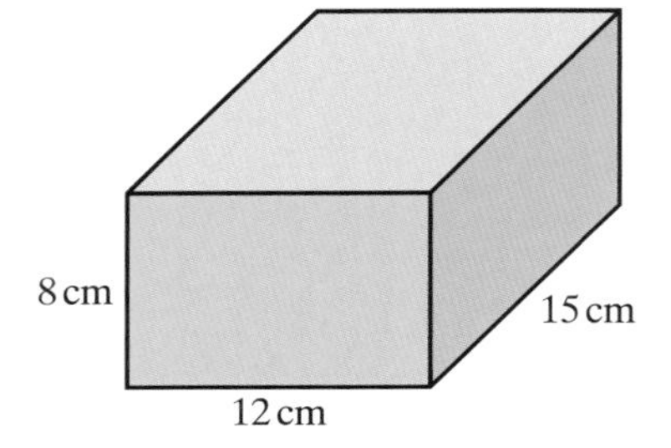

5 The cross section of a steel girder is in the shape of a letter L.
The cross section is shown in the diagram below.

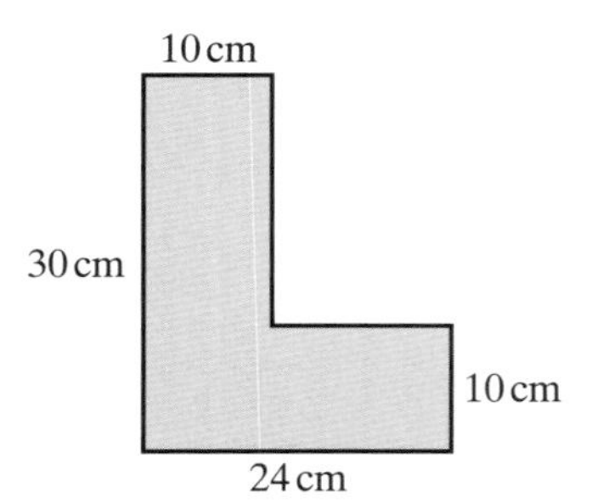

a) Work out the area of the L-shaped cross section.
The girder is 80 cm long.
b) Work out the volume of the girder.

6 A cube measures 12 cm along each side.

a) Work out the volume of the cube.

b) Work out the surface area of the cube.

7 A cuboid measures 15 cm by 20 cm by 30 cm

a) Work out the volume of the cuboid.

b) Work out the surface area of the cuboid.

8 Look at this cuboid.

Tom has worked out:

A, the perimeter of the shaded face

B, the surface area

C, the volume.

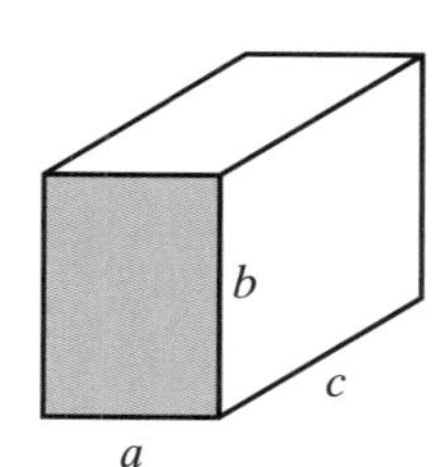

Here are his answers:

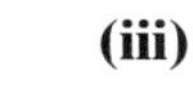

a) Match together A, B and C with Tom's answers.

Here are the formulae he used:

(i) $2ab + 2ac + 2bc$ **(ii)** $2a + 2b$ **(iii)**

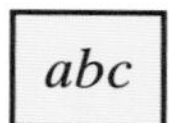

b) Match together A, B and C with Tom's formulae.

9 The diagram shows a sketch of a swimming pool.

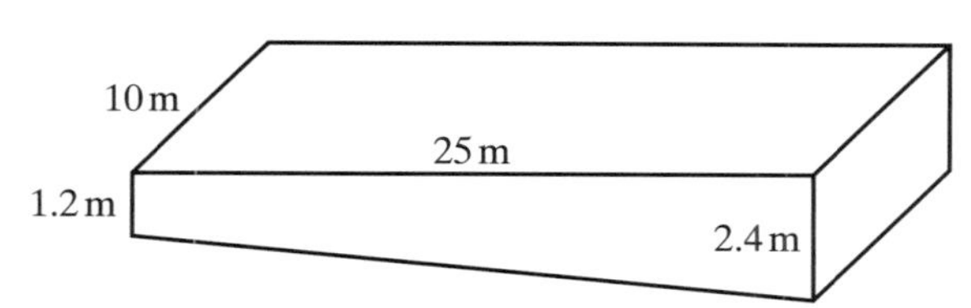

The pool is 1.2 m deep at the shallow end, and 2.4 m deep at the deep end.

The pool is 25 m long, and is 10 m wide.

a) Work out the volume of the pool.

1 cubic metre = 1000 litres

b) Work out the number of litres of water in the pool when it is full.

10 A cube has a volume of 125 cm³

a) Work out the dimensions of the cube.

b) Calculate the surface area of the cube.

11 A cuboid has a volume of 30 cm³

Its dimensions are all integers.

a) Use trial and improvement to work out the dimensions of the cuboid.

b) Calculate the surface area of the cuboid.

REVIEW EXERCISE 14

Don't use your calculator for Questions 1–7.

1 Find: **(i)** the area, **(ii)** the perimeter of each of these shapes.

a)

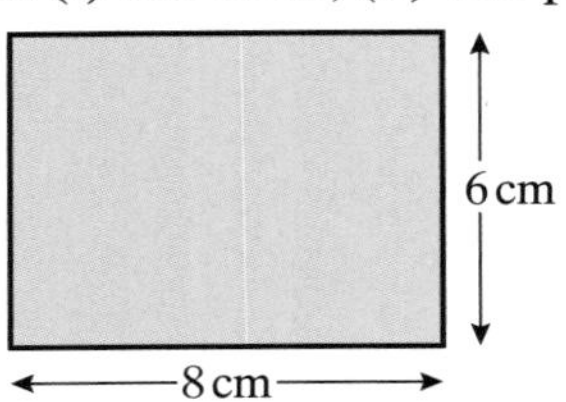

b)

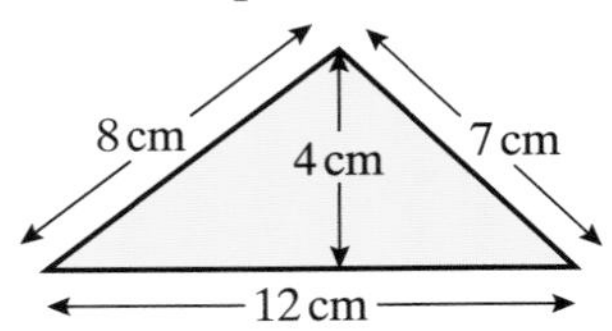

c)

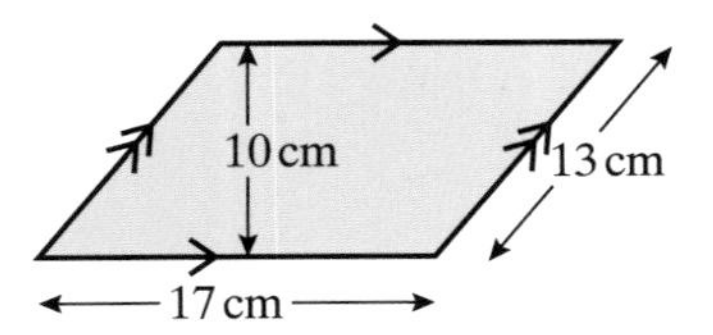

d)

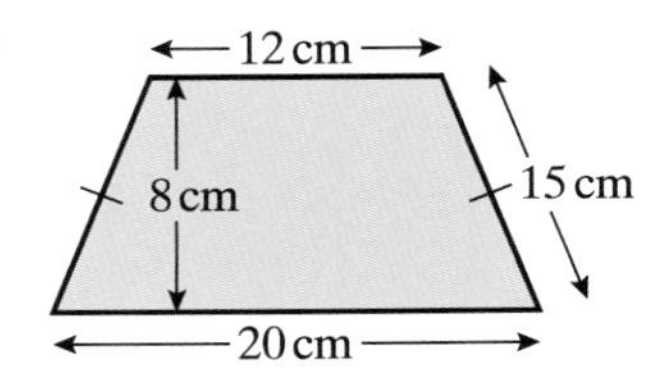

2 Find the area of these shapes:

a)

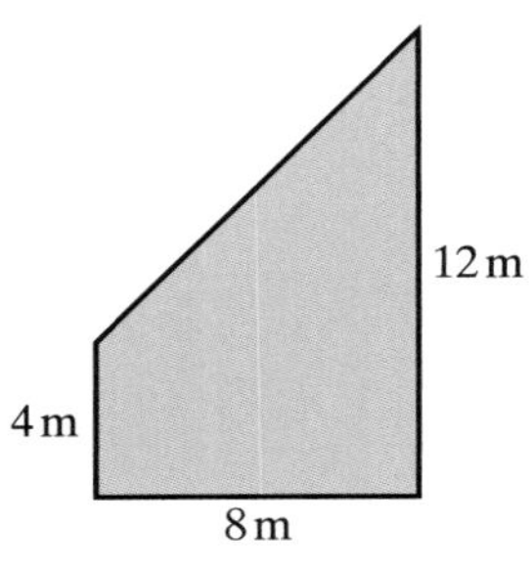

b)

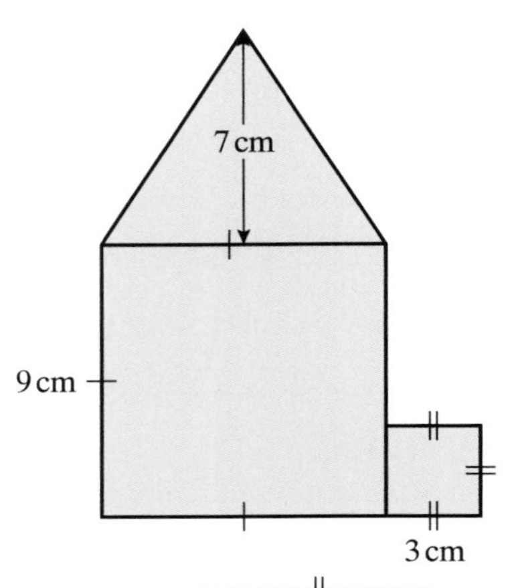

c)

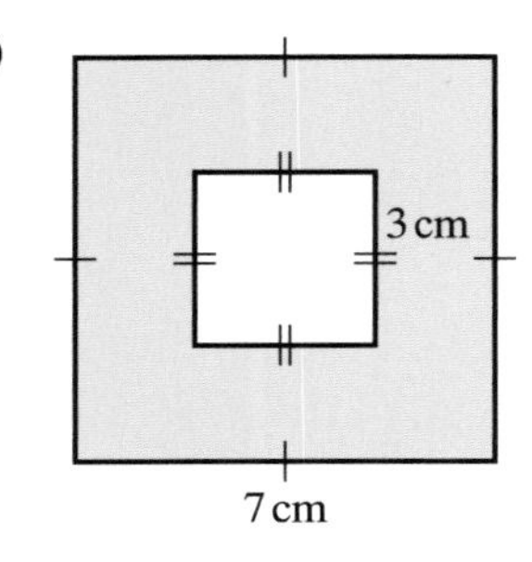

d)

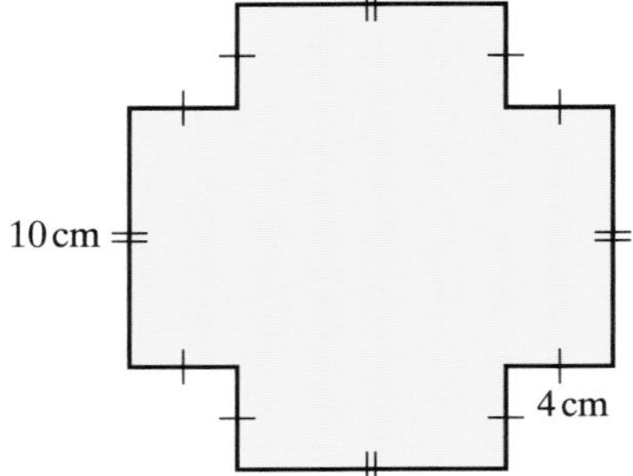

3 Find the volume of these prisms.

a)

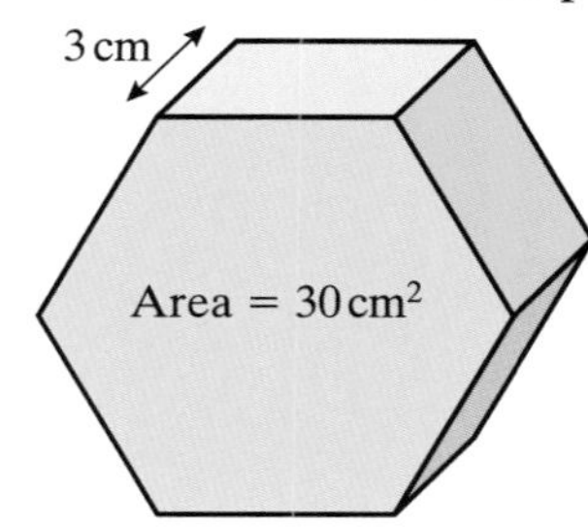

b)

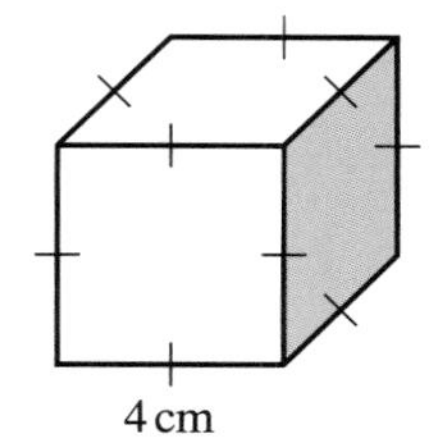

c)

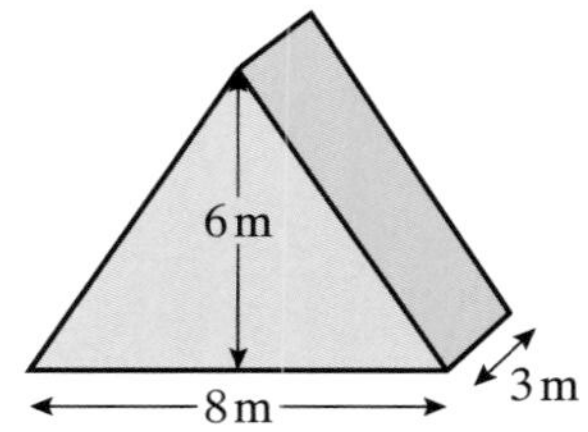

d)

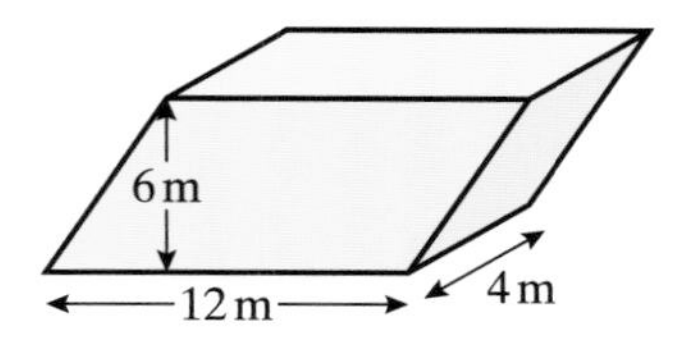

4 The diagram shows a rectangle.

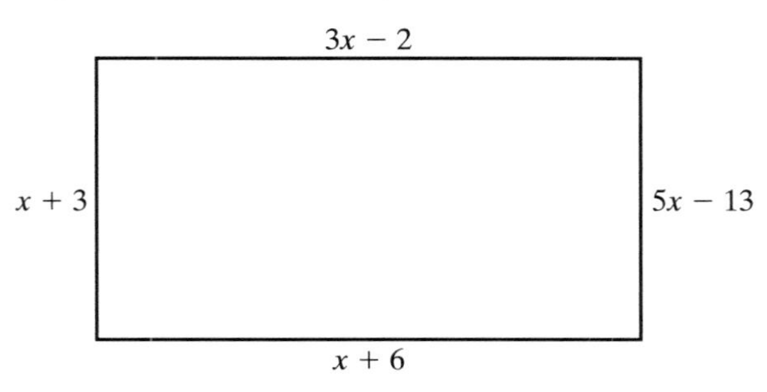

a) Work out the value of x

b) Calculate the perimeter of the rectangle.

c) Work out the area of the rectangle.

5 The diagram shows a wedge in the shape of a triangular prism.

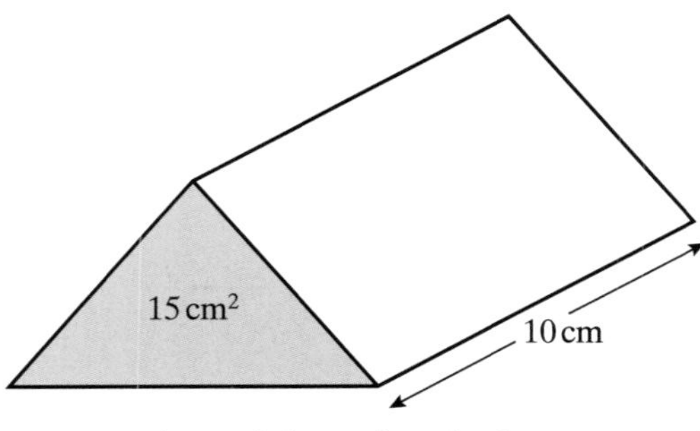

Diagram *not* accurately drawn

The cross section of the prism is shown as a shaded triangle.
The area of the triangle is 15 cm^2
The length of the prism is 10 cm
Work out the volume of the prism. **[Edexcel]**

6 Work out the volume of the triangular prism.
State the units with your answer.

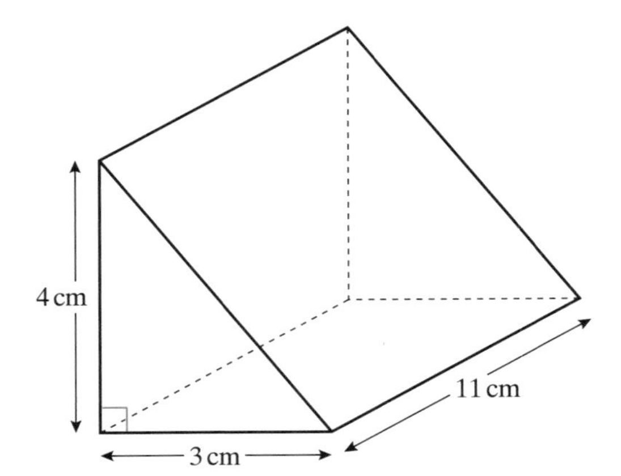

Diagram *not* accurately drawn

[Edexcel]

7 Work out the surface area of the triangular prism.
State the units with your answer.

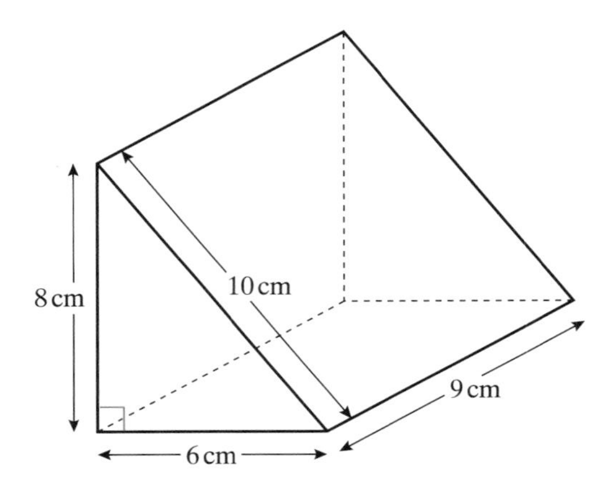

Diagram *not* accurately drawn

[Edexcel]

You can use your calculator for Questions 8–20.

8 A shaded shape has been drawn on the centimetre grid.

a) **(i)** Find the area of the shaded shape.

(ii) Find the perimeter of the shaded shape.

The shaded shape has two lines of symmetry.

b) Copy the shaded shape.
Draw the two lines of symmetry on the shaded shape.

c) Find the volume of this prism.

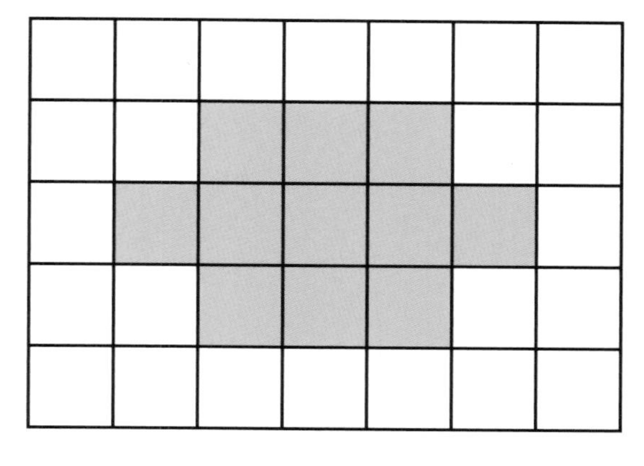

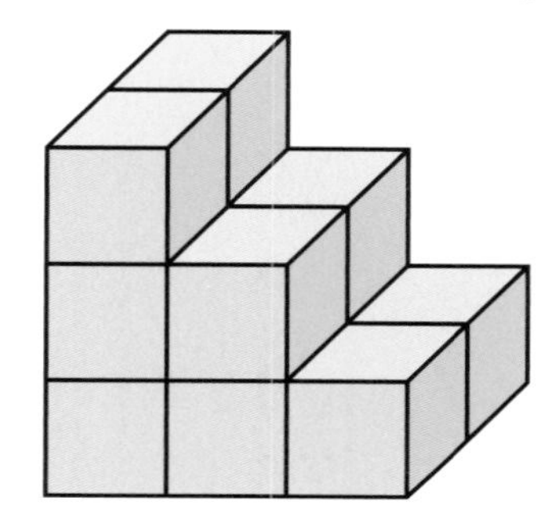

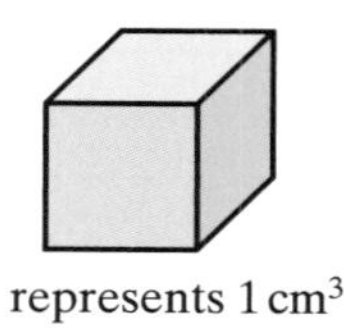

represents 1 cm³

Diagram *not* accurately drawn

[Edexcel]

9 **a)** **(i)** Find the area of the shaded shape.

(ii) Find the perimeter of the shaded shape.

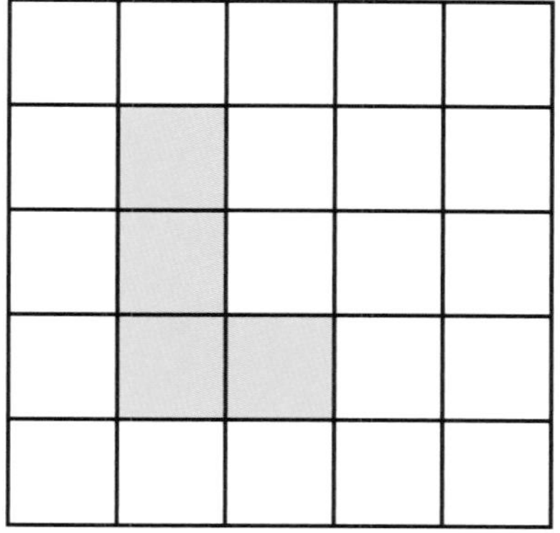

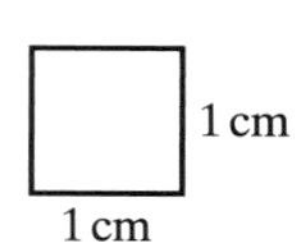

Here is a solid prism made from centimetre cubes.

b) Find the volume of the solid prism.

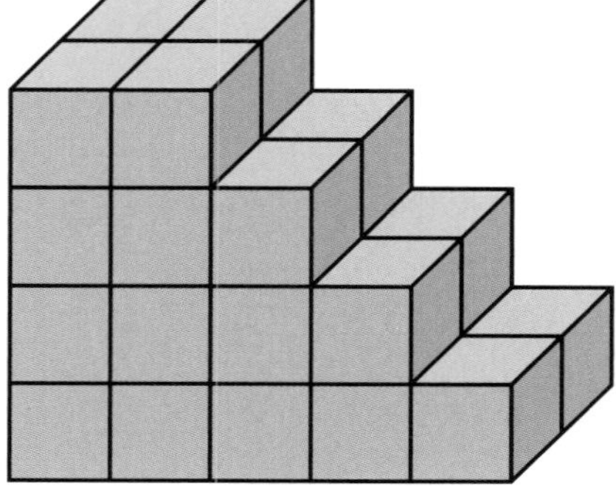

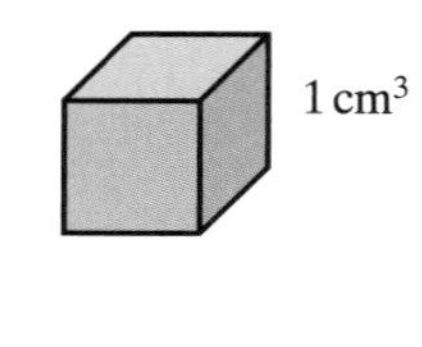

[Edexcel]

10 A cuboid has a volume of 175 cm³
Its length is 2.5 cm and its width is 3.5 cm
Work out the height of the cuboid.

11 Three different rectangles each have an area of 28 cm^2
The lengths of all the sides are whole numbers of centimetres.
For each rectangle work out the lengths of the two sides. **[Edexcel]**

12 The diagram shows a trapezium ABCD.

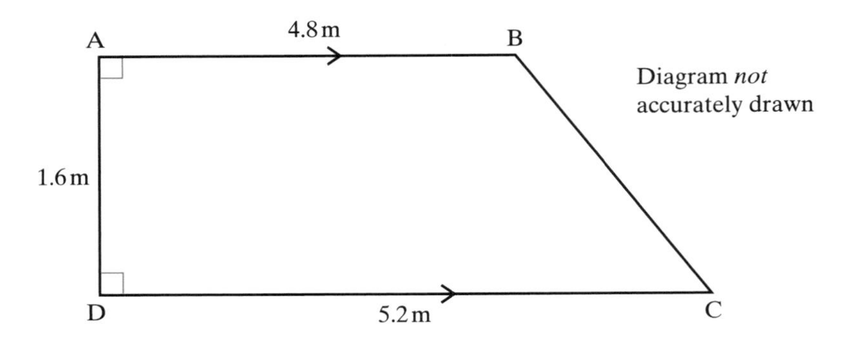

AB is parallel to DC.
AB = 4.8 m, DC = 5.2 m and AD = 1.6 m
Angle BAD = 90° and angle ADC = 90°
Calculate the area of the trapezium. **[Edexcel]**

13 The diagram shows a water tank in the shape of a cuboid.
The measurements of the cuboid are
20 cm by 50 cm by 20 cm

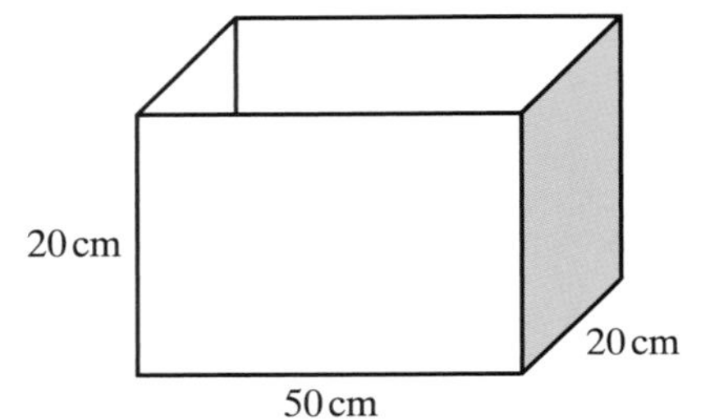

a) Work out the volume of the water tank.

Water is poured into the tank at a rate of 5 litres per minute.
1 litre = 1000 cm^3

b) Work out the time it takes to fill the water tank completely.
Give your answer in minutes. **[Edexcel]**

14 The diagram shows a shape.

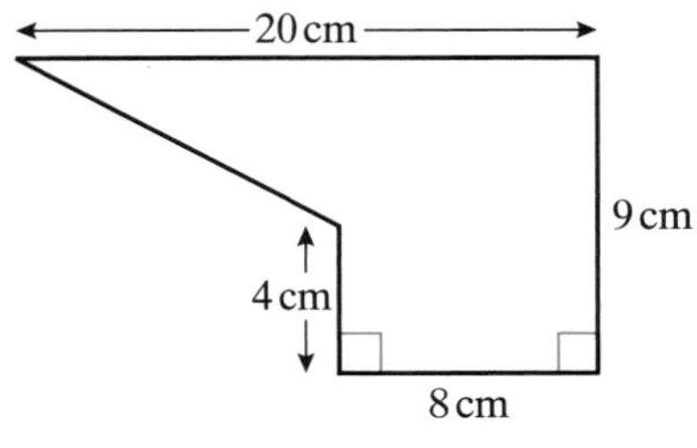

Work out the area of the shape. **[Edexcel]**

15 The diagram shows a prism.

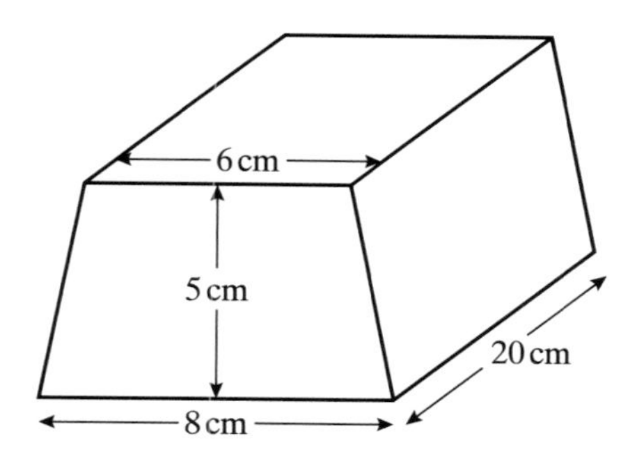

Diagram *not* accurately drawn

The cross section of the prism is a trapezium.
The lengths of the parallel sides of the trapezium are 8 cm and 6 cm
The distance between the parallel sides of the trapezium is 5 cm
The length of the prism is 20 cm

a) Work out the volume of the prism.

The prism is made out of gold.
Gold has a density of 19.3 grams per cm^3

b) Work out the mass of the prism.
Give your answer in kilograms. **[Edexcel]**

16 The lengths, in cm, of the sides of the triangle are $x + 1$, $2x + 5$ and $3x + 2$

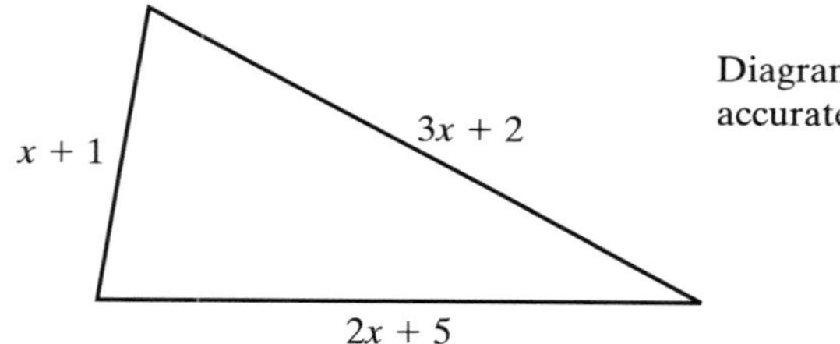

Diagram *not* accurately drawn

a) Write down, in terms of x, an expression for the perimeter of the triangle.
Give your expression in its simplest form.

The perimeter of the triangle is 50 cm

b) Work out the value of x **[Edexcel]**

17 The width of a rectangle is x centimetres.
The length of the rectangle is $(x + 4)$ centimetres.

$x + 4$

x

a) Find an expression, in terms of x, for the perimeter of the rectangle.
Give your expression in its simplest form.

The perimeter of the rectangle is 54 centimetres.

b) Work out the length of the rectangle. **[Edexcel]**

18 The diagrams show a paperweight.

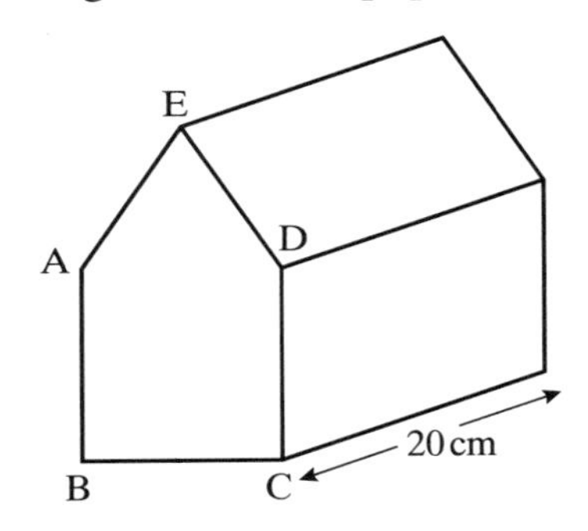

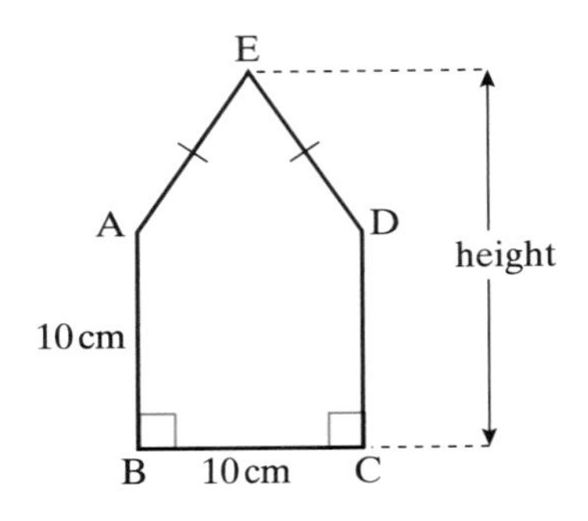

Diagram *not* accurately drawn

ABCDE is a cross section of the paperweight.
AB, BC and CD are three sides of a square of side 10 cm
AE = DE
The area of the cross-section is 130 cm^2

a) Work out the height of the paperweight.

The paperweight is a prism of length 20 cm

b) Work out the volume of the paperweight.
Give the units with your answer. **[Edexcel]**

19 The diagram represents a large tank in the shape of a cuboid.
The tank has a base.
It does not have a top.
The width of the tank is 2.8 metres, the length is 3.2 metres, the height is 4.5 metres.

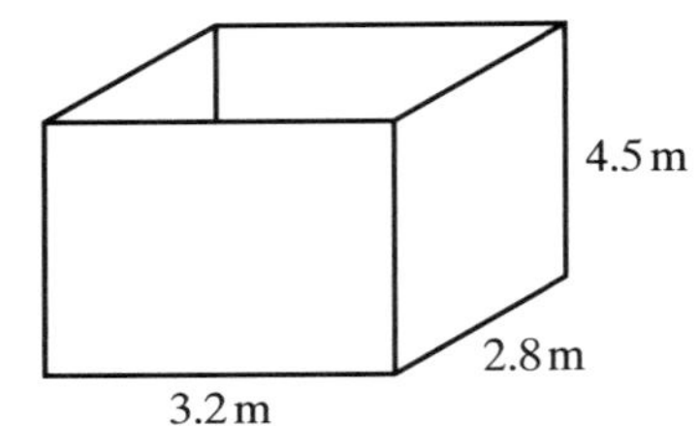

Diagram *not* accurately drawn

The outside of the tank is going to be painted.
1 litre of paint will cover 2.5 m^2 of the tank.
The cost of the paint is £2.99 per litre.
Calculate the total cost of the paint needed to paint the outside of the tank. **[Edexcel]**

20

Diagrams *not* accurately drawn

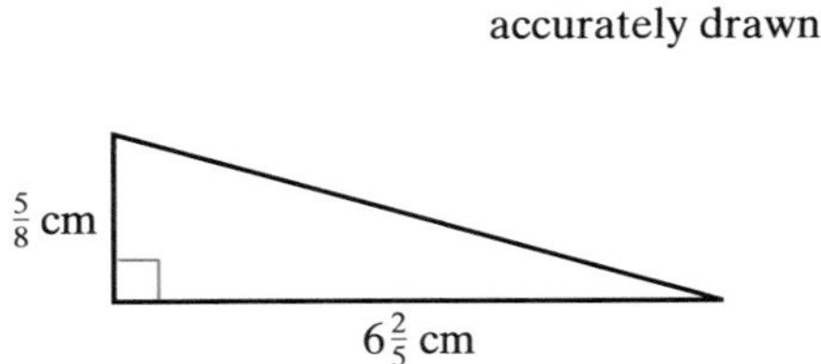

The area of the square is 18 times the area of the triangle.
Work out the perimeter of the square. **[Edexcel]**

KEY POINTS

1 The area of a rectangle = length × width

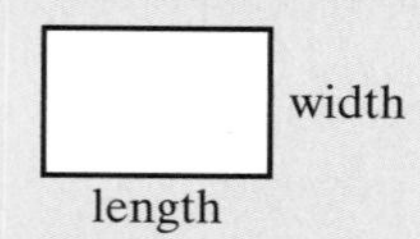

2 The area of a triangle = $\frac{1}{2}bh$

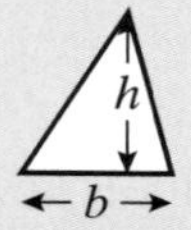

3 The area of a parallelogram = bh

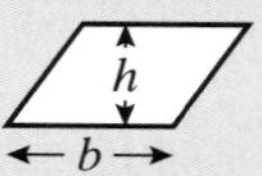

4 The area of a trapezium = $\frac{1}{2}(a + b)h$

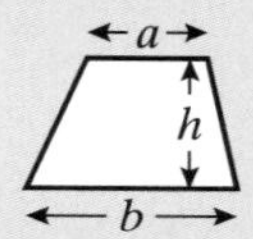

5 The perimeter of a shape = the sum of the lengths of the sides

6 The volume of a cuboid = length × width × height

7 A prism has a constant cross-sectional area.

8 The volume of a prism = cross-sectional area × length

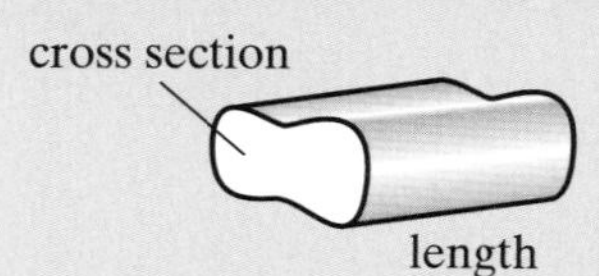

9 To find the surface area of a solid, work out the area of each separate flat surface and add them up.

10 Don't forget to put units in your answers.

Internet Challenge 14

Triangles

Here are some questions and puzzles about triangles.
Try to solve them on your own at first, then use the internet as a check.
Remember to find your answers on more than one site if possible.

1 The word triangle comes from *tri-* (= three) and −angulus (= corner).
From what language do these words originate?

2 What is the 'triangle inequality'?

3 A triangle has perimeter 15 cm, and each of its sides is a whole number of centimetres in length.
How many different triangles like this can be drawn?

4 What is the 'Bermuda triangle'?

5 What is a 'triangulation column' used for?
Where might you expect to find one?

6 How many triangles are there in each of these figures?

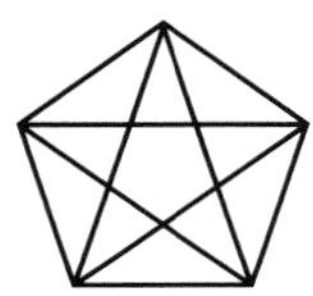

Figure 1

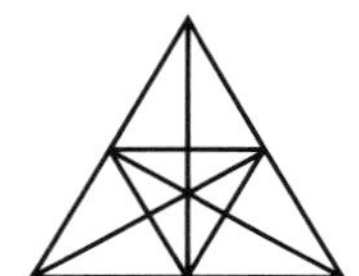

Figure 2

CHAPTER 15

Circles and cylinders

In this chapter you will **learn how to:**

- calculate the circumference and area of a circle
- calculate areas of sectors
- use circle formulae in reverse
- find the surface area and volume of a cylinder
- obtain exact expressions for areas and volumes in terms of π

You will also be **challenged to:**

- investigate measuring the Earth.

Starter: Three and a bit...

1 Use your calculator to work out the value of each of these expressions.
Write down all the figures on your calculator display.
Each answer should be a little over 3.

$3 + \frac{1}{8}$ $\qquad \frac{22}{7}$ $\qquad \sqrt{10}$ $\qquad \frac{333}{113}$

$\sqrt[4]{\left(\frac{2143}{22}\right)}$ $\qquad 3 + \frac{8}{60} + \frac{30}{60^2}$ $\qquad \frac{88}{\sqrt{785}}$

$\frac{355}{113}$ $\qquad \left(\frac{4}{3}\right)^4$ $\qquad \sqrt{2} + \sqrt{3}$

These are all approximations to an important mathematical number called 'pi'.

This is stored on your calculator as a key marked with a π symbol.

2 Use this key to obtain the value of pi correct to as many significant figures as possible, and write it down.

3 Which one of these approximations is the closest?

15.1 Circumference and area of a circle

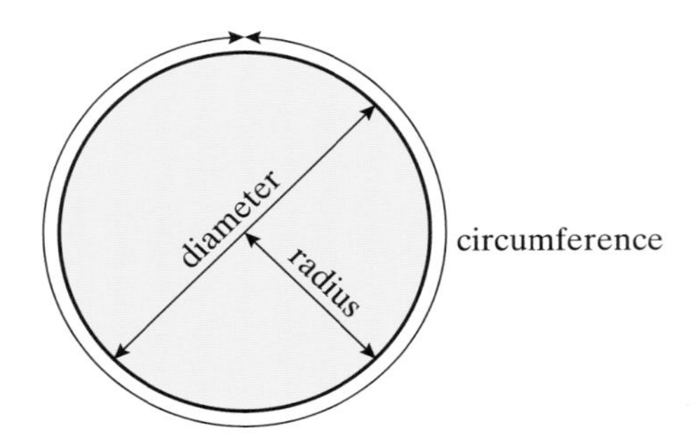

The distance all the way round the outside of a circle is known as its **circumference**.

The circumference of any circle is just over three times its **diameter**.

More precisely, the circumference is 3.1415926… times the diameter.

3.1415926… is known as **pi** (the Greek letter 'p'), written π.

The value of pi is stored in your calculator and can be called up using the π key.

Circumference of a circle = π × diameter $C = \pi d$

Sometimes it is more convenient to work with a circle's **radius**.

The radius is exactly half the diameter.

Circumference of a circle = 2 × π × radius $C = 2\pi r$

EXAMPLE

Find the circumference of these two circles.

a)

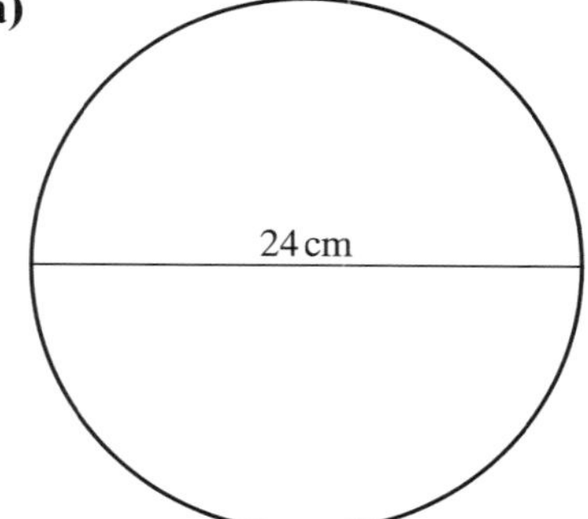

b)

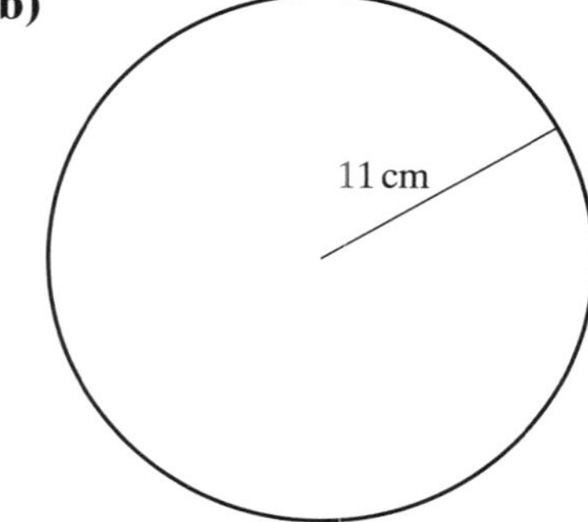

SOLUTION

Round the answer to 3 significant figures in this kind of question…
…but show your unrounded answer first.

a) This circle has diameter $d = 24$ cm

$C = \pi d$

$= \pi \times 24$

$= 75.39822369$

$= \underline{75.4 \text{ cm}}$ to 3 s.f.

b) This circle has radius $r = 11$ cm

$C = 2\pi r$

$= 2 \times \pi \times 11$

$= 69.11503838$

$= \underline{69.1 \text{ cm}}$ to 3 s.f.

The area of a circle is found by using this formula:

> Area of a circle = $\pi \times$ the square of the radius
> $A = \pi r^2$

EXAMPLE

Find the area of these two circles.

a)

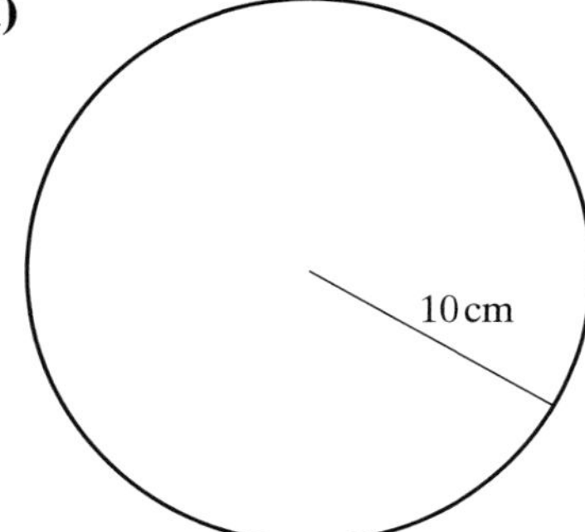

b)

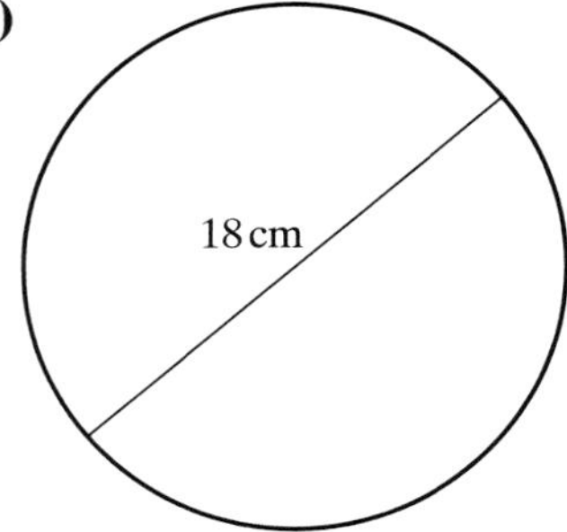

SOLUTION

a) This circle has radius $r = 10$ cm

$A = \pi r^2$

$= \pi \times 10^2$

$= \pi \times 100$

$= 314.1592654$

$= \underline{314\ \text{cm}^2}$ to 3 s.f.

You can just type this straight into your calculator – it knows to work out the square first.

b) This circle has radius $r = 18 \div 2 = 9$ cm

$A = \pi r^2$

$= \pi \times 9^2$

$= \pi \times 81$

$= 254.4690049$

$= \underline{254\ \text{cm}^2}$ to 3 s.f.

Take care to choose the right formula when working with circles.

It might help to remember that the formula containing the *squared term*, πr^2, is used for finding *area*, which is measured in *square units*.

EXERCISE 15.1

Give the answers to each of these problems to an appropriate degree of accuracy.

Make sure you give the units.

1 Find the circumference of these circles:

a)
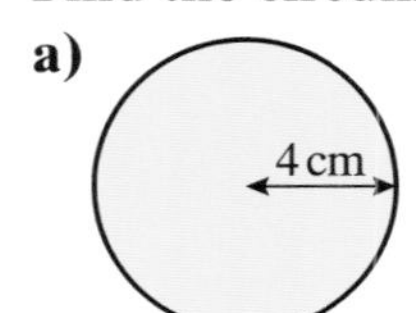

b)
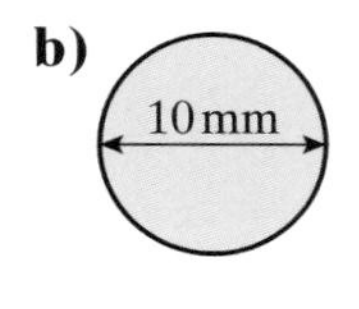

c)
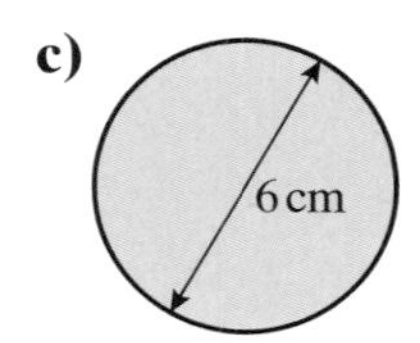

d)
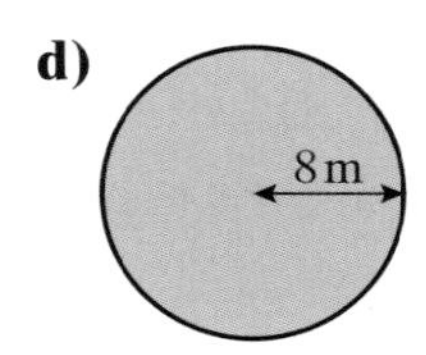

2 Find the area of these circles:

a)
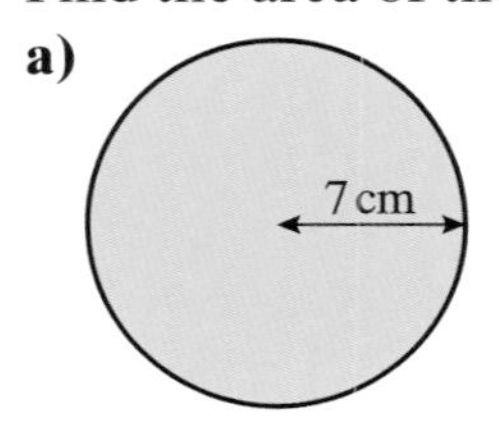

b)
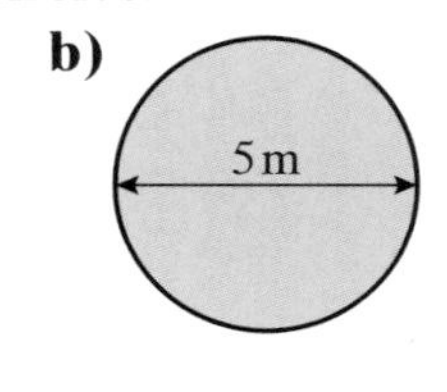

c)
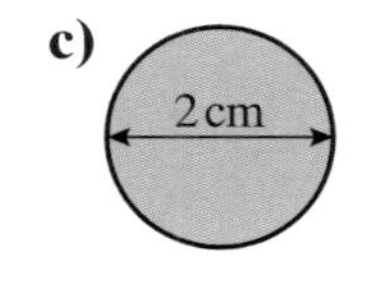

d)
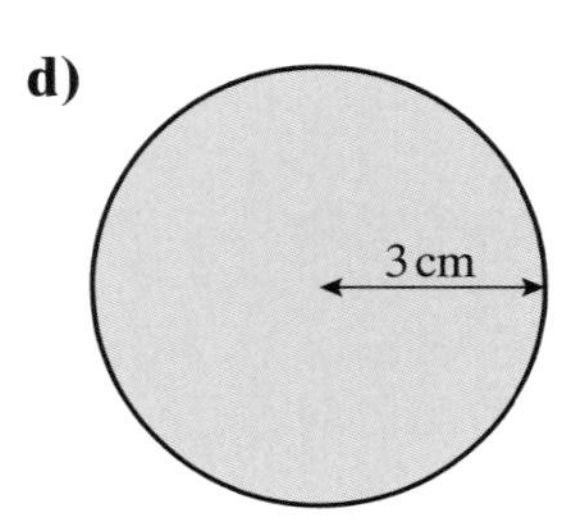

3 A circle has radius 12 mm
Find its circumference.

4 A circle has diameter 22 cm
Find its circumference.

5 A circle has radius 18 cm
Find its area.

6 A circle has diameter 11.5 cm
Find its area.

7 A circle has diameter 7 cm
Find its area.

8 A circle has radius 0.85 m
Find its area.

9 A circle has diameter 250 cm
Find its circumference.

10 A circle has radius 1.06 m
Find its circumference.

15.2 Solving problems involving circles

Some questions may require you to use the circle formulae in combination with other area or perimeter calculations.

EXAMPLE

Marie is riding a unicycle.
The diameter of the wheel is 50 cm

a) How far does she travel when the wheel makes one complete revolution?

b) How many revolutions does the wheel make when Marie cycles 20 m?

SOLUTION

a) When the wheel makes one full turn, the unicycle moves a distance equal to the circumference of the wheel.

Using $C = \pi d$,

$C = \pi \times 50$

$= 157.07\ldots$

$= 157.1$ to 1 d.p.

She travels <u>157.1 cm</u>

b) We need to use the same units, so change 157.1 cm to metres by dividing by 100:

157.1 cm = 1.571 m

We need to work out how many times 1.571 goes into 20:

$20 \div 1.571 = 12.7$

So the wheel makes <u>12.7 revolutions</u>

EXAMPLE

Work out the shaded area of this shape:

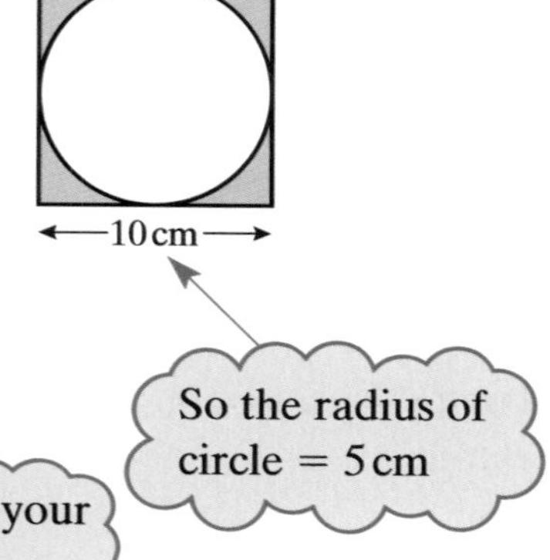

SOLUTION

Shaded area = area of square − area of circle

Area of square $= 10 \times 10 = 100\,\text{cm}^2$

Area of circle $= \pi r^2$

$= \pi \times 5^2$

$= 78.5\ldots\,\text{cm}^2$

Don't round your answer yet.

So the shaded area $= 100 - 78.5\ldots$

$= \underline{21.5\,\text{cm}^2}$ to 3 s.f.

EXERCISE 15.2

Give the answers to each of these problems to an appropriate degree of accuracy.

1 A circular ink spot has a diameter of 66 mm
Calculate its circumference.

2 A face of a one euro coin is a circle of diameter of 23.25 mm
Calculate its area.

3 Emma decides to run around a circular race track.
The radius of the track is 25 metres.
a) Work out the length of one lap of the track.

Emma wants to run at least 5000 metres.
She wants to run a whole number of laps.
b) Work out the minimum number of laps that Emma must run.

4 A circular CD disc is cut from a plastic square of side 12 cm
A central hole of diameter 1.5 cm is then cut from the centre.
Calculate the area of the plastic used to make the CD disc.

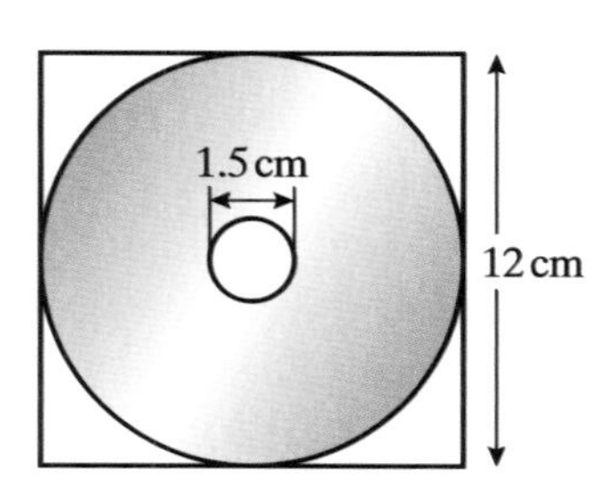

5 The diagram shows a simple 'eclipse viewer' observing aid.
The frame is made of cardboard, and comprises a rectangle with two circular holes cut in it.
The holes are then filled with a reflective safety film that blocks harmful radiation from the Sun.
The holes are each of diameter 4 cm
The rectangle measures 15 cm by 6 cm

4 cm 4 cm 6 cm 15 cm

a) Calculate the area of one of the circular holes.
b) Work out the area of the cardboard frame.

6 The diagram shows an ornamental stained glass window.
The window is a circle, of radius 30 cm
The rectangle measures 48 cm by 36 cm
The glass inside the rectangle is stained blue.
The glass outside the rectangle is stained yellow.
There is a boundary, made of lead, indicated by the heavy black line.
The lead is of negligible thickness.

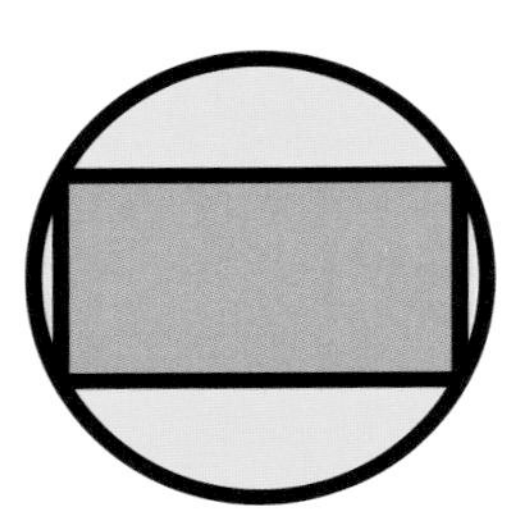

This means you can ignore the thickness of the lead.

a) Work out the length of the lead boundary.
Give your answer correct to the nearest centimetre.
b) Work out the area of the blue glass.
c) Work out the area of the yellow glass.
Give your answer correct to the nearest square centimetre.

7 Peter is rolling a hoop with a diameter of 60 cm
a) How far, in metres, does the hoop travel when Peter rolls it one complete revolution?
b) How far, in metres, does the hoop travel when Peter rolls it 10 complete revolutions?
c) How many revolutions does the hoop make when Peter rolls it 50 m?

8 The diagram shows a metal washer.
It is made from a circle of radius 6 mm
A smaller circle of radius 3 mm is then removed from the centre and discarded.
a) Calculate the area of the large circle.
b) Calculate the area of the smaller circle.
c) Hence find the area of the washer.

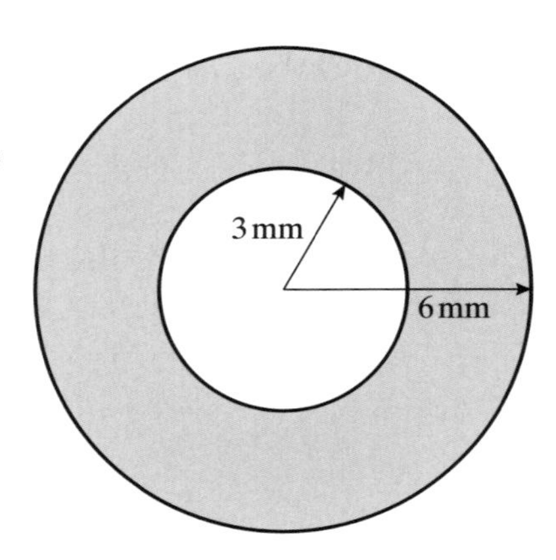

15.3 Sectors of a circle

Half a circle is called a **semicircle**.

Circles can be sliced up into quarters or into other sized fractions of a complete circle.
These are called **sectors**.

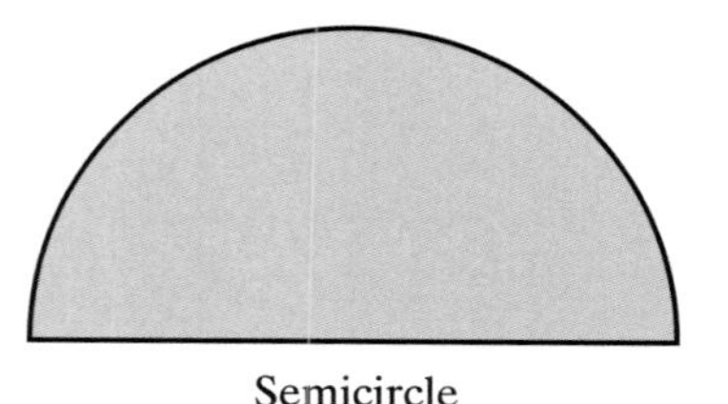

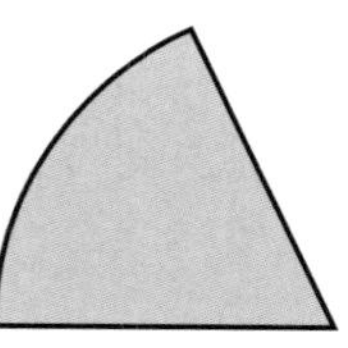

Sector

The curved boundary along the edge of a sector is called an **arc**.

You can find the length of an arc by calculating the corresponding fraction of the circumference of a circle.

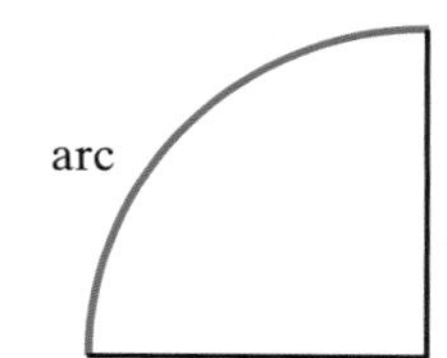

The area of a sector can be found in a similar way.

A line which touches a circle exactly once is called a **tangent**.

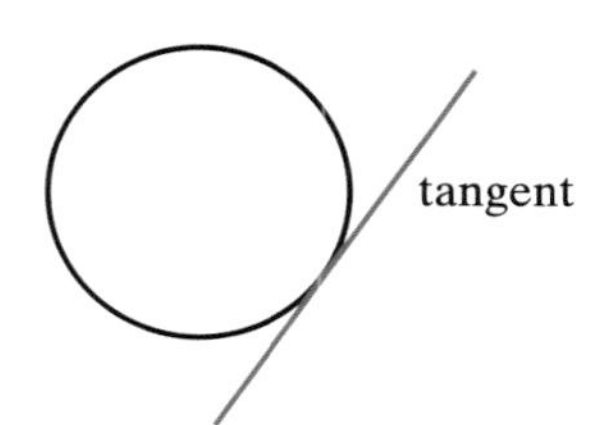

A line which cuts a circle into two (but is not the diameter) is called a **chord**.

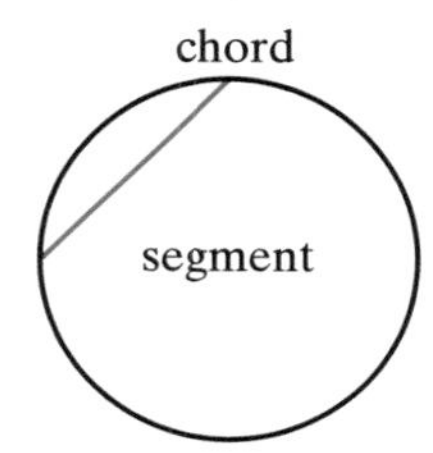

A chord cuts the circle into two **segments**.

EXAMPLE

Calculate: **a)** the perimeter, **b)** the area of this sector of a circle.

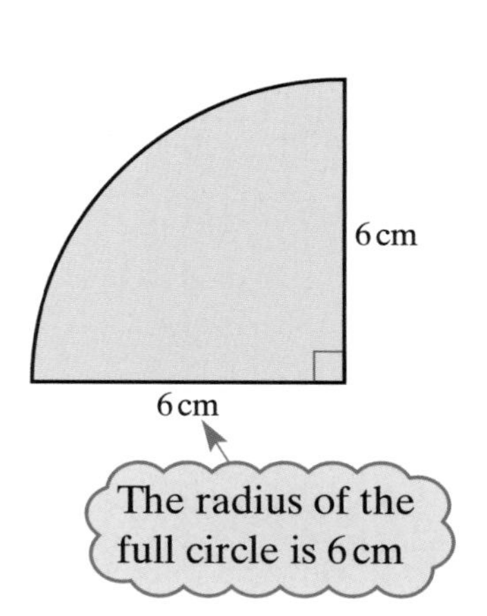

SOLUTION

a) The sector is one quarter of a circle.

The circumference of the full circle would be
$2 \times \pi \times 6 = 37.69911184\ldots$

The arc length of the sector is $37.69911184 \div 4 = 9.42$ cm

So the perimeter is $9.42 + 6 + 6 = \underline{21.43 \text{ cm}}$ to 3 s.f.

b) The area of the full circle would be $\pi \times 6^2 = 113.097\ldots \text{ cm}^2$

The area of the sector is $113.097\ldots \div 4 = \underline{28.3 \text{ cm}^2}$ to 3 s.f.

EXAMPLE

The diagram shows an ornamental flowerbed.
It is in the shape of a rectangle, with semicircles at each end.
The rectangle is of length 2.8 metres.
Each semicircle has radius 1.1 metre.

a) Calculate the perimeter of the flower bed.

The gardener is going to put edging around the flowerbed.
Edging costs £2.50 per metre.

b) Work out how much the edging will cost.

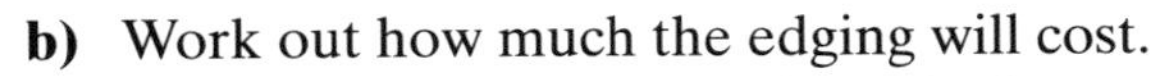

c) Calculate the area of the flowerbed.

The gardener plans to add fertiliser to the flowerbed.
1 bag of fertiliser will be sufficient for 0.8 square metres of flowerbed.

d) How many bags of fertiliser will the gardener need to buy?

SOLUTION

a) The two semicircles are equivalent to a single circle with radius $r = 1.1$

$$C = 2\pi r$$
$$= 2 \times \pi \times 1.1$$
$$= 6.911\ldots$$

The two rectangular edges are 2.8 metres each, so the total perimeter is given by:

$$P = 2.8 + 2.8 + 6.911\ldots$$
$$= 12.511\ldots$$

So the perimeter is 12.5 m to 3 s.f.

b) Cost of edging = 12.51… × £2.50
= £31.28

c) The area of the semicircles is equivalent to the area of a single circle with $r = 1.1$

$$A = \pi r^2$$
$$= \pi \times 1.1^2$$
$$= 3.801\ldots$$
$$= 3.80\text{ m}^2 \text{ to 3 s.f.}$$

The rectangular part measures 2.8 m by 2.2 m
So its area is:

$$A = 2.8 \times 2.2$$
$$= 6.16\text{ m}^2$$

So the total area is $3.80 + 6.16 = 9.96\text{ m}^2$

d) 1 bag of fertiliser is sufficient for 0.8 m^2
Therefore the gardener needs:

$$9.96 \div 0.8 = 12.45 \text{ bags}$$

So the gardener needs to buy 13 bags

EXERCISE 15.3

Give the answers to each of these problems to an appropriate degree of accuracy.

1 Calculate: **(i)** the perimeter, **(ii)** the area of each sector.

a) 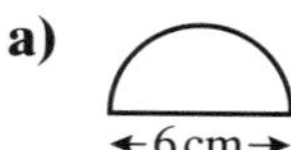

b) 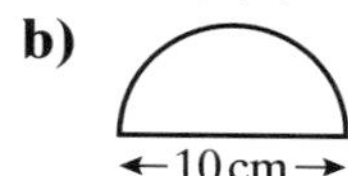

c) 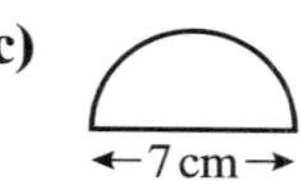

d) 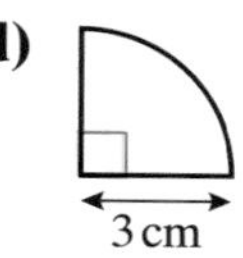

e) 4 cm

f) 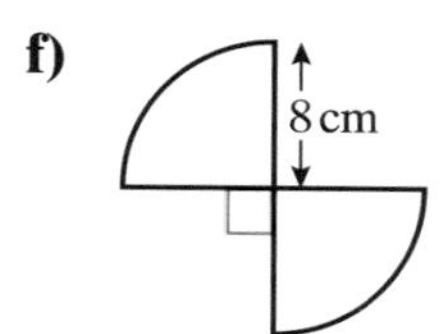

g) 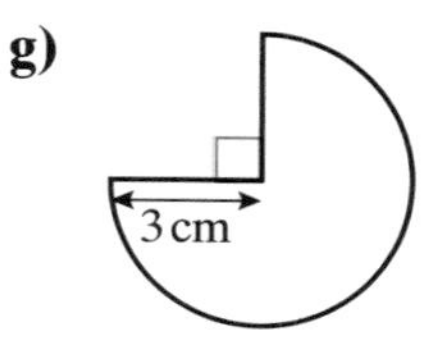

h) 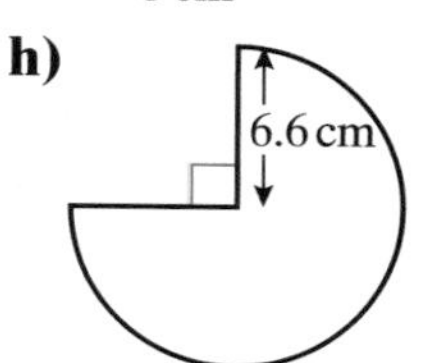

2 A pizza of diameter 12 inches is to be shared between four people.
They cut it into four equal sectors.

a) Work out the angles at the centre of each sector of pizza.
b) Work out the area of one sector.

3 The diagram shows a running track.
It is made up of two straight sections, and two semicircular ends.
The dimensions are marked on the diagram.

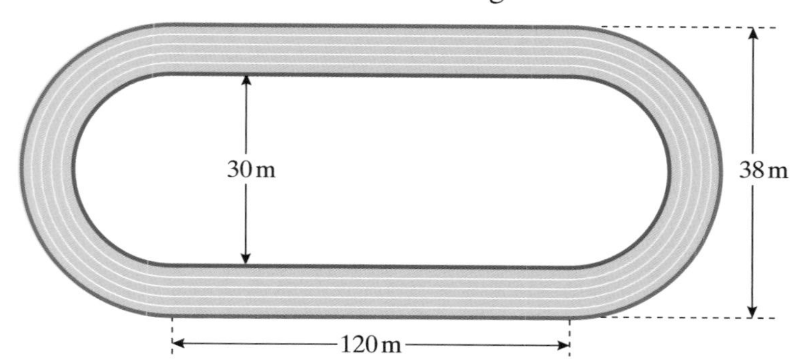

Steve runs around the **outside boundary** of the track, marked with a **red line**.
Seb runs around the **inside boundary**, marked in **blue**.
They each run one lap of the track.

a) Work out the length of the **outside boundary** of the track.
b) Work out the length of the **inside boundary** of the track.
c) How much further does Steve run?

4 Work out the area of the shaded segment.

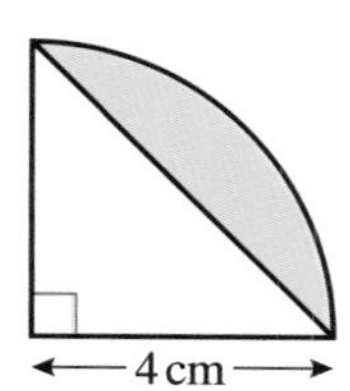

15.4 Circumference in reverse

Suppose you want to find the radius or diameter of a circle so that it will have a given circumference.

Then you would need to apply the formulae for circumference in reverse.

EXAMPLE

A circular hula-hoop has a circumference of 240 cm

Find its radius correct to the nearest centimetre.

SOLUTION

Let the radius be r metres.

$$2 \times \pi \times r = 240$$

$\div 2$ (both sides)

$$\pi \times r = 120$$

$\div \pi$ (both sides)

$$r = 38.1\ldots \text{ cm}$$

The radius is 38 cm to the nearest centimetre.

EXERCISE 15.4

Give the answers to each of these problems to an appropriate degree of accuracy.

1 A circle has circumference 15.5 cm
Find its diameter.

2 A circle has circumference 18.8 cm
Find its diameter.

3 A circle has circumference 10 m
Find its diameter.

4 A circle has circumference 40 mm
Find its radius.

5 A circle has circumference 22 cm
Find its radius.

6 A circle has circumference 52 m
Find its radius.

7 A circle has circumference 14 cm
Find its radius.

8 A circle has circumference 8 m
Find its radius.

9 The diagram shows a running track.
The ends are semicircles of radius x metres.
The straights are of length 35 metres each.
The total distance around the outside of the track is 100 metres.

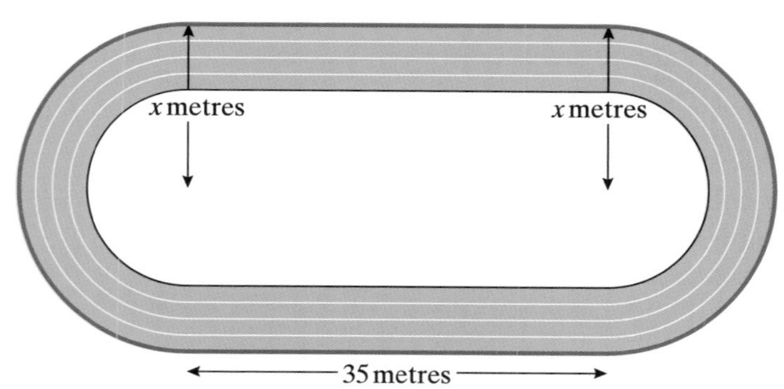

Calculate the value of x

15.5 Surface area and volume of a cylinder

You can make a hollow cylinder by rolling up a rectangular sheet of paper.

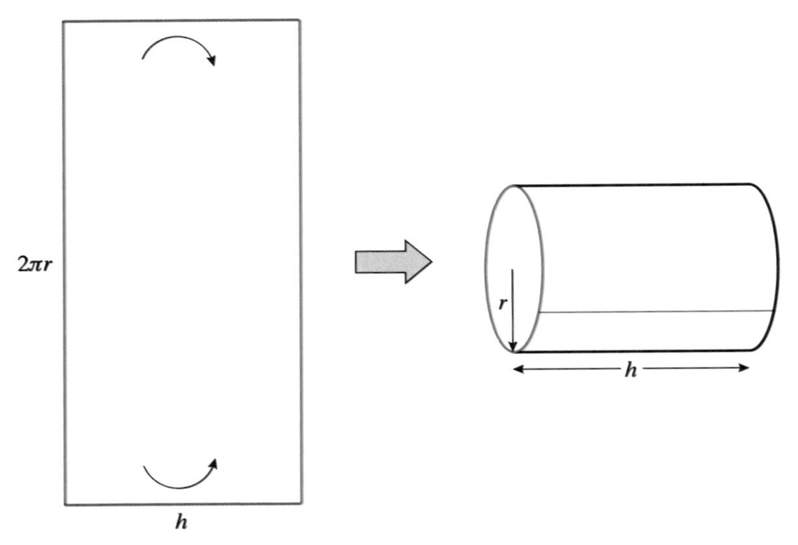

Suppose the cylinder has height h and radius r

Then the distance marked in red on the diagram is the circumference of each end of the cylinder, which is $2\pi r$

The **curved surface area** of the cylinder has the same area as the original rectangle.

This is $2\pi r$ times h, giving the formula:

> Curved surface area of a cylinder $= 2\pi rh$

A cylinder can be thought of as a prism with a circular base.

So its volume is found by multiplying the cross-sectional area (πr^2) by the length (h) to obtain the formula:

> Volume of a cylinder $= \pi r^2 h$

EXAMPLE

A metal pipe is in the form of a cylinder.
It is 1.5 metres long and 22 centimetres in diameter.

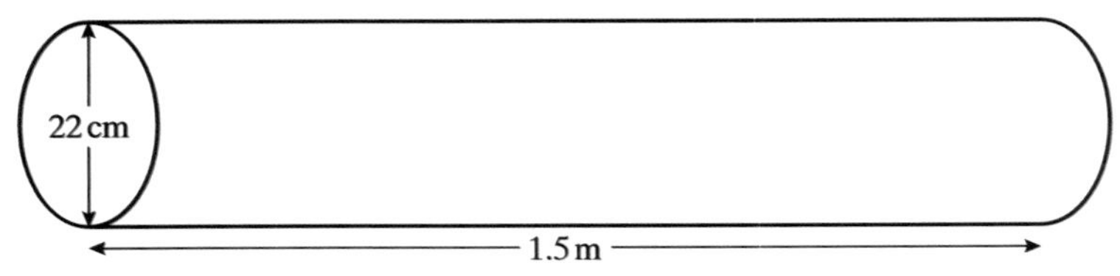

a) Calculate the curved surface area of the pipe, in square centimetres.
b) Work out the volume of the cylindrical pipe, in cubic centimetres.

SOLUTION

You need to change 1.5 m to centimetres, so:

1.5 m = 150 cm

$r = 11$ and $h = 150$

a) Curved surface area $= 2\pi rh$

$= 2 \times \pi \times 11 \times 150$

$= 10\,367.25\ldots$

$= \underline{10\,400 \text{ cm}^2}$ to 3 s.f.

b) Volume $= \pi r^2 h$

$= \pi \times 11^2 \times 150$

$= 57\,019.90\ldots$

$= \underline{57\,000 \text{ cm}^3}$ to 3 s.f.

The volume of a solid can be used to work out its mass, if you know the density of the material from which it is made.

Density is often measured in grams per cubic centimetre or in kilograms per cubic metre (see Section 11.5).

mass = volume × density

EXAMPLE

The diagram shows a steel cylinder.
It has a radius of 10 centimetres and is 2 centimetres thick.

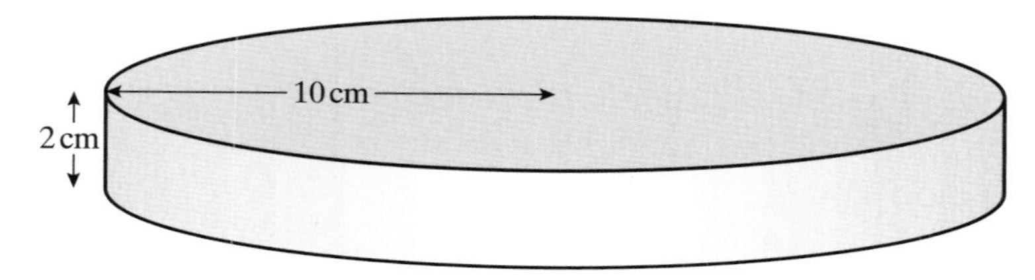

a) Work out the volume of the cylinder.

The steel has a density of 7.6 grams per cubic centimetre.

b) Work out the mass of the cylinder.

SOLUTION

a) The cylinder has $r = 10$ and $h = 2$

$V = \pi r^2 h$

$= \pi \times 10^2 \times 2$

$= 628.3\ldots$

$= \underline{628 \text{ cm}^3}$ to 3 s.f.

b) Mass of cylinder = volume × density

$= 628.31\ldots \times 7.6$

$= 4775.22\ldots$

$= \underline{4780 \text{ grams}}$ to 3 s.f.

EXERCISE 15.5

Give the answers to each of these problems to an appropriate degree of accuracy.

1 A cylinder has radius 12 cm and height 19 cm
Find its volume.

2 A cylinder has radius 5 cm and height 2 cm
Find its curved surface area.

3 A cylinder has diameter 22 cm and height 8 cm
a) Find its volume. **b)** Find its curved surface area.

4 A cylinder has radius 9 cm and height 12 cm
Find its volume.

5 The diagram shows a hollow cylinder.
Work out the curved surface area of the cylinder.

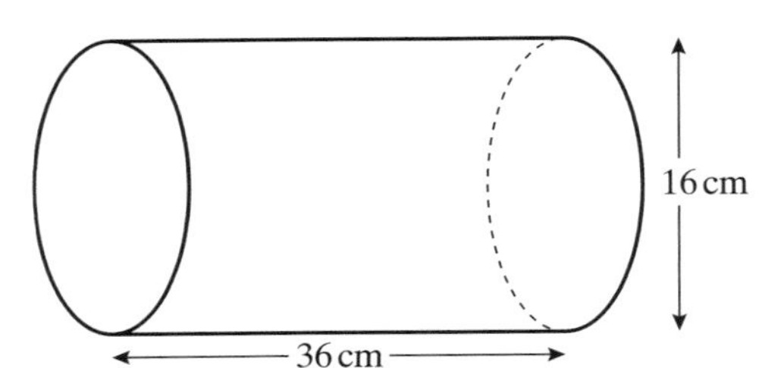
16 cm
36 cm

6 A hollow cylindrical pipeline has an internal diameter of 15 cm
The pipeline is 120 metres in length.
a) Work out the volume of the pipeline.
Give your answer in cm^3
b) $1000\ cm^3 = 1$ litre
Express the volume of the pipeline in litres.

7 A biscuit tin is in the shape of a cylinder.
It has radius 9 cm and height 14 cm
Work out the volume of the cylinder.

8 A sweet packet is in the shape of a hollow cardboard cylinder.
The inside diameter of the cylinder is 2.5 cm and it has a height of 15 cm
a) Work out the volume of the cylinder.

The sweets have a volume of $1.5\ cm^3$ each.
b) Show that the packet cannot contain as many as 50 sweets.

9 Jamie and Joe have been working on an exercise about cylinders.
They have to find the volume of a cylinder with diameter 14 cm and height 24 cm

The volume is 3690 cm³ correct to 3 s.f.
Jamie
The volume is 14 800 cm³ correct to 3 s.f.
Joe

a) Work out who is right.
b) Suggest what mistake has been made by the person who is wrong.

15.6 Exact calculations using pi

So far in this chapter, you have used the pi key on your calculator.

Although this is very convenient, it does introduce a slight inaccuracy to your work.

You can make exact statements about areas and volumes of circles and cylinders, by leaving π in both your working and your final answer.

Some examination questions will instruct you to do this.

EXAMPLE

The diagram shows a circular washer.
It is made from a metal circle of radius 4 mm, with a circular hole of radius 2 mm removed.
Work out the area of the washer.
Leave your answers in terms of π

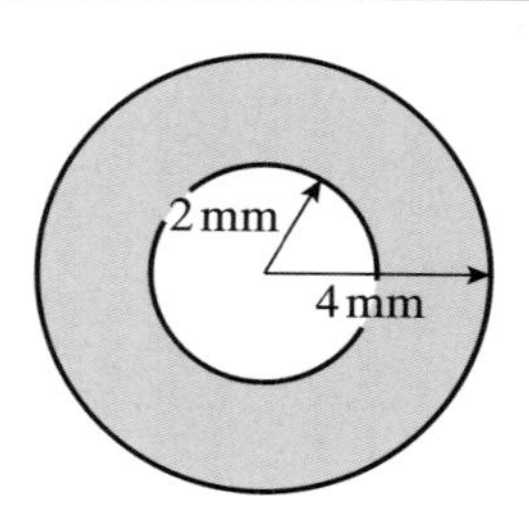

SOLUTION

The area of the larger circle is:

$$\pi r^2 = \pi \times 4^2 = \pi \times 16 = 16\pi \text{ mm}^2$$

The area of the smaller circle is:

$$\pi r^2 = \pi \times 2^2 = \pi \times 4 = 4\pi \text{ mm}^2$$

So the area of the washer $= 16\pi - 4\pi$

$= \underline{12\pi \text{ mm}^2}$

EXAMPLE

A cylinder has height 8.5 cm and radius 4 cm
Work out **a)** its curved surface area, **b)** its volume.
Leave your answers in terms of π

SOLUTION

Just multiply all the numbers except π together. Leave π in the answer.

For this cylinder, $h = 8.5$ and $r = 4$

a) Curved surface area $= 2\pi rh$

$= 2 \times \pi \times 4 \times 8.5$

$= 2 \times \pi \times 34$

$= \pi \times 68$

$= \underline{68\pi \text{ cm}^2}$

b) Volume $= \pi r^2 h$

$= \pi \times 4^2 \times 8.5$

$= \pi \times 16 \times 8.5$

$= \pi \times 136$

$= \underline{136\pi \text{ cm}^3}$

EXERCISE 15.6

1 A circle has diameter 20 cm
Work out:
a) its circumference **b)** its area.
Leave your answers in terms of π

2 A circle has radius 11 cm
Work out:
a) its circumference **b)** its area.
Leave your answers in terms of π

3 A cylinder has radius 12 cm and height 8 cm
a) Find the curved surface area of the cylinder.
b) Find the volume of the cylinder.
Leave your answers in terms of π

4 A circle has circumference 24π centimetres.
a) Find the exact radius of the circle.
b) Find the exact area of the circle.
Leave your answer in terms of π

5 A circle has a circumference of 10π centimetres.
a) Find the exact radius of the circle.
b) Find the exact area of the circle.
Leave your answer in terms of π.

6 A cylinder has volume $300\pi\ \text{cm}^3$
It has radius 10 cm
Work out its height.

7 A cylinder has volume $250\pi\ \text{cm}^3$
It has radius 5 cm
Work out its height.

8 The diagram shows a quarter of a circle.
The radius is 16 cm
a) Find the shaded area in terms of π
b) Find an exact expression for the perimeter of the shape.

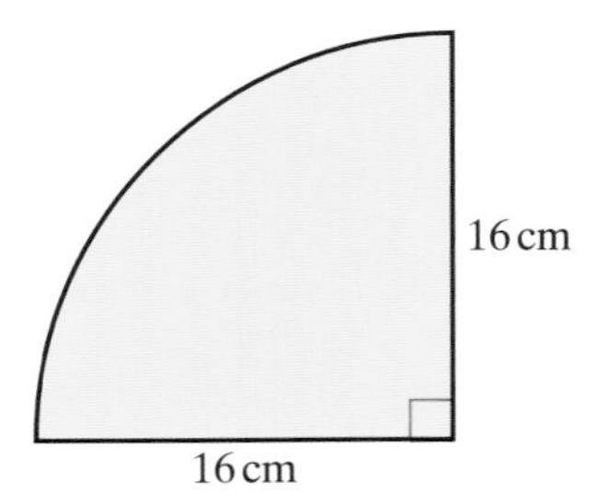

9 The diagram shows an ornamental design.
It is in the shape of a square, with semicircles on each of the four sides.
The square is of side 12 cm

a) Find the area of one of the semicircles.
Leave your answer in terms of π

b) Hence find an exact expression for the area of the ornamental design.

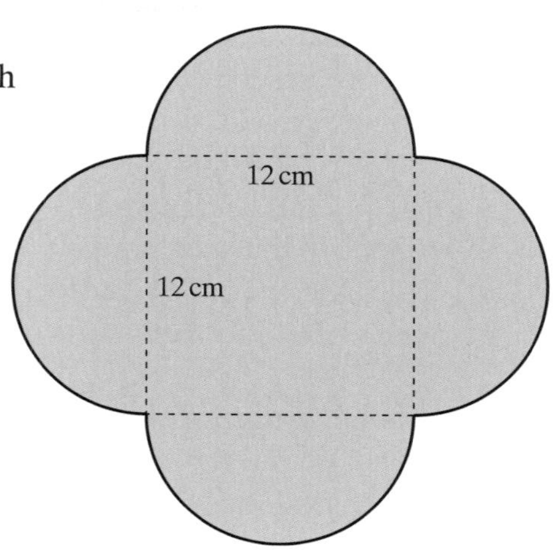

10 The diagram shows two cylinders.
Cylinder A has diameter 6 cm and height 8 cm
Cylinder B has diameter 8 cm and height 6 cm

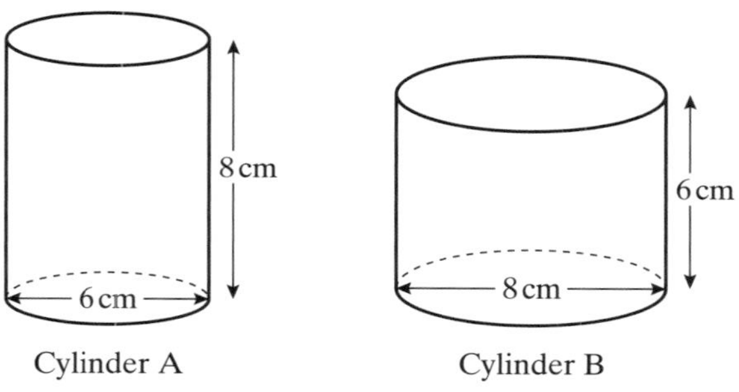

a) Show that both cylinders have exactly the same curved surface area.

b) Work out the volume of each cylinder, leaving your answers in terms of π
Which cylinder has the larger volume?

REVIEW EXERCISE 15

Give the answers to each of these problems to an appropriate degree of accuracy.

1 A circle has radius 28 cm
Work out its area.

2 A circle has diameter 90 mm
Work out its circumference.

3 A circle has radius 1.9 cm
Work out its circumference.

4 A circle has diameter 64 mm
Work out its area.

5 A circle has a radius of 10 cm
 a) Find its circumference in terms of π
 b) Find its area in terms of π

6 A closed cylinder has a radius of 10 cm and a height of 15 cm
 a) Calculate the area of one of its circular ends.
 b) Calculate the curved surface area of the cylinder.
 c) Hence find the total surface area of the cylinder.

7 A circle has a circumference of 15.71 cm
 a) Calculate the radius of this circle.
 b) Hence find the area of the circle.

8 The diagram shows a rectangle inscribed in a circle.
AB = 5 cm, BC = 12 cm and AC = 13 cm
The line segment AC is a diameter of the circle.
Work out the size of the shaded area.

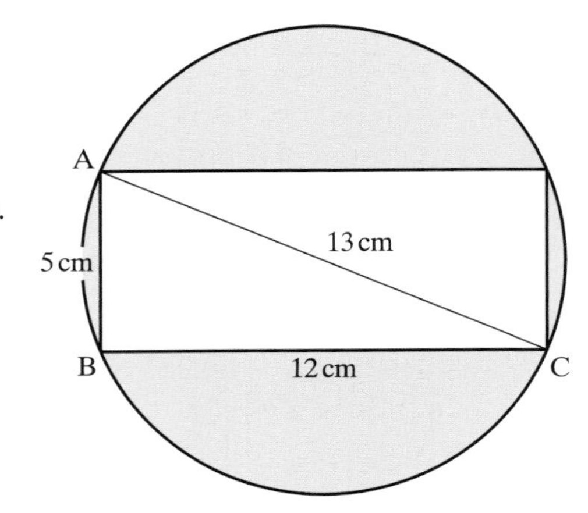

9 The radius of a circle is 5.1 m

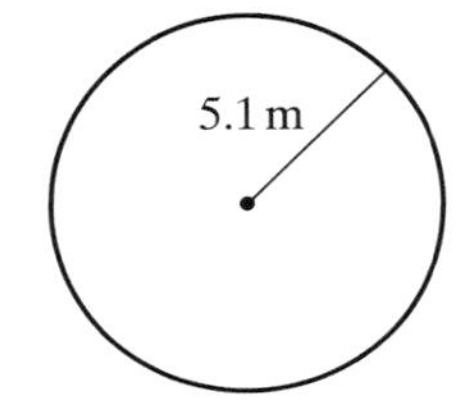

Diagram *not* accurately drawn

Work out the area of the circle.
State the units of your answer. **[Edexcel]**

10 A circle has a radius of 32 cm
Work out the circumference of the circle.
Give your answer correct to the nearest centimetre. **[Edexcel]**

11 A ten pence coin has a diameter of 2.45 cm
Work out the circumference of the coin.

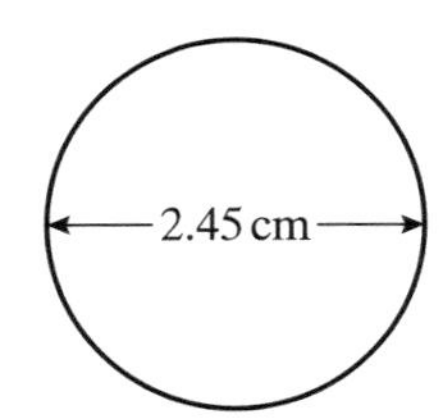

Give your answer in cm correct to 1 decimal place. **[Edexcel]**

12 A circle has a radius of 3 cm

a) Work out the area of the circle.
Give your answer correct to 3 significant figures.

Diagram *not* accurately drawn

A semicircle has a diameter of 9 cm

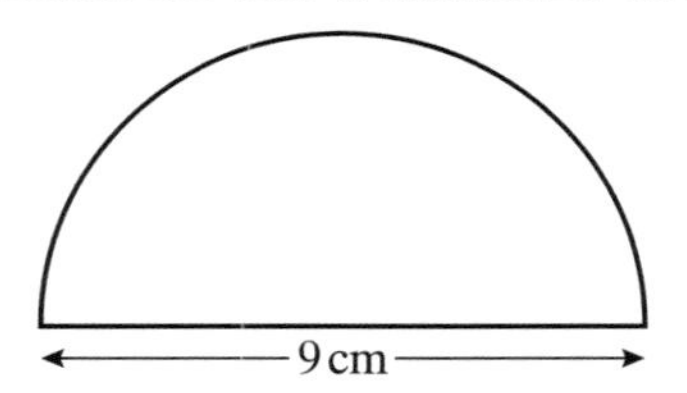

Diagram *not* accurately drawn

b) Work out the perimeter of the semicircle.
Give your answer correct to 3 significant figures.

[Edexcel]

13 The diagram shows a semicircle.
The diameter of the semicircle is 15cm
Calculate the area of the semi-circle.
Give your answer correct to 3 significant figures.

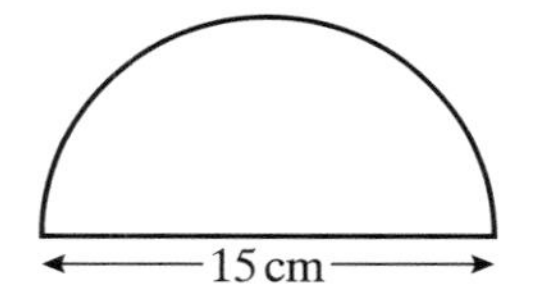

Diagram *not* accurately drawn

[Edexcel]

14 The diagram shows a right-angled triangle ABC and a circle.
A, B and C are points on the circumference of the circle.
AC is a diameter of the circle.
The radius of the circle is 10 cm
AB = 16 cm and BC = 12 cm

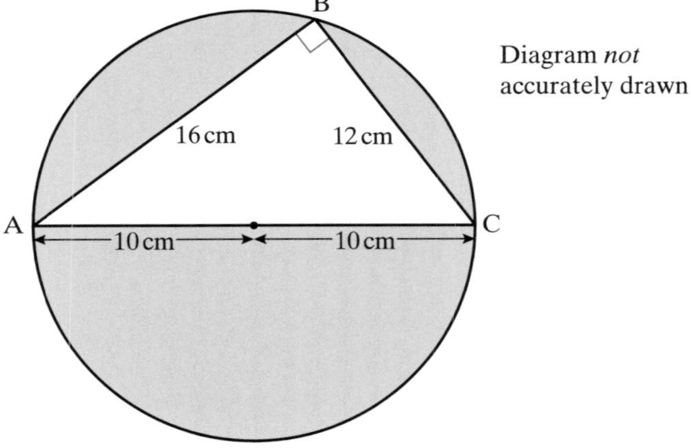

Diagram *not* accurately drawn

Work out the area of the shaded part of the circle.
Give your answer correct to the nearest cm^2

[Edexcel]

15 A can of drink is in the shape of a cylinder.
The can has a radius of 4 cm and a height of 15 cm
Calculate the volume of the cylinder.
Give your answer correct to 3 significant figures.

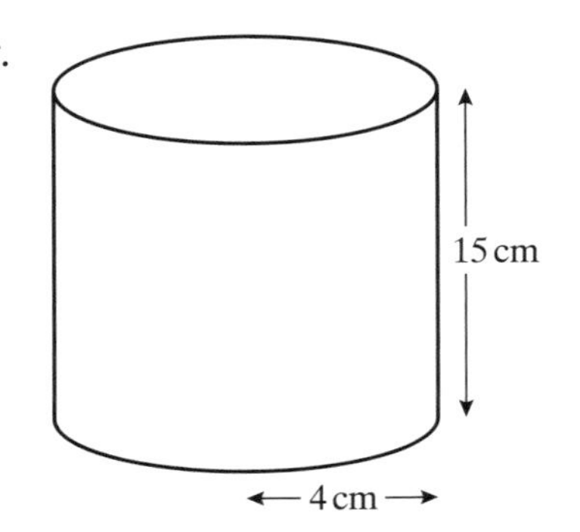

Diagram *not* accurately drawn

[Edexcel]

16 An ice hockey puck is in the shape of a cylinder with a radius of 3.8 cm, and a thickness of 2.5 cm
It is made out of rubber with a density of 1.5 grams per cm^3

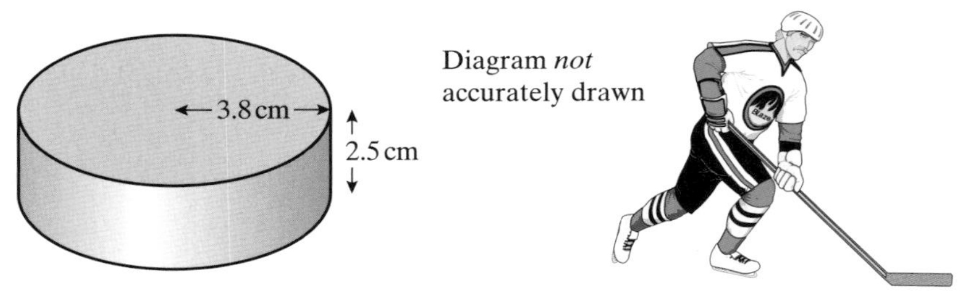

Work out the mass of the ice hockey puck.
Give your answer correct to three significant figures. **[Edexcel]**

17 The diagram shows a shape, made from a semicircle and a rectangle.

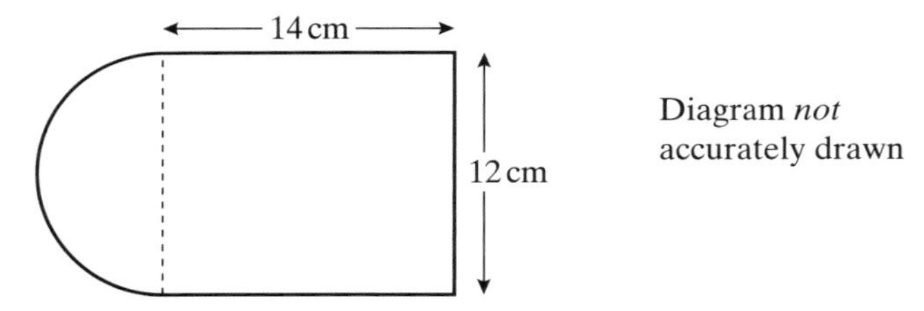

The diameter of the semicircle is 12 cm
The length of the rectangle is 14 cm
Calculate the **perimeter** of the shape.
Give your answer correct to 3 significant figures. **[Edexcel]**

KEY POINTS

1. The circumference of a circle is given by:

   $C = 2\pi r$ or $C = \pi d$

2. The area of a circle is given by:

   $A = \pi r^2$

3. To work out the area of a semicircle:
   - find the area of the whole circle
   - divide it by 2

4. To work out the area of a quarter circle:
   - find the area of the whole circle
   - divide it by 4

5. The arc length of a sector can be found in a similar way.

6. If a question asks you to find the perimeter of a sector, remember to include the two radii as well as the curved arc.

7. A line which touches a circle exactly once is called a tangent.

8. A line which cuts a circle into two (but is not the diameter) is called a chord.

9. The volume of a cylinder of radius r and height h is given by:

   $V = \pi r^2 h$

10. The curved surface area of the cylinder is given by:

    $A = 2\pi rh$

11. Make sure that you know how to use your calculator's π key correctly.

12. Answers to calculations will normally need to be rounded off – an examination question will tell you how many significant figures or decimal places are required.

13. Remember that some examination questions will ask you to leave your answers as exact expressions in terms of π.

When reading examination questions, take care to check whether you have been told the radius or the diameter of the circle.

Internet Challenge 15

Measuring the Earth

A few hundred years ago many people thought the Earth was flat.

They feared you might fall off the edge if you travelled too far from home!

Most people now accept that the Earth is roughly spherical, with a diameter of about 12 800 km.

Use the internet to help research the answers to these questions about the Earth.

1 What observational evidence can you find to support the claim that the Earth is roughly spherical?

2 What organisation claims to have been 'deprogramming the masses since 1547'?

3 Find an accurate value for the Earth's equatorial diameter.
Use this figure to calculate the Earth's circumference (around the equator).

4 Find an accurate value for the Earth's polar diameter.
Use this figure to calculate the Earth's circumference from pole to pole.

5 Find the definition of a Great Circle.
Is the equator a Great Circle?

6 Who was the first person to circumnavigate the globe, that is, travel right round the Earth?
How long did the journey take, and when was it completed?

7 Who first circumnavigated the world pole to pole?
When?

8 Some adventurous sailors take part in round the world yacht races.
How far do they typically travel?
Do you think they really do travel around the world, in the strictest sense?

9 What is the origin of our word 'geometry'?

10 The size of the Earth was first measured accurately by Eratosthenes, around 200 BC.
Find out as much as you can about Eratosthenes and the methods he used.
You might want to collect your findings into a poster for your classroom, or prepare a *Powerpoint* presentation for your class.

CHAPTER 16

Pythagoras' theorem

In this chapter you will learn how to:

- use Pythagoras' theorem to test whether triangles are right angled
- use Pythagoras' theorem to find an unknown side in a right-angled triangle.

You will be challenged to:

- investigate Pythagorean triples.

Starter: Finding squares and square roots on your calculator

When you multiply a number by itself, you are finding its **square**.

For example, 3 squared is 9 because $3 \times 3 = 9$

This is usually written $3^2 = 9$

To find the square of 3.1, you would probably prefer to use a calculator:

3.1 [x^2] $= 9.61$

The reverse process of squaring is called **square rooting**, or finding the **square root**.

This is much harder than squaring and usually requires the use of a calculator with a square root [$\sqrt{}$] key.

You may find that your calculator screen fills with decimal figures; if so, it is usual to round the answer off to 3 significant figures.

1 Look at these numbers.
Work out the square of each one.
Several of them can be done without a calculator, but you may use a calculator for the harder ones.

4 7 2.5 1.2 0.8 13 6 16

2 Look at these numbers.
Work out the square root of each one.
If the answers are not exact then you should round off to 3 significant figures.

13 10 16 22.5 12.25 64 130 121

3 Use your calculator to find the square roots of these numbers, to 3 significant figures where necessary.

8 9 10 11 12 13 14 15 16 17

4 Why is it possible to find the square roots of some whole numbers without using a calculator?

16.1 Introducing Pythagoras' theorem

Pythagoras' theorem concerns right-angled triangles.

Suppose you have a right-angled triangle with sides of lengths a, b and c, with c being the longest side.
The longest side is called the **hypotenuse**.

Pythagoras' theorem states that:

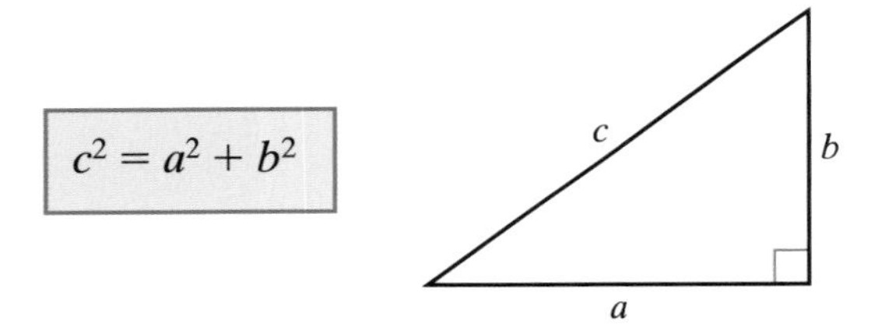

For example, if $a = 4$ cm and $b = 3$ cm then c would be 5 cm

This is because:

$$5^2 = 3^2 + 4^2$$

since $25 = 9 + 16$

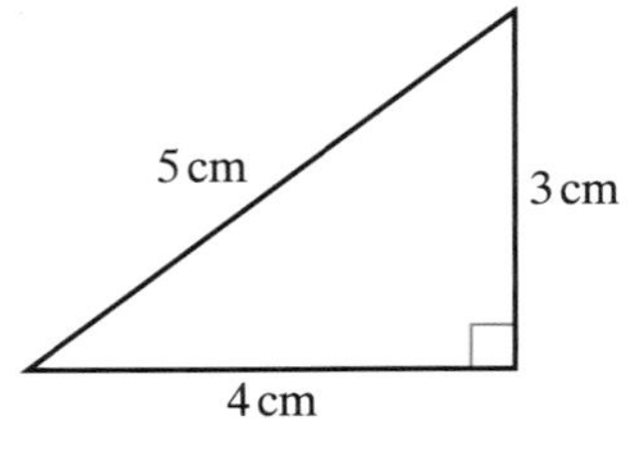

In fact, Pythagoras' theorem only works in right-angled triangles.
So you can use it to check whether a triangle actually is right angled or not.

EXAMPLE

Use Pythagoras' theorem to check whether each of these triangles is right angled or not.

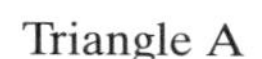

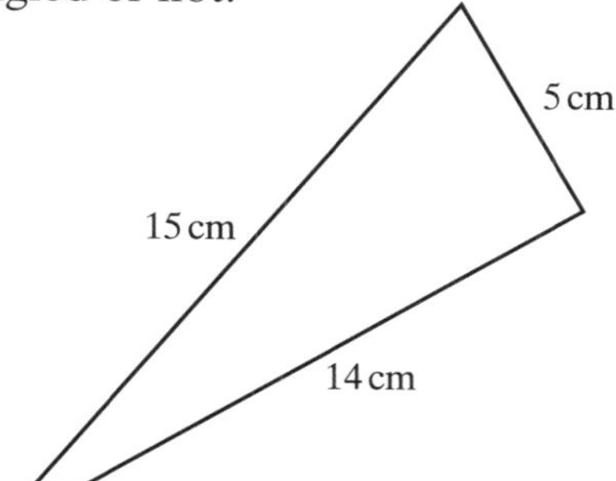

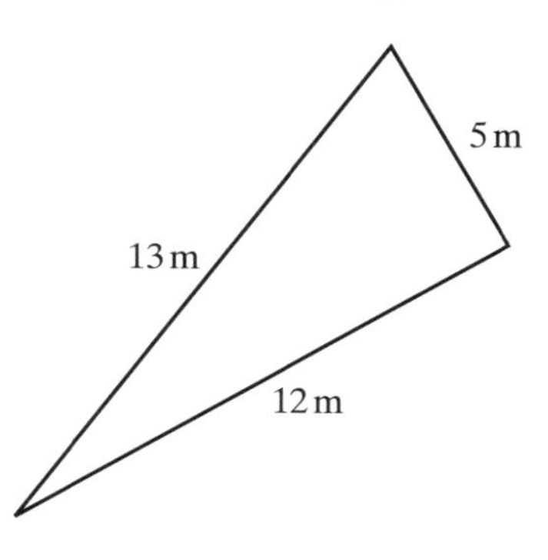

SOLUTION

Call the longest side 'c' and the other two sides 'a' and 'b'

In triangle A, $c = 15, a = 14$ and $b = 5$

Now $\quad c^2 = 225$

and $\quad a^2 + b^2 = 14^2 + 5^2 = 196 + 25 = 221$

Since $225 \neq 221$ then triangle A is not right angled

In triangle B, $c = 13, a = 12$ and $b = 5$

Now $\quad c^2 = 169$

and $\quad a^2 + b^2 = 12^2 + 5^2 = 144 + 25 = 169$

Since c^2 and $a^2 + b^2$ are equal (169) then triangle B is right angled

EXERCISE 16.1

1 Look at these triangles, and use Pythagoras' theorem to decide whether they are right angled or not.
The diagrams are not drawn to scale.

a)

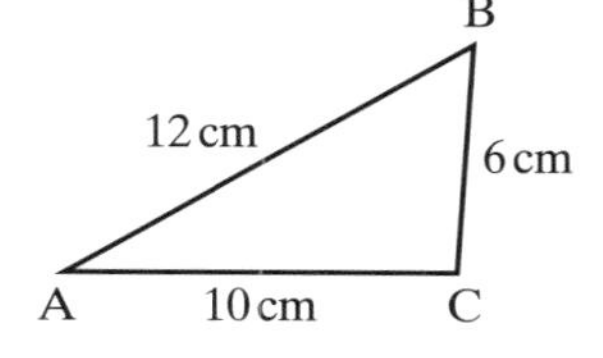

b)

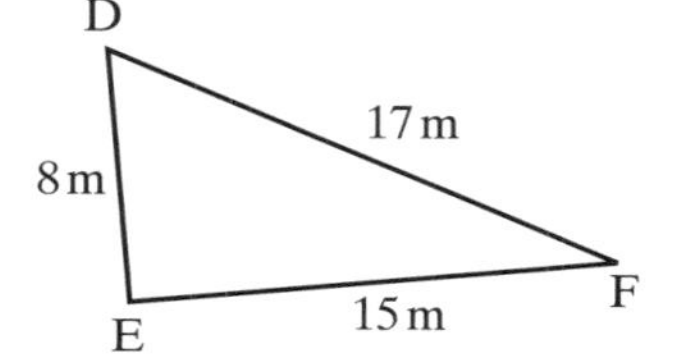

c)

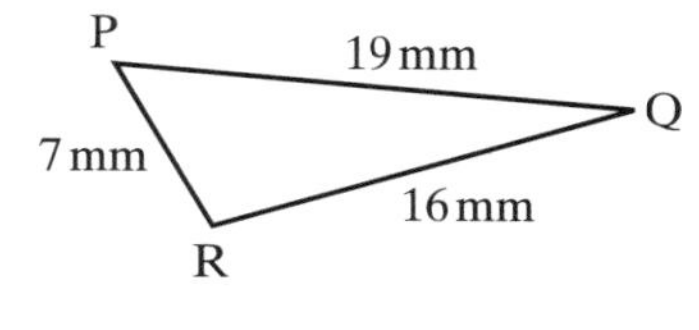

2 For each of the triangles described below, use Pythagoras' theorem to decide whether it is right angled.
If so, name the angle at which the right angle is located.

a) AB = 8 cm, BC = 6 cm, CA = 2.5 cm
b) AB = 7.5 cm, BC = 4.5 cm, CA = 6 cm
c) AB = 12 mm, BC = 13 mm, CA = 5 mm
d) DE = 10.1 cm, EF = 7.1 cm, FD = 7.2 cm
e) DE = 12 m, EF = 16 m, FD = 20 m
f) PQ = 3.3 cm, QR = 5.8 cm, RP = 4.5 cm
g) PQ = 6 km, QR = 7 km, RP = 8 km
h) RS = 26 mm, ST = 24 mm, RT = 10 mm

16.2 Using Pythagoras' theorem to find a hypotenuse

In this section, we will be working with triangles that are known to be right angled, and will use Pythagoras' theorem to find the length of the **hypotenuse**.

Remember, this is the longest side, and is always located directly opposite the right angle.

To find the hypotenuse (c) you:

> **SQUARE** the two shorter sides (a and b),
> **ADD** the results together, and then
> **SQUARE ROOT** your answer.

This can be written formally as:

$$c = \sqrt{a^2 + b^2}$$

EXAMPLE

Calculate the length c, of the side AB in the triangle.

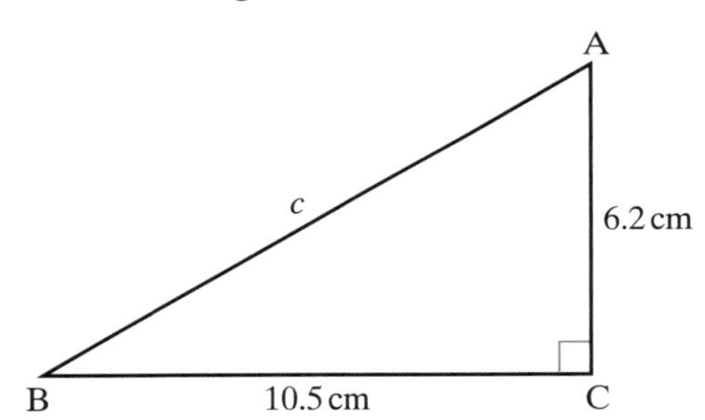

SOLUTION

Pythagoras' theorem tells us that:

$$c^2 = 6.2^2 + 10.5^2$$
$$= 38.44 + 110.25$$
$$= 148.69$$

So $\sqrt{148.69} = 12.193\,850\,91$

$$= \underline{12.2} \text{ (3 s.f.)}$$

SQUARE the two shorter sides.

ADD them together.

SQUARE ROOT the answer.

EXERCISE 16.2

1 Find the length of the hypotenuses represented by the letters a to f below.
Give your answers to an appropriate degree of accuracy.

a)

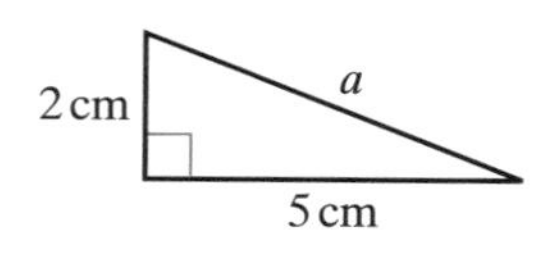

b)

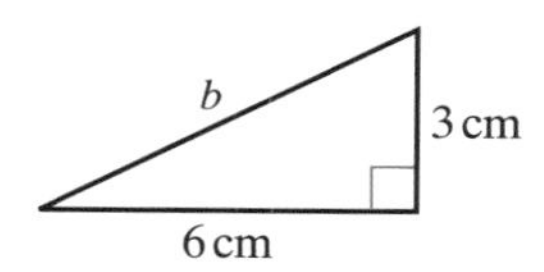

c)

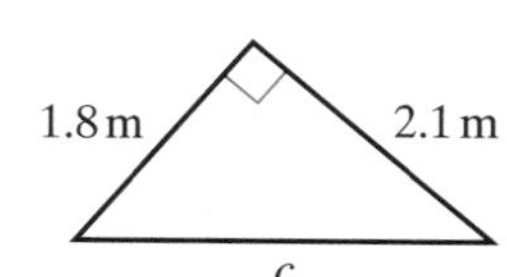

d)

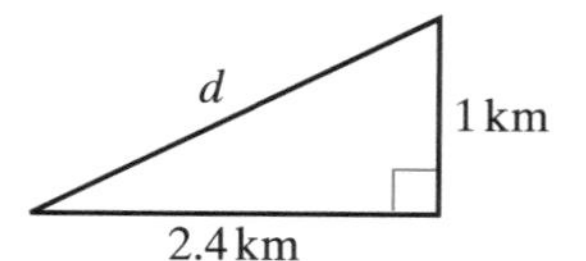

e)

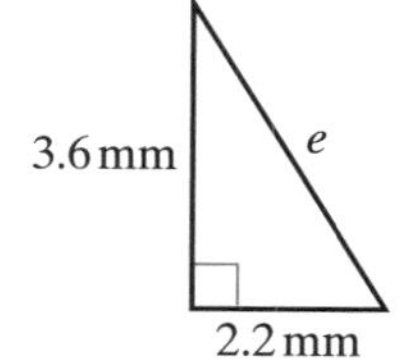

f)

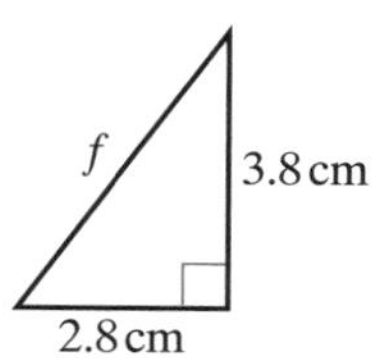

2 Find the length of the diagonal of each rectangle.

a)

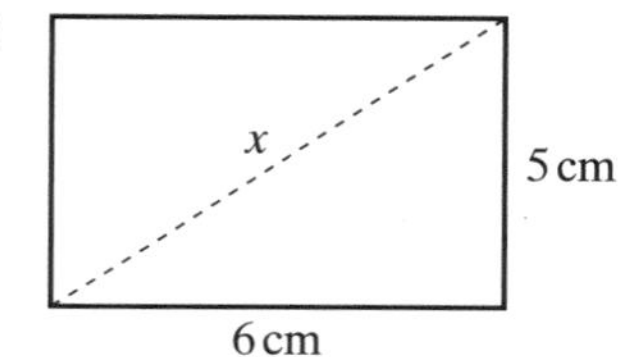

b)

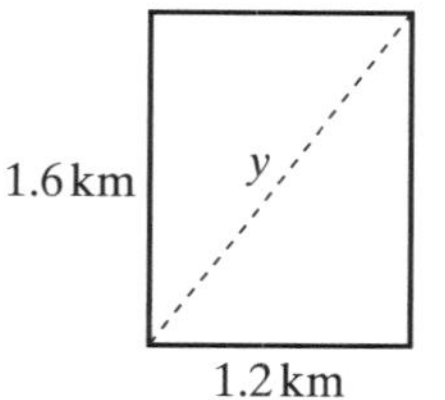

c)

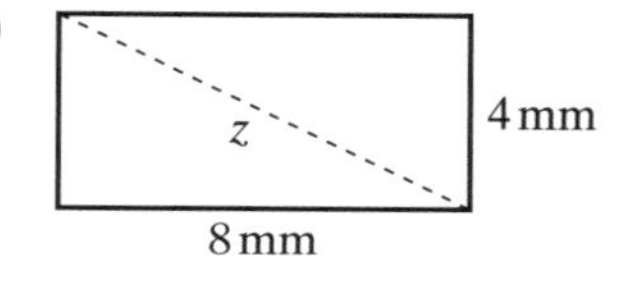

3

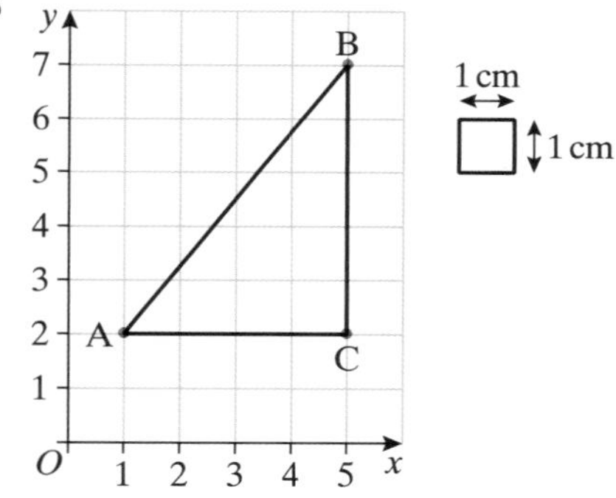

a) Find the length of **(i)** AC, **(ii)** BC.
b) Hence find AB.

4

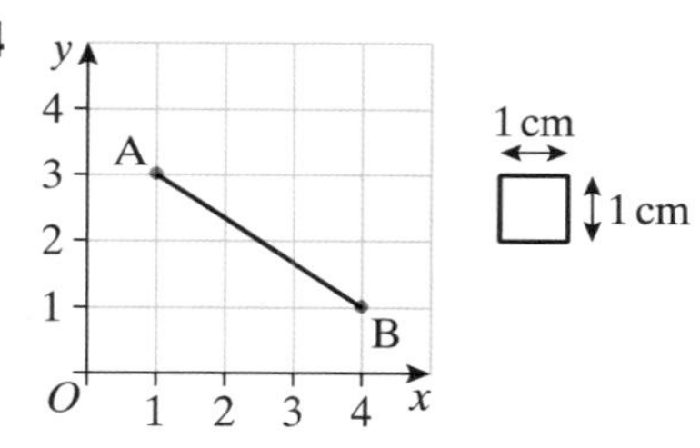

Find the length of AB.

16.3 Using Pythagoras' theorem to find one of the shorter sides

The method used above can be adapted when the unknown side is not the hypotenuse.

In this case, the calculation requires a subtraction instead of an addition.

To find one of the shorter sides (say a) you:

> **SQUARE** the other two sides (c and b),
> **SUBTRACT** the smaller number from the larger, and then
> **SQUARE ROOT** your answer.

This can be written formally as:

$$a = \sqrt{c^2 - b^2}$$

EXAMPLE

Find the value of y in the right-angled triangle.

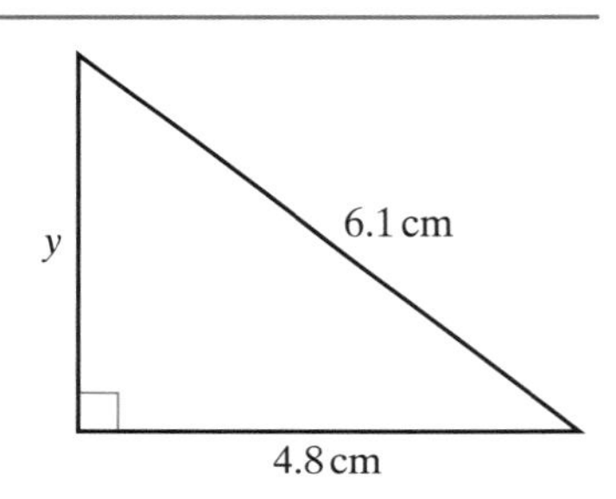

SOLUTION

$6.1^2 = 37.21$ ← SQUARE

$4.8^2 = 23.04$

$y^2 = 37.21 - 23.04$ ← SUBTRACT

$= 14.17$

So $y = \sqrt{14.17}$ ← SQUARE ROOT

$= 3.764306045$

$= \underline{3.76\text{ cm}}$ (3 s.f.) ← Don't forget the units (cm) in your answer.

EXERCISE 16.3

1 Find the length of the sides marked by the letters a to f below.
Give your answers to 3 significant figures where appropriate.

a)

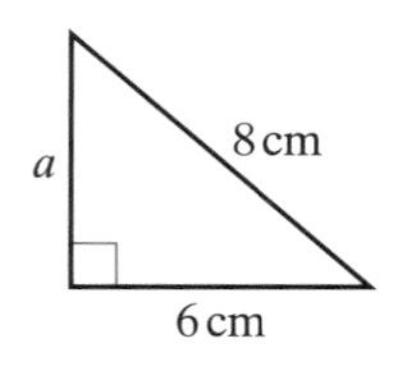

b)

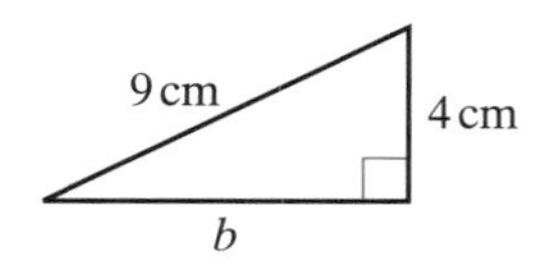

c)

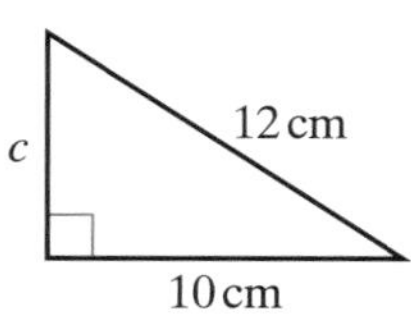

d)

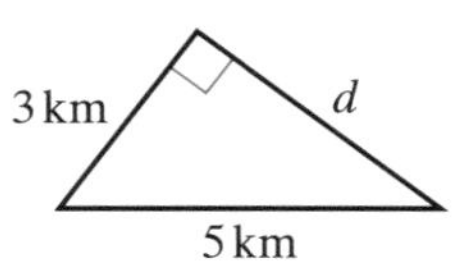

e)

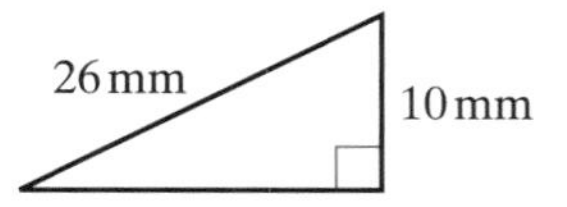

f)

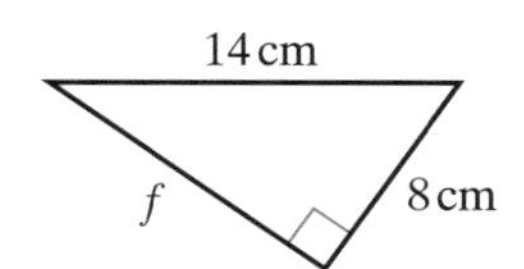

2 A rectangle has length 24 cm and width x cm
Its diagonal is of length 25 cm
Find the value of x

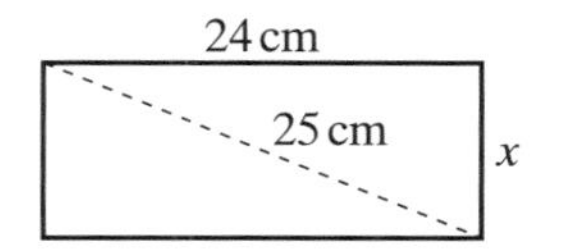

3 A ship sails due North for 12 km, then turns and sails due East for y km
It ends up 16 km in a direct straight line from its start point.
Find the value of y

16.4 More Pythagoras

You need to be able to decide whether you are finding the hypotenuse or one of the shorter sides.

EXERCISE 16.4

1 Find the length of the sides marked by the letters a to f below.
Give your answers to an appropriate degree of accuracy.

a)

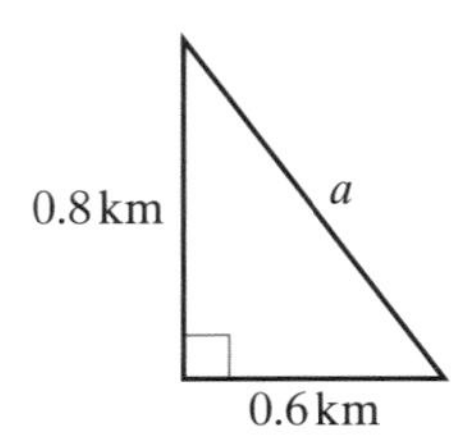

b)

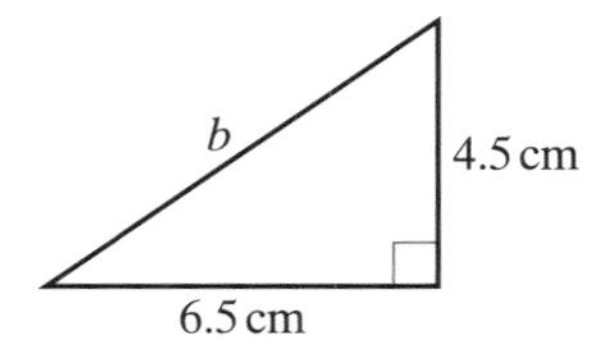

c)

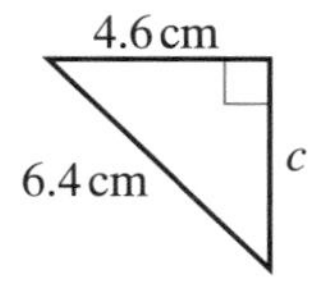

d)

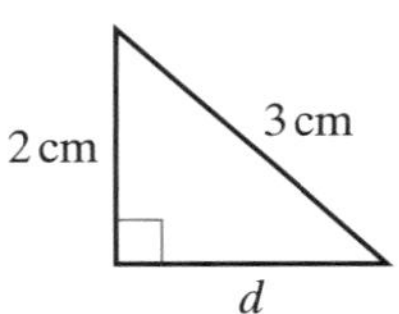

e)

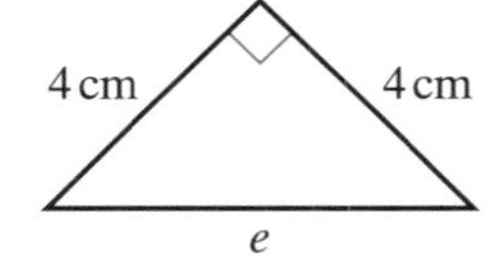

f)

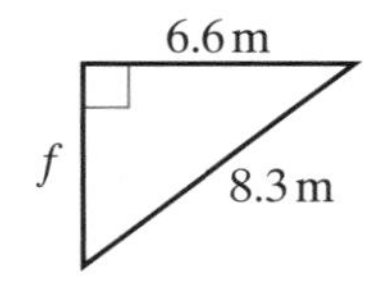

2 Find the length of the diagonals in these squares.

a)
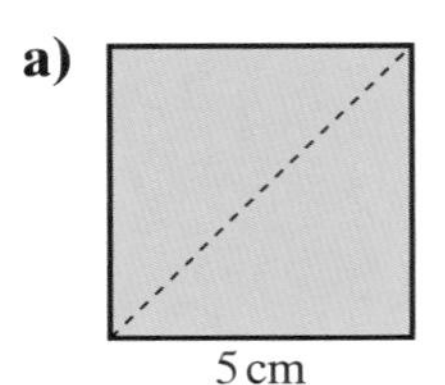

b)
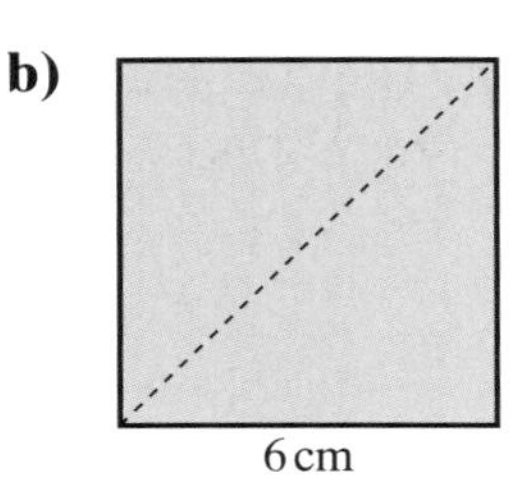

3 **a)** Work out the height of this isosceles triangle.
b) Find the area of the triangle.

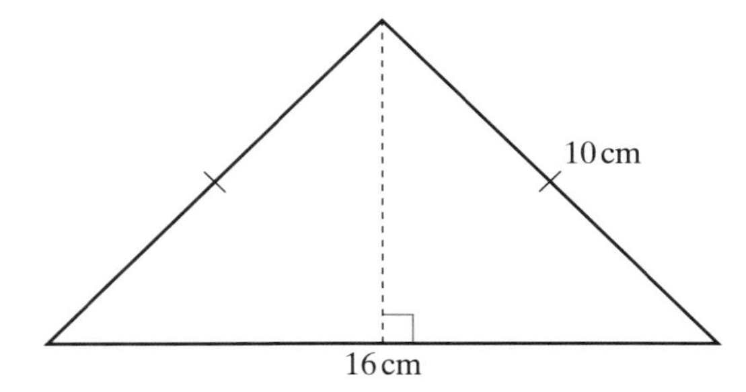

4 The diagram shows a rectangular field with a footpath running diagonally across it.
a) Work out the length of the footpath.

Michelle walks from A to B and then from B to C.
Mohammed walks from A to C along the footpath.
b) How much farther does Michelle walk than Mohammed?

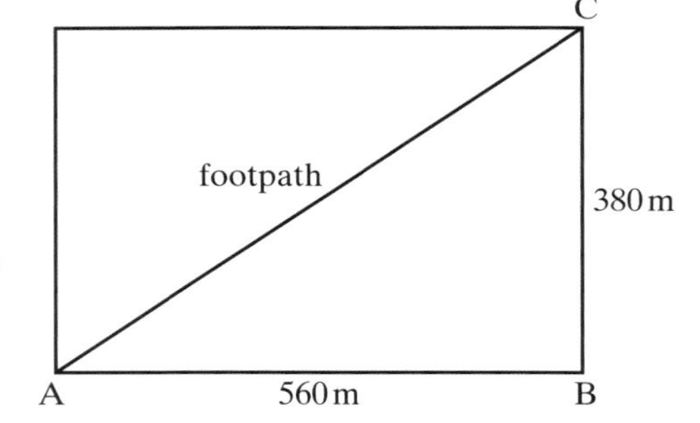

5 A man places a 13 m long ladder against a wall.
The foot of the ladder is 5 m from the wall.
How high from the ground is the top of the ladder?

REVIEW EXERCISE 16

1 Find the missing lengths, denoted by the letters a to f
Round your answers to an appropriate degree of accuracy.

a)
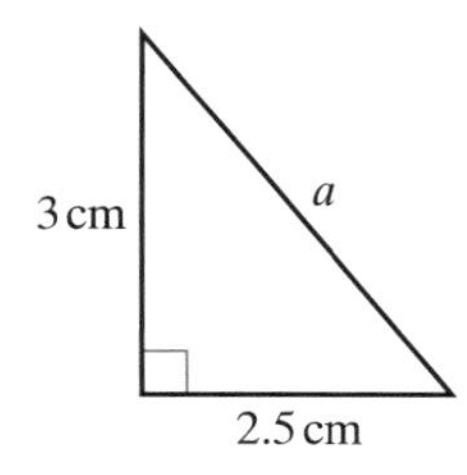

b)
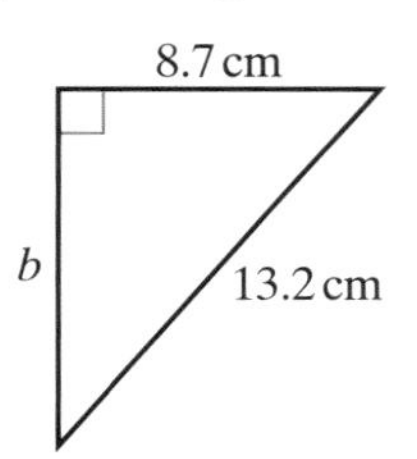

c)
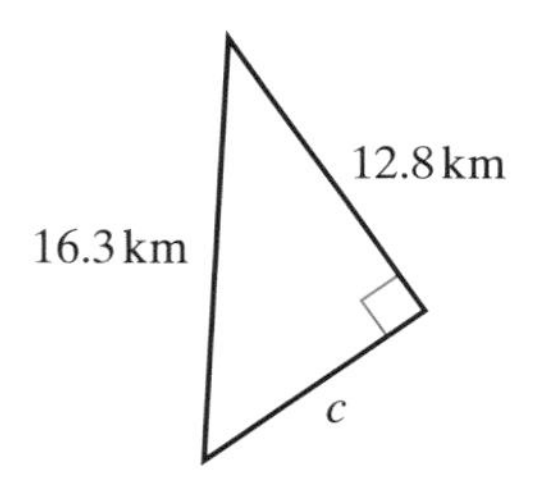

d)
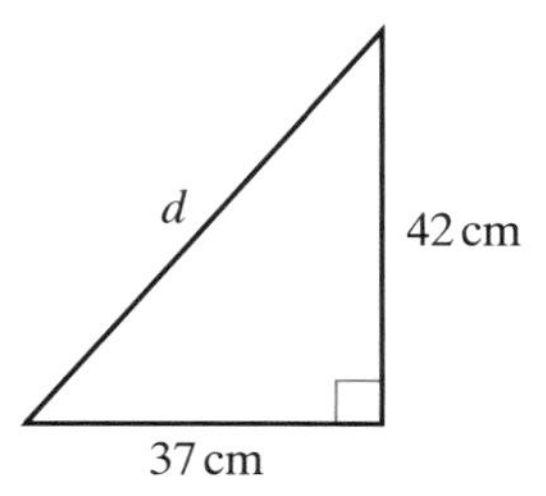

e)
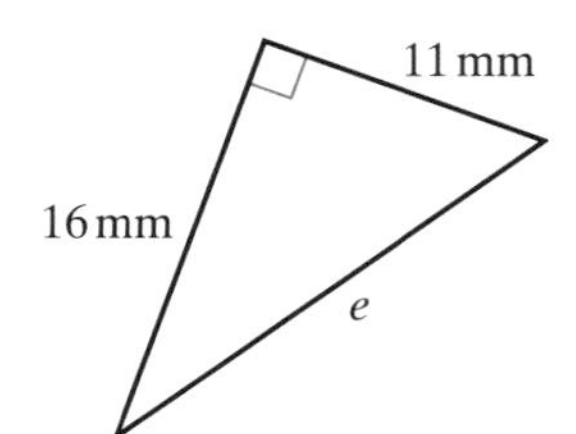

f)
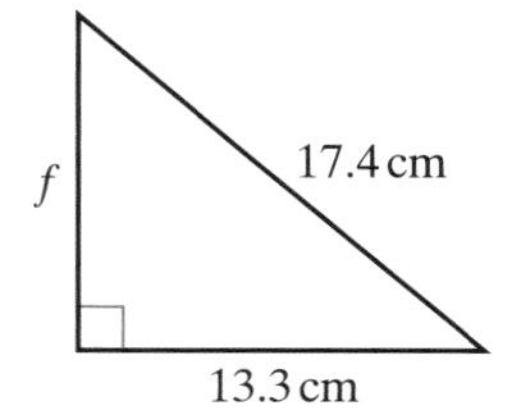

2 The diagram (right) shows two connected right-angled triangles.

a) Write down the exact value of x, without using a calculator.

b) Use your calculator to find the value of y, giving your answer correct to 2 decimal places.

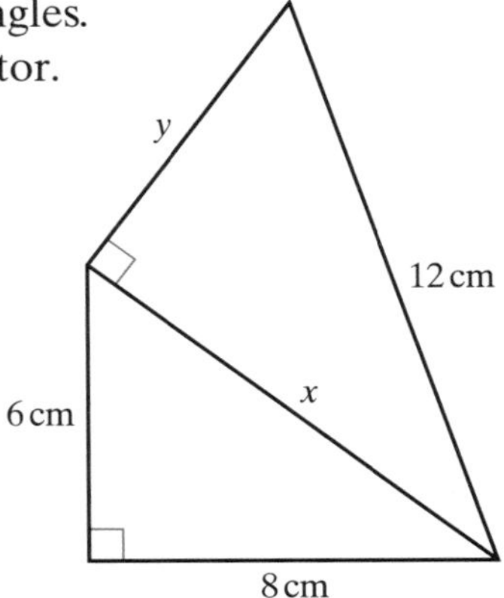

3 The diagram below shows 2 points A and B on a centimetre grid.

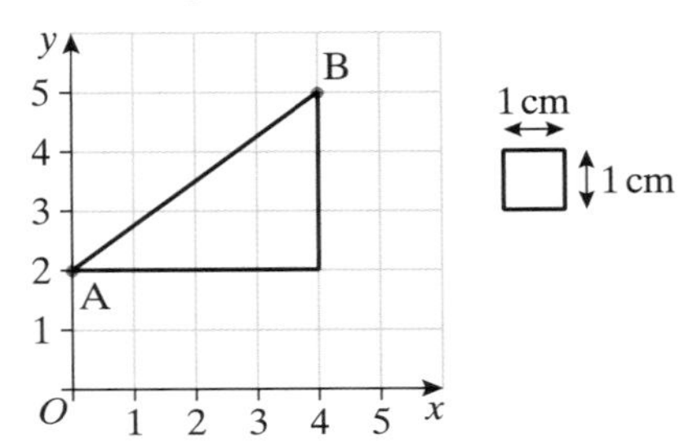

Find the length of AB.

4 The diagram shows a sketch of a triangle.

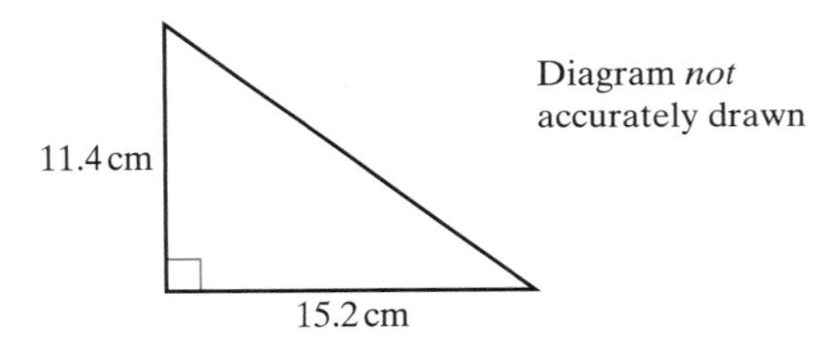

a) Work out the area of the triangle.
State the units of your answer.

b) Work out the perimeter of the triangle. **[Edexcel]**

5 ABCD is a rectangle.
AC = 17 cm and AD = 10 cm

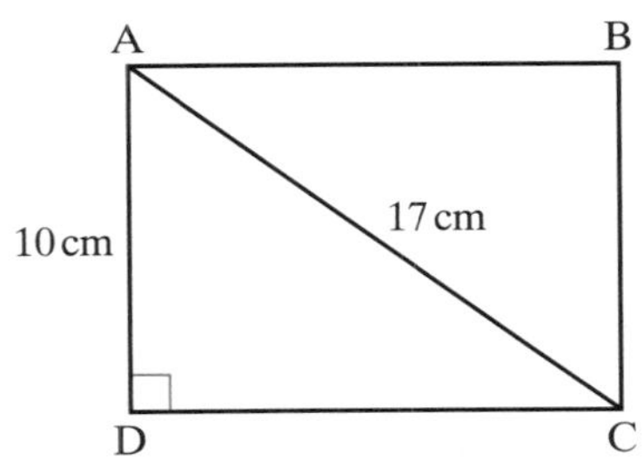

Diagram *not* accurately drawn

Calculate the length of the side CD.
Give your answer correct to 1 decimal place. **[Edexcel]**

6 ABC is a right-angled triangle.
AC = 5 m and CB = 8.5 m

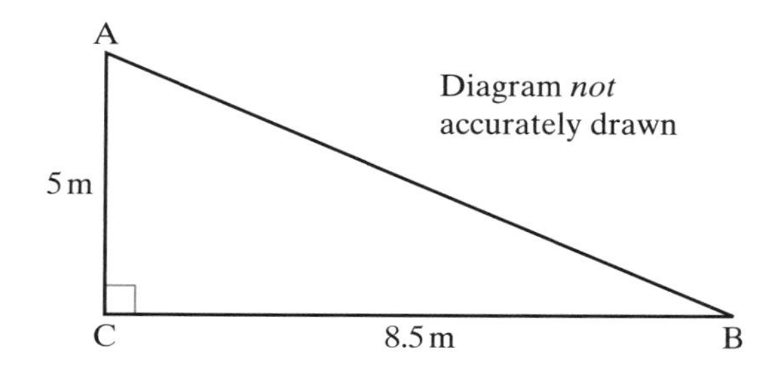

a) Work out the area of the triangle.
b) Work out the length of AB.
Give your answer correct to 2 decimal places. **[Edexcel]**

KEY POINTS

1. For a right-angled triangle with sides a, b and c, where c is the longest side, then
   $$c^2 = a^2 + b^2$$
   This is Pythagoras' theorem.
2. Pythagoras' theorem works in reverse:
   if $c^2 = a^2 + b^2$ then the triangle must be right-angled.
3. To find the hypotenuse (longest side): $c = \sqrt{a^2 + b^2}$
   - square the two shorter sides
   - add the results together
   - square root your answer.
4. To find one of the shorter sides: $a = \sqrt{c^2 - b^2}$
   - square the other two sides
   - subtract the smaller number from the larger
   - square root your answer.
5. Pythagoras problems sometimes involve two stages.
   Do not round off answers to multi-stage problems until all the calculations have been completed.
6. It is useful to draw the right-angled triangle you are using at each stage of your calculations.

Internet Challenge 16

Investigating Pythagorean triples

Probably the most well-known right-angled triangle has sides in the ratio of 3 : 4 : 5, and is known as the (3, 4, 5) triangle.

The numbers 3, 4 and 5 form a Pythagorean 'triple' (or 'triplet').

This means that they are whole numbers satisfying $a^2 + b^2 = c^2$

Another Pythagorean triple is (5, 12, 13)

Here are some questions about Pythagorean triples.
You may use the internet to help you research some of the answers.

1 Find c such that $(8, 15, c)$ is a Pythagorean triple.

2 Find all the Pythagorean triples in which each number does not exceed 25

3 How can we easily know that (6, 8, 10) and (9, 12, 15) are Pythagorean triples without doing any calculations?

4 Are there any patterns or formulae for generating these triples?

5 Are there infinitely many Pythagorean triples?

6 Are there any Pythagorean quadruples – positive whole numbers a, b, c, d such that $a^2 + b^2 + c^2 = d^2$?

7 Find out as much as you can about Fermat's Last Theorem.
Has it been proved yet?

CHAPTER 17

Transformations

In this chapter you will **learn how to:**

- carry out simple reflections, translations, rotations and enlargements
- use combinations of these transformations
- decide whether two shapes are congruent or similar.

You will be **challenged to:**

- investigate some geometric definitions.

Starter: Monkey business

Nine monkeys have fallen out of their tree.

They are not all the same shape and size.

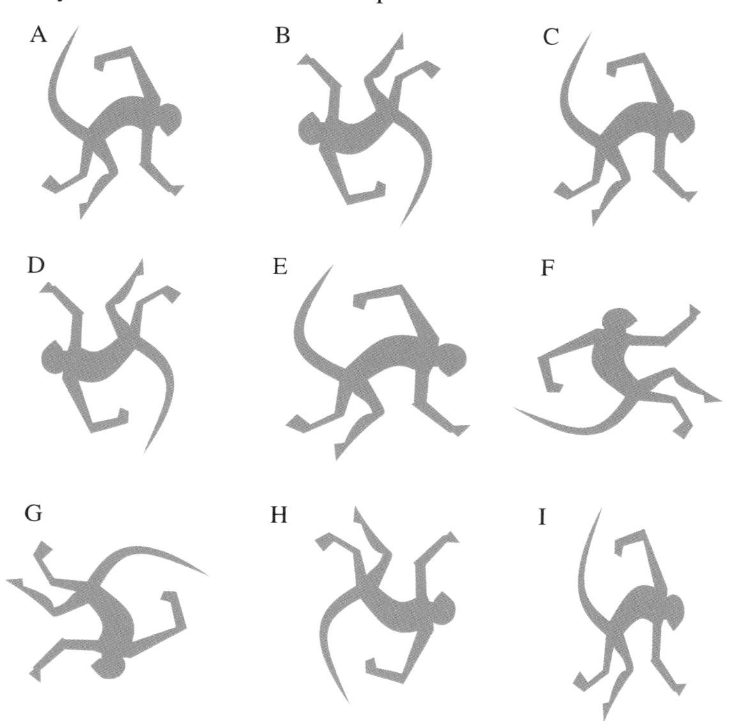

Pick out the monkeys that are the same shape and size.

These monkeys are **congruent**.

17.1 Symmetry

Many shapes possess **mirror symmetry**, or **reflection symmetry**.

Such 2-D shapes have a mirror line, and this divides the shape into two matching halves, one being a mirror image of the other.

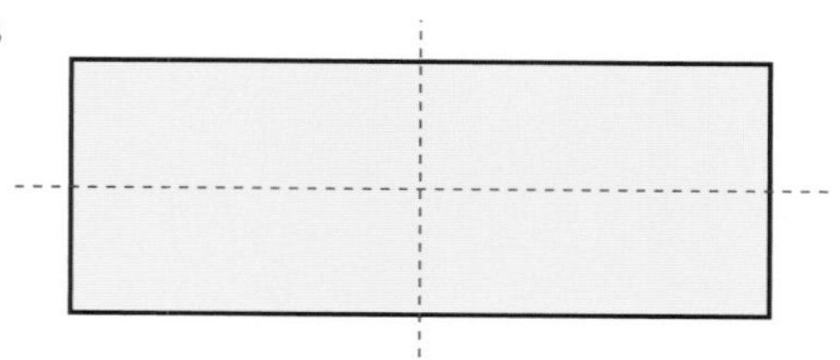

A rectangle has two lines of symmetry.

The matching halves are exactly the same shape and size.

We say they are **congruent**.

3-D shapes have a **plane of symmetry** instead.

Again, the shape divides into two **congruent** halves.

This 3-D shape has two planes of symmetry:

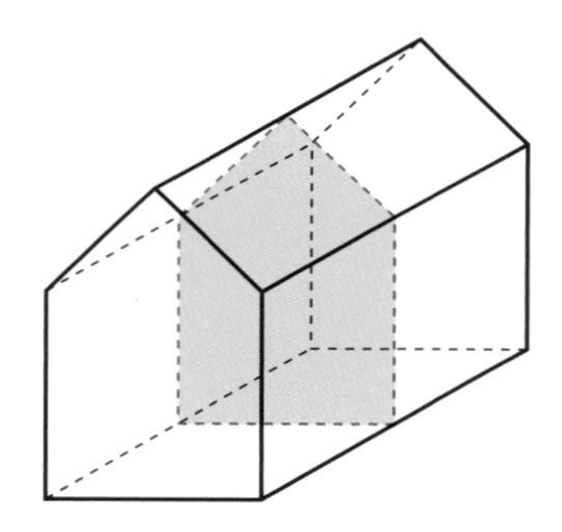

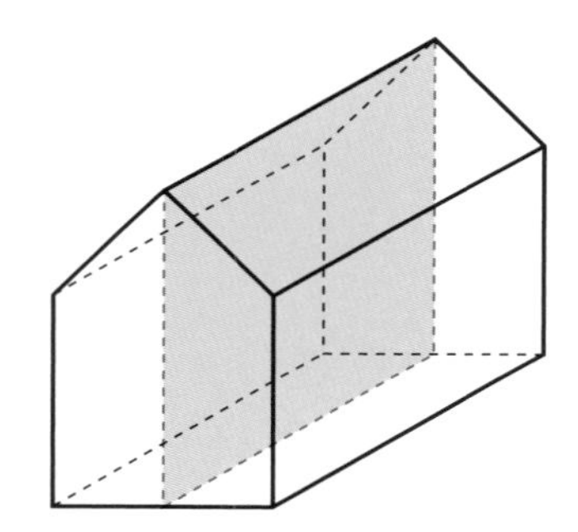

EXERCISE 17.1

1 **(i)** Draw all the lines of symmetry on a copy of the following shapes.

(ii) Write down how many lines of symmetry each shape has.

a)

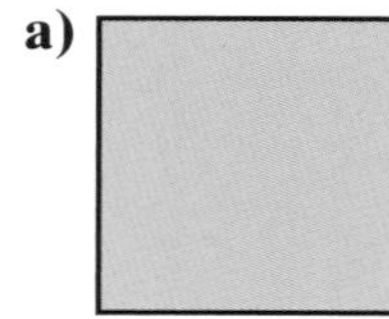

Square

b)

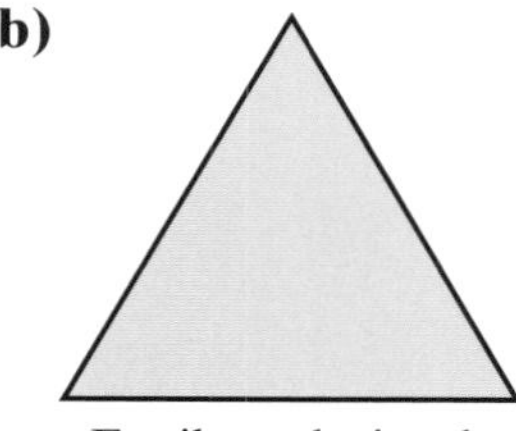

Equilateral triangle

c)

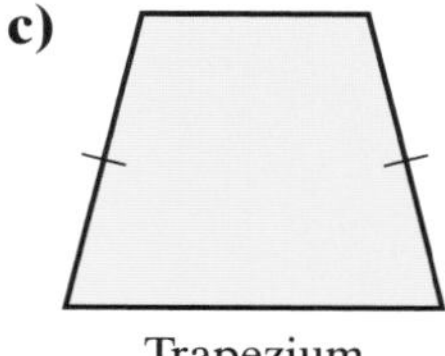

Trapezium

d)

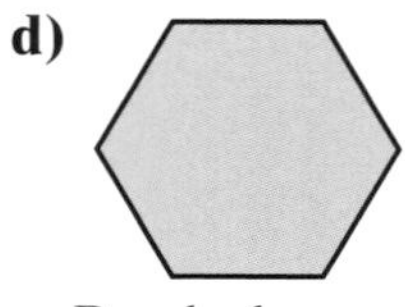

Regular hexagon

e)

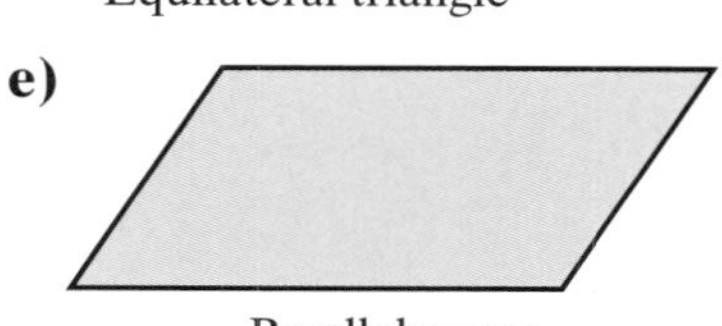

Parallelogram

f)

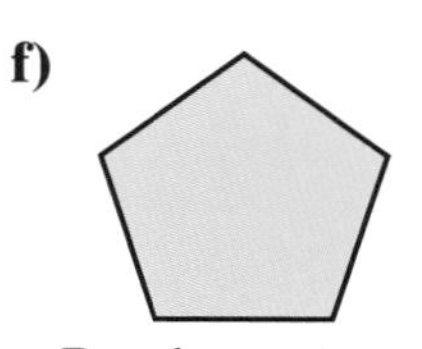

Regular pentagon

2 (i) Draw all the lines of symmetry on a copy of the following shapes.

(ii) Write down how many lines of symmetry each shape has.

a)

b)

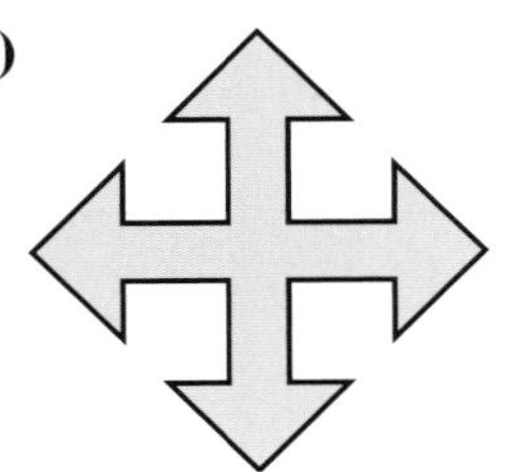

c)

d)

3 Shade in squares on a copy of the following grids so that the resulting pattern is symmetrical about the lines shown.

a)

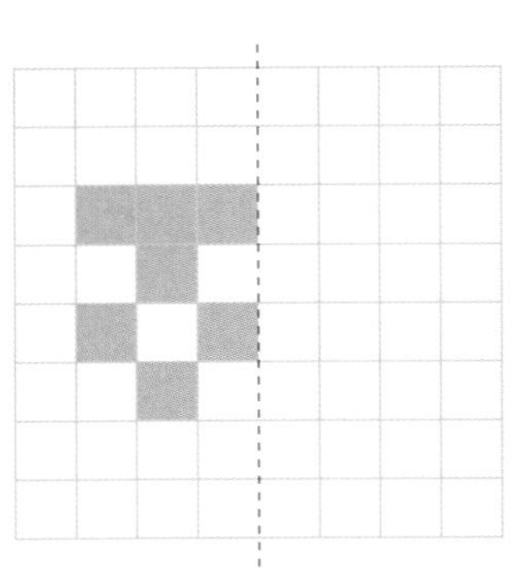

b)

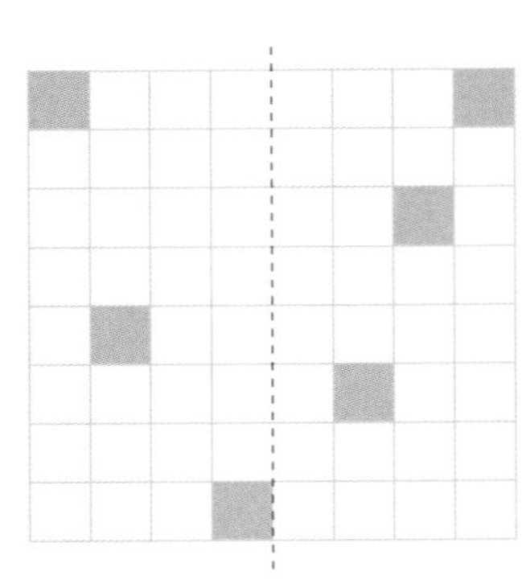

c)

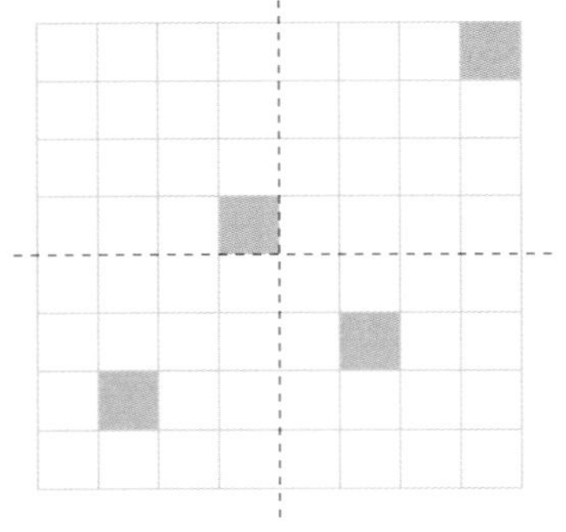

d)

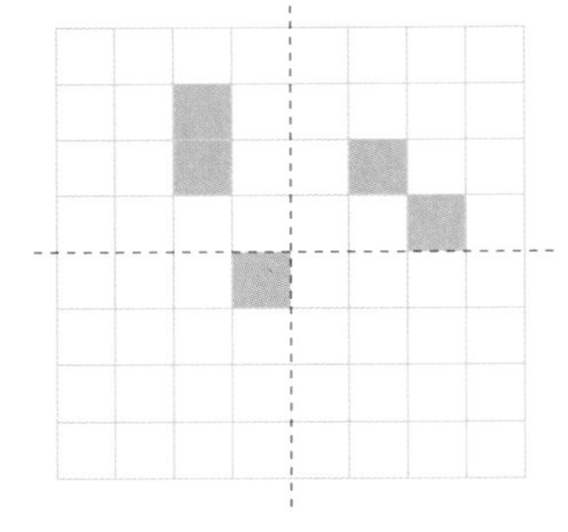

e)

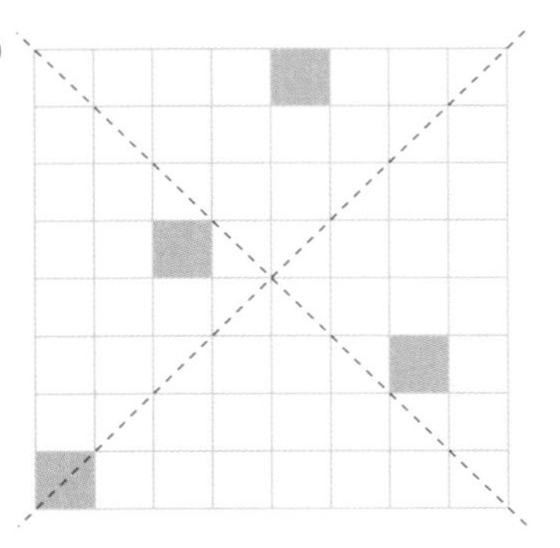

f)

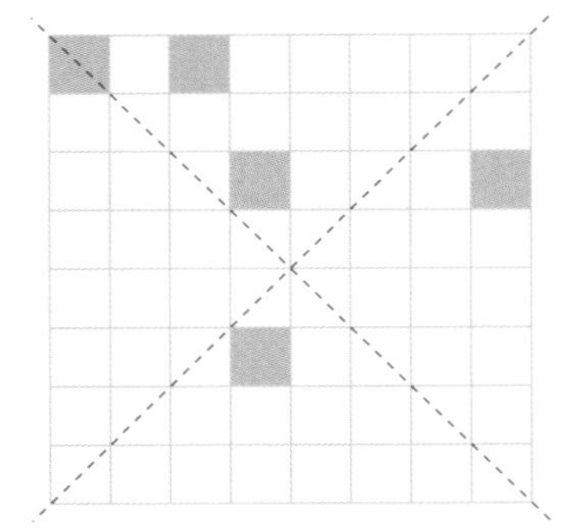

4 Sketch in the planes of symmetry on a copy of the following shapes.

a)

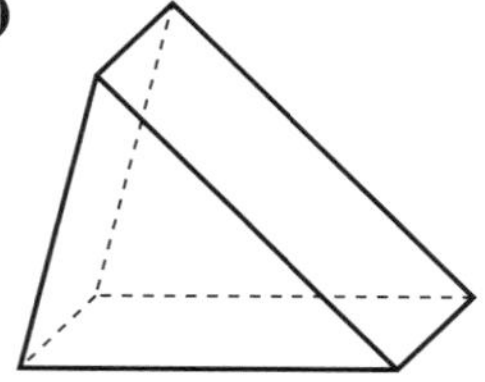

b)

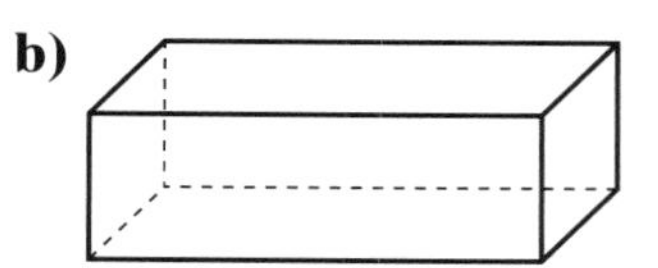

c)

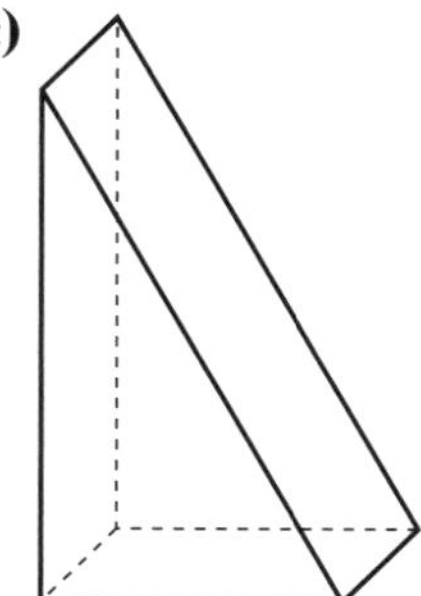

d)

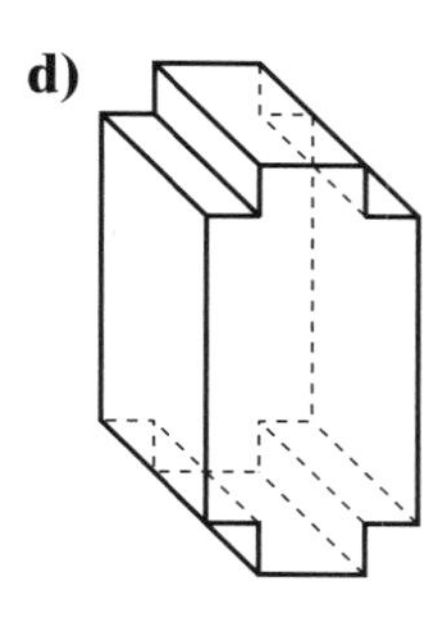

5 Which of these shapes are **congruent**?

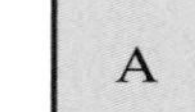

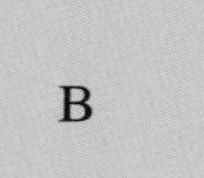

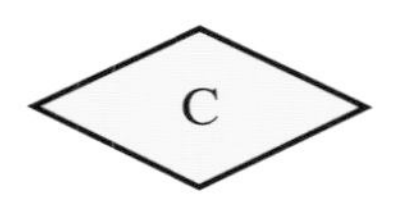

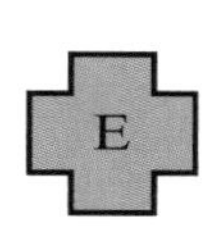

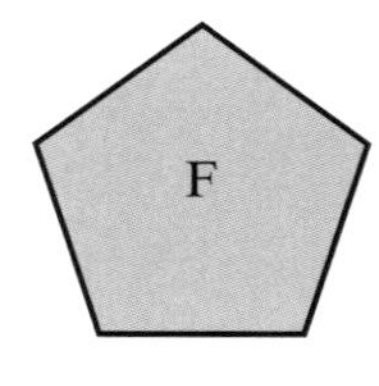

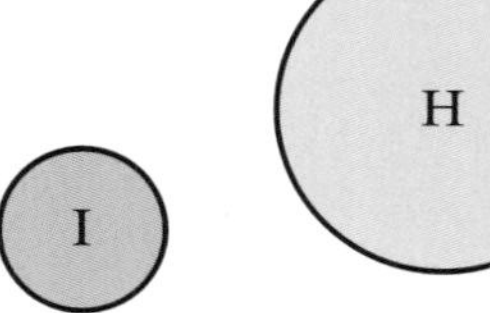

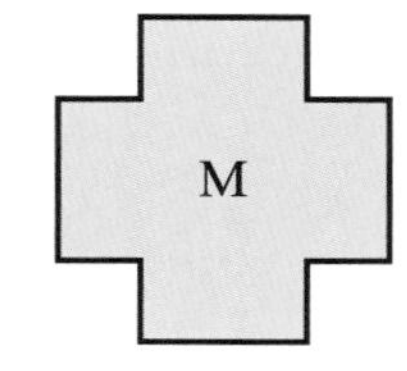

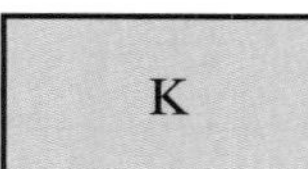

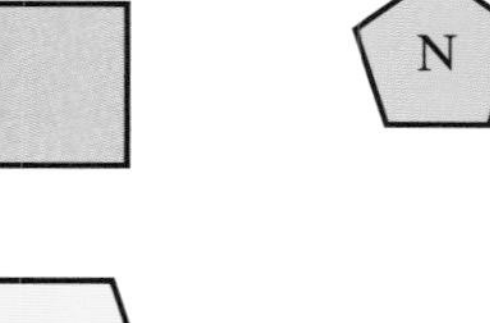

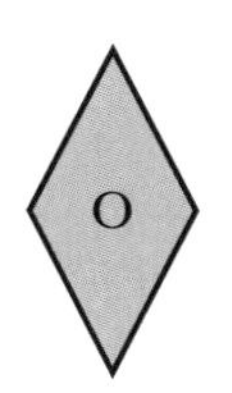

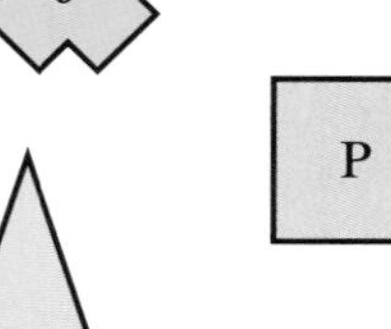

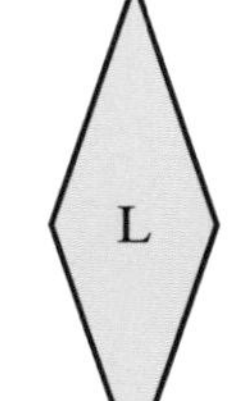

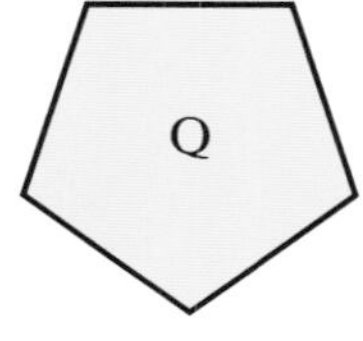

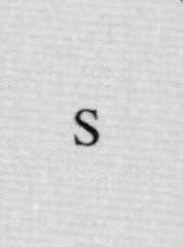

17.2 Reflections on a coordinate grid

Remember how to write down the equations of horizontal and vertical lines (see Section 10.4).

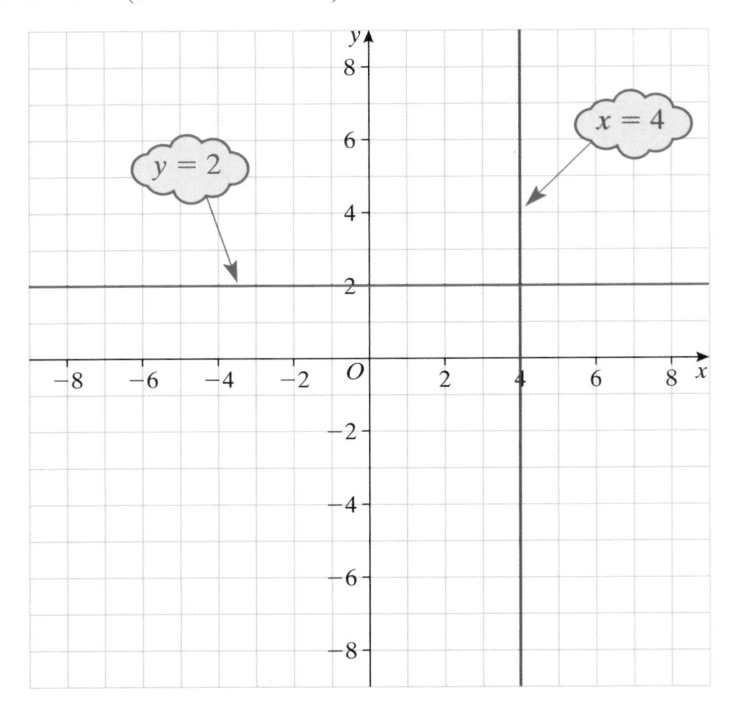

You also need to know the following two lines:

You can make a symmetric 2-D shape by reflecting a given shape in a **mirror line**.

This is usually done using a squared coordinate grid.

The mirror line has an equation.

The mirror line might be:

- horizontal (e.g. $y = 3$)
- vertical (e.g. $x = -2$)
- diagonal (e.g. $y = x$)

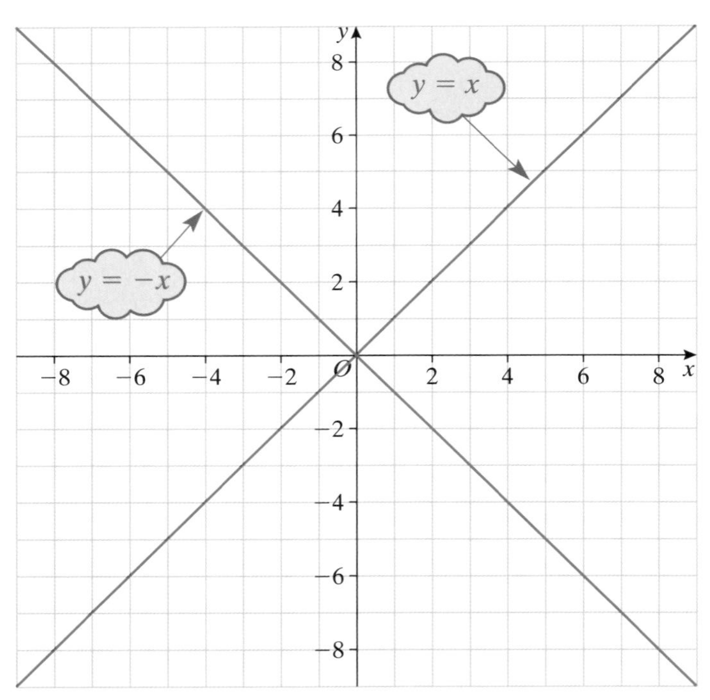

EXAMPLE

Reflect the given shape in the line $x = 5$

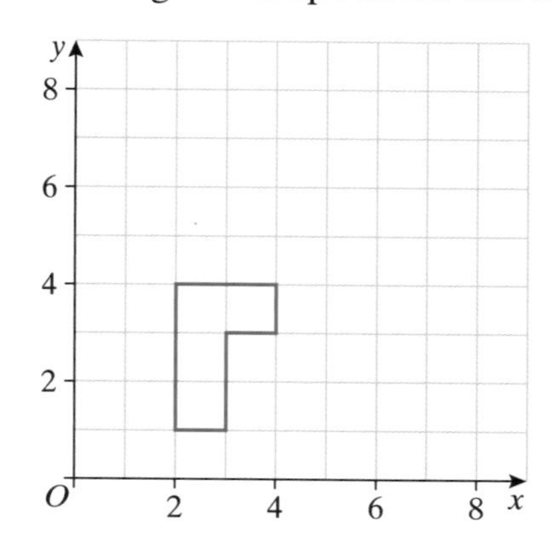

SOLUTION

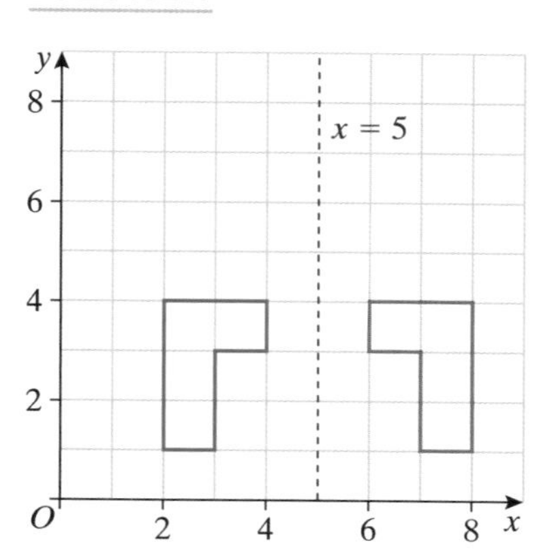

Problems involving a diagonal mirror line can be more difficult to visualise.
It helps to rotate your book so that the mirror line is vertical.

EXAMPLE

The diagram shows a triangle P.
The triangle has been reflected in a mirror line to form an image Q.

a) Draw the mirror line on the diagram. **b)** Write down the equation of the mirror line.

SOLUTION

a)

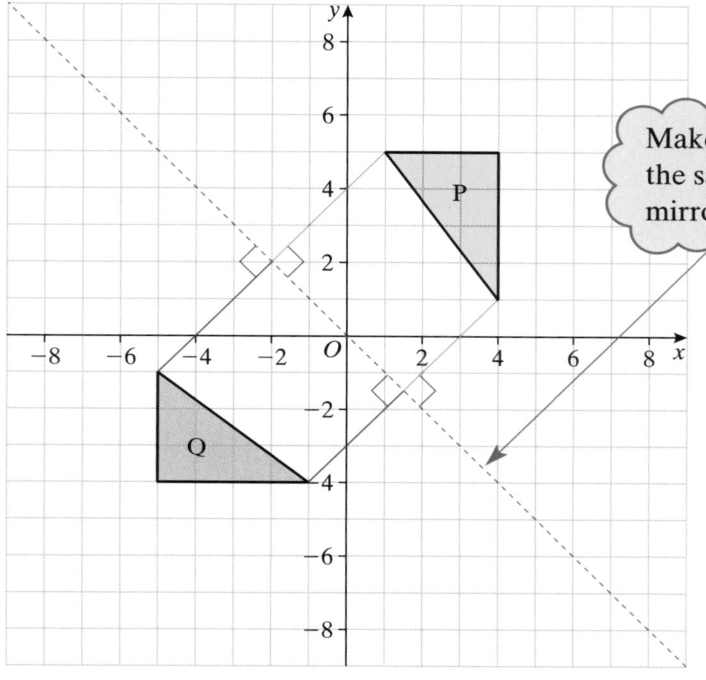

b) $y = -x$

Make sure that Q is exactly the same distance from the mirror line as P.

P and Q are **congruent**.

EXERCISE 17.2

In Questions 1–7, on a copy draw the reflection of the given shape in the mirror line shown.

Label the mirror line with its equation in each case.

1

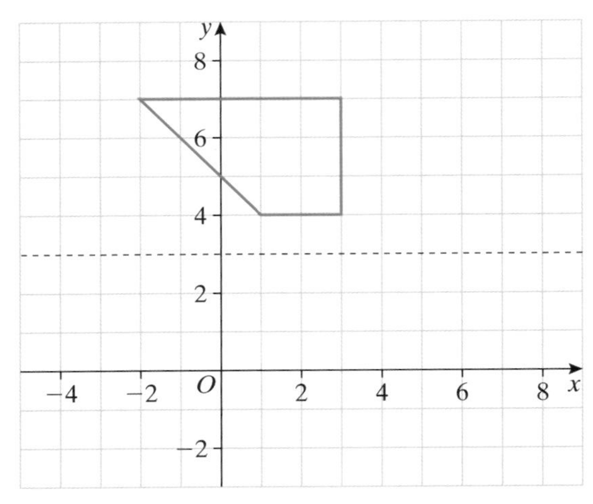

2

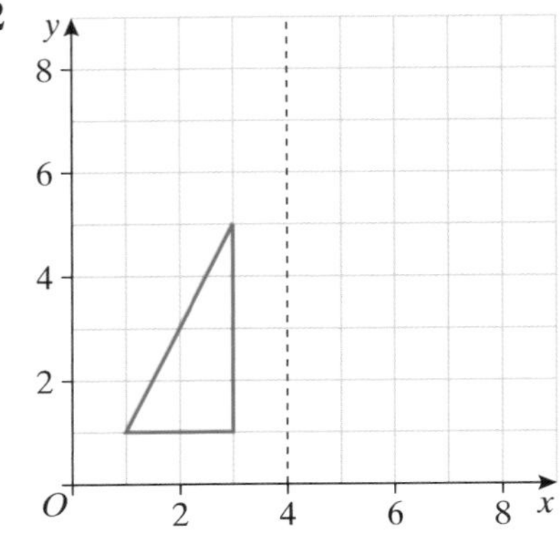

3

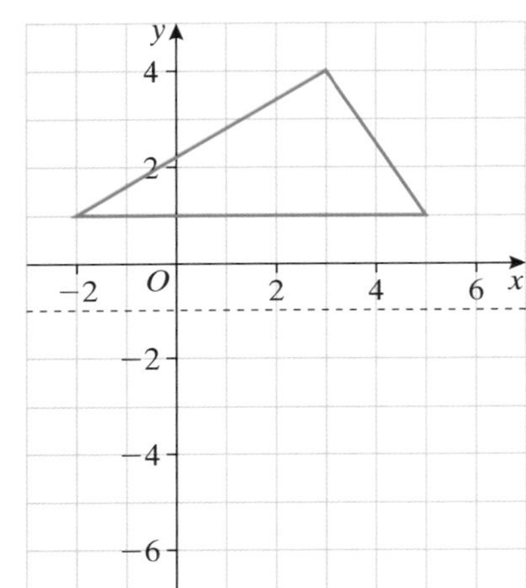

4

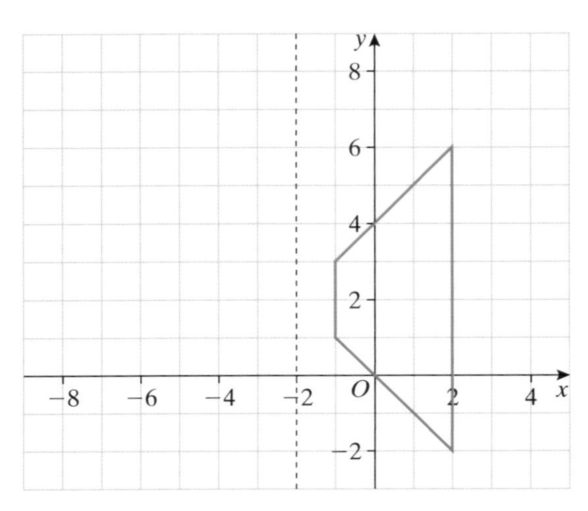

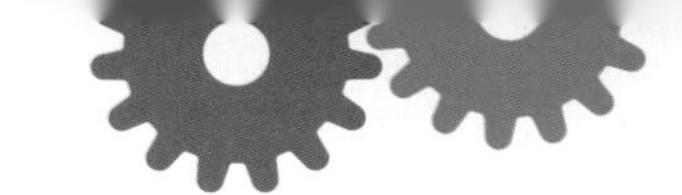

5

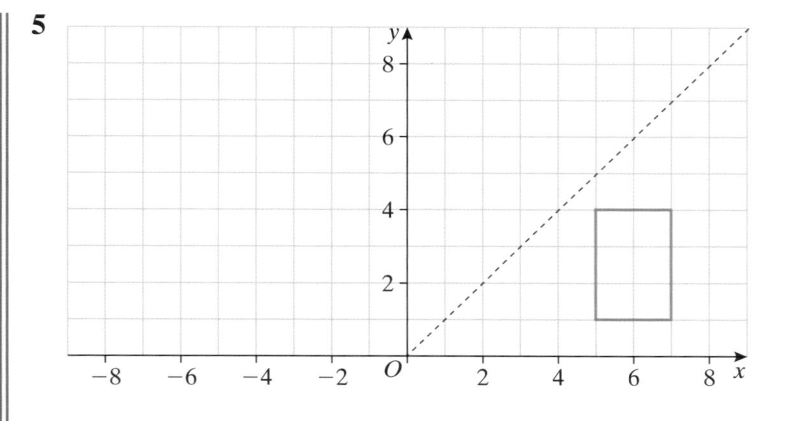

6

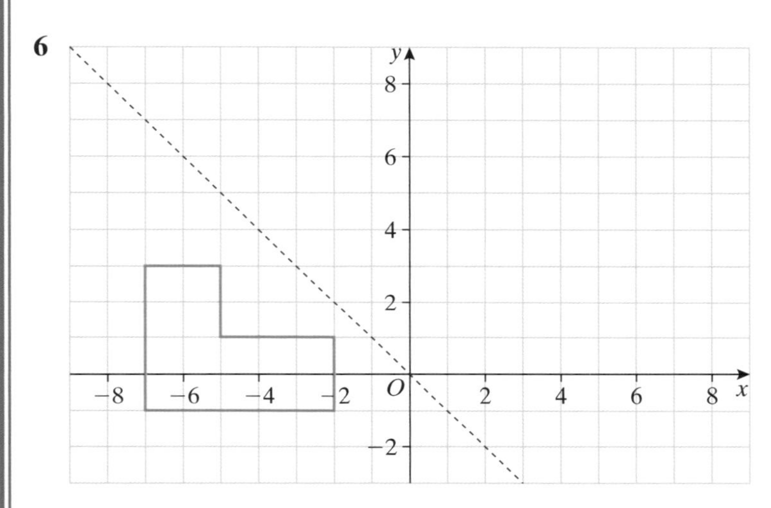

7

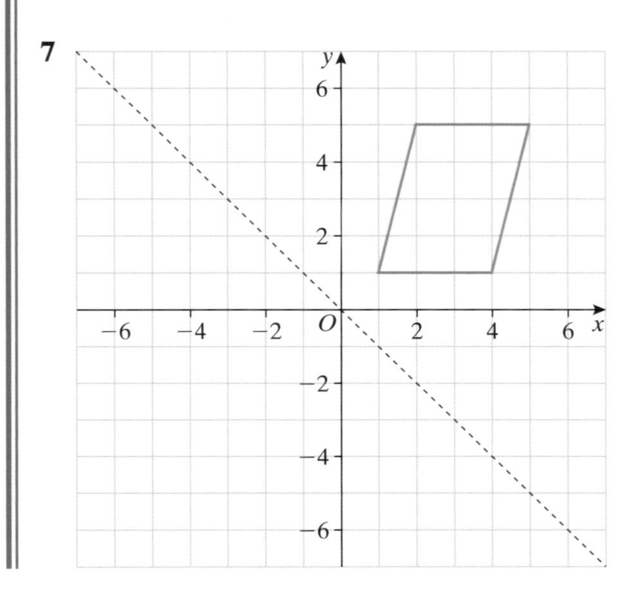

8 The diagram shows a triangle S and its mirror image T.

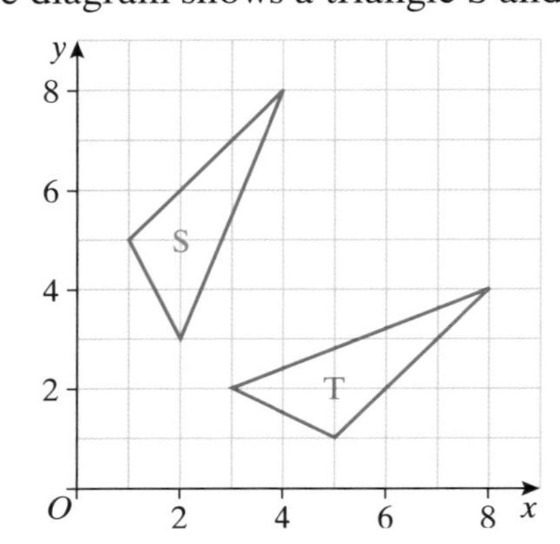

a) On a copy, draw the mirror line that has been used for the reflection.

b) Write down the equation of the mirror line.

9 The diagram shows a triangle S and its mirror image T.

a) On a copy, draw the mirror line that has been used for the reflection.

b) Write down the equation of the mirror line.

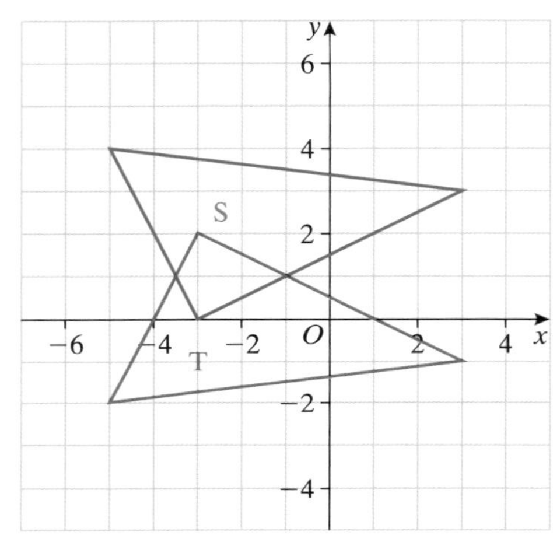

10 The diagram shows a letter L shape, labelled X.
The shape is to be reflected in a mirror line.
Part of the reflection has been drawn on the diagram.

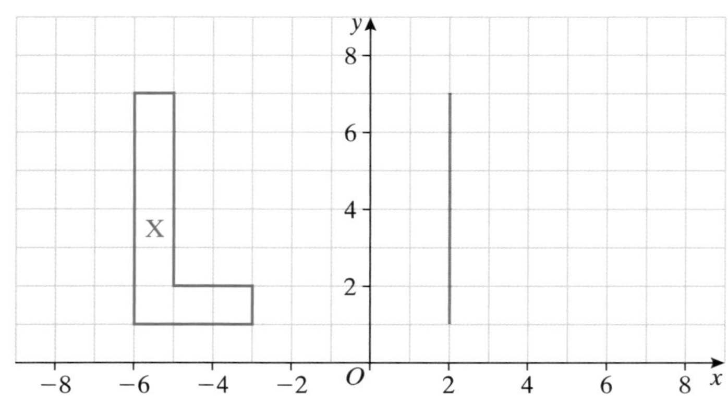

a) Complete the drawing to shown the image.
Label it Y.

b) Mark the mirror line, and give its equation.

11 The diagram shows six triangles A, B, C, D, E and F. The six triangles are all congruent to each other.

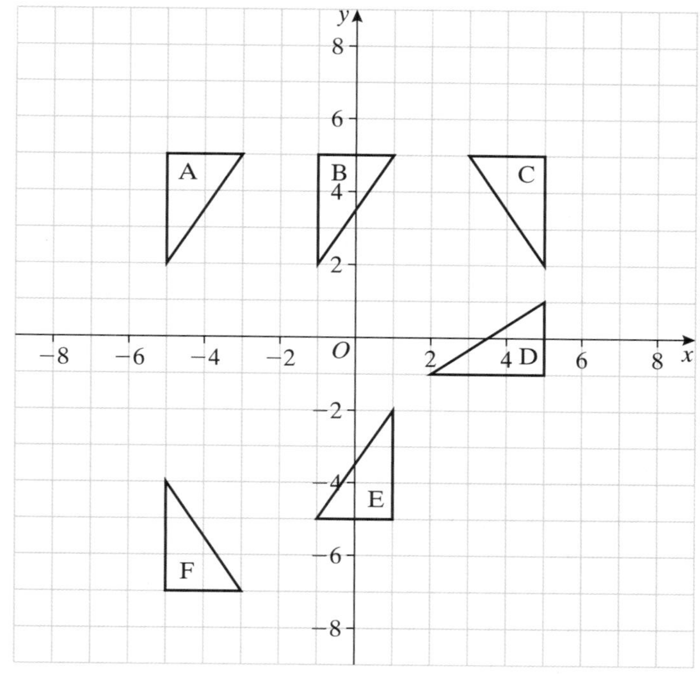

a) Explain the meaning of the word *congruent*.

b) Triangle A can be reflected to triangle F. State the mirror line that achieves this.

c) Triangle C is reflected to another triangle using a mirror line $x = 2$ Which one?

d) Triangle D can be reflected to triangle B using a mirror line. Give the equation of this line.

e) Triangle D can be reflected to triangle E using a mirror line. Give the equation of this line.

12 A triangle T is reflected in a mirror line to form an image, triangle U. Then triangle U is reflected in the same mirror line to form an image, triangle V. What can you say about triangle T and triangle V? Use a copy of the grid to help you.

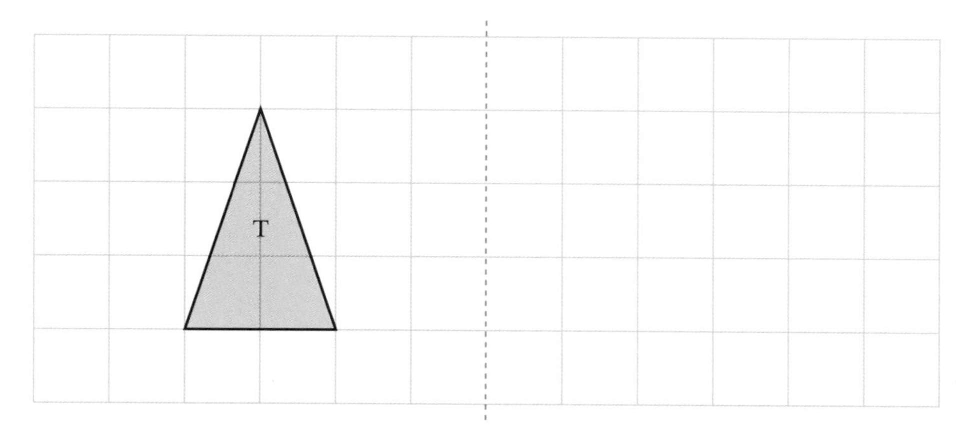

17.3 Translations

A **translation** is when you slide an object left or right and/or up or down.

You need to say how far the object is to be moved in each of the x and y directions.

This can be written as two numbers in a **column vector**:

$\begin{pmatrix} 5 \\ 2 \end{pmatrix}$ means a translation of 5 units to the right and two units up;

$\begin{pmatrix} -2 \\ 0 \end{pmatrix}$ means a translation of 2 units to the left.

EXAMPLE

The grid shows a shape labelled S.

a) Translate the shape S by the vector $\begin{pmatrix} 4 \\ -2 \end{pmatrix}$

Label the resulting shape T.

b) Translate the shape T by the vector $\begin{pmatrix} 5 \\ 0 \end{pmatrix}$

Label the resulting shape U.

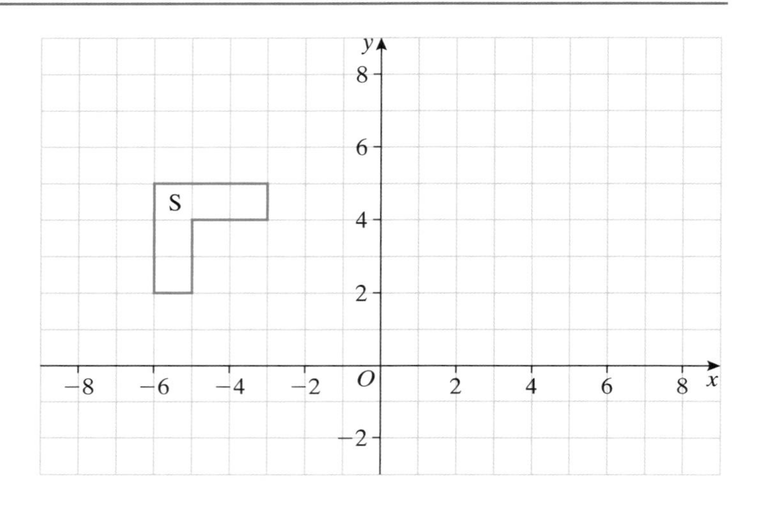

SOLUTION

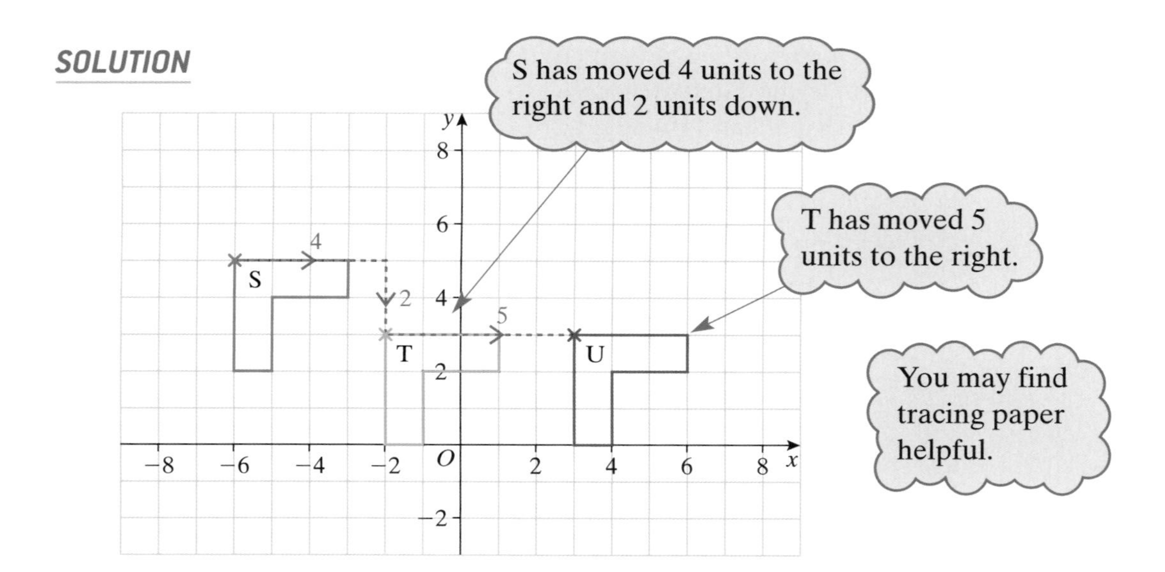

EXAMPLE

The grid shows two rectangles L and M.

a) Describe the transformation that maps L onto M.

b) Describe the transformation that maps M onto L.

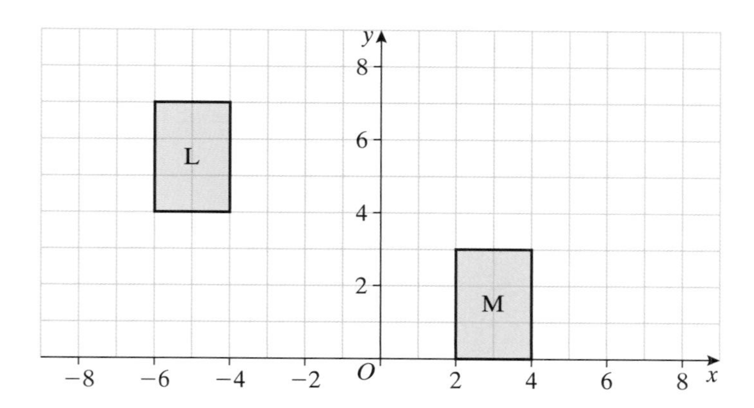

SOLUTION

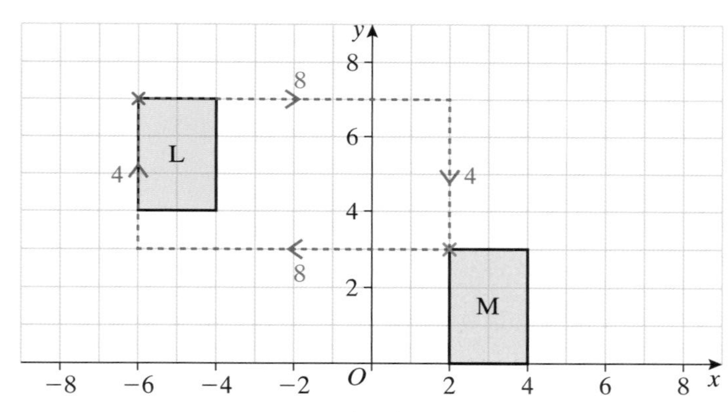

L and M are **congruent**.

a) A translation by the vector $\begin{pmatrix} 8 \\ -4 \end{pmatrix}$

b) A translation by the vector $\begin{pmatrix} -8 \\ 4 \end{pmatrix}$

EXERCISE 17.3

In Questions 1–4, describe:

a) the translation that takes A onto B

b) the translation that takes B onto A.

1

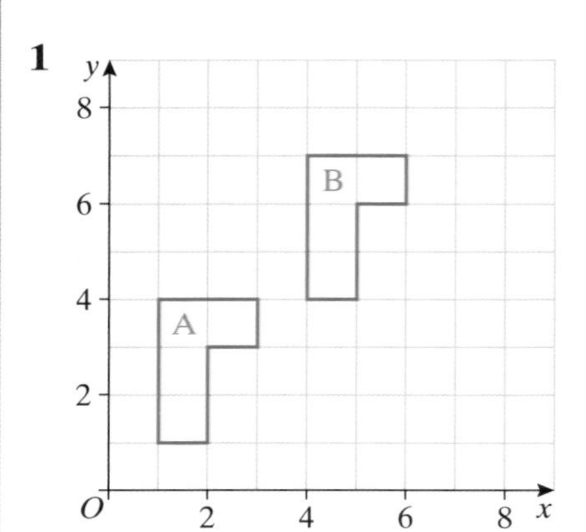

2

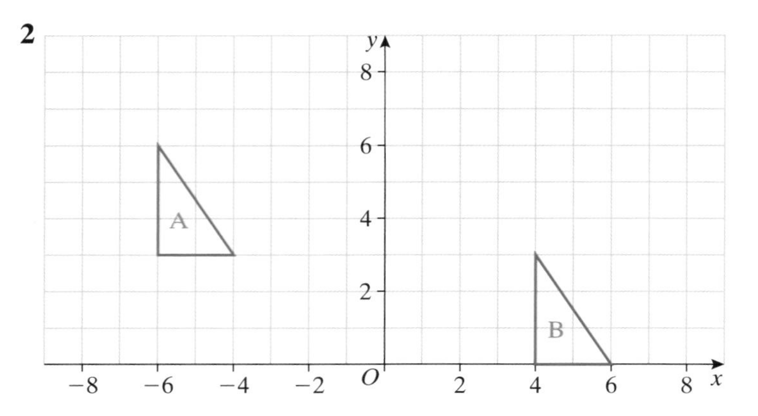

3

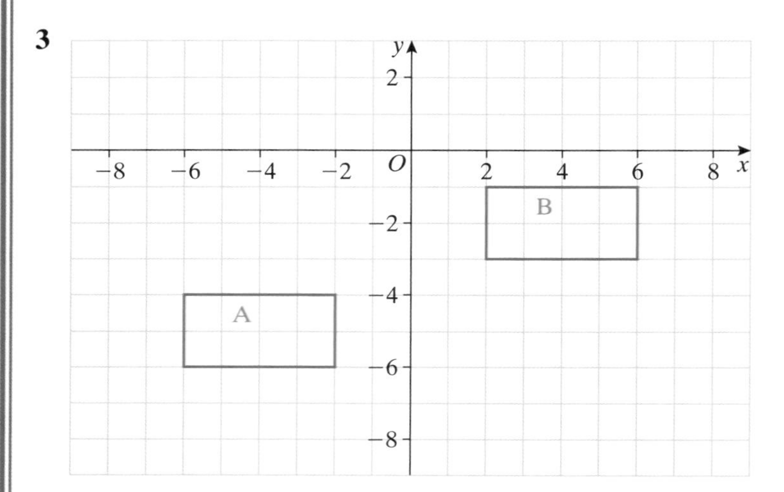

4

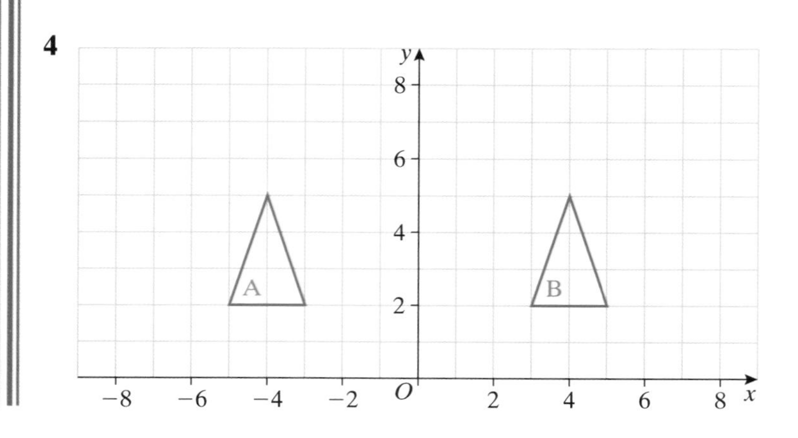

5 **a)** Translate triangle A by the vector $\begin{pmatrix} -4 \\ -2 \end{pmatrix}$

Label the image B.

b) Translate triangle B by the vector $\begin{pmatrix} 0 \\ -5 \end{pmatrix}$

Label the image C.

c) Translate triangle C by the vector $\begin{pmatrix} 4 \\ 0 \end{pmatrix}$

Label the image D.

d) Describe the single translation that would take triangle D onto triangle A.

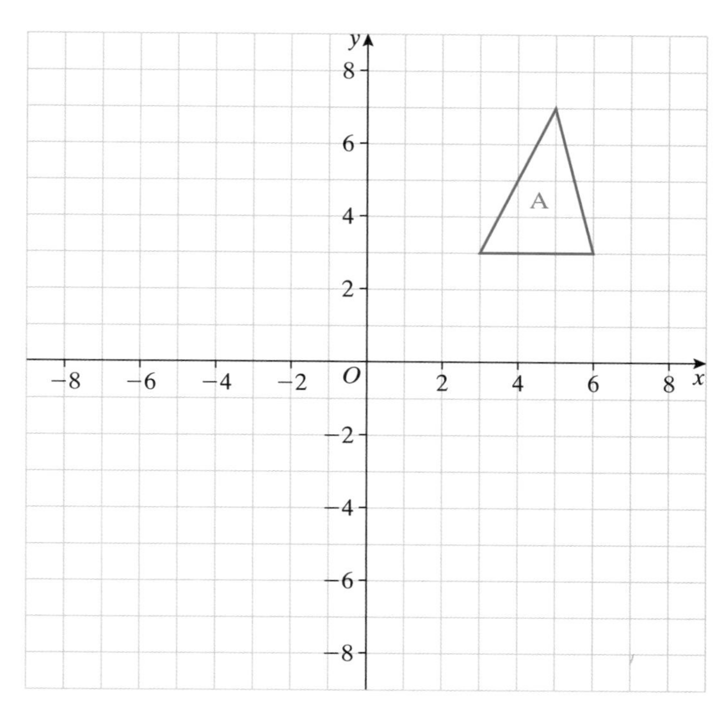

6

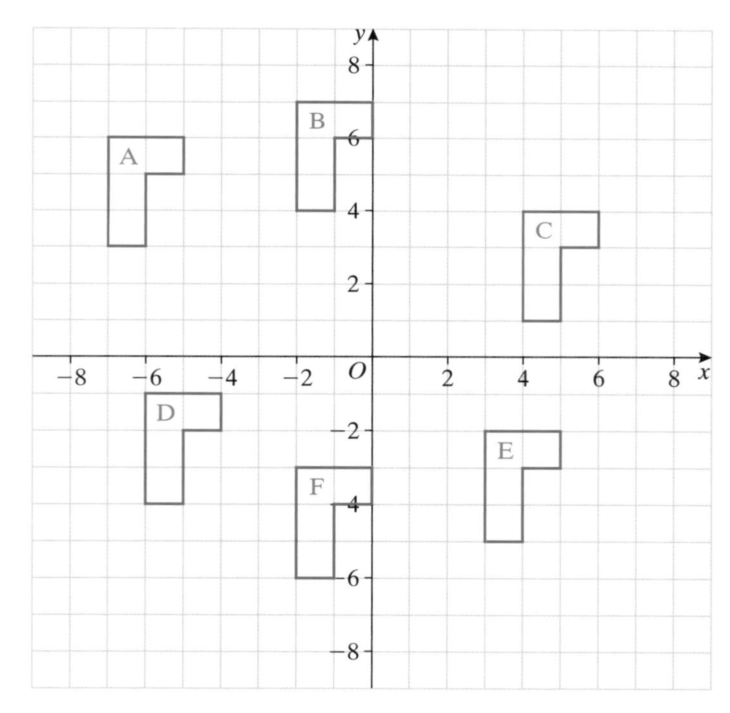

Describe the translation that takes Γ-shape from:

a) A onto B **b)** B onto F **c)** D onto E **d)** E onto A

e) E onto C **f)** C onto D **g)** A onto F **h)** B onto D

17.4 Rotational symmetry

When a rectangle is rotated (turned) through 360°, it looks the same on two occasions.

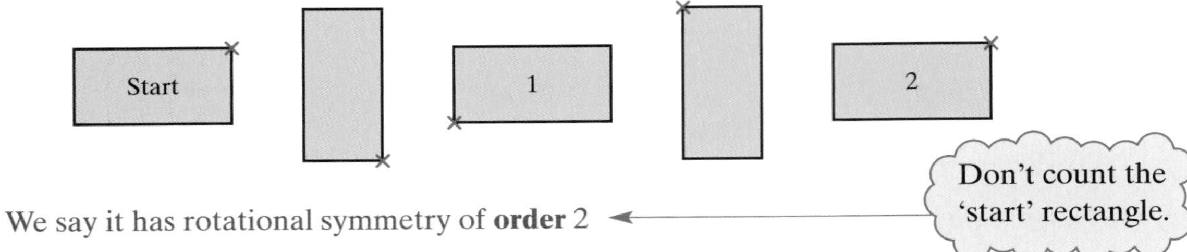

Don't count the 'start' rectangle.

We say it has rotational symmetry of **order** 2

The **order of rotational symmetry** is the number of times a shape looks the same when it is rotated one full turn.

EXAMPLE

Write down the order of rotational symmetry for each of these shapes.

a)

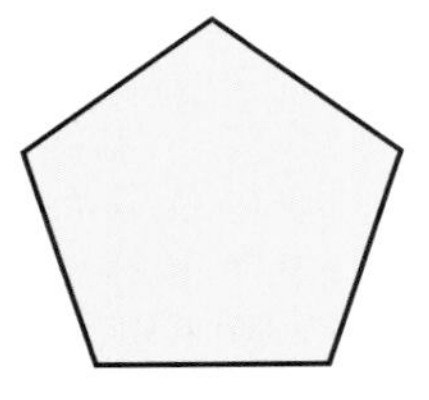

b)

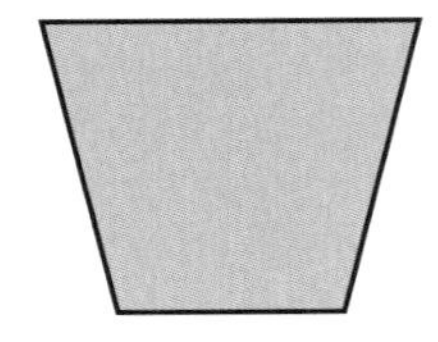

SOLUTION

a)

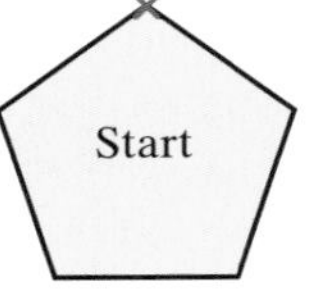

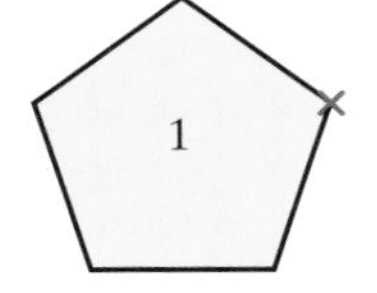

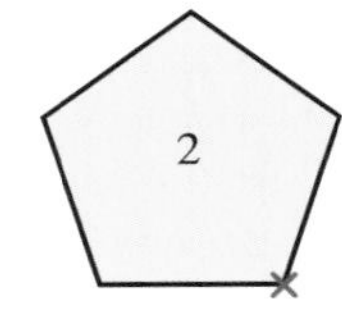

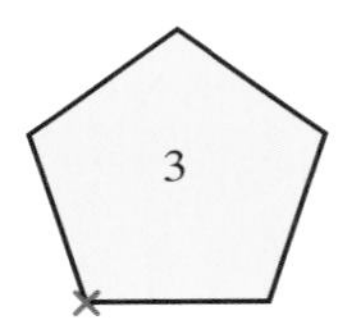

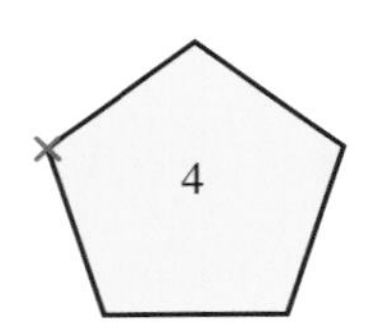

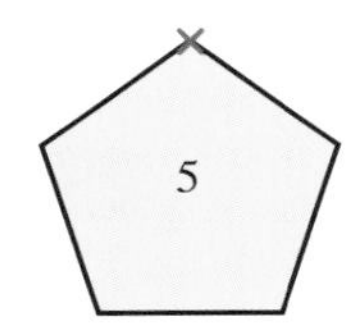

The pentagon has rotational symmetry of order 5

b)

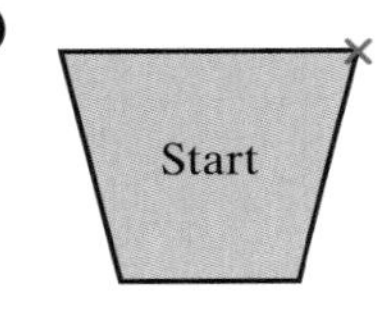

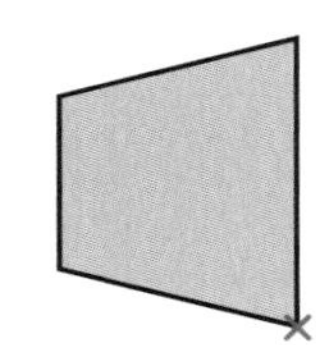

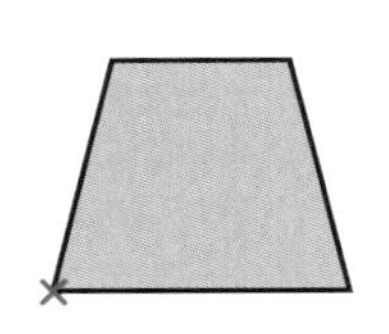

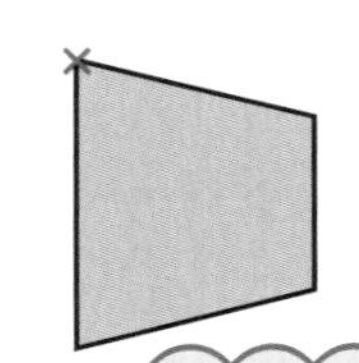

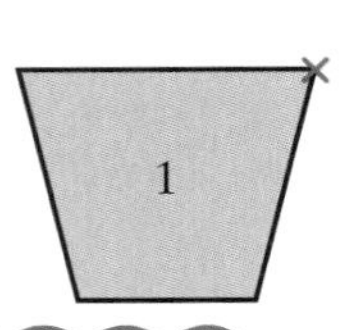

The trapezium doesn't have rotational symmetry

Sometimes we say this is rotational symmetry of order 1

The order of rotational symmetry of a regular polygon is the same as the number of sides it has.

EXERCISE 17.4

1 Write down the order of rotational symmetry of each of these shapes.

a) 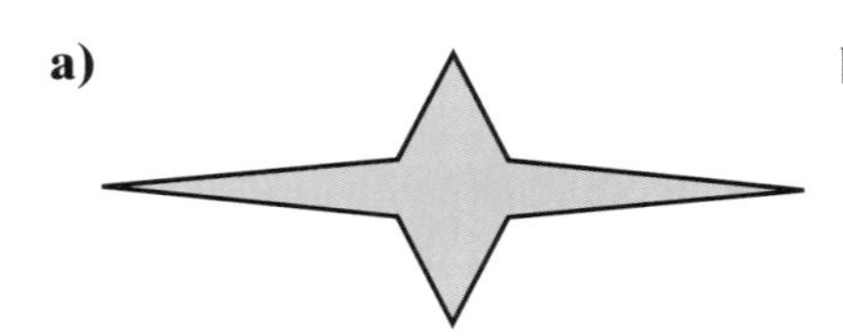

b)

c)

d) 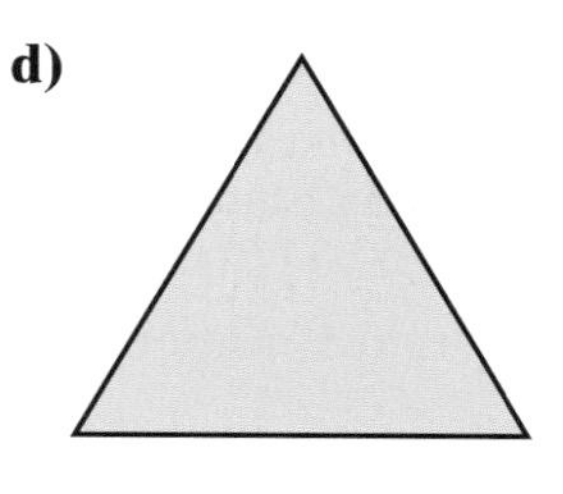

e) 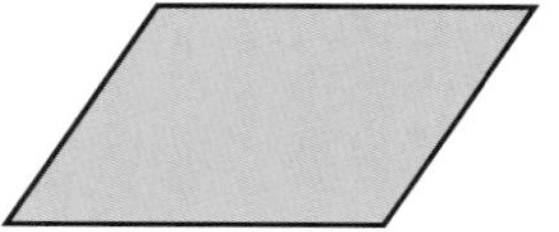

f) 

g) 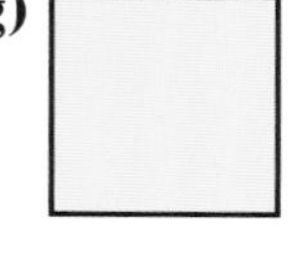

h) 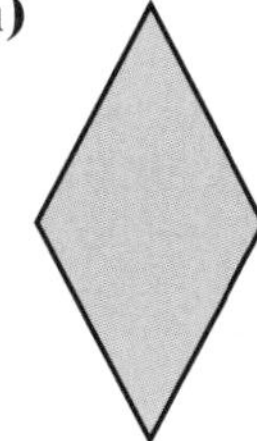

2 Complete a copy of this shape so that it has two lines of symmetry and rotational symmetry of order 2

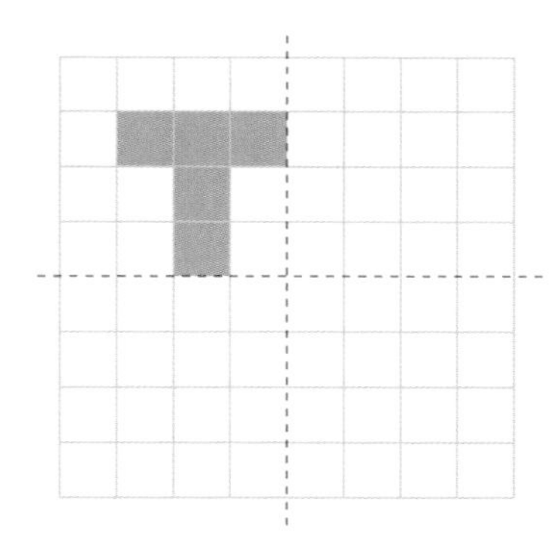

3 Add one square to a copy of this shape so it has rotational symmetry of order 2

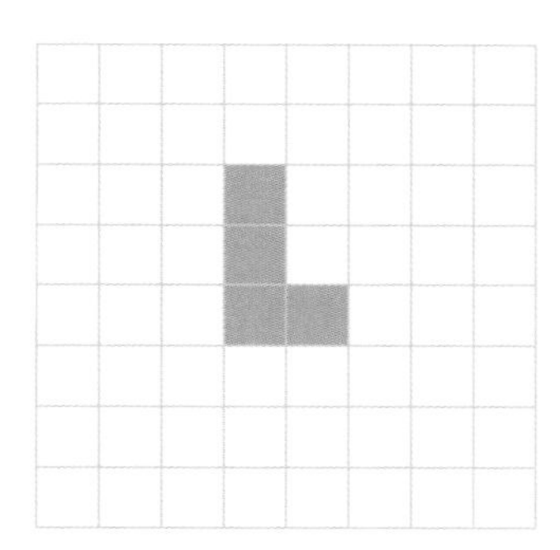

4 Add three squares to a copy of this shape so it has rotational symmetry of order 4

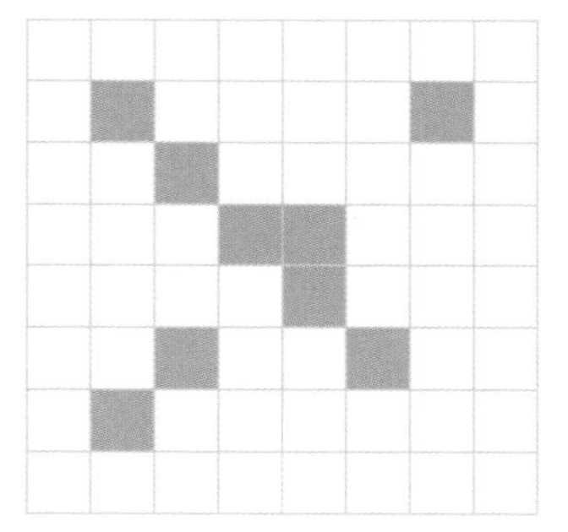

17.5 Rotations

A shape can be turned to face in a different direction, while remaining the same shape and size – this is known as **rotation**.

An imaginary point acts as a pivot for the rotation – this is the **centre of rotation**.

You must remember to state:

- the size of the turn, or **angle of rotation**,
- whether it is **clockwise** or **anticlockwise**
- and the coordinates of the **centre of rotation**.

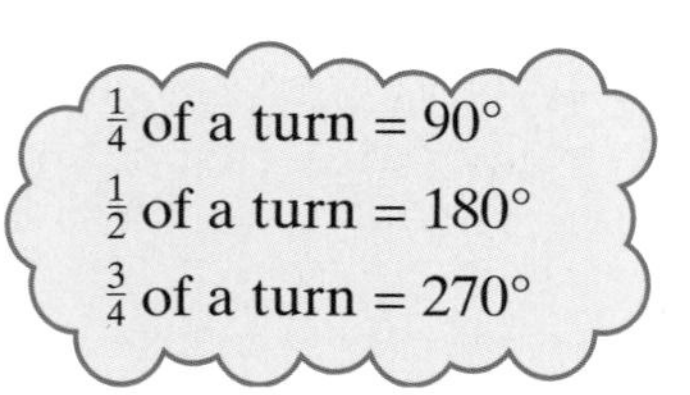

$\frac{1}{4}$ of a turn = 90°
$\frac{1}{2}$ of a turn = 180°
$\frac{3}{4}$ of a turn = 270°

EXAMPLE

The diagram shows a rectangle labelled S.

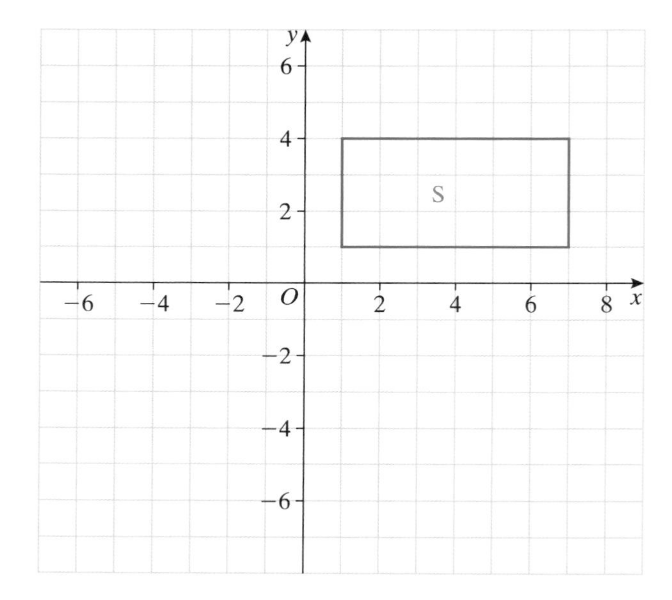

a) Rotate shape S through 90° clockwise, about the origin O.
Label the resulting shape T.

b) Now rotate the shape T through 180° about O.
Label the resulting shape U.

c) Describe a single rotation that would take S directly to U.

SOLUTION

a)

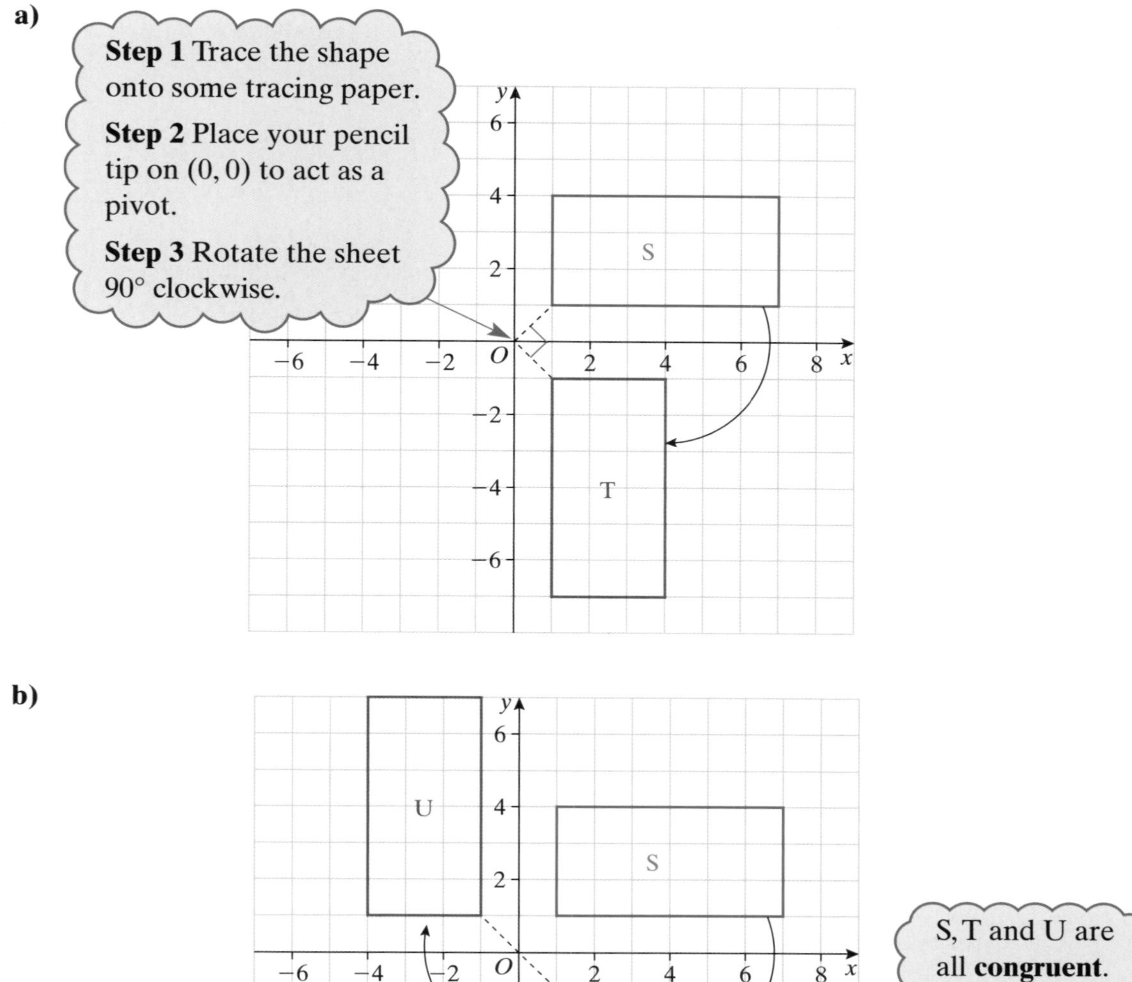

Step 1 Trace the shape onto some tracing paper.
Step 2 Place your pencil tip on (0, 0) to act as a pivot.
Step 3 Rotate the sheet 90° clockwise.
y
6
4
2
−6
−4
−2
O
2
4
6
8
x
−2
−4
−6
S
T

b)

S, T and U are all **congruent**.

The direction of this second rotation was not given in the question … because 180° clockwise and 180° anticlockwise are exactly the same.

c) U can be obtained directly from S by a 90° rotation anticlockwise about O

Rotations are often performed with the point $(0, 0)$ as the centre of rotation, but they can be done about other centres.

The **origin**.

EXAMPLE

The diagram shows a triangle M drawn on a grid.

Rotate the triangle M through $\frac{1}{4}$ of a turn anticlockwise about the point P at $(1, 0)$

Label this new triangle N.

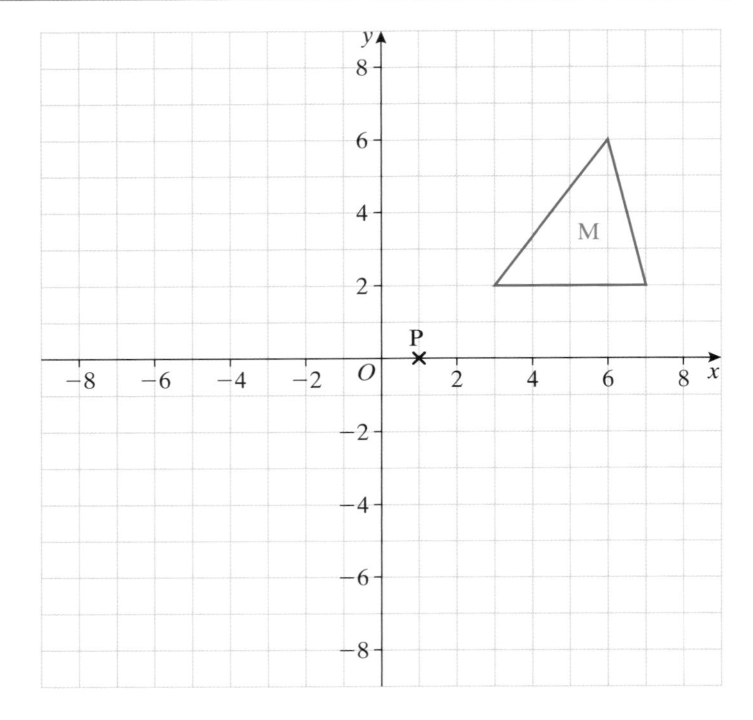

SOLUTION

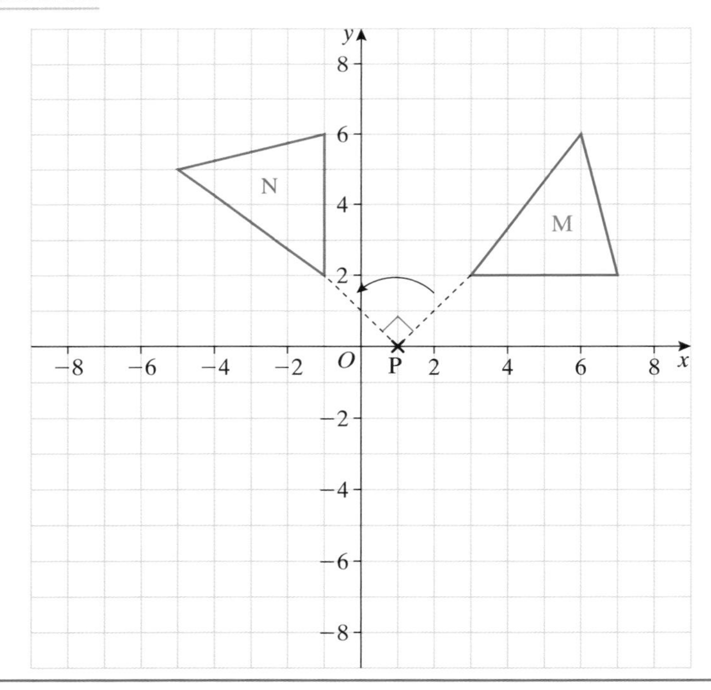

EXERCISE 17.5

Use copies of the coordinate grids with *x* and *y* axes numbered from −8 to 8

1 Rotate the trapezium shape 90° clockwise about O.

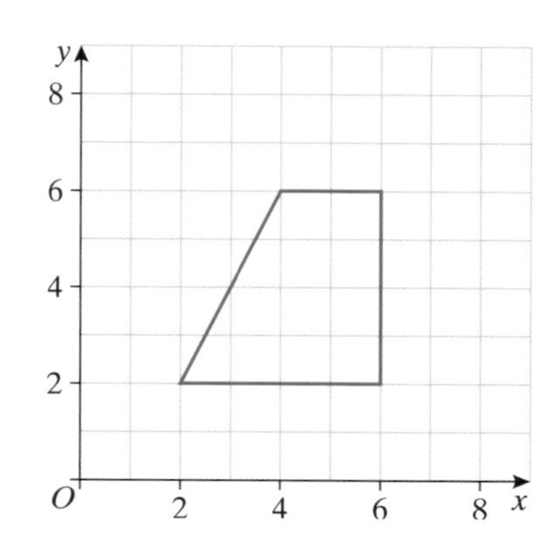

2 Rotate the shape by 180° about O.

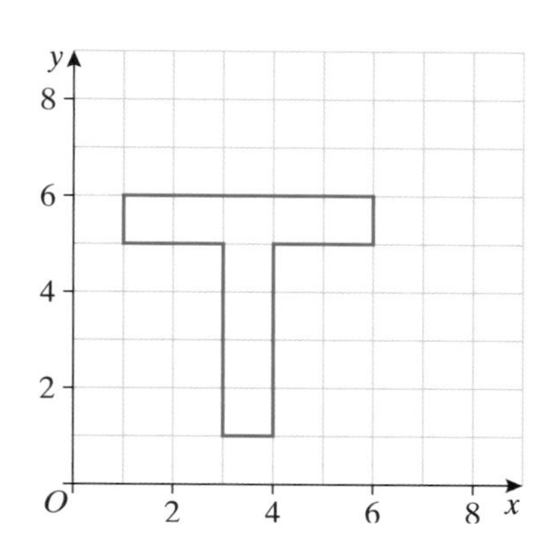

3 a) Rotate the triangle T1 $\frac{1}{4}$ of a turn anticlockwise about O. Label the result T2.

b) Rotate T2 $\frac{1}{2}$ a turn about O. Label the result T3.

c) Describe the single rotation that takes T1 directly to T3.

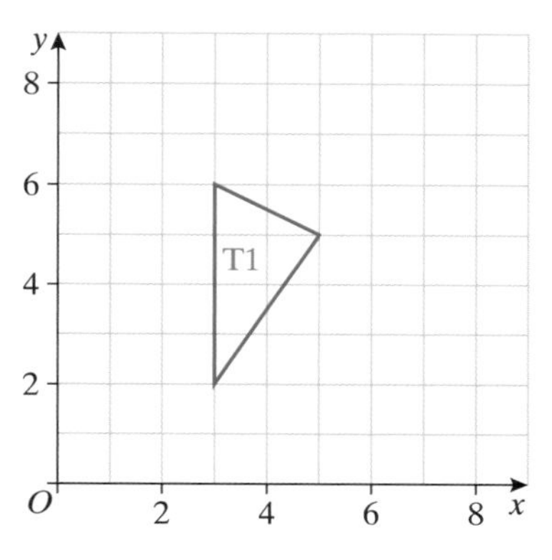

4 a) Rotate shape A 90° anticlockwise about (1, 0) Label the result B.

b) Rotate shape B by 180° about (0, 0) Label the result C.

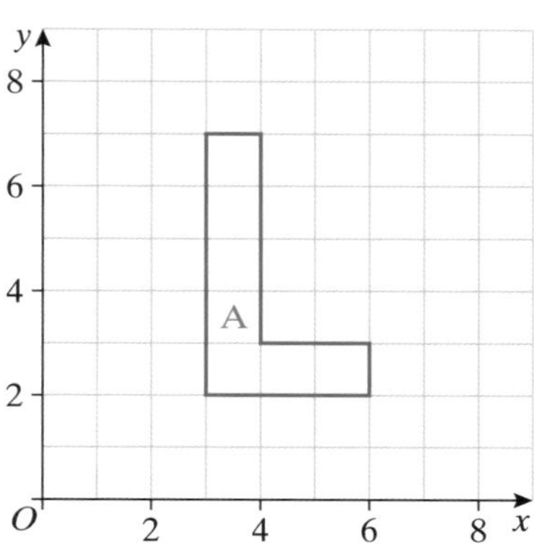

5

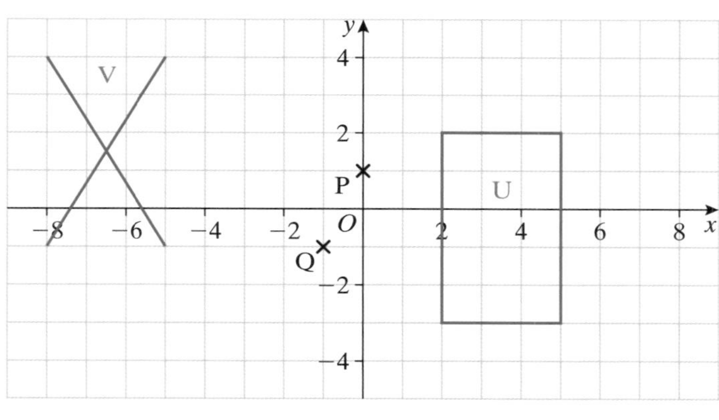

a) Rotate shape U 90° anticlockwise about point P (0, 1)

b) Rotate shape V 90° clockwise about point Q (−1, −1)

6 a) Rotate the triangle 90° clockwise about (1, 1)

b) Now rotate both the new triangle and the original one by 180° about (1, 1)

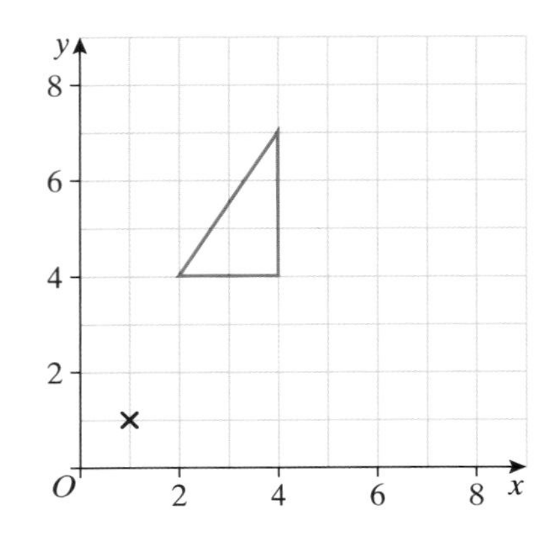

7 The diagram shows an object A and its image B after a rotation.

a) Write down the size and direction of the angle of rotation.

b) Write down the coordinates of the centre of rotation.

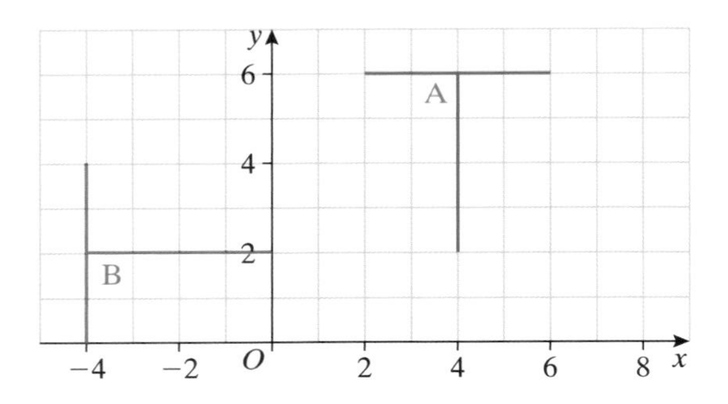

17.6 Combining transformations

There are three transformations that preserve congruence.

So the shape or size of an object does not change.

They are:

- reflection
- translation
- rotation

You may be required to combine two transformations together.

EXAMPLE

a) Reflect the given triangle T1 in the line $x = -4$, and label the result T2.

b) Reflect T2 in the line $x = 1$, and label the result T3.

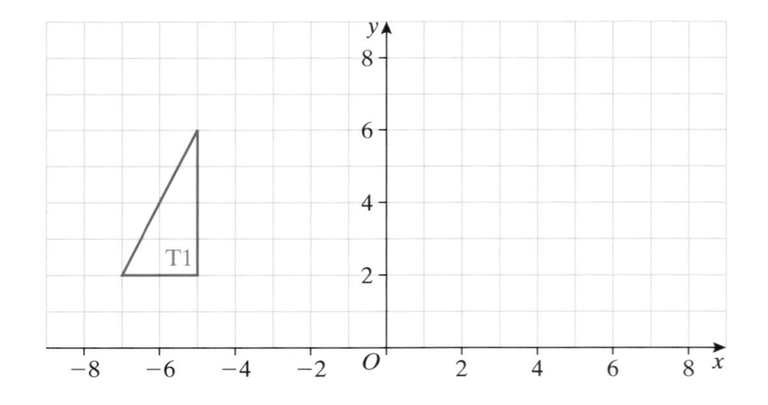

SOLUTION

a)

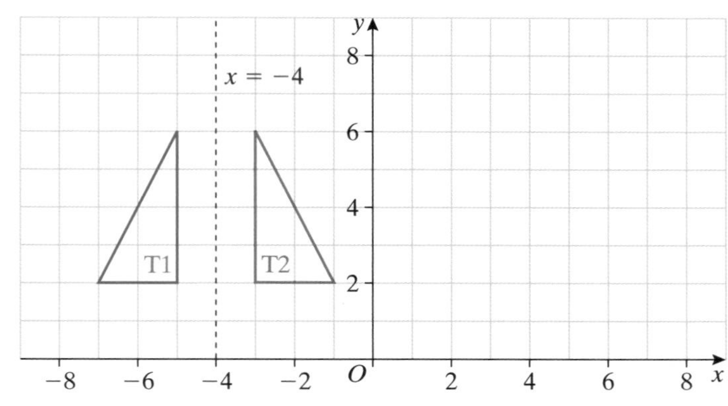

b)

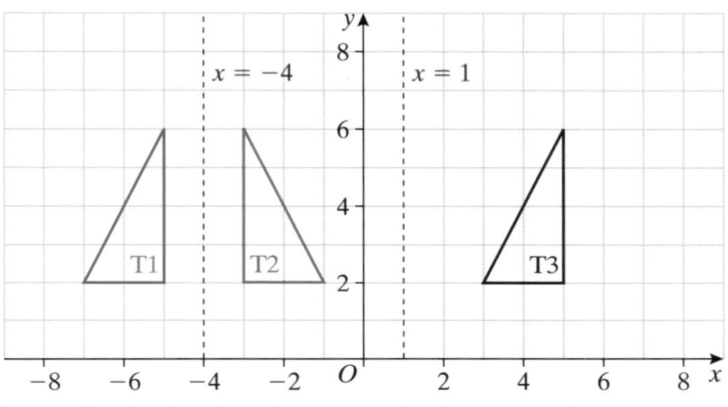

EXERCISE 17.6

1 a) On a copy, reflect triangle S in the line $x = -1$
Label the new triangle T.

b) Reflect triangle T in the x axis.
Label the new triangle U.

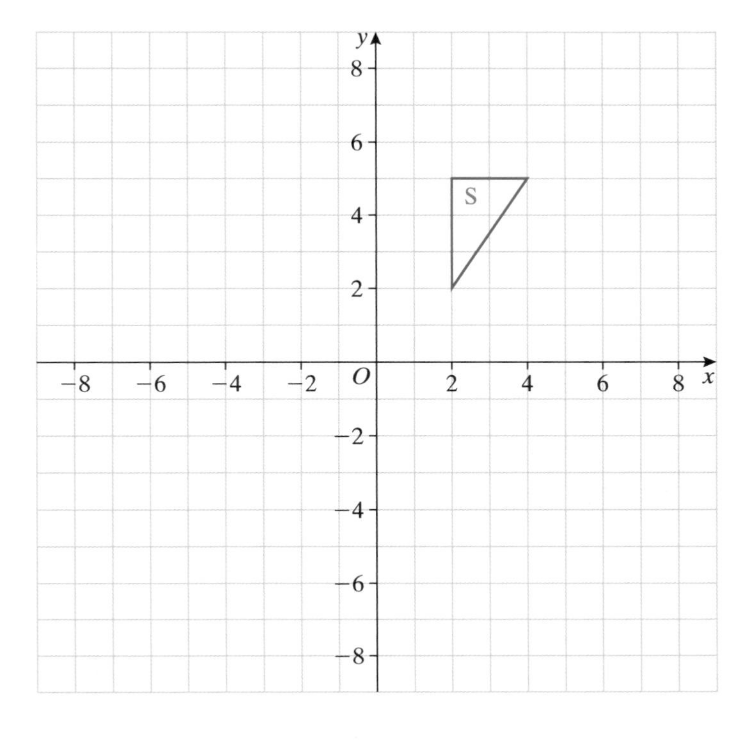

2 The diagram shows a triangle, T.

a) On a copy, translate triangle T by $\begin{pmatrix} -6 \\ 0 \end{pmatrix}$

Label its image triangle U.

b) Rotate triangle U by 180° about O. Label the resulting triangle V.

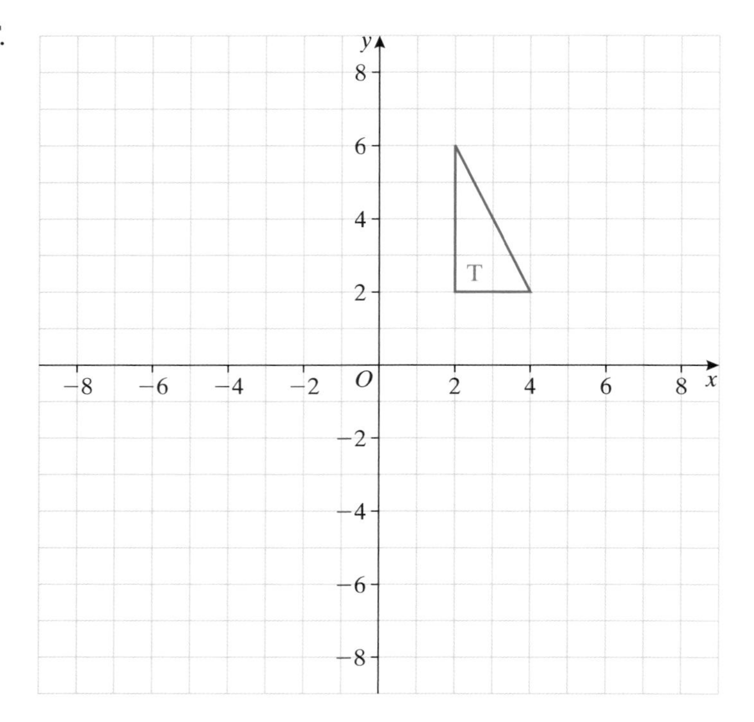

3 The diagram shows a triangle, S.

a) On a copy, reflect triangle S in the y axis. Label this image triangle T.

b) Reflect triangle S in the line $y = 1$ Label this image triangle U.

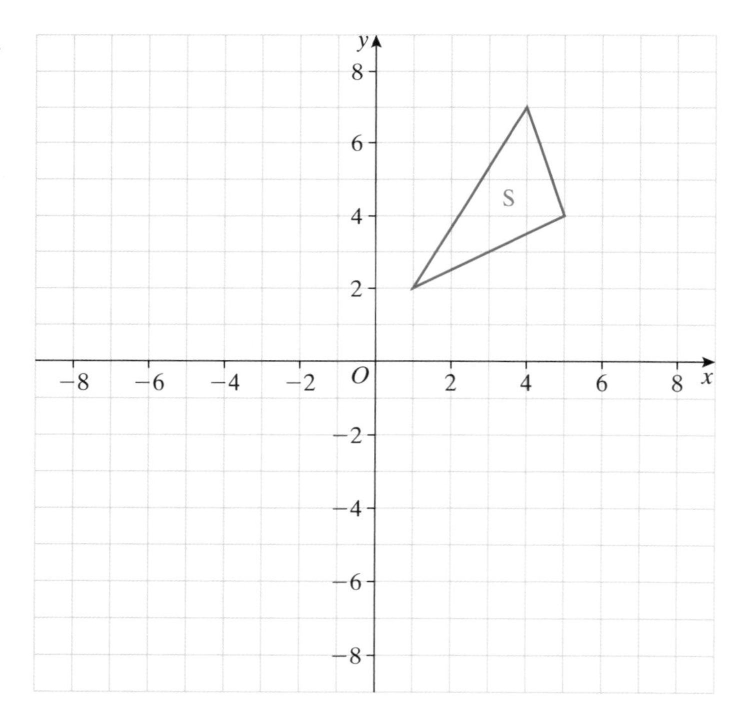

4 The diagram shows a set of points that make a letter F-shape. The shape is labelled F1.

a) On a copy, reflect the shape F1 in the x axis. Label the result F2.

b) Reflect F2 in the line $y = x$. Label the result F3.

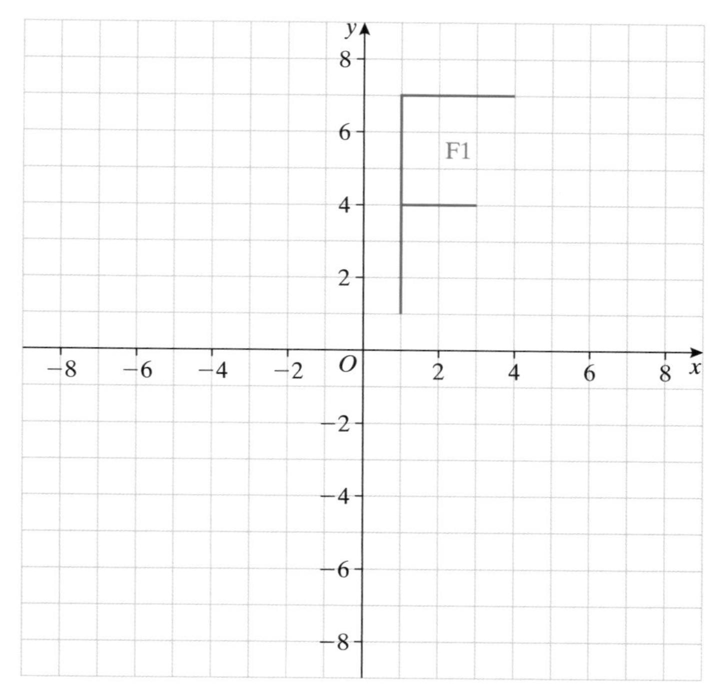

5 The diagram shows a quadrilateral F.

a) Rotate quadrilateral F through 90° anticlockwise about O. Label the result G.

b) Rotate quadrilateral G through 90° clockwise about (4, −4) Label the result H.

c) Describe the single transformation that would take shape H onto shape F.

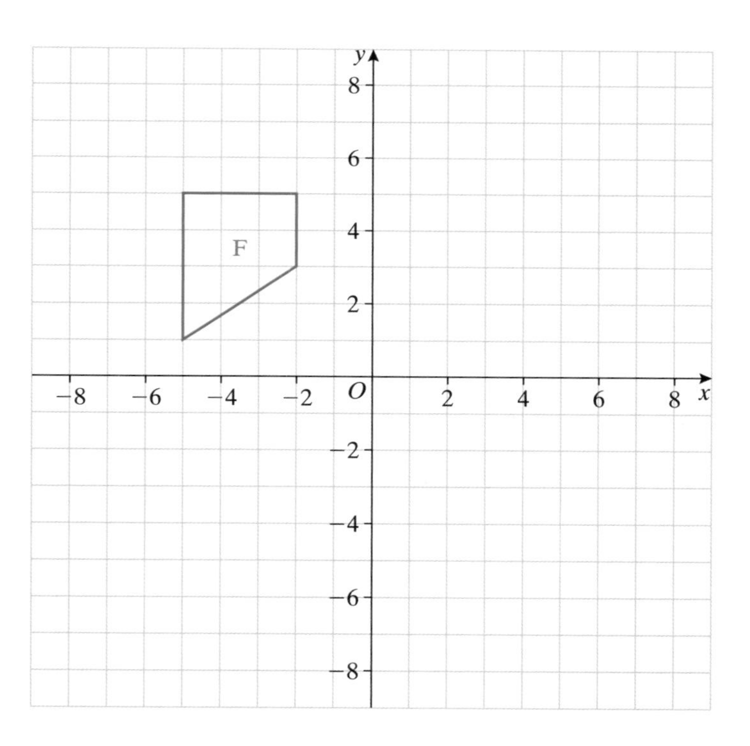

6 The diagram shows a triangle A.

a) On a copy, rotate triangle A by 90° anticlockwise about (0, 3). Label this image B.

b) Rotate triangle A by 180° about the origin O. Label this image C.

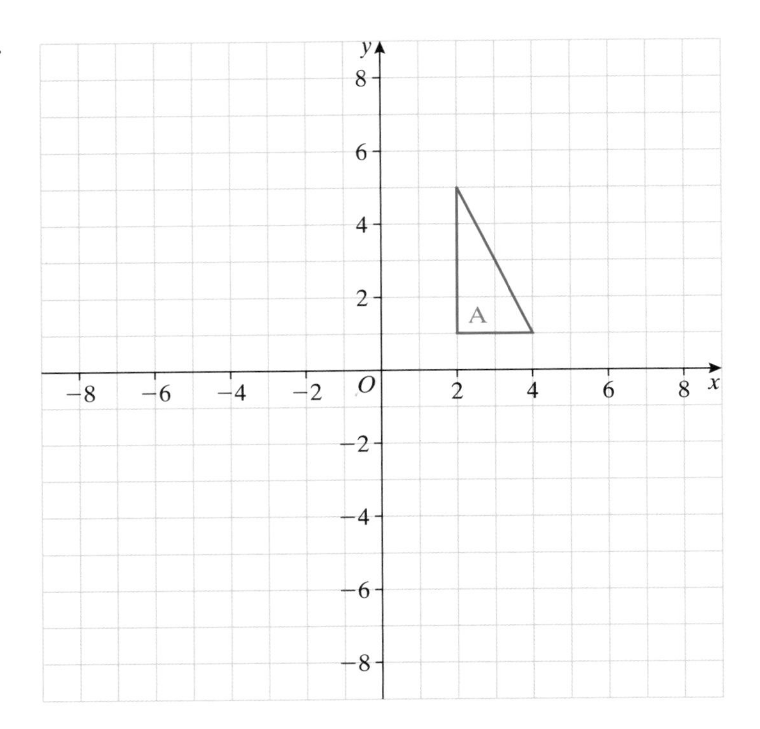

17.7 Enlargements

Look at these two P-shapes drawn on a centimetre grid.

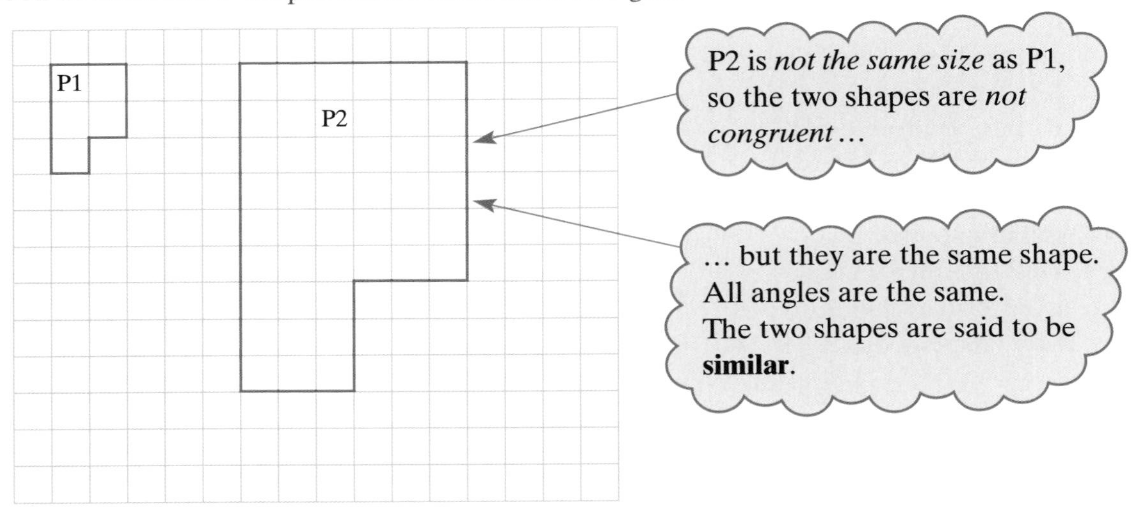

P1
P2
P2 is *not the same size* as P1, so the two shapes are *not congruent*…
… but they are the same shape. All angles are the same. The two shapes are said to be **similar**.

P2 is an enlargement of P1. It is 3 times larger.

We say, 'P2 is an enlargement of P1, scale factor 3'

This means that every side of P1 has been made 3 times longer.

In mathematics we also say that P1 is an enlargement of P2 (even though it is smaller!).

We would say, 'P1 is an enlargement of P2, scale factor $\frac{1}{3}$'

EXAMPLE

a) Find the **(i)** perimeter, **(ii)** area of P1.

b) Find the **(i)** perimeter, **(ii)** area of P2.

c) How many times larger is the **(i)** perimeter, **(ii)** area of P2 than of P1?

SOLUTION

a) **(i)** The perimeter of P1 $= 2 + 2 + 1 + 1 + 1 + 3 =$ 10 cm

(ii) The area of P1 is 5 cm^2

b) **(i)** The perimeter of P2 $= 6 + 6 + 3 + 3 + 3 + 9 =$ 30 cm

(ii) The area of P2 is 45 cm^2

c) **(i)** The perimeter of P2 is $3\times$ larger than P1

(ii) The area of P2 is $9 \times$ larger than P1

When a shape is enlarged, the perimeter increases by the same factor as the scale factor.

So when the scale factor is 2 the perimeter is doubled.

When the scale factor is $\frac{1}{2}$ the perimeter is halved.

When a shape is enlarged by a scale factor:

- between 0 and 1, then the shape gets smaller;
- greater than 1, then the shape gets larger;
- of 1, then the shape stays the same size.

The area increases by the square of the scale factor.

EXAMPLE

Enlarge this shape by a scale factor $\frac{1}{2}$

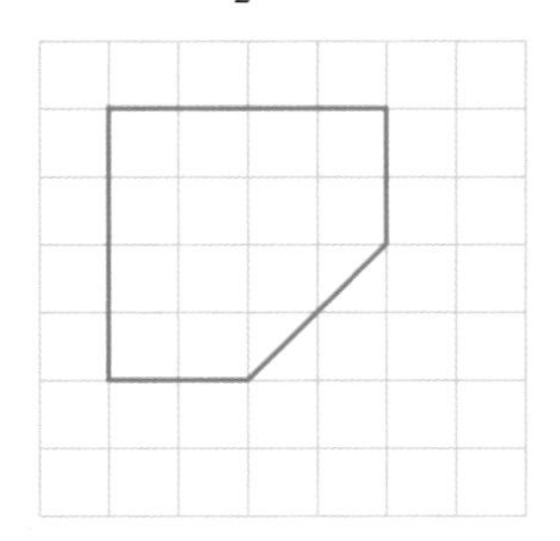

SOLUTION

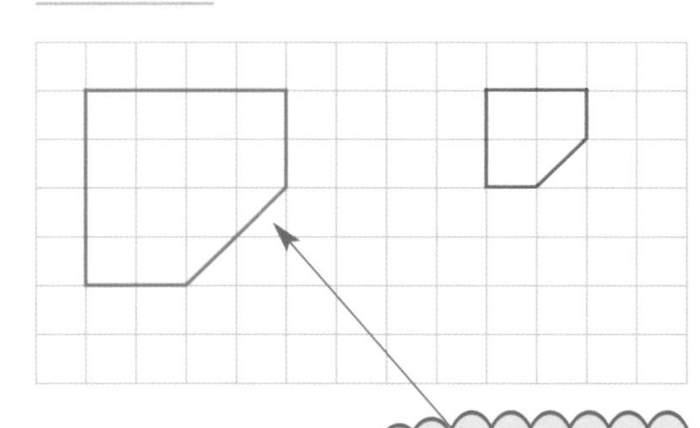

Take care with diagonal lines. It is best to enlarge the other lines first.

Two shapes are **similar** when one is an enlargement of the other.

EXAMPLE

Which of these rectangles are:

a) congruent

b) similar?

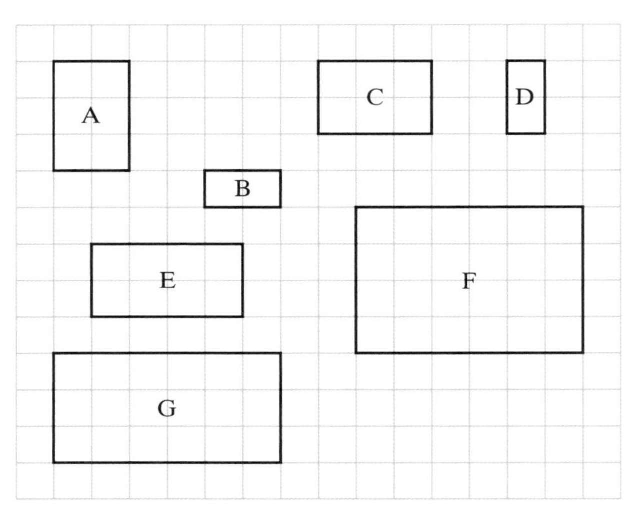

SOLUTION

a) Rectangles A and C are the same shape and size so they are congruent.

Rectangles B and D are the same shape and size so they are congruent.

b) Congruent shapes are also similar.

Each side of rectangle E is twice the length of the sides in B.

Each side of rectangle G is $3 \times$ the length of the sides in B.

So B, D, E and G are similar.

Each side of rectangle F is twice the length of the sides in rectangle A.

So A, C and F are similar.

EXERCISE 17.7

1 Enlarge each shape by scale factor 2

a)

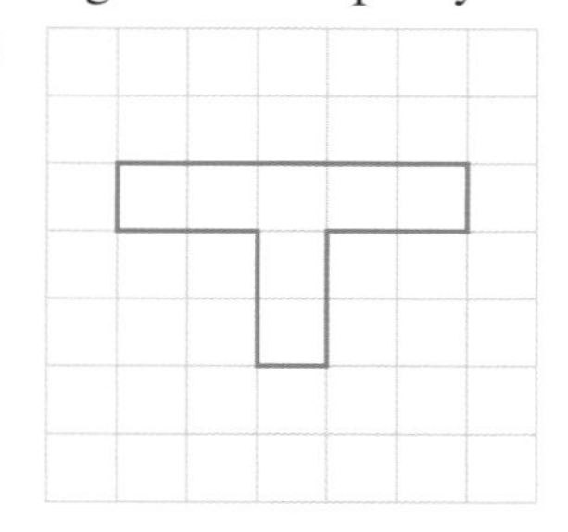

b)

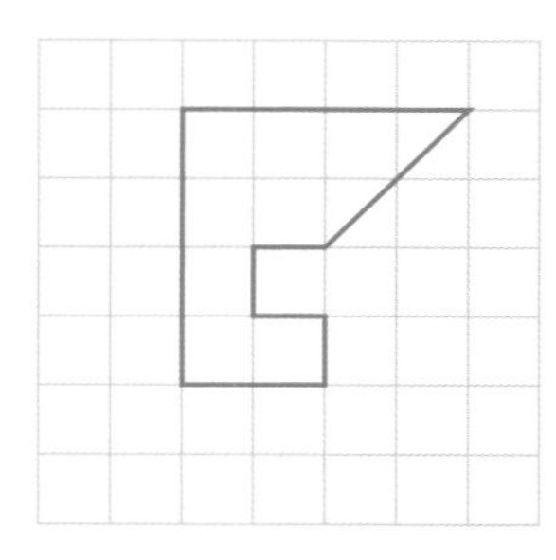

2 Enlarge the following shapes by scale factor 3

a)

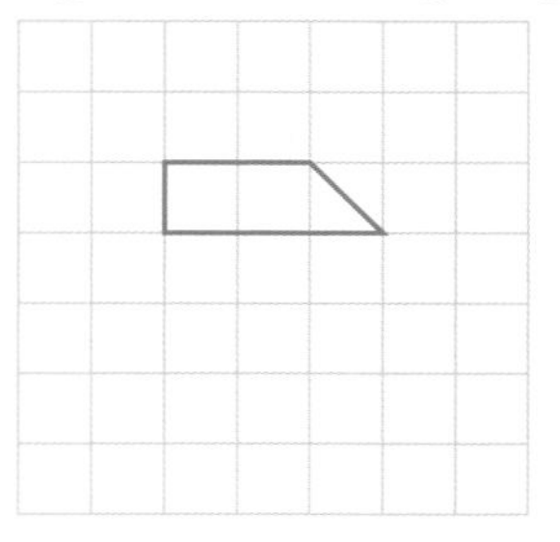

b)

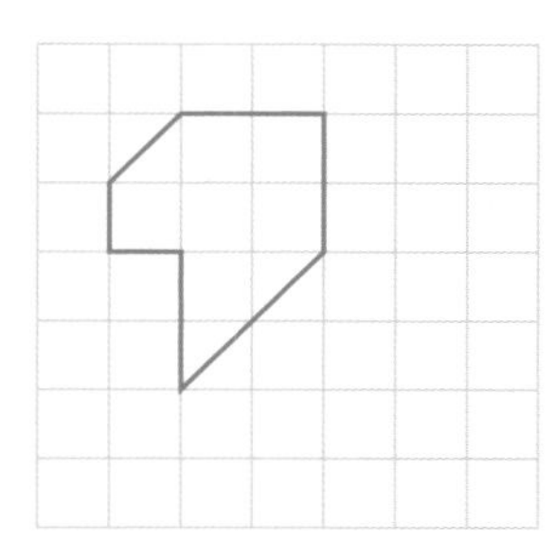

3 Enlarge the following shapes by scale factor $\frac{1}{2}$

a)

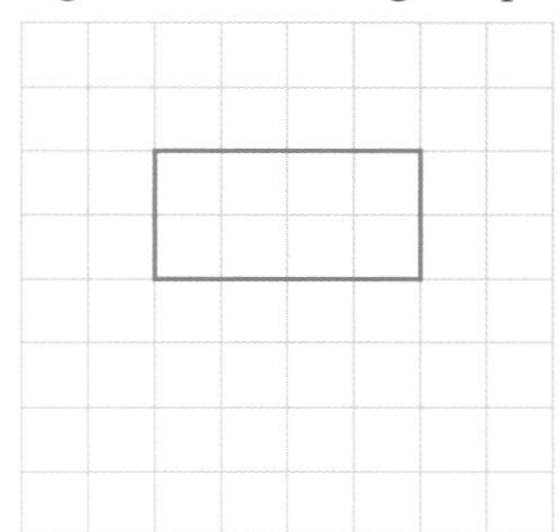

b)

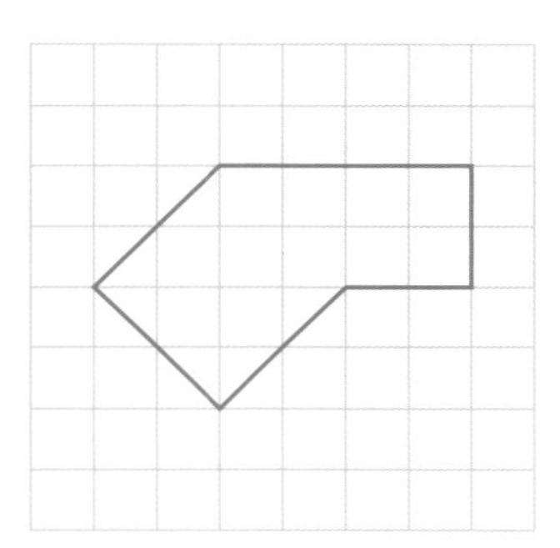

4 Enlarge this shape by scale factor:

a) $\frac{1}{2}$

b) $1\frac{1}{2}$

c) 2

d) $2\frac{1}{2}$

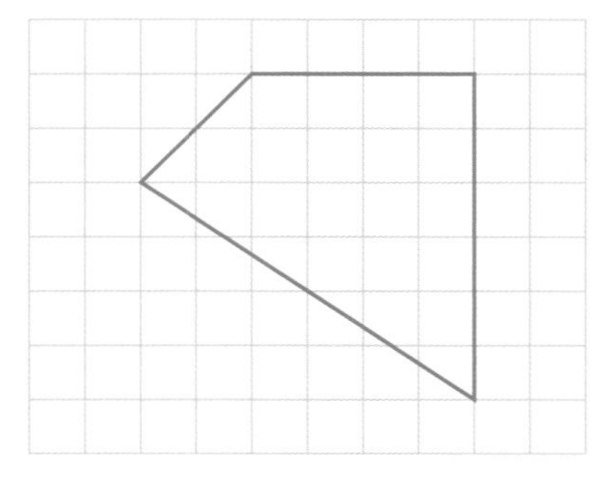

5 A shape has a perimeter of 20 cm
Write down the perimeter of the shape when it is enlarged by a scale factor of:

a) 3 **b)** 2 **c)** $1\frac{1}{2}$ **d)** $\frac{1}{2}$

6 Write down which of these rectangles are: **a)** similar **b)** congruent.

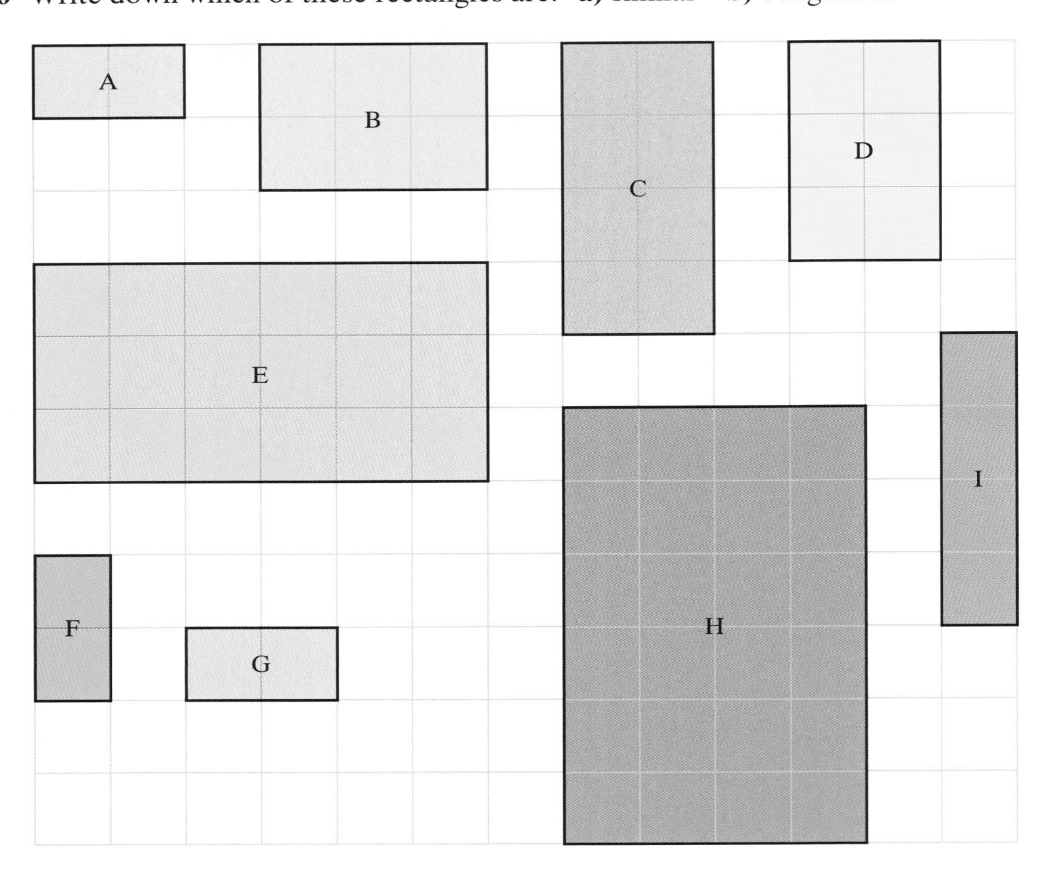

7 Which of these triangles are similar?

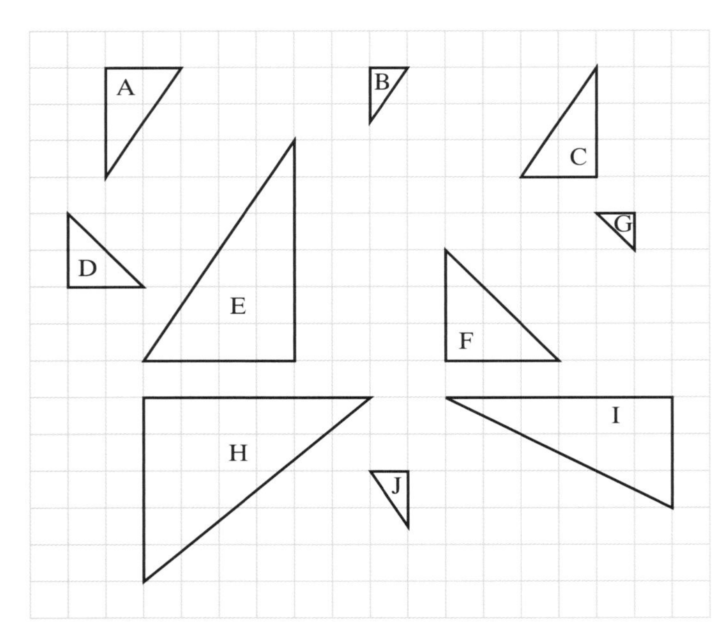

8 Are any two squares similar?
Give a reason for your answer.
Use this diagram to help you.

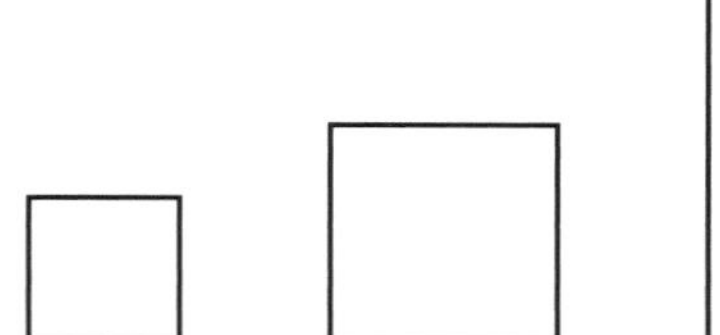

9 Are any two circles similar?
Give a reason for your answer.
Use this diagram to help you.

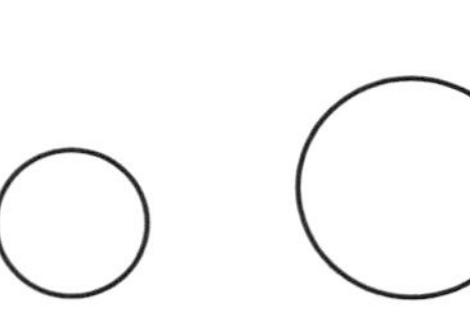

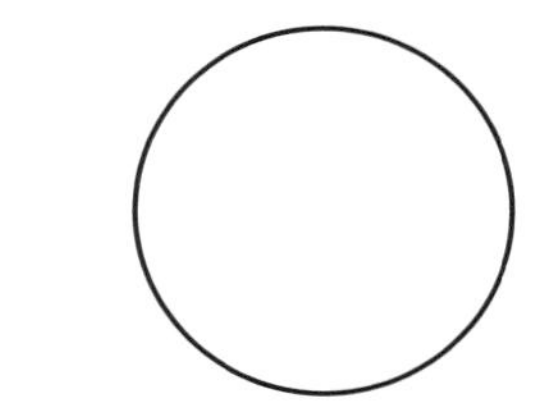

10 Are any two rectangles similar?
Give a reason for your answer.
Use this diagram to help you.

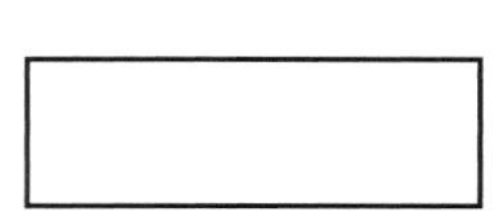

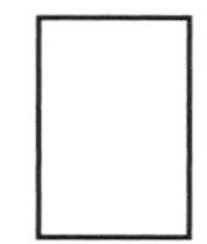

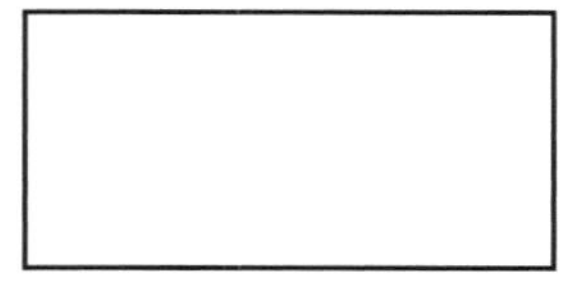

11 A quadrilateral has angles of 100°, 90°, 110° and 60°
Write down the angles in the quadrilateral when it is enlarged by a scale factor of: **a)** $\frac{1}{2}$ **b)** 2

12 **a)** Find the **(i)** perimeter, **(ii)** area of this rectangle.

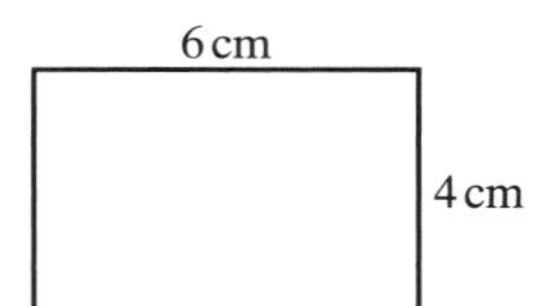

The rectangle is enlarged by scale factor 2.

b) Write down the **(i)** perimeter, **(ii)** area of the enlarged rectangle.

13 **a)** Find the volume of this cuboid.

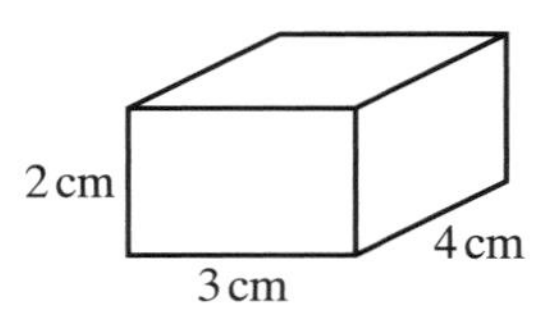

The cuboid is enlarged by scale factor 2.

b) Work out the volume of the enlarged cuboid.

17.8 Centre of enlargement

In the previous section you enlarged shapes and drew the enlargement anywhere you liked on the page.

In this section you will enlarge shapes on a coordinate grid.

You need to make sure that the shape ends up in the right place on the grid.

You are given the coordinates of a point, called the **centre of enlargement**, which tells you where to draw the enlarged shape.

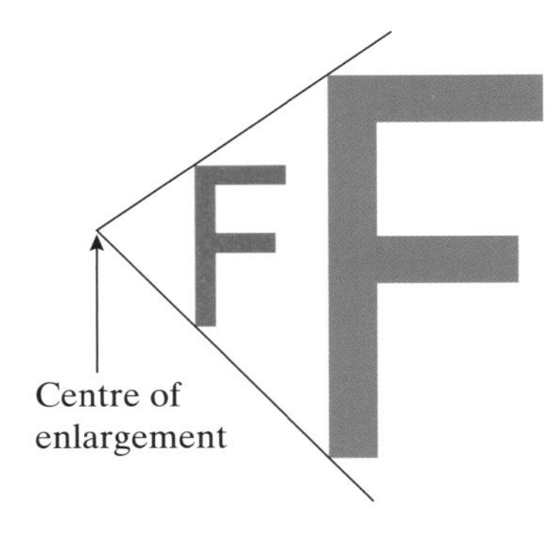

Scale factor > 1

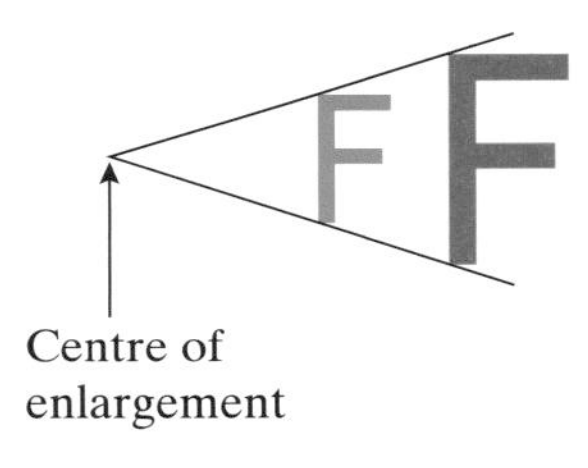

$0 <$ scale factor < 1

The following example shows you how to use centres of enlargement.

EXAMPLE

The diagram shows a letter F-shape and two points, P and Q.

a) Using P as the centre of enlargement, draw a letter F to scale factor 2

b) Using Q as the centre of enlargement, draw a letter F to scale factor $\frac{1}{2}$

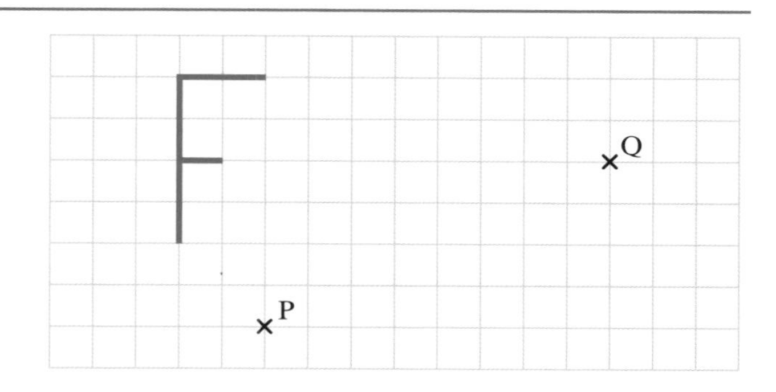

SOLUTION

a)

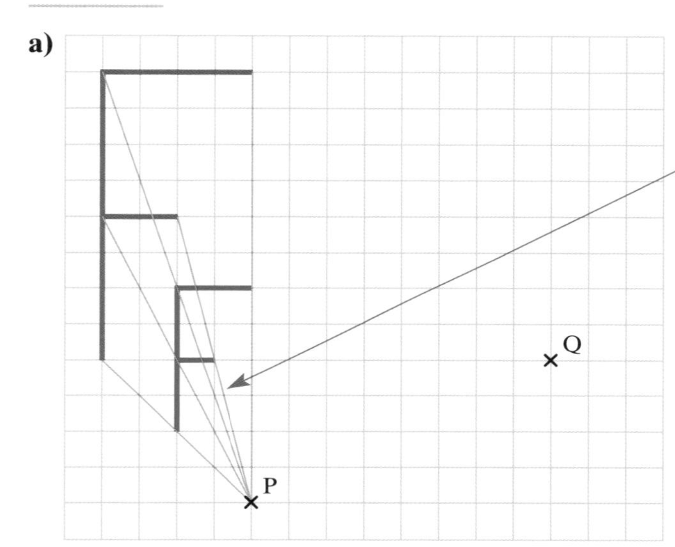

Draw rays from P to each corner of the original F shape. Then extend these rays so they are twice their original length (factor is $\times 2$).
The rays will give the corners of the enlarged shape.

b)

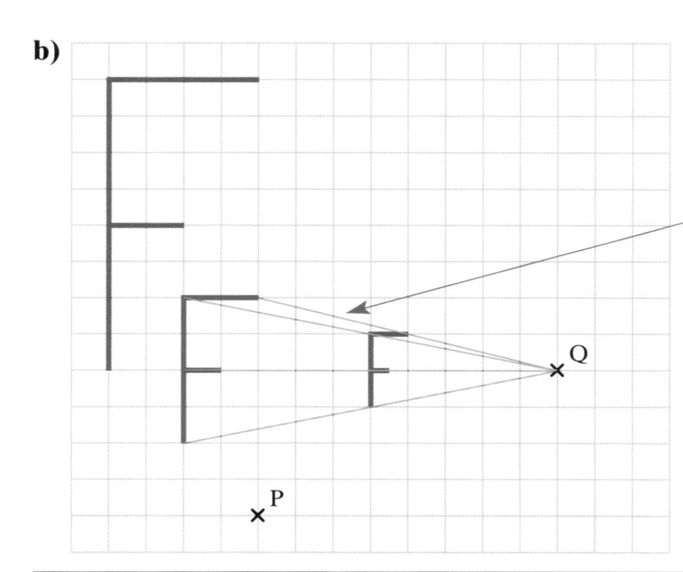

Again draw rays from Q to each corner of the original F shape. Then go half way along the rays (factor is $\times \frac{1}{2}$), to find the corners of the enlar ged’ shape (which is actually smaller than the original).

EXERCISE 17.8

1 The diagram shows a shape A.

a) On a copy enlarge shape A by scale factor 2, centre P. Label the new shape B.

b) Enlarge shape A by scale factor 3, centre P. Label the new shape C.

c) State whether shapes B and C are:

(i) congruent

(ii) similar.

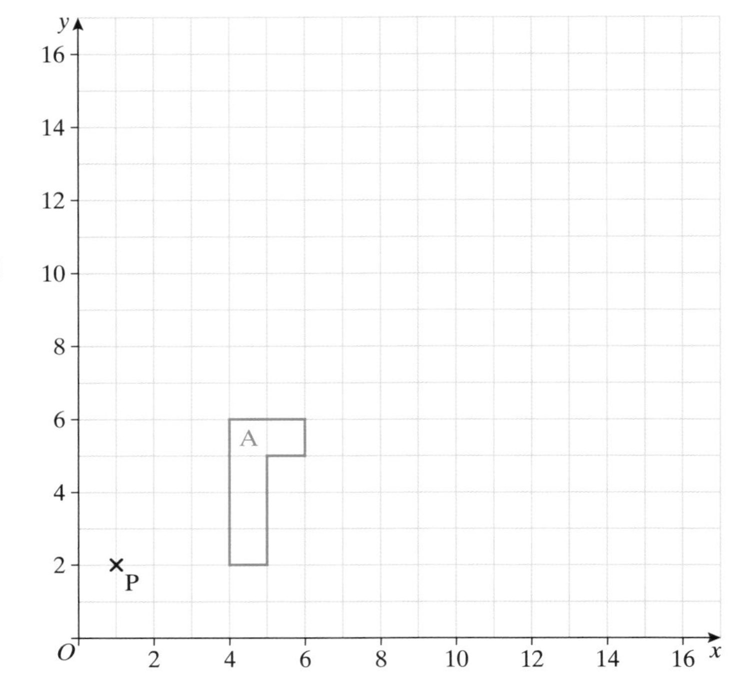

2 The diagram shows a triangle.

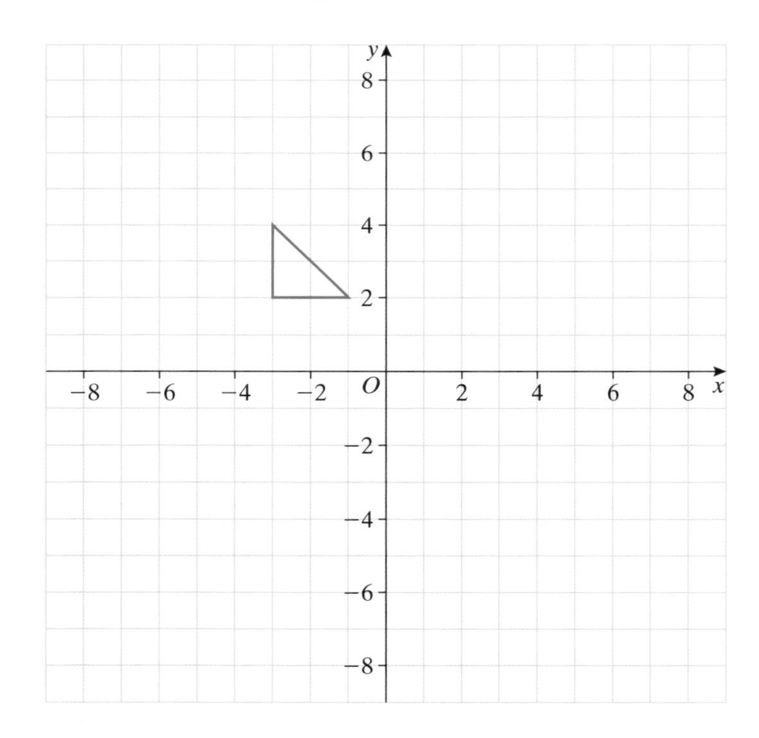

On a copy, enlarge the shape by scale factor $2\frac{1}{2}$, using centre $(-7, 6)$

3 The diagram shows a shape A and two centres, P and Q, marked with crosses.

a) On a copy enlarge shape A, with scale factor 2, centre P.
Label the result B.

b) Enlarge shape B, with scale factor $\frac{1}{2}$, centre Q.
Label the result C.

c) State whether shapes A and B are:
(i) congruent **(ii)** similar.

d) Are shapes A and C congruent?

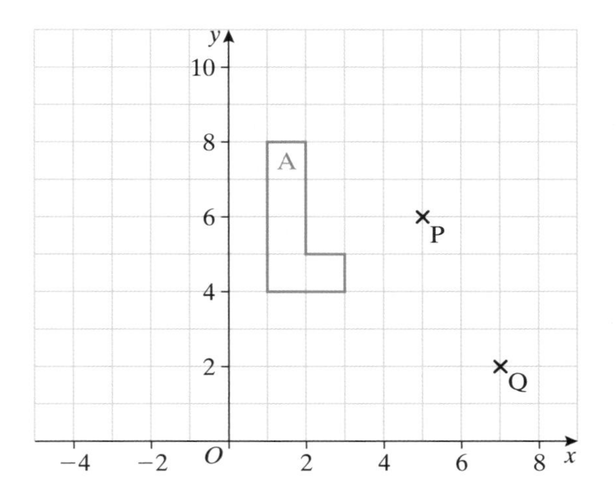

4 The diagram shows object A and its image B after an enlargement.

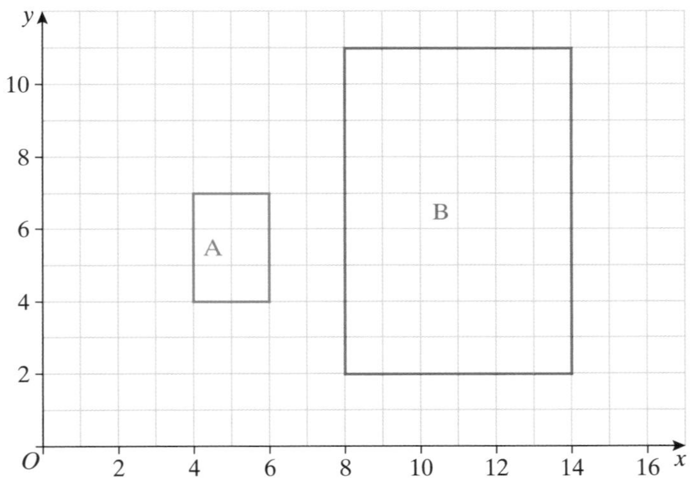

State the scale factor for the enlargement.

REVIEW EXERCISE 17

1 Here are four road signs.

A

B

C

D

Two of these road signs have one line of symmetry.

a) Write down the letters of each of these two road signs.

Only one of these four road signs has rotational symmetry.

b) **(i)** Write down the letter of this road sign.
(ii) Write down its order of rotational symmetry.

[Edexcel]

2 A shaded shape is shown on the grid of centimetre squares.

a) Work out the perimeter of the shaded shape.

b) Work out the area of the shaded shape.

c) On a copy, reflect the shape in the mirror line. **[Edexcel]**

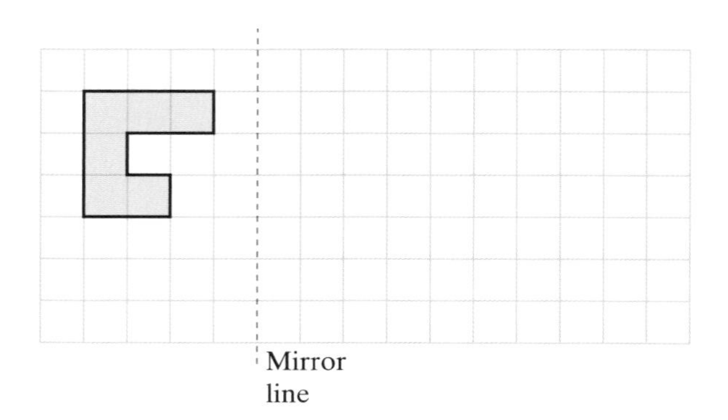

3 Here are some numbers:

11 16 18 36

68 69 82 88

From these numbers, write down a number which has:

a) exactly **one** line of symmetry

b) **two** lines of symmetry and rotational symmetry of order 2

c) rotational symmetry of order 2 but no lines of symmetry. **[Edexcel]**

4 Here is a rectangle.
The length of the rectangle is 4 cm
The width of the rectangle is 3 cm

a) **(i)** Work out the area of the rectangle.

(ii) Work out the perimeter of the rectangle.

The rectangle is enlarged by a scale factor of 5

b) Write down the length and width of the enlarged rectangle.

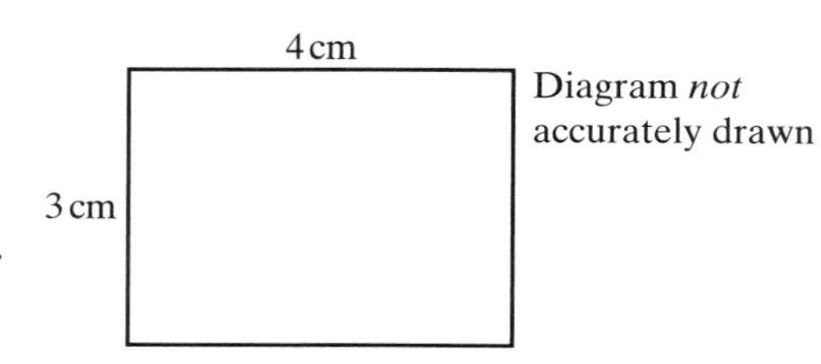

[Edexcel]

5 **a)** Copy the diagram.
Shade one square so that the shape has exactly one line of symmetry.

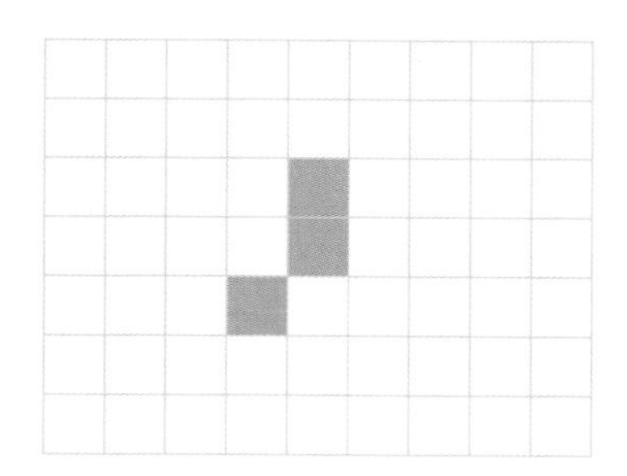

b) Copy the diagram.
Shade one square so that the shape has rotational symmetry of order 2

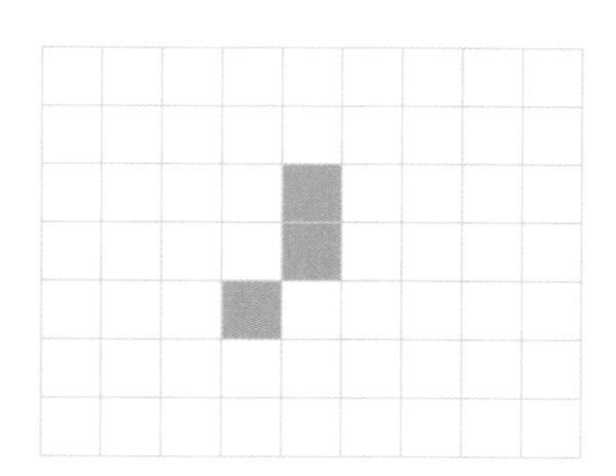

[Edexcel]

6 A triangle is shown on the grid below.
On a copy of the grid, draw the reflection of the triangle in the y axis.

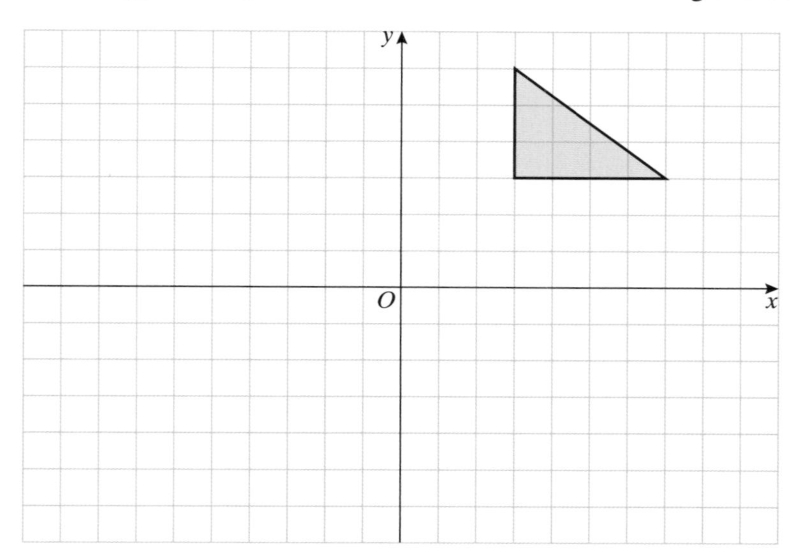

[Edexcel]

7 Triangle B is a reflection of triangle A.

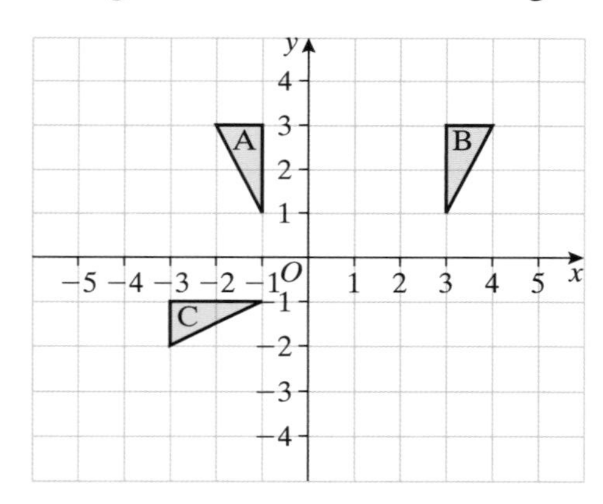

a) On a copy of the grid, draw the mirror line for this reflection.
b) Write down the equation of the mirror line. [Edexcel]

8 Triangle A and triangle B have been drawn on a grid.
a) On a copy, reflect triangle A in the line $x = 3$
Label this triangle C.
b) Describe fully the single transformation which will map triangle A onto triangle B.

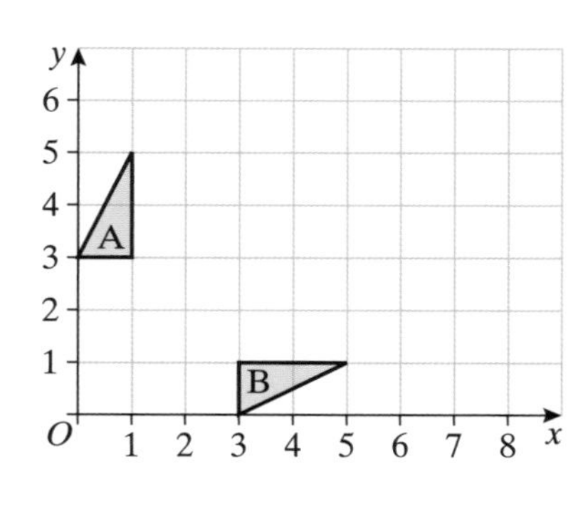

[Edexcel]

9 On a copy of the grid, enlarge the shaded triangle by a scale factor 2, centre O.

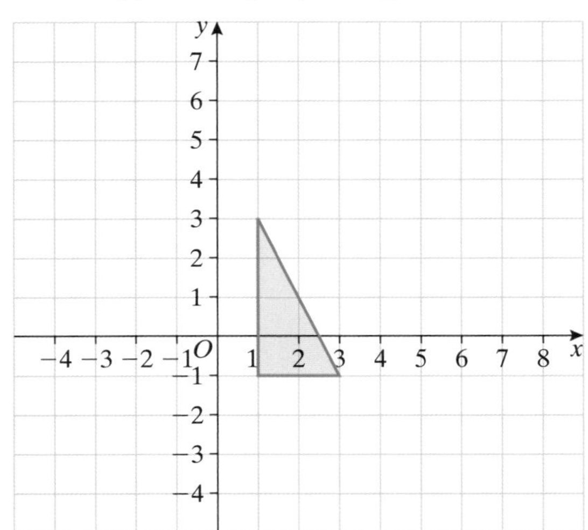

[Edexcel]

10 A shape has been drawn on a grid of centimetre squares.

a) Work out the area of the shape. State the units with your answer.

b) On a copy of the grid, enlarge the shape with a scale factor of 2

[Edexcel]

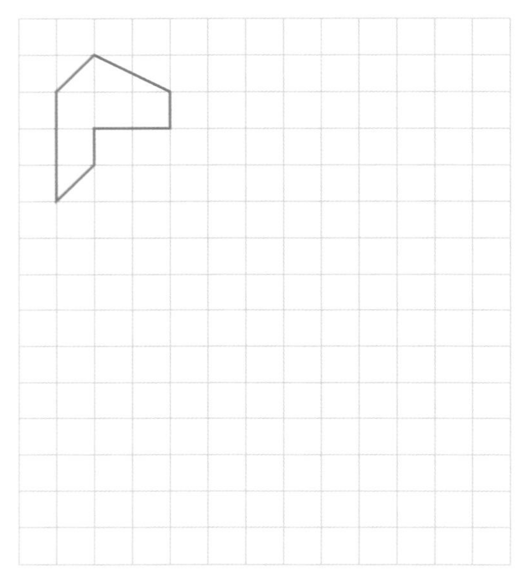

11

A

a) On the grid, rotate triangle A 180° about O. Label your new triangle B.

b) On a copy of the grid, enlarge triangle A by scale factor $\frac{1}{2}$, centre O. Label your new triangle C.

{Edexcel}

12 On a copy of the grid, enlarge the shaded triangle by a scale factor $1\frac{1}{2}$, centre P.

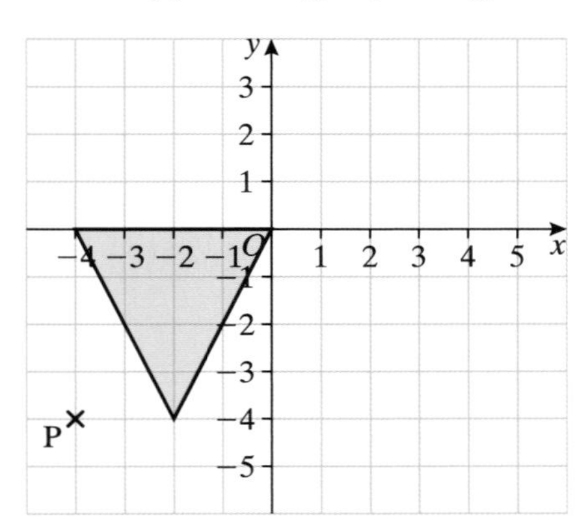

[Edexcel]

13

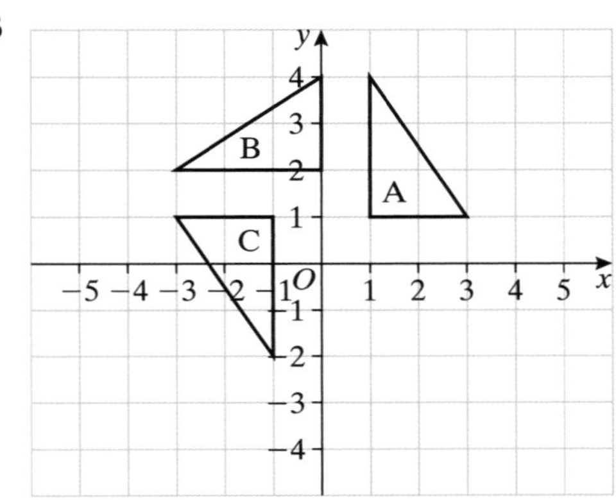

Shape A is rotated 90° anticlockwise, centre (0, 1), to shape B.
Shape B is rotated 90° anticlockwise, centre (0, 1), to shape C.
Shape C is rotated 90° anticlockwise, centre (0, 1), to shape D.
On a copy of the grid, mark the position of shape D. [Edexcel]

14

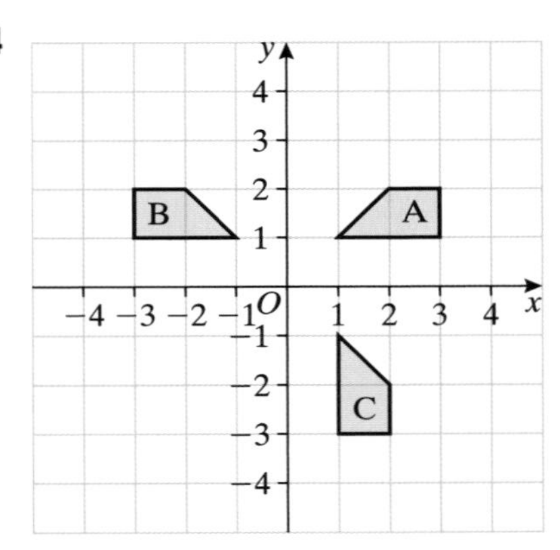

a) Describe fully the single transformation which takes shape A on to shape B.
b) Describe fully the single transformation which takes shape A on to shape C.

[Edexcel]

15 Shape A is shown on the grid.
Shape A is enlarged, centre $(0, 0)$, to obtain shape B.
One side of shape B has been drawn for you.

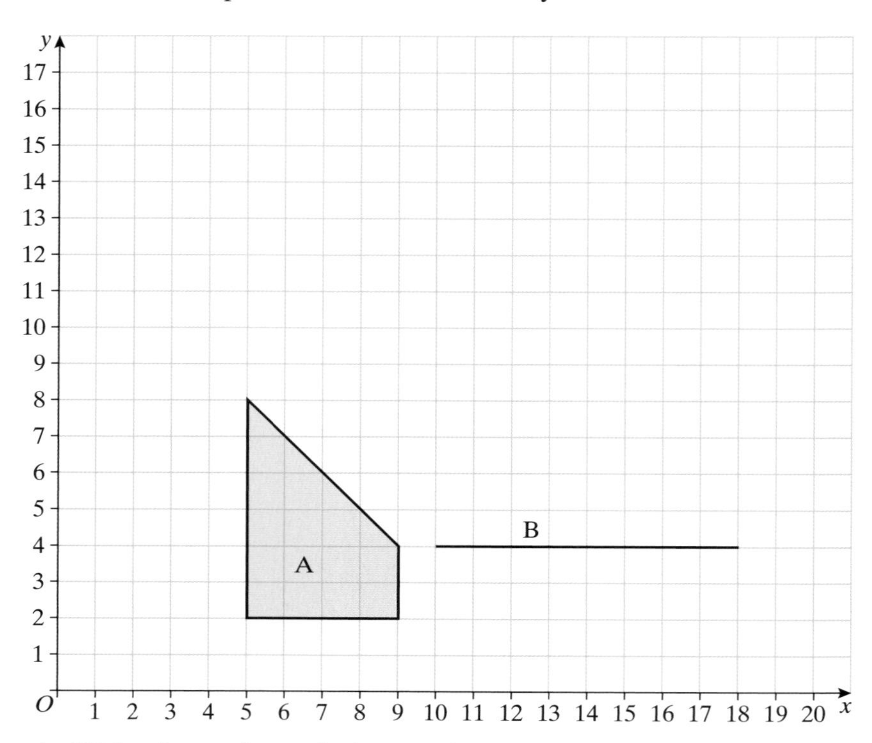

a) Write down the scale factor of the enlargement.

b) On a copy of the grid, complete shape B.

The shape A is enlarged by scale factor $\frac{1}{2}$, centre $(5, 16)$ to give the shape C.

c) On a copy of the grid, draw shape C. **[Edexcel]**

KEY POINTS

1. Shapes may have reflection symmetry and/or rotational symmetry.
2. The order of rotational symmetry tells you how many times the shape looks the same when it is rotated 360°
3. A reflection is specified by a mirror line.
4. A rotation is specified by
   - a centre of rotation
   - an angle of rotation
   - and a direction (clockwise or anticlockwise).
5. A translation can be expressed in vector form – for example,
   $\begin{pmatrix} 2 \\ -3 \end{pmatrix}$ means 2 right and 3 down.
6. Two shapes are congruent if they are exactly the same shape and size.
7. Reflections, rotations and translations all preserve congruence.
8. An enlargement is specified by a centre of enlargement and a scale factor.
   - Scale factors bigger than 1 make the image bigger.
   - Scale factors between 0 and 1 reduce the image.
   - Enlargements do not preserve congruence.
9. Two shapes are similar when one shape is an enlargement of the other.
10. When solid objects are enlarged by a scale factor, their perimeters increase by the same ratio.
    The angles remain the same.

Internet Challenge 17

Geometrical definitions

Mathematicians like to attach precise meanings to certain words – these are definitions.

In these sentences the letters of the key words have been replaced with □ symbols.

Find the missing word in each case.
(You will know some of these already, but you may need to look up some of the less well-known ones on the internet.)

1 An □□□□□□□□□□□ is a mathematical solid with 20 faces.

2 A □□□□□□□□ is the simple name for a circular prism.

3 If two objects are the same shape and size they are said to be □□□□□□□□□.

4 If two shapes are alike in shape but one is larger than the other, they are said to be mathematically □□□□□□□.

5 Z-angles are, more properly, called □□□□□□□□□ angles.

6 □□□□□□□□ lines never touch; they remain at a constant distance apart.

7 A □□□□□ is a solid object in the form of a perforated ring (like ring doughnut).

8 The interior angles of an □□□□□□□ add up to 1080°.

9 A □□□□□□□□□□ is exactly half of a sphere.

10 The highest point of a pyramid is known as its □□□□.

11 A pyramid with a triangular base is called a □□□□□□□□□□□.

12 □□□□□□□ is the correct mathematical name for a 'diamond' with four equal sides.

13 An angle of one-sixtieth of a degree is called a □□□□□□ of □□□.

14 An angle of 57.296° is called one □□□□□□.

15 The diagram below shows a □□□□□□□□□ cone
This is also a □□□□□□□.

CHAPTER 18

Constructions and loci

In this chapter you will **learn how to**:

- construct triangles from given information
- carry out standard compass constructions on line segments
- solve locus problems
- sketch 3-D shapes on isometric paper
- draw and use nets
- draw and interpret plans and elevations of 3-D shapes.

You will also be **challenged to**:

- investigate polyhedra.

Starter: Round and round in circles

Use a sharp pencil, a pair of compasses and a ruler to make this drawing.

You might want to colour the diagram in when you have finished it.

Here are two ideas:

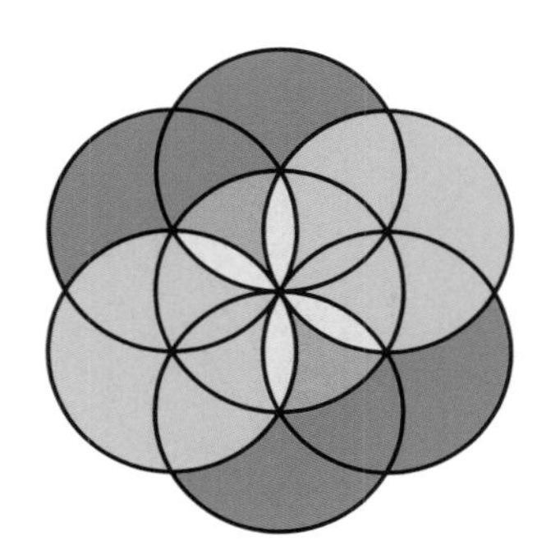

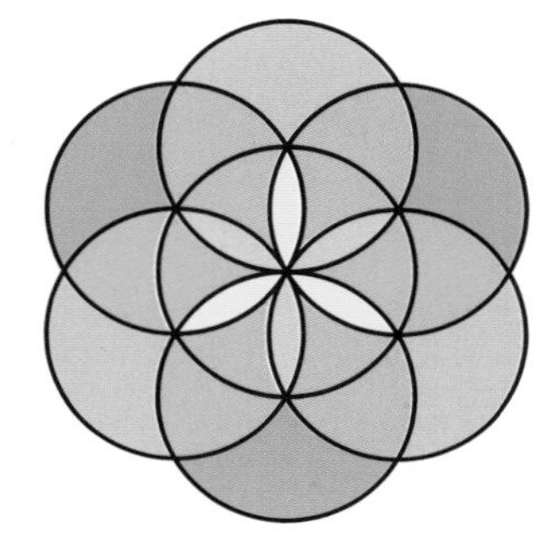

Now try designing your own circle patterns.

18.1 Constructing triangles from given information

There are four ways of constructing a triangle given three pieces of information.

EXAMPLE

A triangle PQR has side PQ = 9 cm
PR = 5 cm and angle QPR = 55°

The diagram is not accurately drawn.

Use a ruler and protractor to construct this triangle.

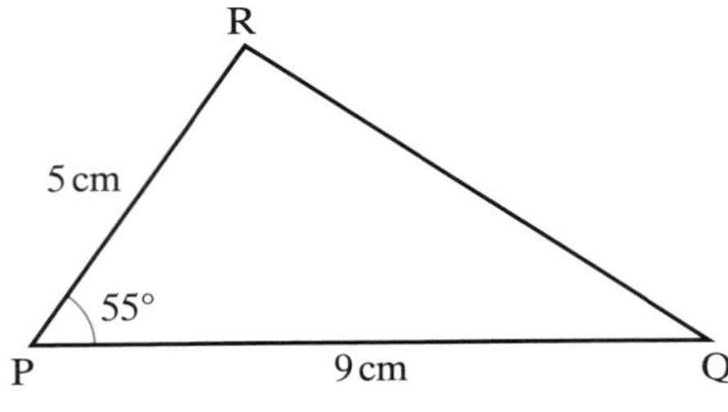

SOLUTION

Step 1 Draw the line PQ.

Step 2 Use your protractor to make an angle of 55°

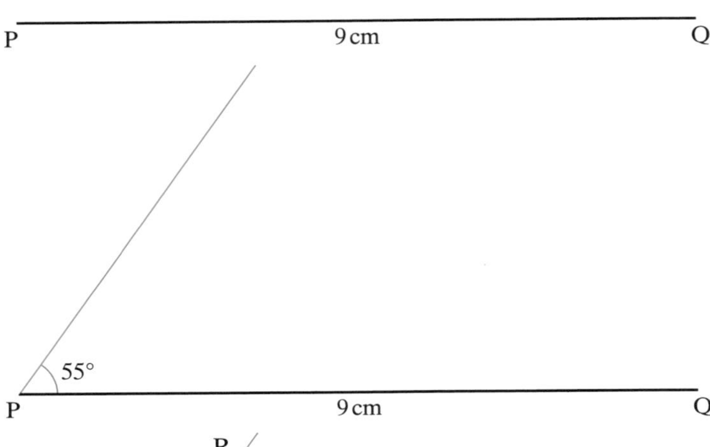

Step 3 Measure off length 5 cm along the new line. Mark this point R.

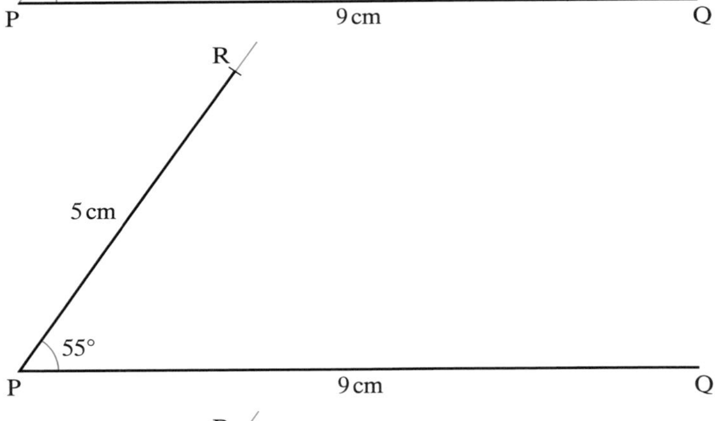

Step 4 Join R and Q.

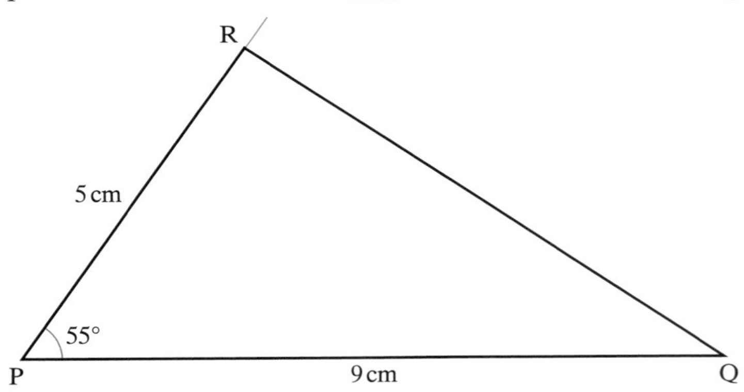

EXAMPLE

A triangle ABC has side AB = 8 cm, angle BAC = 40° and angle ABC = 70°

The diagram is not accurately drawn.

Use a ruler and protractor to construct this triangle.

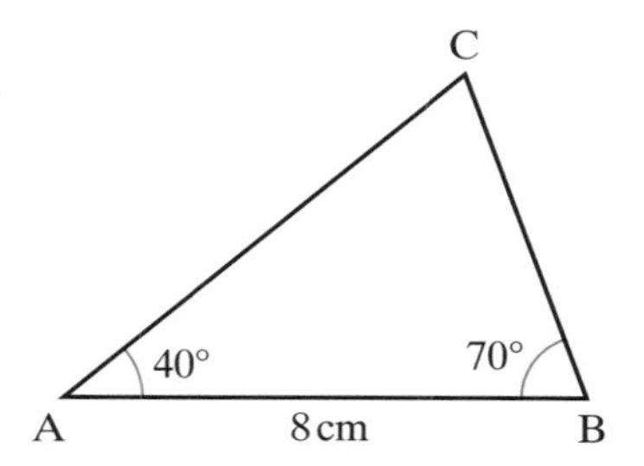

SOLUTION

Step 1 Draw the line AB, of length 8 cm
Use your protractor to measure an angle of 40° at A.
Add a long line from A at an angle of 40°

Step 2 Draw a long line at an angle of 70° from B.

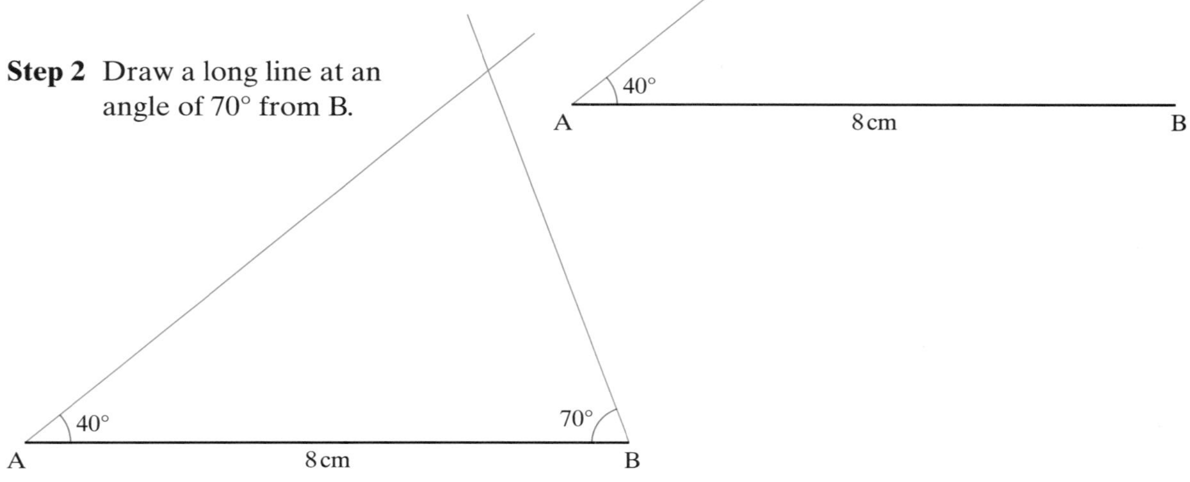

Step 3 These two lines must intersect (meet) at C so that the diagram may be completed.

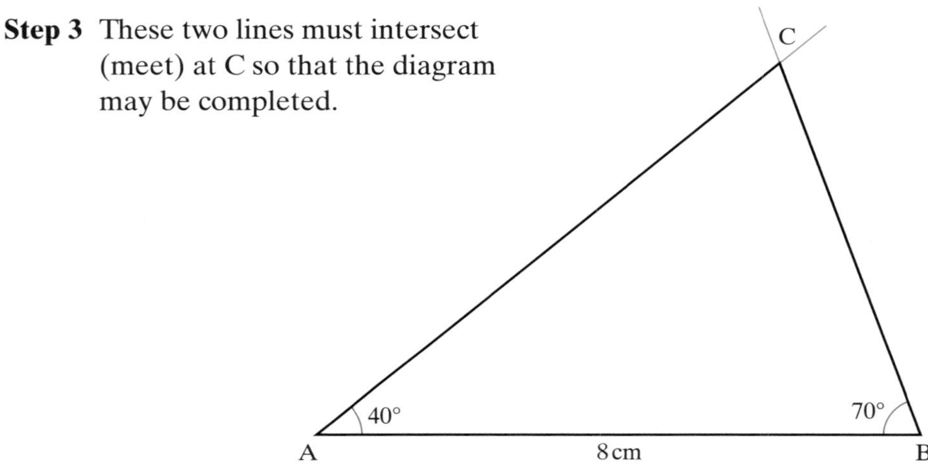

EXAMPLE

A triangle LMN has sides
LM = 6 cm, LN = 7 cm and MN = 8 cm

The diagram is not accurately drawn.

Use ruler and compasses to construct this triangle.

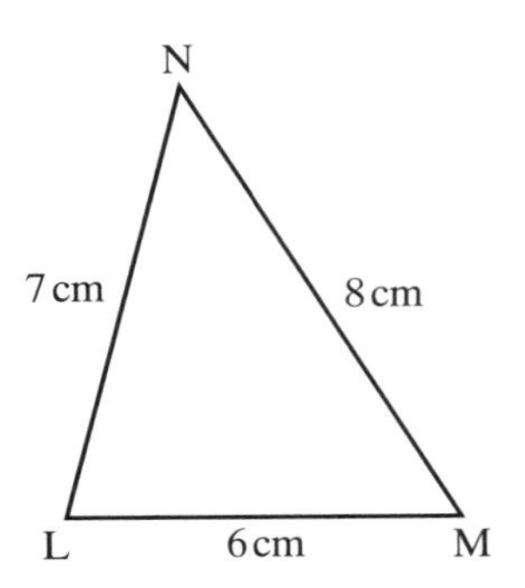

SOLUTION

Step 1 Draw a line LM, of length 6 cm

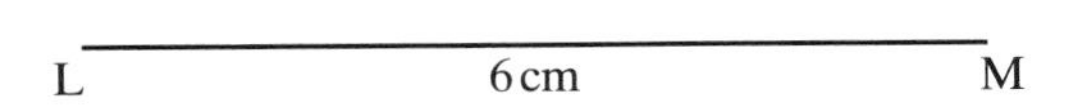

Step 2 Draw an arc of radius 7 cm from L.
Draw an arc of radius 8 cm from M.

Step 3 These arcs must intersect at N so that the construction can be completed.

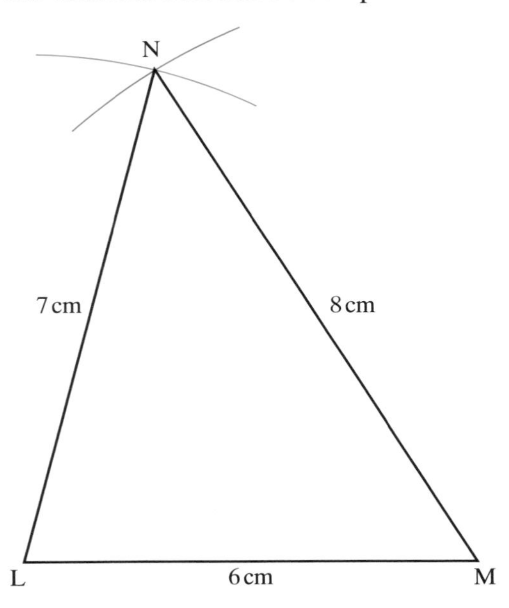

In an examination, make sure you leave all your construction lines showing because if you delete them you cannot score all the marks available.

EXAMPLE

In triangle ABC you are given that
AB = 7 cm, BC = 5 cm and angle CAB = 40°

Use ruler and compasses and a protractor to construct this triangle.

Show that there are two different solutions based on the given information.

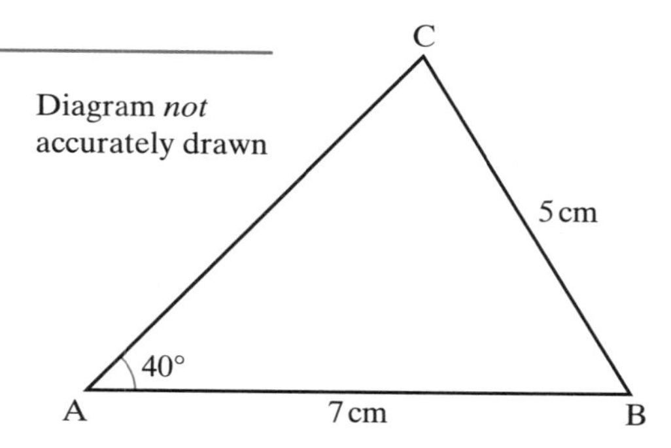

SOLUTION

Step 1 Draw a line AB of length 7 cm
Draw a long line from A at an angle of 40°

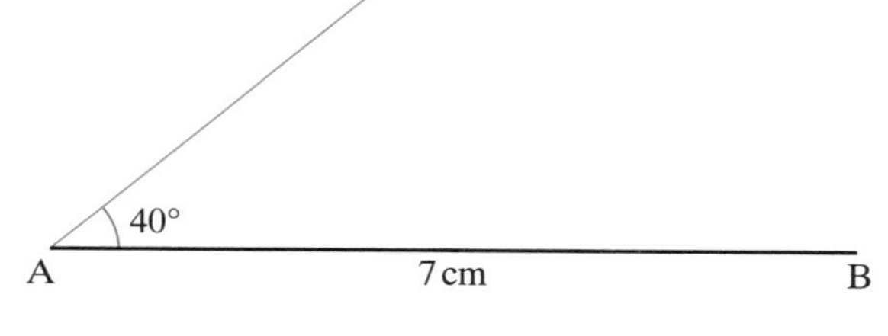

Step 2 Open your compasses to a radius of 5 cm
Draw an arc centred on B.

This arc intersects the line from A in two places.
So there are two ways of completing the construction.

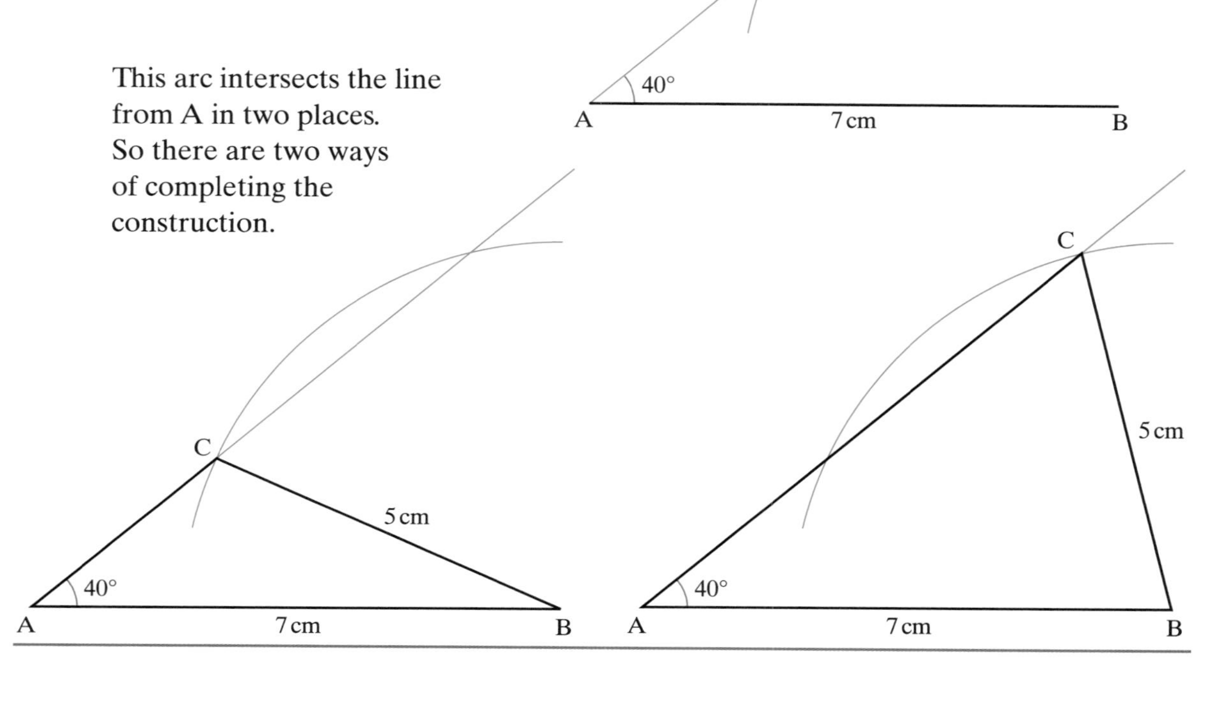

The following exercise gives you some practice at making accurate drawings of triangles.

Be sure to leave your construction lines visible, so that your teacher (or the examiner!) can follow your methods clearly.

EXERCISE 18.1

1 Use ruler, protractor and compasses to construct these triangles. The diagrams are not accurately drawn.

a)

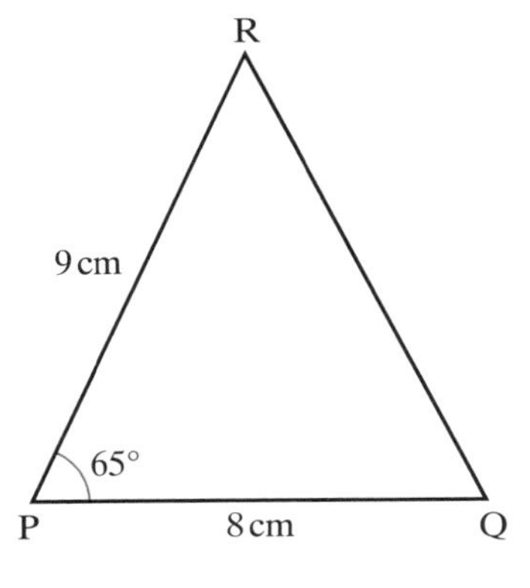

b)

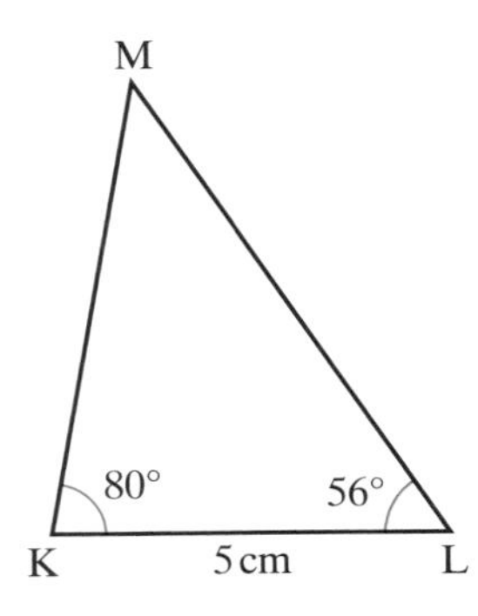

c)

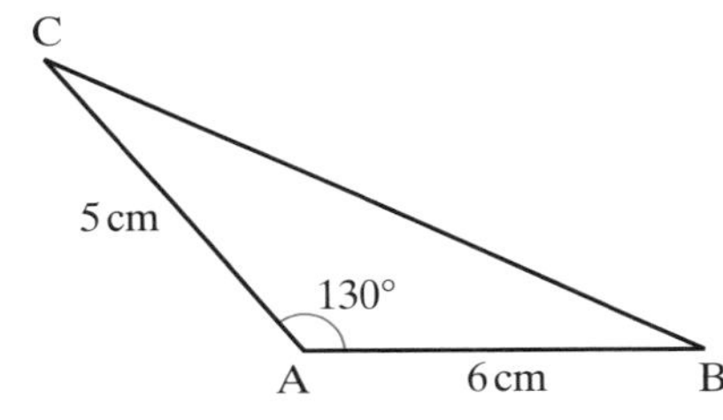

d)

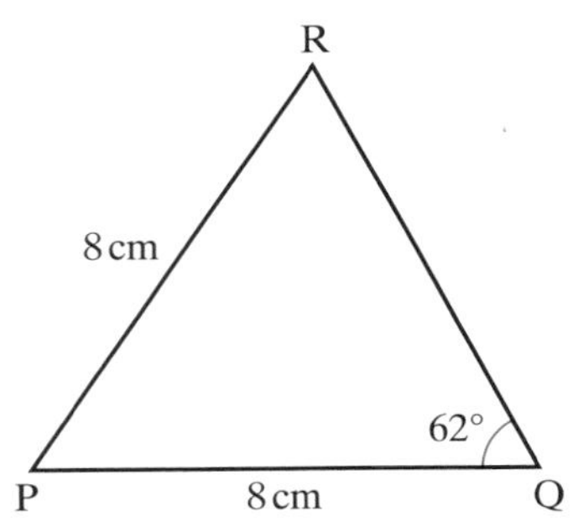

e)

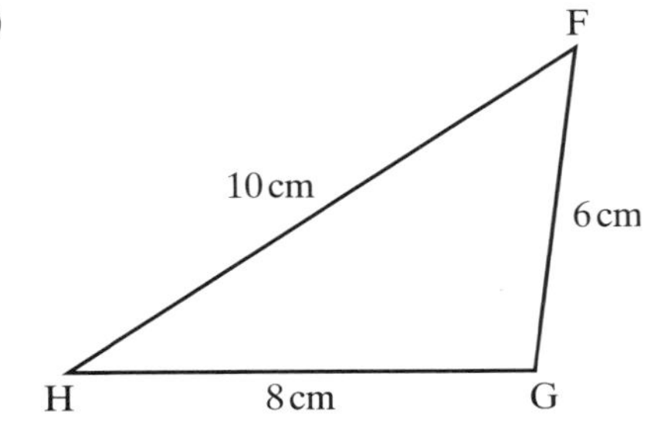

f)

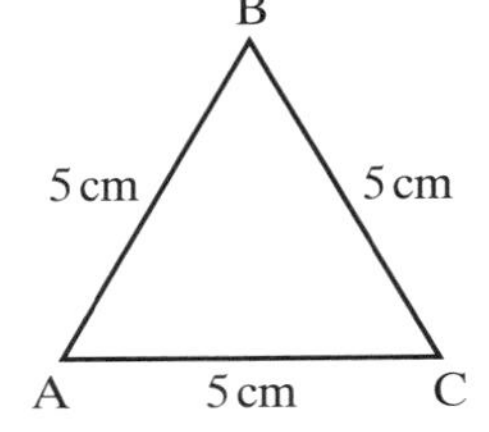

2 The sketch shows a triangle with AB = 85 mm, BC = 55 mm, AC = 70 mm

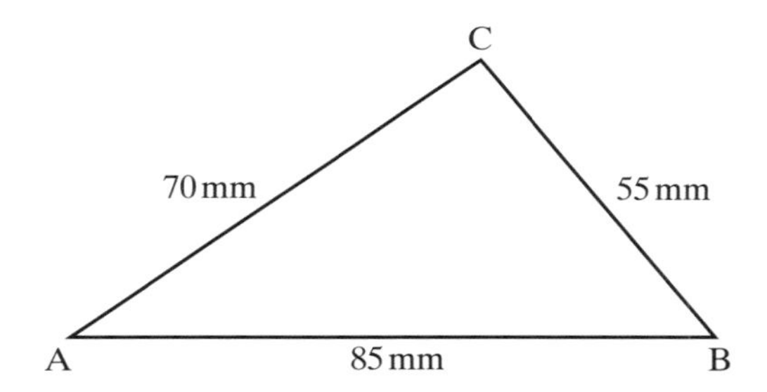

Diagram *not* accurately drawn

Make an accurate diagram of the triangle.

3 The sketch shows a triangle with AB = 140 mm, AC = 122 mm and angle ACB = 90°

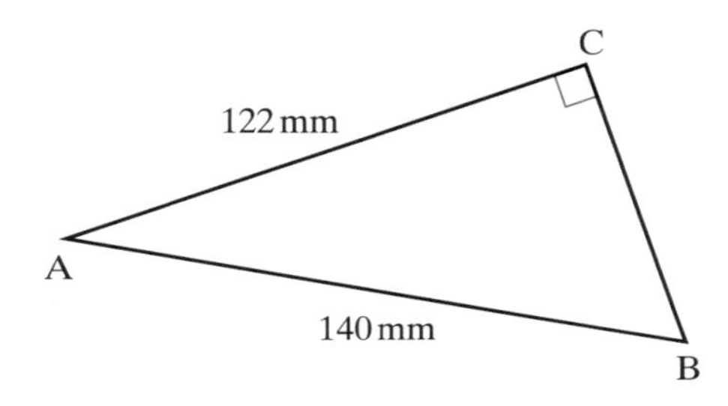

Diagram *not* accurately drawn

Make an accurate diagram of the triangle.

4 Draw triangle RST with RT = 8.5 cm, RS = 6.5 cm, angle RTS = 45°
Use ruler and compasses to construct this triangle.
Show that there are two different solutions based on the given information.

5 Using compasses, try to make an accurate drawing of triangle PQR with sides PQ = 10 cm, QR = 5 cm, RP = 4 cm
What difficulty do you encounter?

18.2 More constructions

There are four constructions that you need to be able to do.

Exam questions will expect you to do these with a pair of compasses and a ruler.

You should leave any construction lines plainly visible.

1 Bisecting an angle

'Bisecting' means to cut in half.

EXAMPLE

Use ruler and compasses to construct the angle bisector of the angle Q shown in the diagram.

60°

Q

SOLUTION

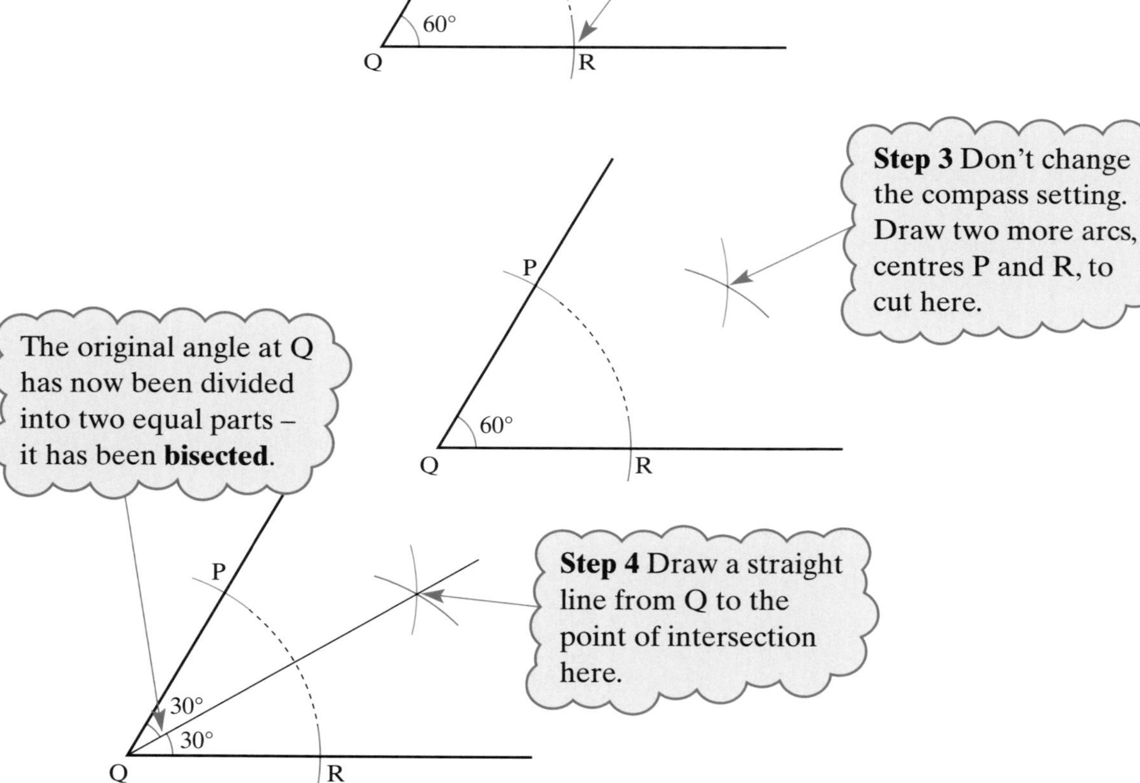

Step 1 Using compasses, draw an arc with centre Q, so it cuts one of the lines. Call this point P.
Step 2 Don't change the compass setting. Draw a second arc with centre Q, to cut the other line. Call this point R.
P
60°
Q
R
Step 3 Don't change the compass setting. Draw two more arcs, centres P and R, to cut here.
P
60°
Q
R
The original angle at Q has now been divided into two equal parts – it has been **bisected**.
Step 4 Draw a straight line from Q to the point of intersection here.
P
30°
30°
Q
R

2 Constructing the perpendicular bisector of a line

Two lines which are at right angles to each other are **perpendicular**.

EXAMPLE

Use ruler and compasses to construct the perpendicular bisector of the line segment shown in the diagram.

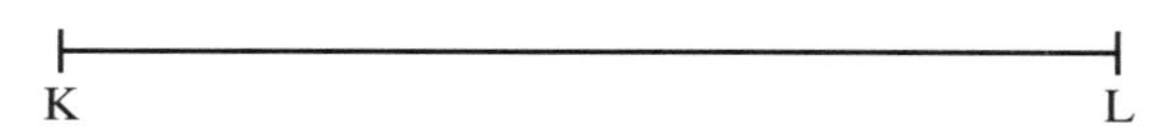

K
L

This means to draw a line at right angles to KL passing through the midpoint of KL.

SOLUTION

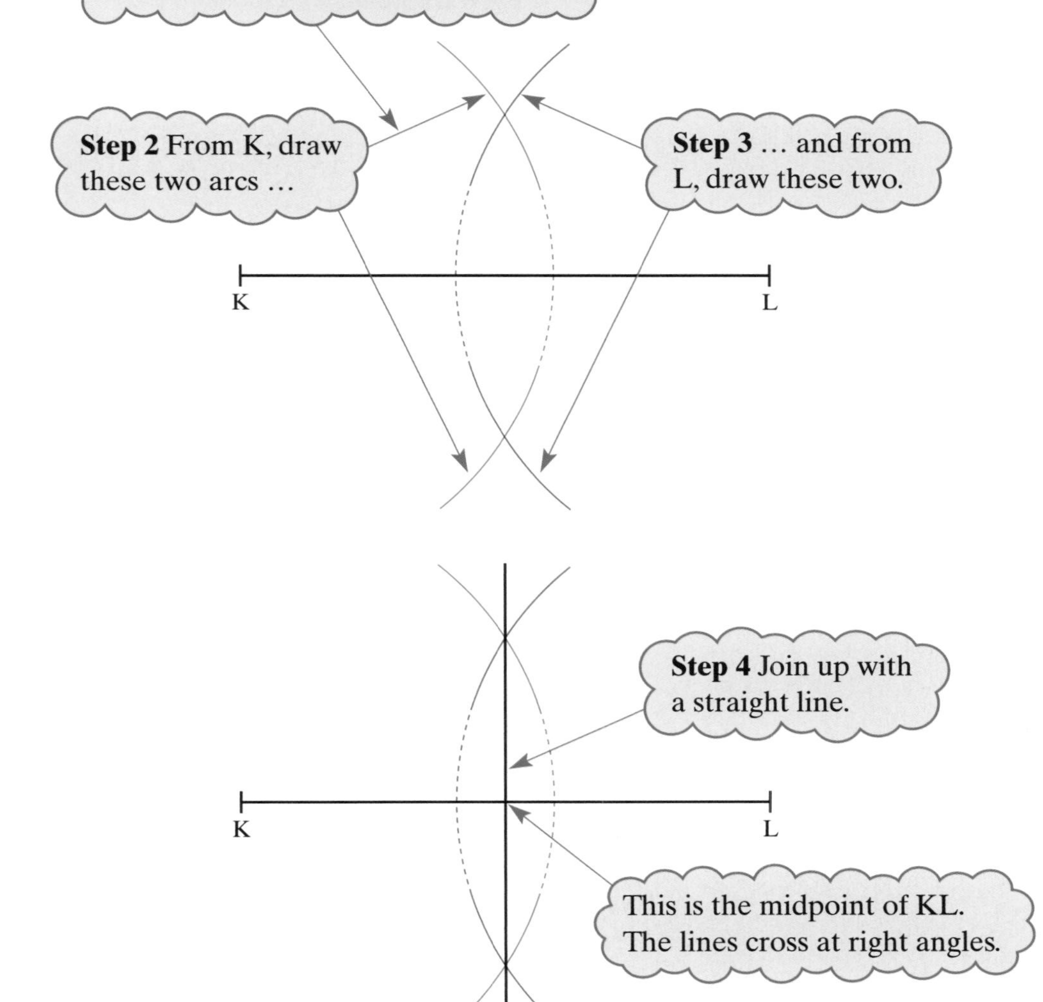

Step 1 Open the compasses to more than half the distance from K to L.
Step 2 From K, draw these two arcs …
Step 3 … and from L, draw these two.
K
L
Step 4 Join up with a straight line.
K
L
This is the midpoint of KL. The lines cross at right angles.

3 Dropping a perpendicular from a point to a line

EXAMPLE

Use ruler and compasses to draw a line from P perpendicular to the line segment AB shown in the diagram.

SOLUTION

Step 1 Construct two arcs centred at P, to cut the line at Q and R.

Step 2 Make two further arcs, centred at Q and R, to intersect at S.

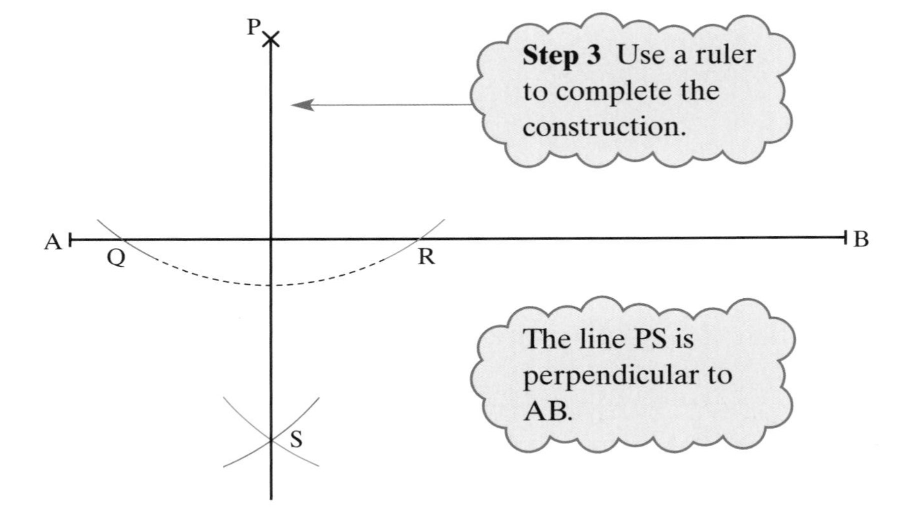

P
Step 3 Use a ruler to complete the construction.
A
Q
R
B
The line PS is perpendicular to AB.
S

4 Constructing a perpendicular from a point on a line

EXAMPLE

Use ruler and compasses to draw a line from P perpendicular to the line segment AB shown below.

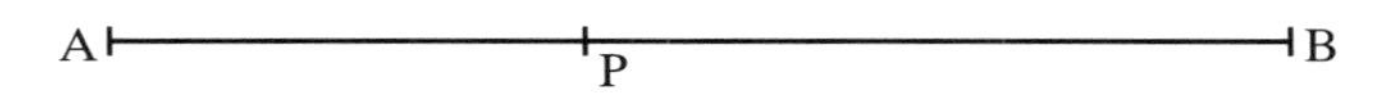

SOLUTION

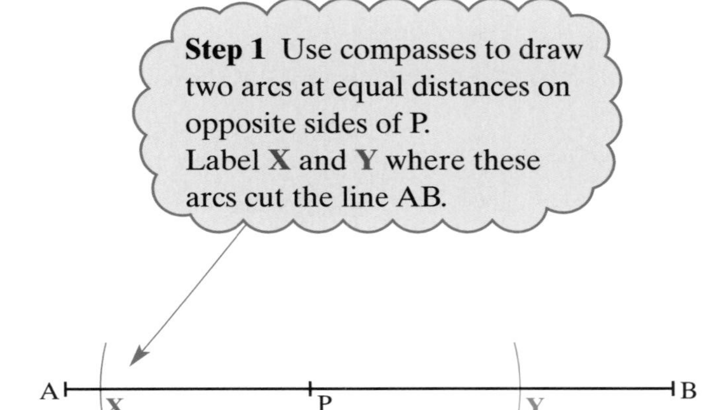

Step 1 Use compasses to draw two arcs at equal distances on opposite sides of P.
Label **X** and **Y** where these arcs cut the line AB.
A
X
P
Y
B

Step 2 Complete the solution by constructing the perpendicular bisector of XY (see method 2 page 435)

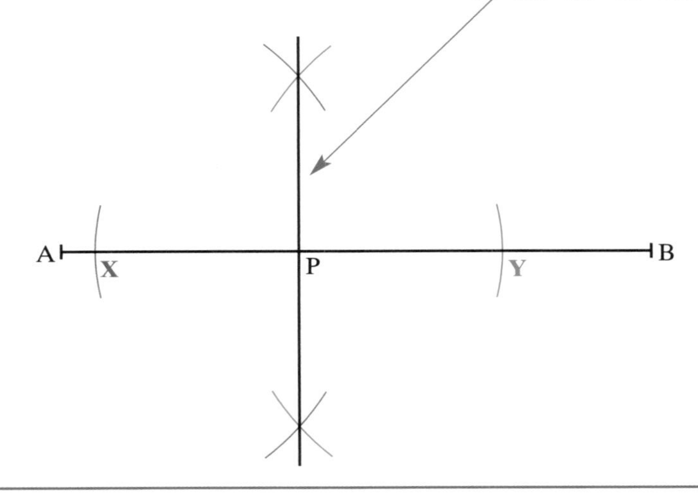

EXERCISE 18.2

1 Use ruler and compasses to construct the perpendicular bisectors of these line segments.

a) 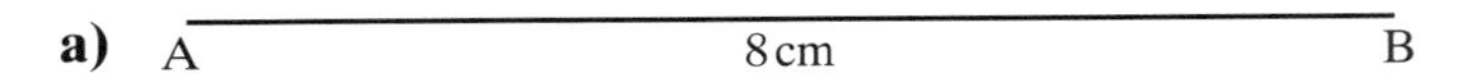

b) A 10 cm B

c) A 6 cm B

d) A 13 cm B

e) A 7.5 cm B

2 Use ruler and compasses to construct the bisectors of these angles.

a)

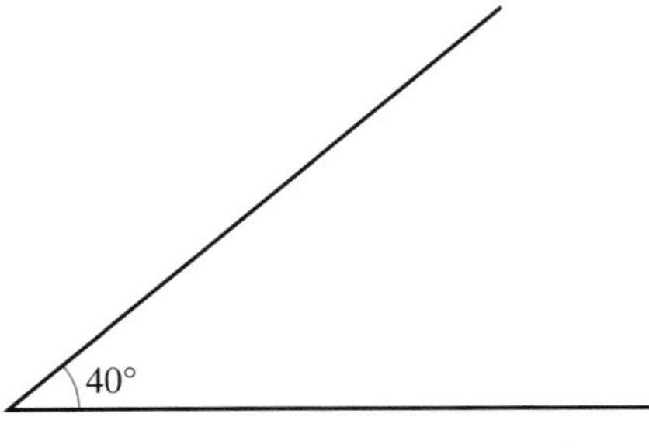

b)

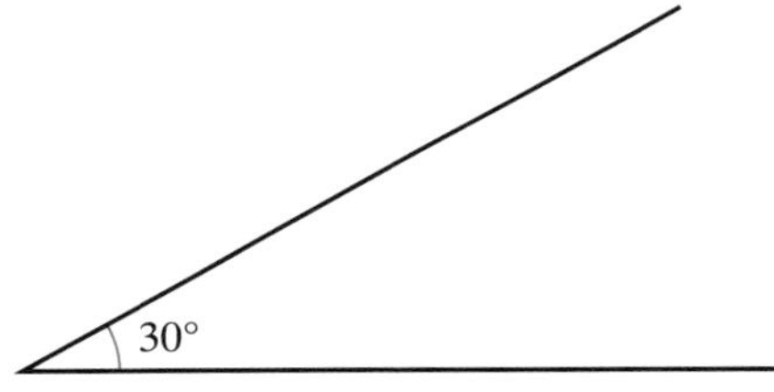

c)

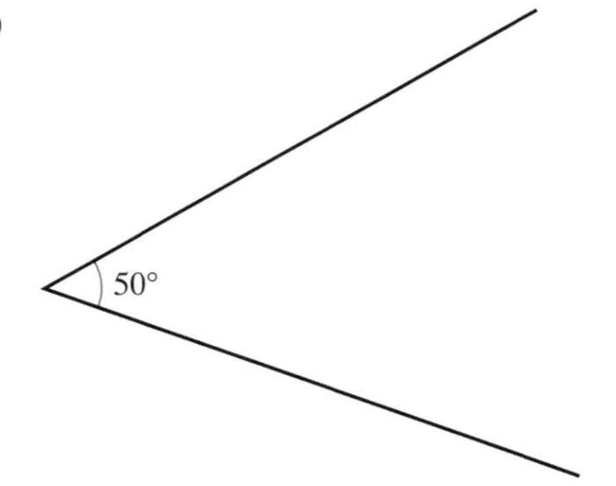

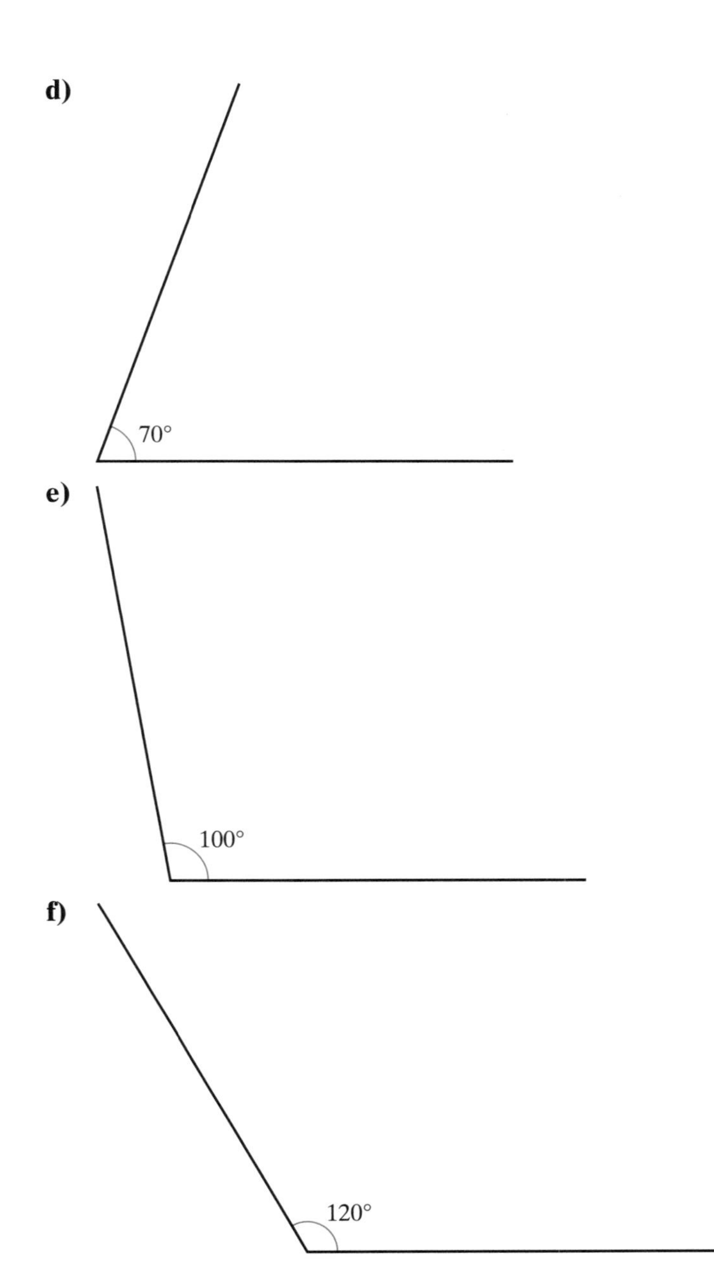

d)
70°
e)
100°
f)
120°

3 Use ruler and compasses to construct the perpendicular from P to the line segment AB.

4 Use ruler and compasses to construct the line from P perpendicular to the line segment AB.

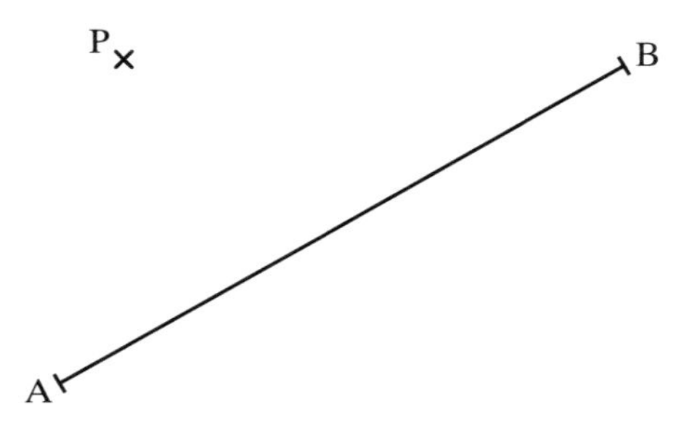

5 Use ruler and compasses to construct a line through P perpendicular to AB on a copy of each of these line segments.

a) 

b) 

c)

B

P

A

6 a) Construct a triangle ABC with sides AB = 10 cm, BC = 14 cm, CA = 12 cm

b) (i) Construct the perpendicular bisector of the side AB.

(ii) Construct the perpendicular bisector of the side BC.

(iii) Construct the perpendicular bisector of the side CA.

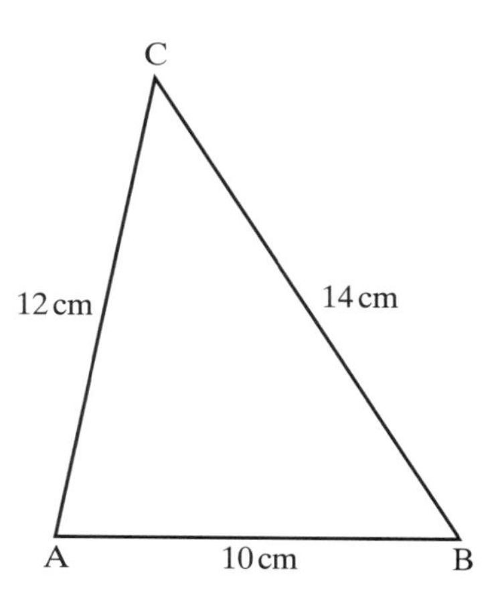

Your three bisectors should all meet at a single point.
Label this point X.

c) Draw a circle, centre X, that passes through A.
What do you notice?

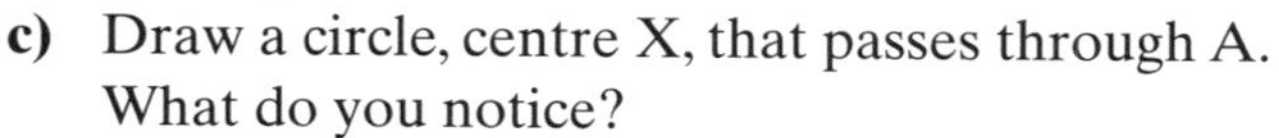

7 a) Construct a triangle PQR with sides PQ = 9 cm, QR = 13 cm, RP = 11 cm

b) (i) Bisect the angle PQR.

(ii) Bisect the angle QRP.

(iii) Bisect the angle RPQ.

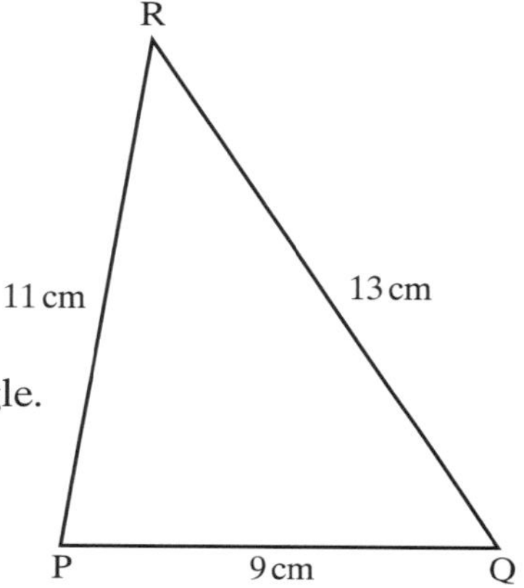

The three bisectors should all meet at a single point.
Label this point Y.

c) Draw a circle, centre Y, that touches **one side** of the triangle.
What do you notice?

8 Follow this method to construct an angle of exactly 30° without using a protractor.

Step 1 Draw a line 8 centimetres long.

Step 2 Use this line as one side of an equilateral triangle.
Use compasses and a ruler to construct the other two sides.
You now have three angles of 60°

Step 3 Choose one of the angles and bisect it.

9 Follow this method to construct an angle of exactly 45° without using a protractor.

Step 1 Draw a line 8 centimetres long.

Step 2 Construct the perpendicular bisector of your line.
You now have four angles of 90°.

Step 3 Choose one of the angles and bisect it.

18.3 Loci

When a point P moves in the plane, subject to a certain rule, it traces out a path, known as a **locus** (*plural*, **loci**).

Here are some loci that you need to know.

Firstly, when P is a fixed distance from a given point you get a **circle**.

For example, imagine a goat tethered to a post T.

The green circle shows the path of the goat walking around the post.

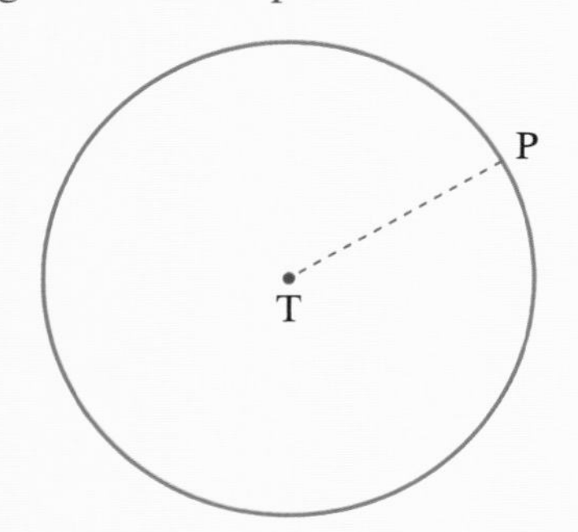

When P is equidistant from two points you get a **perpendicular bisector.**

For example, the two points A and B in the diagram below represent dangerous rocks.

The green line shows the path that a ship would sail so that it would always be exactly the same distance from A as it is from B.

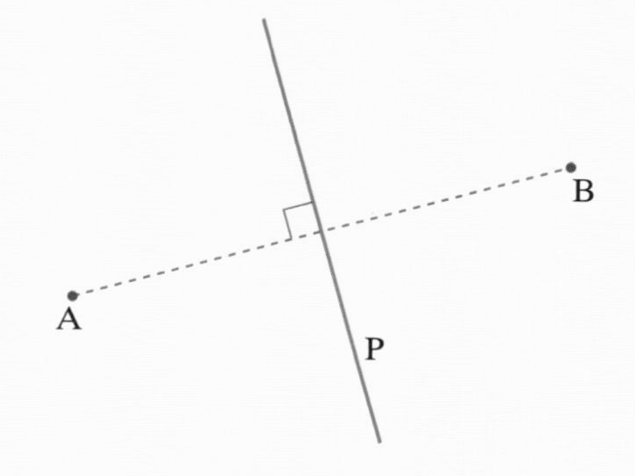

When P is a fixed distance from a line segment you get a **rectangle with semicircular ends**.

For example, imagine walking around the line L keeping exactly 3 m away from it.

The green shape shows the path you would take.

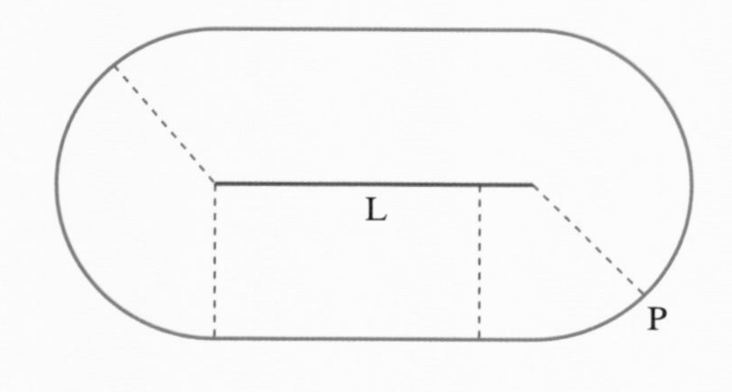

When P is equidistant from two line segments you get an **angle bisector**.

For example, the green line shows the path you would take to stay exactly the same distance from line L as you are from line M.

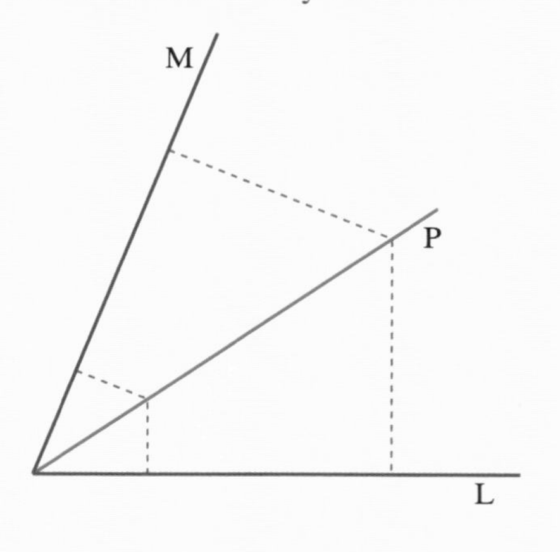

You might find that the following helps you to remember the loci:

one point → circle		one line → 'running track'
two points → perpendicular bisector		two lines → angle bisector

EXAMPLE

A ship sails from an island to the mainland.
Throughout the journey the ship remains at equal distances from the two lighthouses P and Q.
Draw the path of the ship's journey.

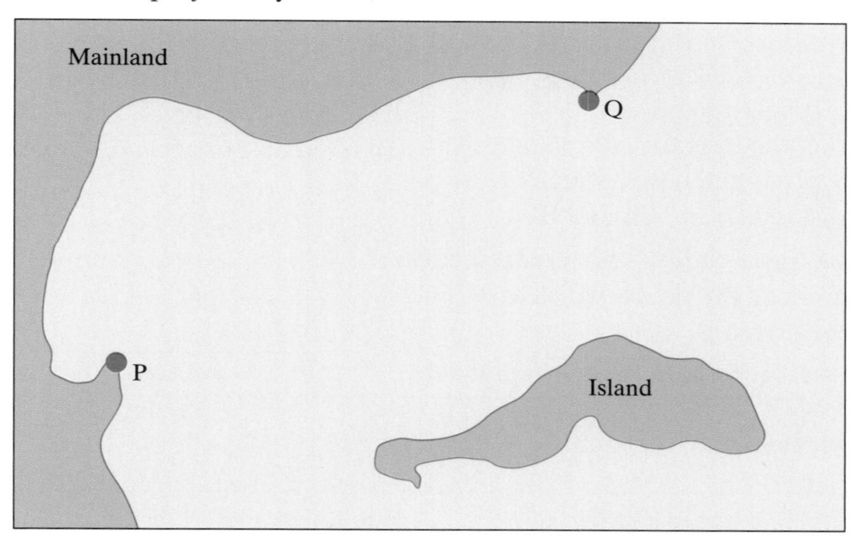

SOLUTION

Using compasses, construct the perpendicular bisector of PQ:

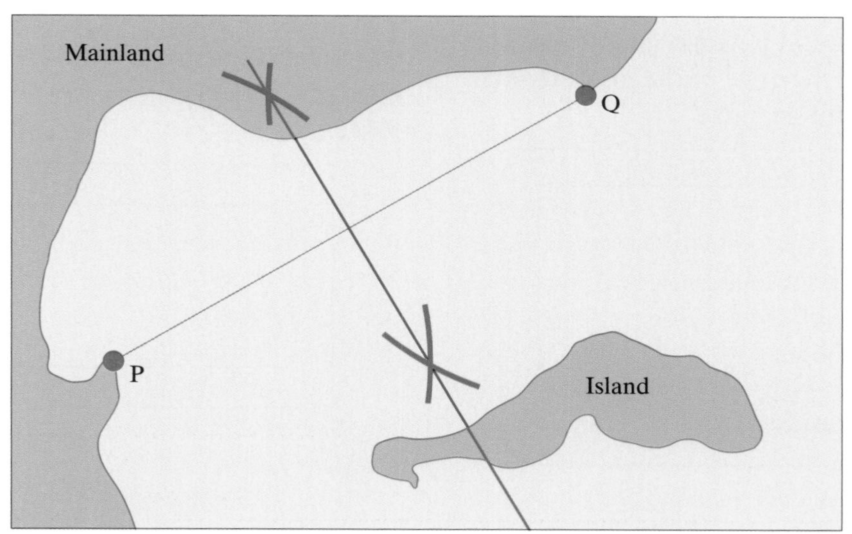

EXAMPLE

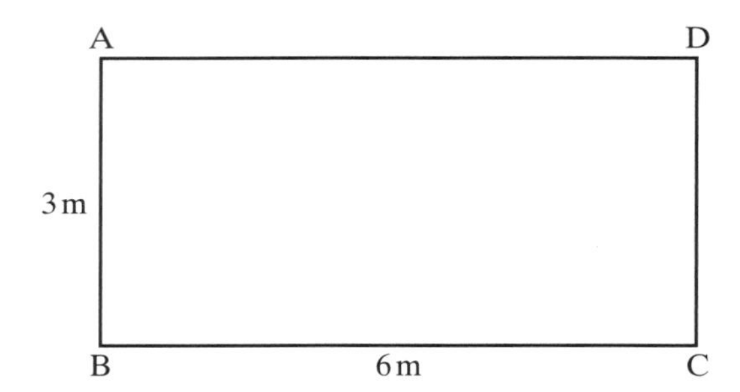

The diagram shows a rectangular garden.
AB = CD = 3 metres,
AD = BC = 6 metres.

A gardener wants to plant a tree in the garden.
The tree must be the same distance from AB as from BC.
The tree must be 3.5 metres from B.
Mark the position of the tree on the diagram.

SOLUTION

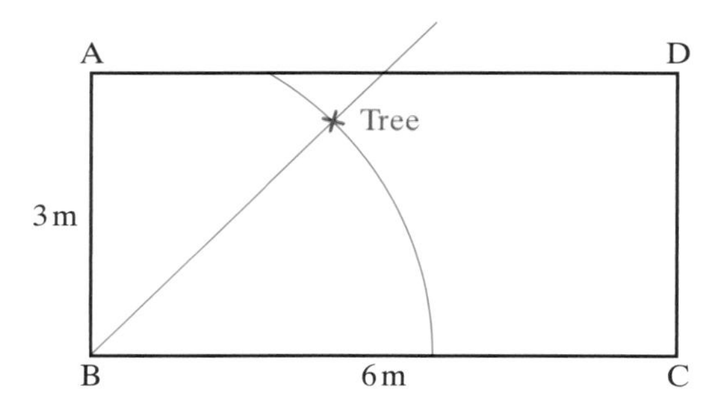

Step 1 Construct the bisector of the angle ABC. This shows all the points which are an equal distance from AB and BC.

Step 2 Draw an arc of radius 3.5 m centred at B. This shows all the points which are 3.5 metres from B.

Step 3 So the tree lies where the two loci meet.

Here are two more loci.

When *P* is a fixed distance outside a rectangle you get a **larger rectangle with arcs**.

For example, the blue rectangle is a swimming pool. The purple rectangle shows the path you would take when you walk around the edge of the pool staying 3 m away from the edge.

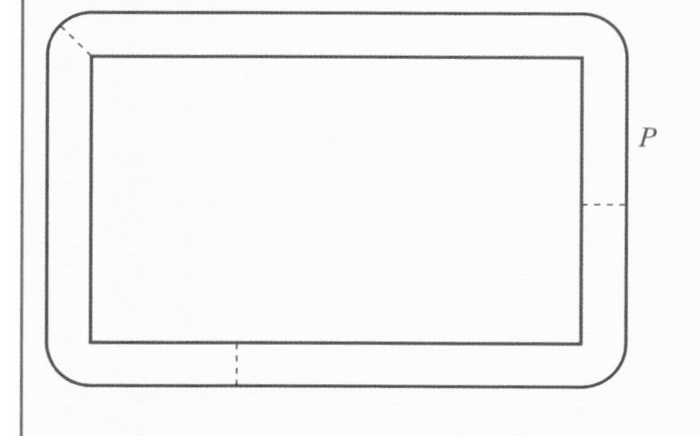

When *P* is a fixed distance inside a rectangle you get a **smaller rectangle**.

For example, the blue rectangle is a fence going around a field.

The purple rectangle shows the path you would take when you walk around the edge of the field staying 3 m away from the fence.

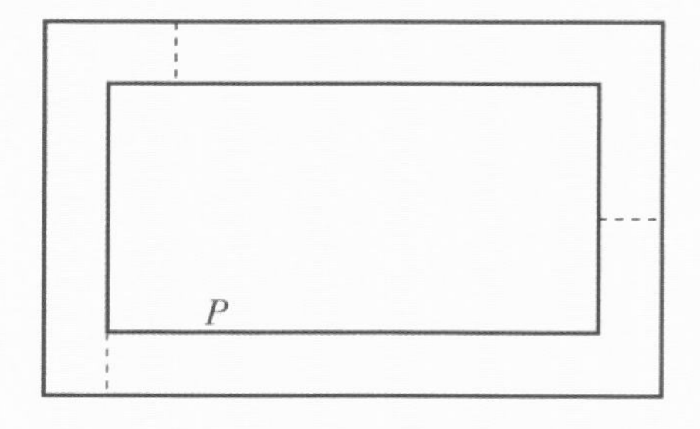

EXERCISE 18.3

In all your answers to this exercise, you should leave all your construction lines showing.

1 Draw the locus of all the points which are exactly 4 cm away from the following:

a) •

b) ______________________________ 10 cm

2 a) Mark a point A on your paper.
b) Draw the locus of the points which are exactly 3 cm away from your point A.
c) Shade the region which contains all the points *less than* 3 cm away from A.

3 a) Draw a line 6 cm long on your paper.
b) Draw the locus of the points which are exactly 2 cm away from your line.
c) Shade the region which contains all the points *less than* 2 cm away from your line.

4 a) Mark two points, A and B, 8 cm apart on your paper.

A •   • B

b) Draw the locus of the points which are equidistant from A and B.
c) Shade the side of the locus which contains the points nearer to A than to B.

5 a) Mark two points, A and B, 6 cm apart on your paper.

A •   • B

b) Draw the locus of the points which are equidistant from A and B.
c) Shade the side of the locus which contains the points nearer to B than to A.

6 a) Use a protractor to draw two lines, AB and AC at 40° to each other.

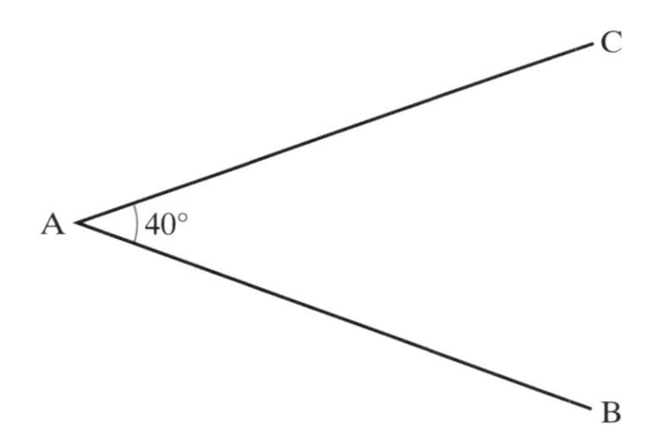

b) Draw the locus of the points which are equidistant from the lines AB and AC.
c) Shade the side of the locus which contains the points nearer to AB than to AC.

7 a) Use a protractor to draw two lines, AB and AC at 60° to each other.

b) Draw the locus of the points which are equidistant from the lines AB and AC.

c) Shade the side of the locus which contains the points nearer to AC than to AB.

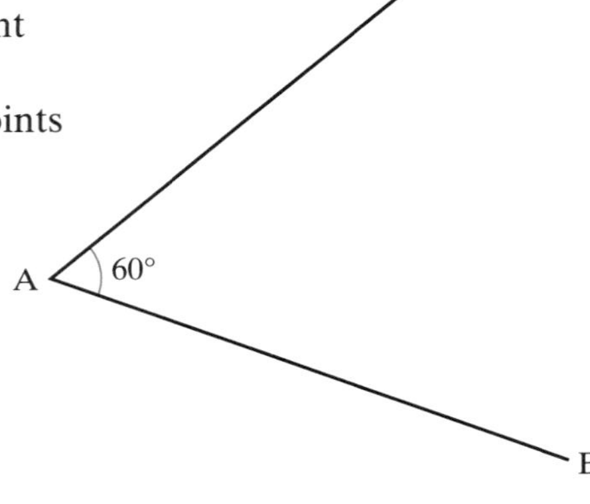

8 In the diagram, the rectangle represents a garden pond.
Josef wants to make a path that is 1 m away from the edge of the pond.
Draw Josef's path on a copy of the diagram.

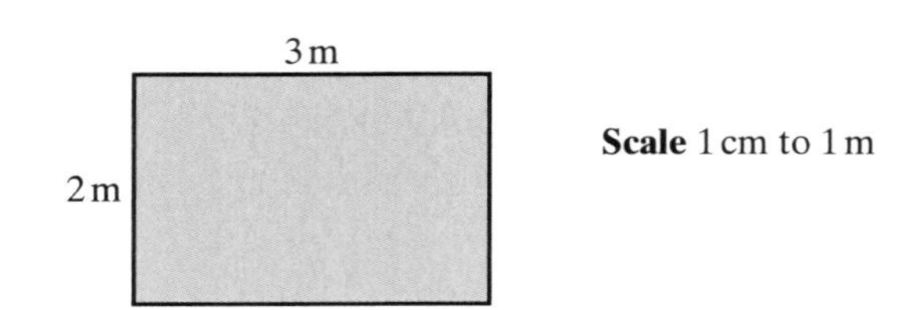

9 In the diagram, the rectangle represents a garden.

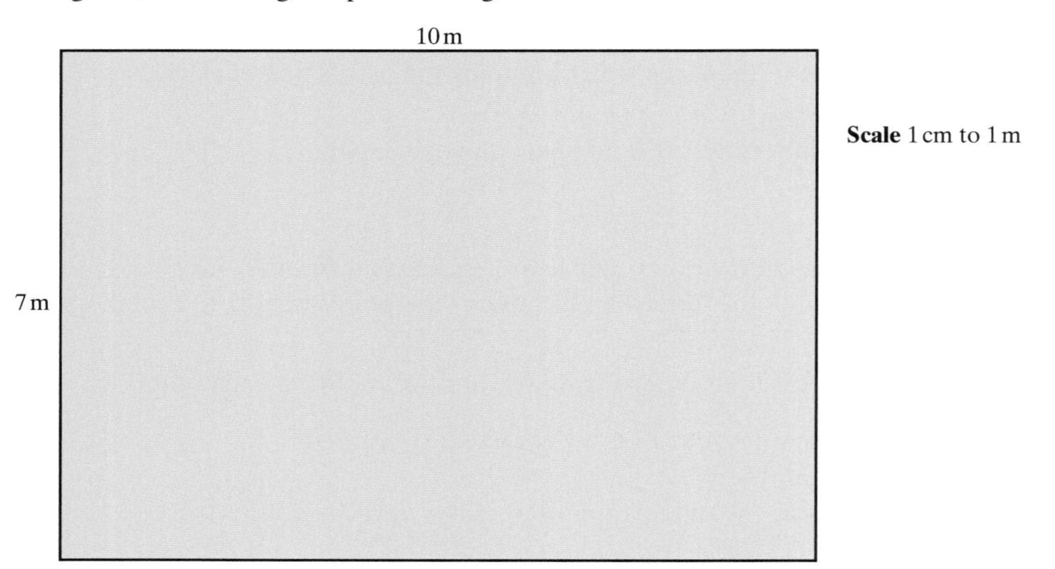

Joanne wants to make a path around her garden.
She wants the path to be 1 m wide and to touch the fence.
The path is *inside* Joanne's garden.
Draw Joanne's path on a copy of the diagram.

10 In the diagram, the rectangle ABCD represents an allotment.

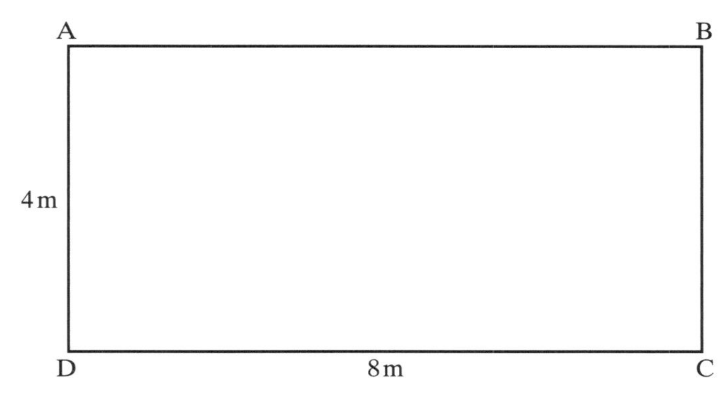

Scale 1 cm to 1 m

It is to be divided up into three regions:

The region formed by all points within 3 metres of AD will be planted with potatoes.

The region formed by all points within 4 metres of B will be planted with cabbages.

The rest of the allotment will be planted with carrots.

Indicate each of the three regions on a copy of the diagram.

11 The diagram shows an animal enclosure at a zoo.

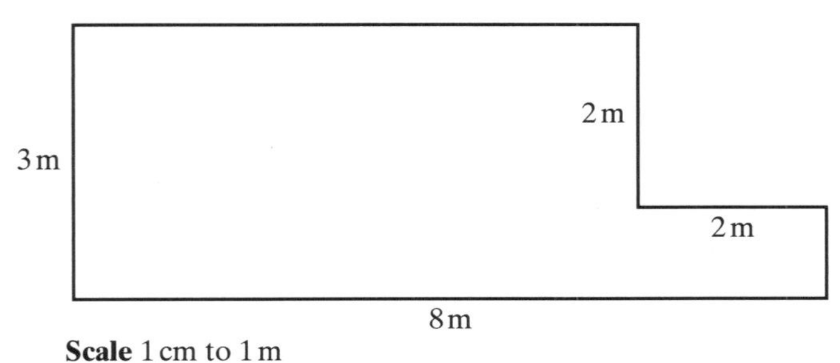

Scale 1 cm to 1 m

The zoo management want to erect a safety fence around the outside of the enclosure.
They want the public to have a good view of the animals, without getting dangerously close.
They decide that the fence should come to within 1 metre of the nearest point of the enclosure.
On a copy of the diagram, make a drawing to show the position of the safety fence.

12 The diagram shows a rectangular garden.
A gravel path is to be laid.

The centre line of the path runs diagonally from A to C.
All parts of the garden within 1 metre of this centre line are to be gravelled.

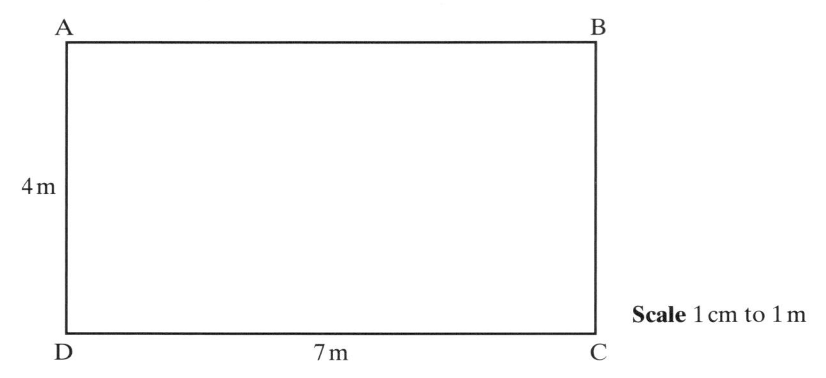

Scale 1 cm to 1 m

a) On a copy of the diagram, mark the centre line of the path.
b) Shade the region to be gravelled.

A rockery is to be planted in the rectangular plot.
The rockery will occupy all points within 1.5 metres of the corner B.
c) Shade the area to be planted as the rockery.

13 The diagram shows a rectangular flowerbed.
A gardener is going to add fertiliser to the flowerbed.
He adds fertiliser to the entire flowerbed except those points within 50 cm of its perimeter.
The gardener also puts turf around the *outside* of the flowerbed.
He turfs the region of all points within 50 cm of the *outside* perimeter of the flowerbed.

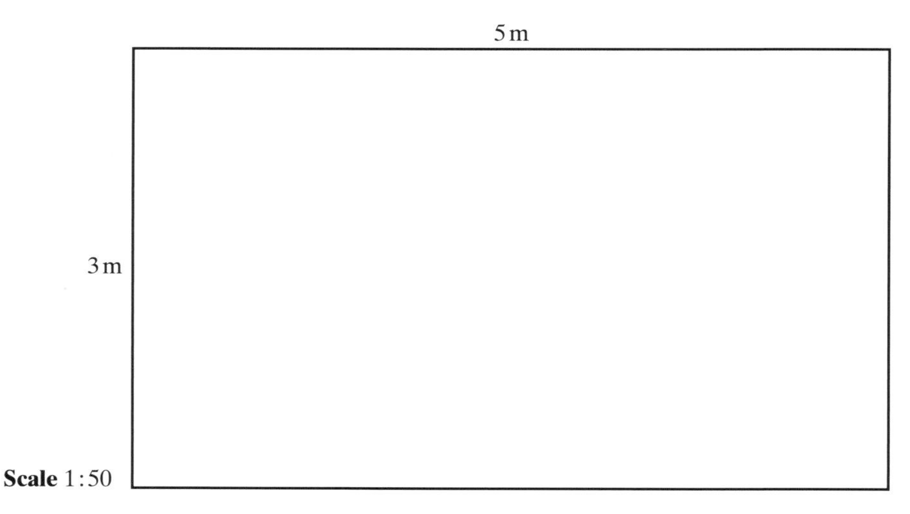

Scale 1:50

a) On a copy of the diagram, shade the region to which fertiliser is applied.
b) Shade the region that is turfed.

14 In a town there are three mobile phone transmitters, all belonging to the same phone network.
When people move around the town, their phones automatically connect to the nearest transmitter.
The diagram shows the outline of the town, and the three transmitters, A, B, and C.

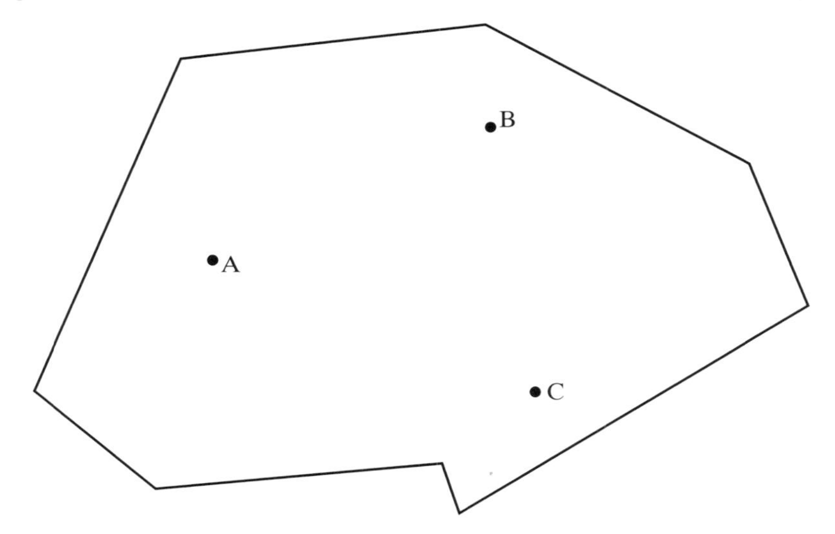

a) On a copy of the diagram, use compasses and a ruler to construct the locus of all points that are equidistant from A and B.
b) Repeat the construction using B and C; and again, using A and C.
c) Hence divide the town into three service regions, one for each transmitter.
Use coloured pencils to mark the regions distinctly.

18.4 2-D representations of 3-D shapes

You can represent a 3-D object in 2-D (on paper) in several different ways.

1 Sketch

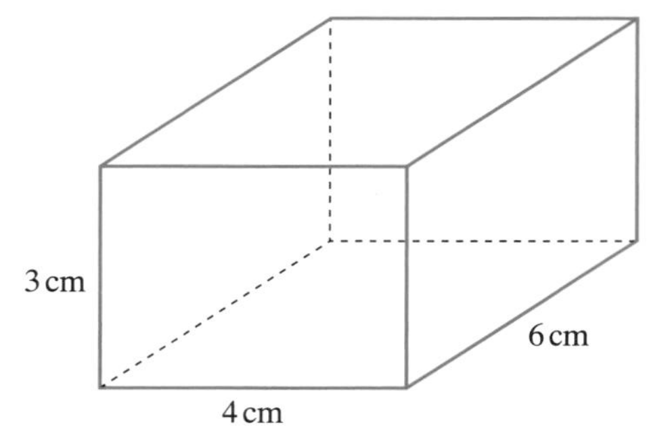

2 Isometric drawing

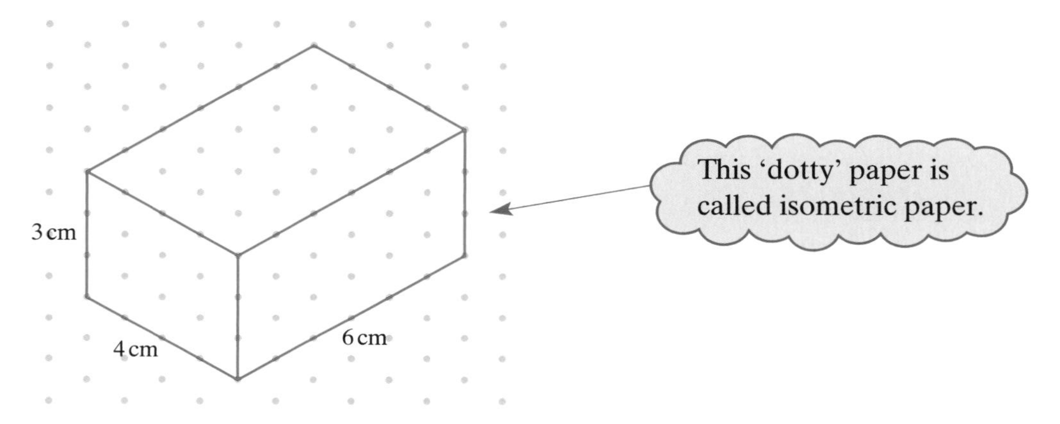

3 A net

Imagine the cuboid wrapped in paper.

When the paper is unfolded it might look like this:

This is called a **net**.

You can add tabs to every other edge so that the net can be folded up into the 3-D shape.

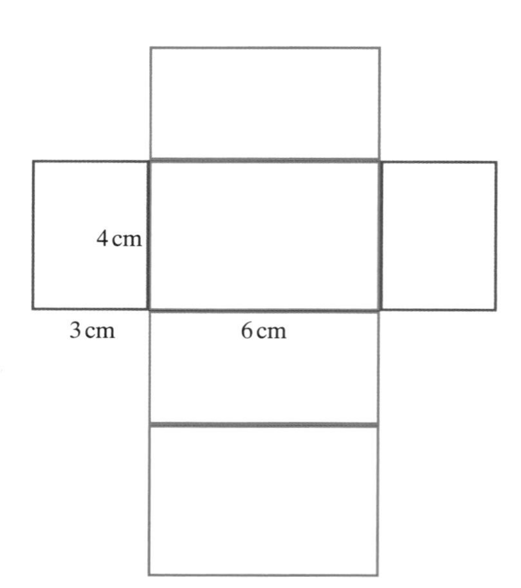

4 Plans and elevations

Another way is to draw a **plan** and **elevations** of the object. This is where you draw a 2-D view of the object from three different directions:

- above (plan view)
- from the side (side elevation)
- from the front (front elevation).

These three views give you all the information you need about a 3-D shape.

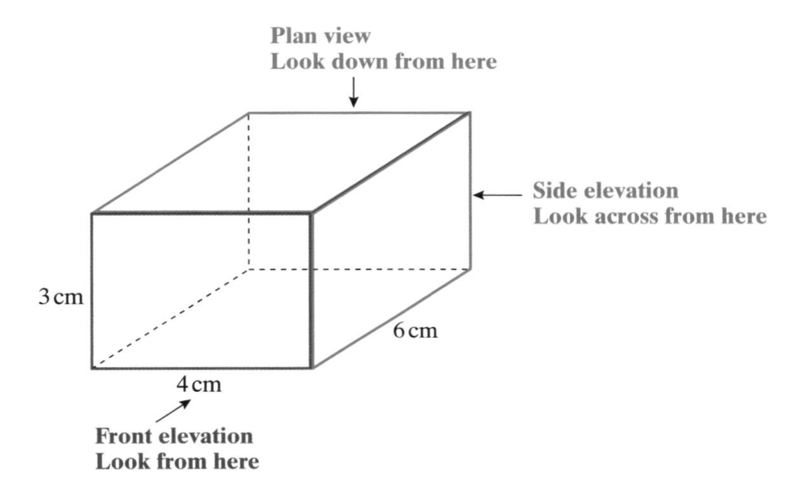

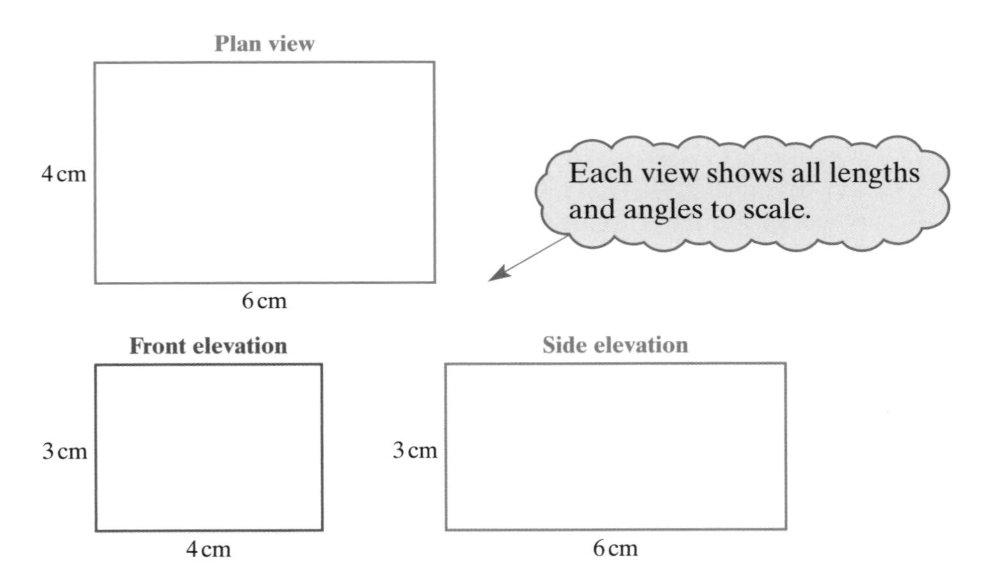

EXAMPLE

Draw a plan, front elevation and side elevation of this 3-D shape.

SOLUTION

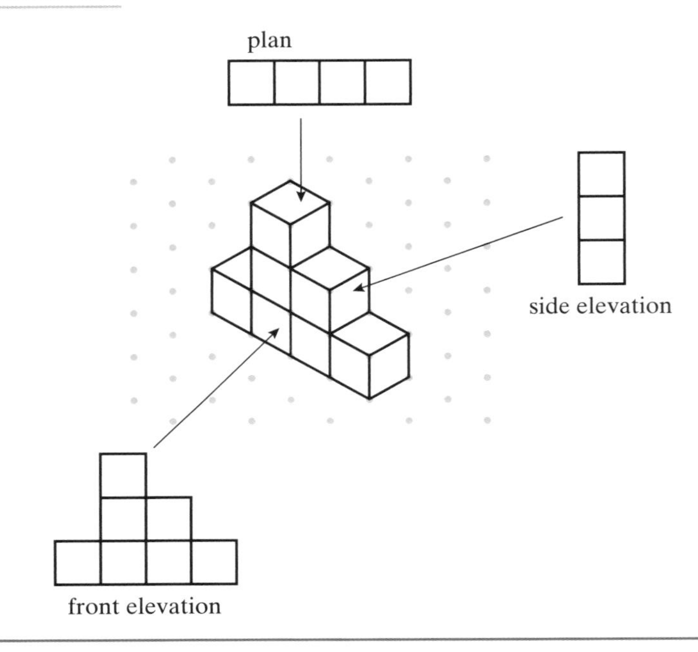

Note: If we draw the side elevation from the other side, one of the 'steps' will be hidden.

To show this we use a dotted line like this:

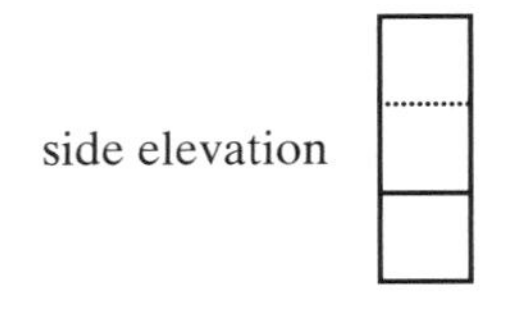

EXAMPLE

Sketch this solid shape from its plan, side and front elevations.

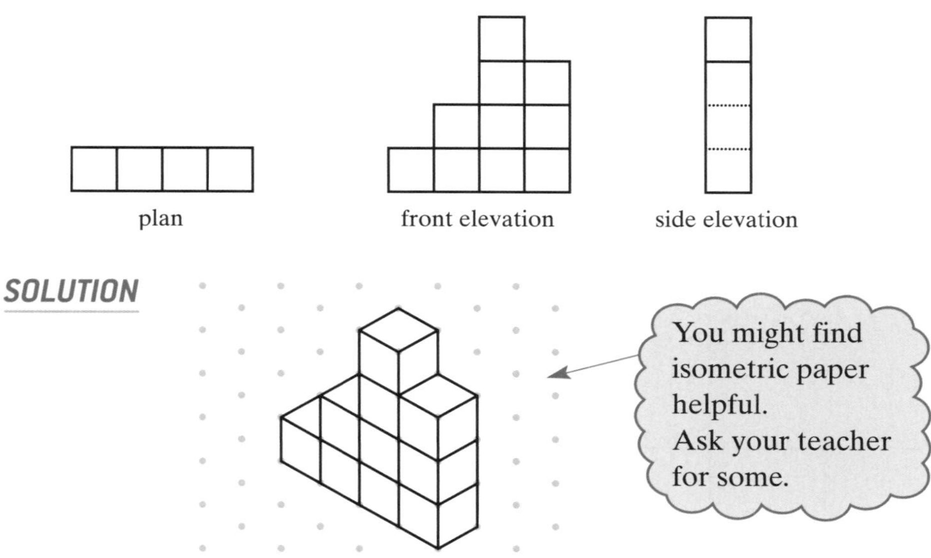

plan
front elevation
side elevation
You might find isometric paper helpful. Ask your teacher for some.

SOLUTION

EXAMPLE

a) Sketch a net of this triangular prism:

b) Sketch the plan view, side and front elevations of the triangular prism.

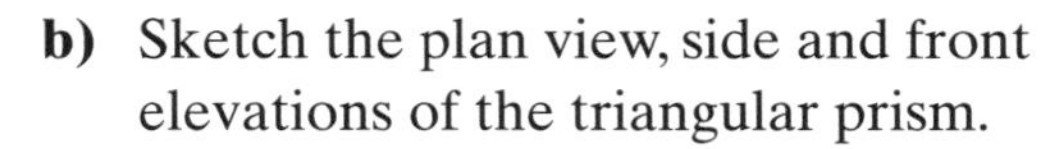

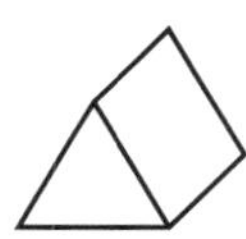

SOLUTION

a)

b)

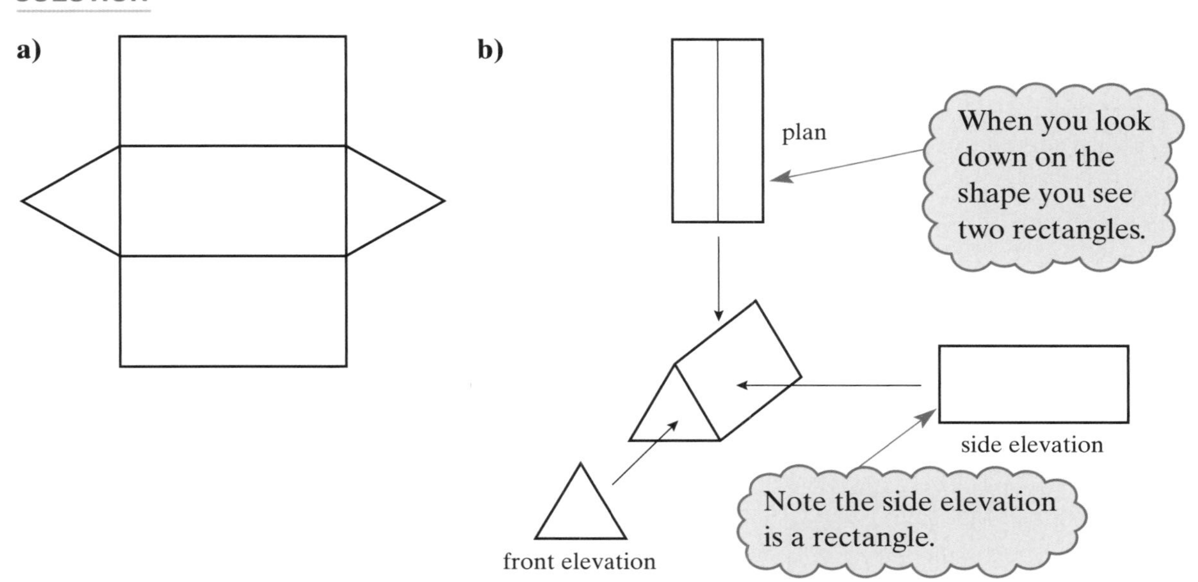

plan
When you look down on the shape you see two rectangles.
side elevation
front elevation
Note the side elevation is a rectangle.

EXERCISE 18.4

1 Match together these 3-D solids with their nets.

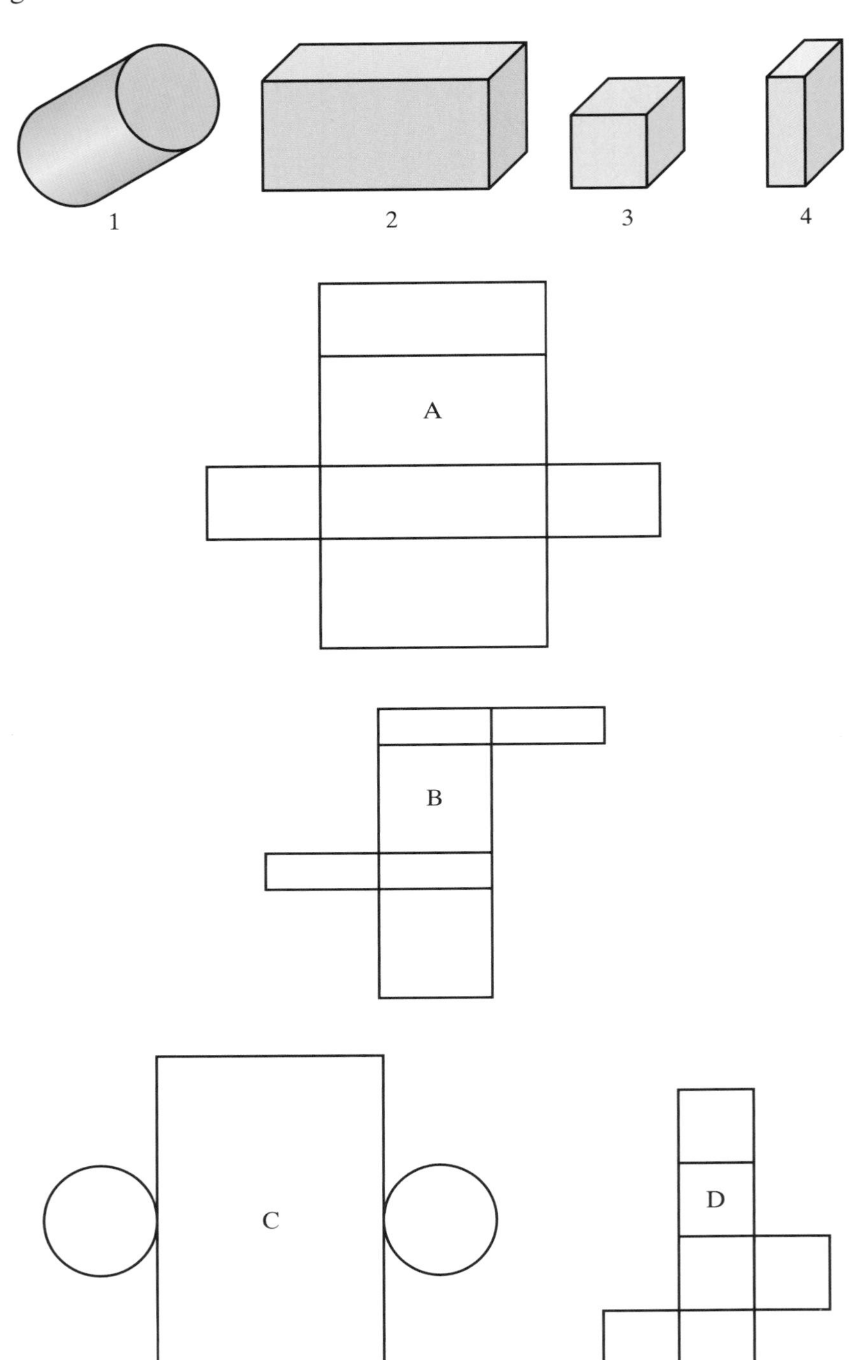

2 Which of these are nets of a cube?

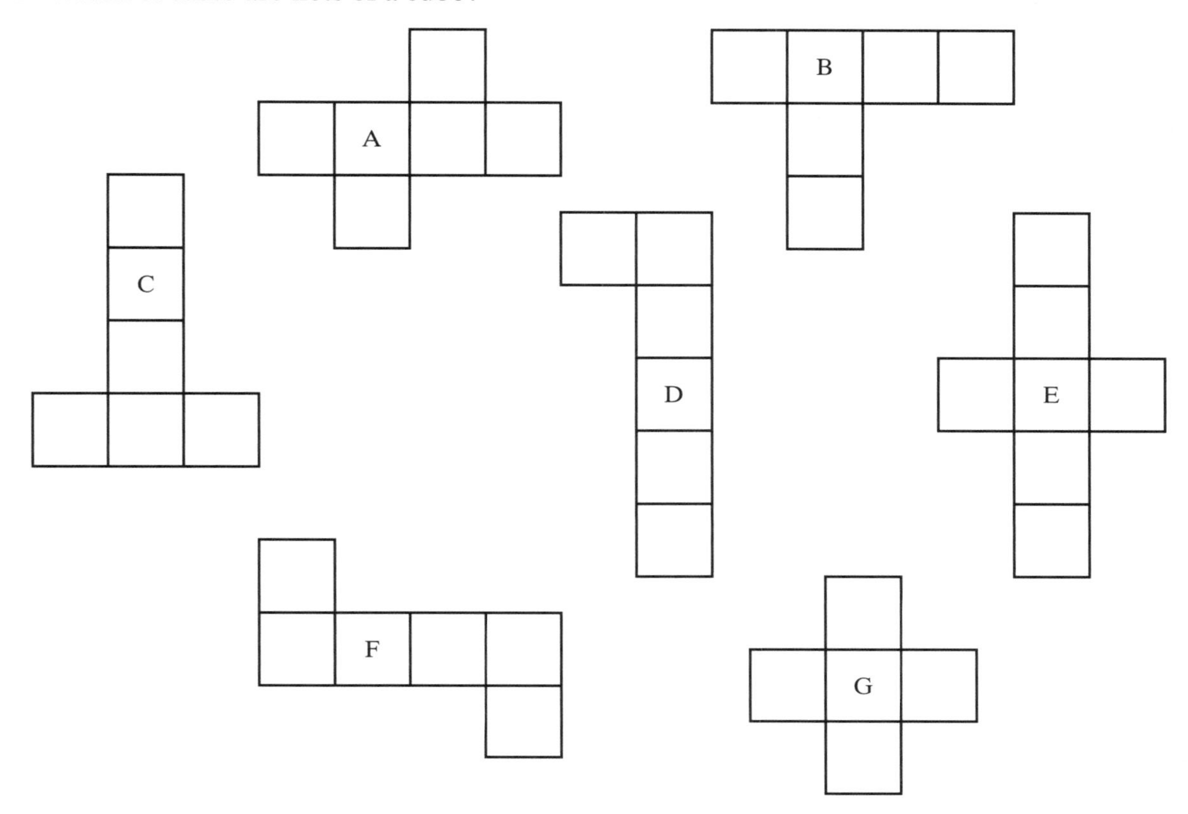

3 On squared paper draw accurate nets of these solids.

a)

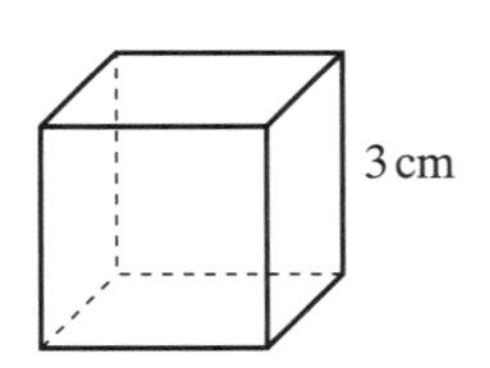

Cube

b)

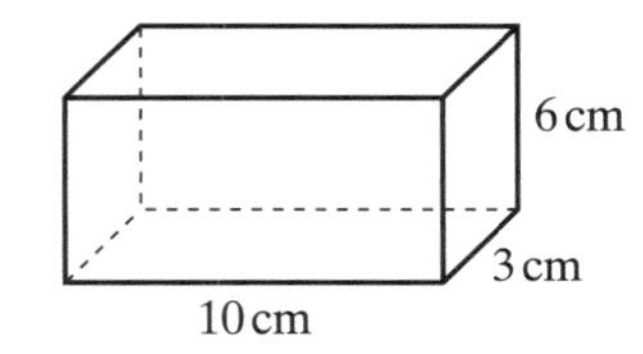

Cuboid

c)

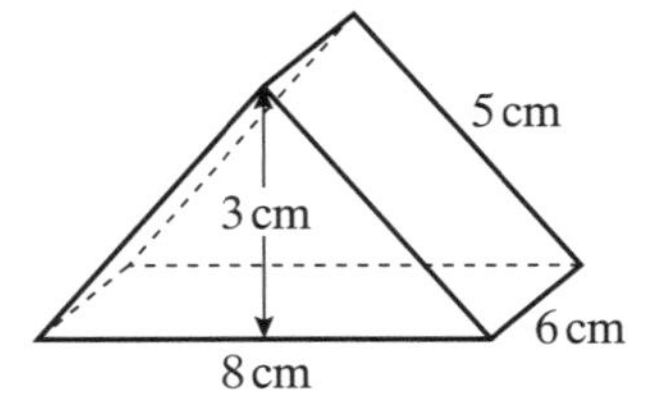

Triangular prism

d)

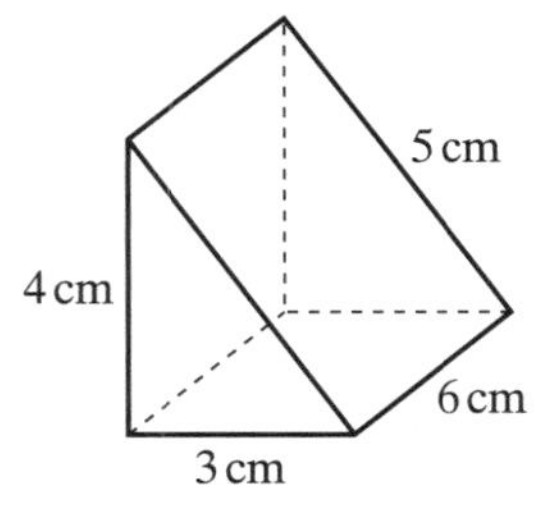

Triangular prism

4 Sketch the 3-D shapes made by these nets.

a)

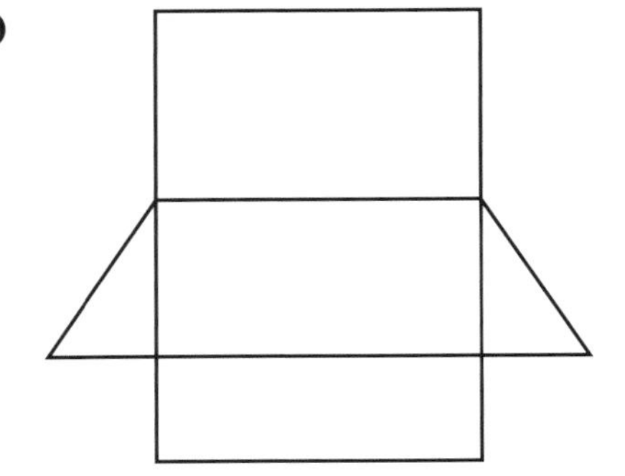

b)

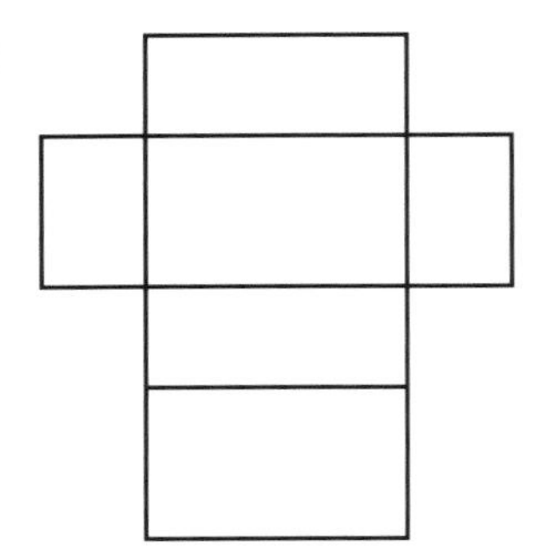

c)

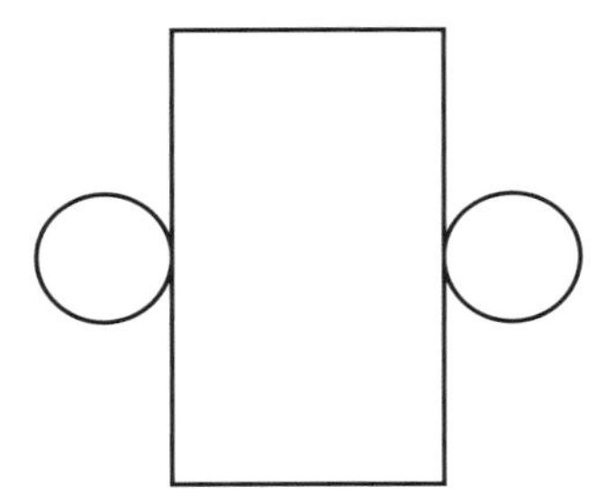

d)

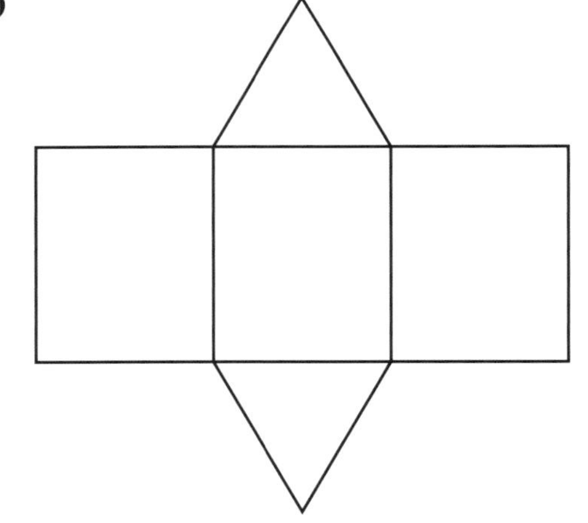

5 The isometric drawing shows some cubes forming an L-shape.

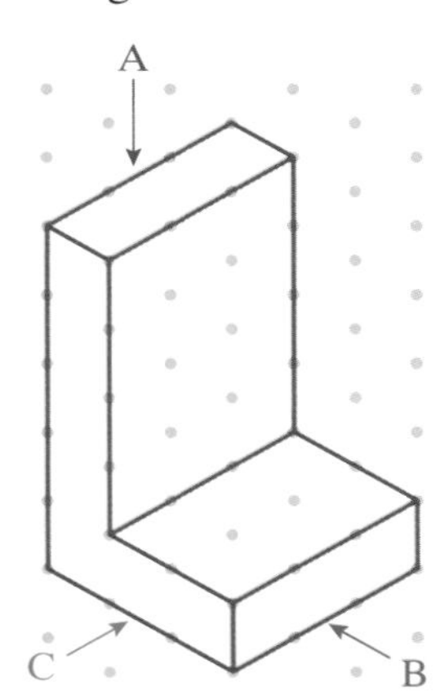

a) Copy the shape on to isometric paper.

b) On squared paper draw a sketch of:

(i) a plan view as seen from A

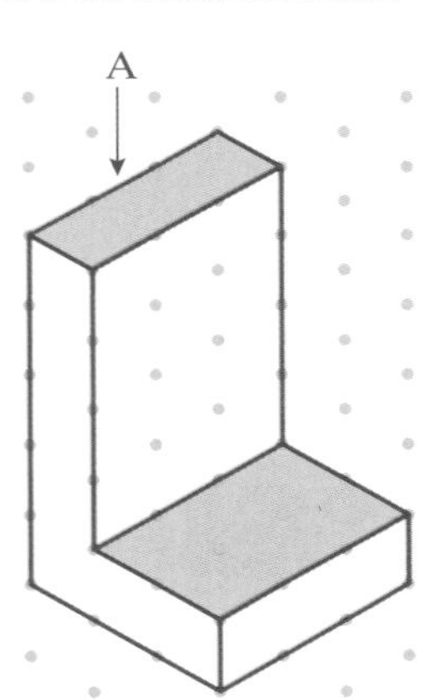

(ii) a front elevation, as seen from B

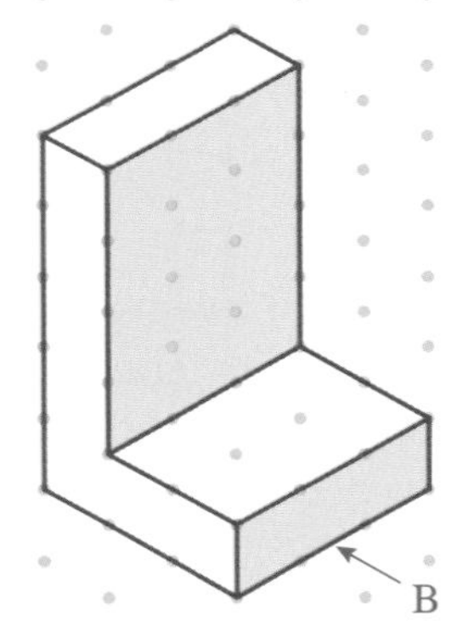

(iii) a side elevation as seen from C.

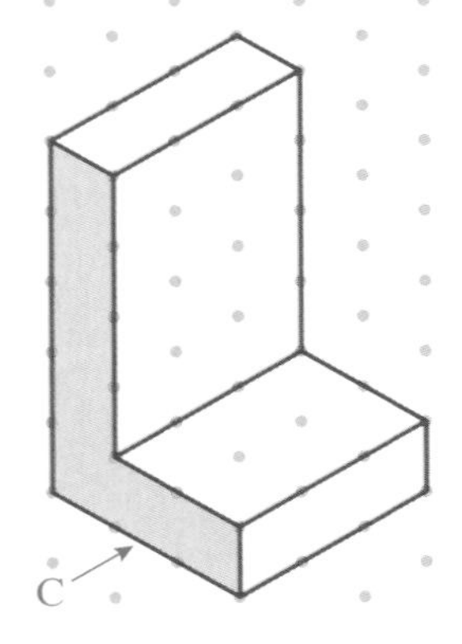

6 **(i)** Copy each of these 3-D shapes on to isometric paper.
(ii) On squared paper draw an accurate plan, front elevation and side elevation of each solid.
(iii) Write down the volume of each shape.

a)

b)

c)

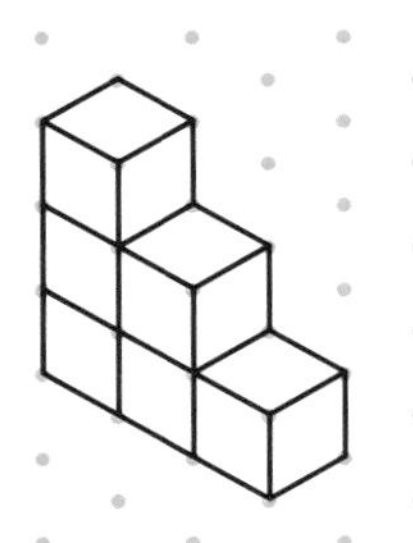

d)

e)

f)

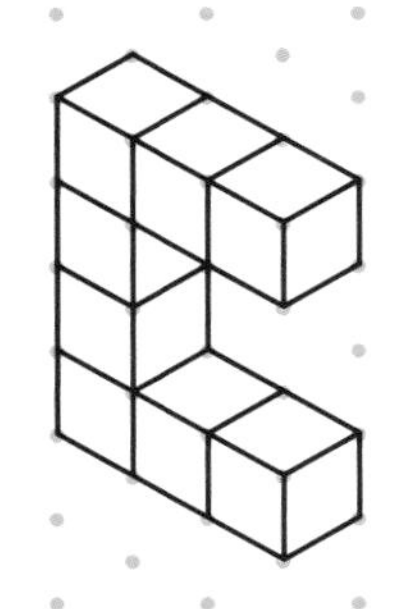

7 Here is a tin of sweets.
The length of the label around the tin is 18.8 cm
a) Draw an accurate net of the tin.
b) Draw accurate plan, front and side elevations of the tin.

8 **(i)** Sketch each of these 3-D shapes onto isometric paper.
(ii) On squared paper draw an accurate plan, front elevation and side elevation of each solid.

a)

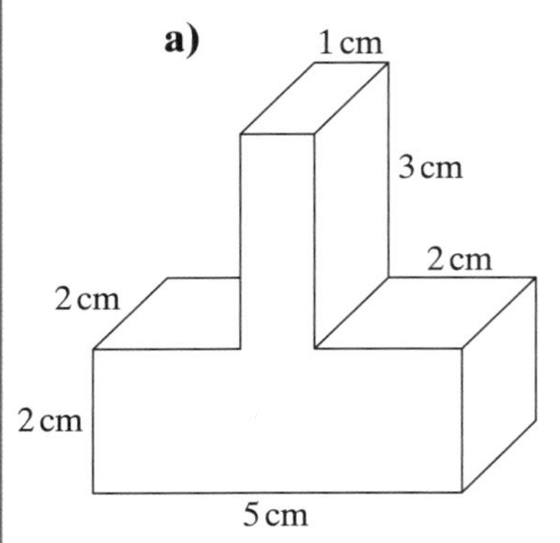

b)

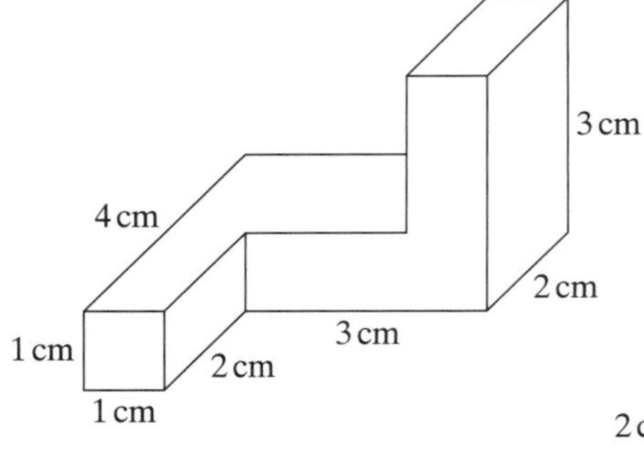

c)

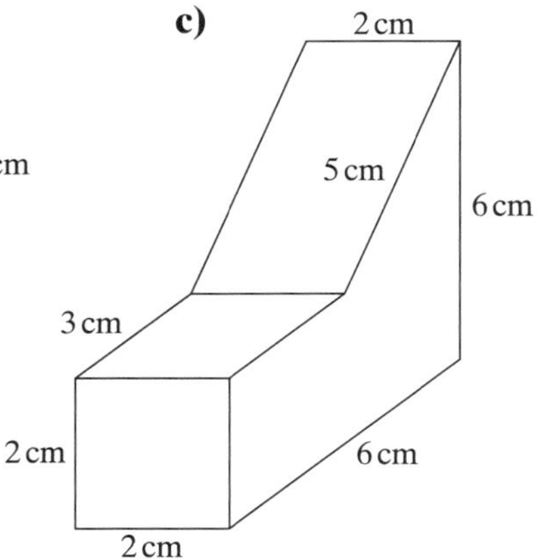

9 A pyramid has a square base whose sides are 6 cm long.

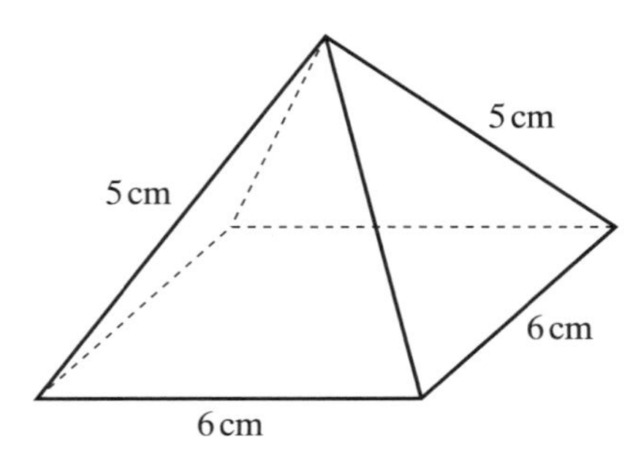

The triangular faces have sides 5 cm, 5 cm and 6 cm

a) Sketch a plan view of the pyramid.

b) Sketch a net for the pyramid.

10 Damini and Jonty have been building shapes with centimetre cubes on a square grid.

The diagrams show plan views of their shapes.

The numbers 1, 2, 3 tell you how many cubes are stacked on top of each square.

Damini

2	1
1	1

A

Jonty

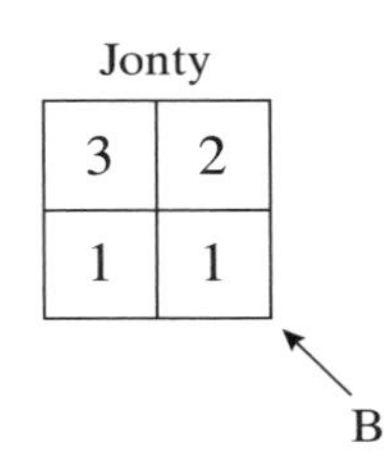

a) Draw a front elevation to show how Damini's shape appears seen from direction A.

b) Make an isometric drawing of Jonty's shape, viewed from direction B.

11 Look at this cube.

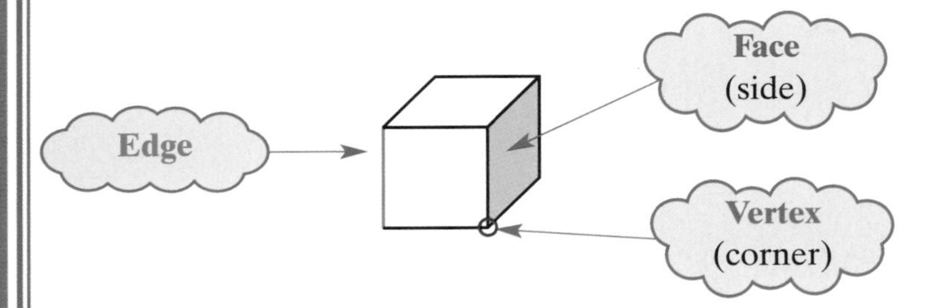

a) How many: **(i)** faces, **(ii)** vertices, **(iii)** edges does it have?

b) Copy and complete this table.

Solid	Number of faces	Number of vertices	Number of edges	Number of faces + number of vertices	Number of edges + 2

c) What do you notice?

REVIEW EXERCISE 18

Don't use a calculator for Questions 1–8.

1 a) On the grid, draw a line from the point C perpendicular to the line AB.

b) Sketch a cylinder.

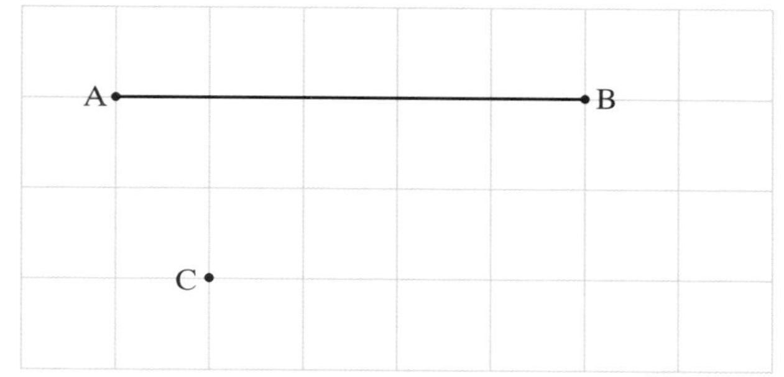

[Edexcel]

2 a) Draw a line 12 cm long.

b) Find the point that is halfway along the line you have drawn. Mark it with a cross.

c) On squared centimetre paper, draw a rectangle that has a length 6 cm and width 4 cm

[Edexcel]

3 Here are the plan, front elevation and side elevation of a 3-D shape.

Draw a sketch of the 3-D shape.

[Edexcel]

plan

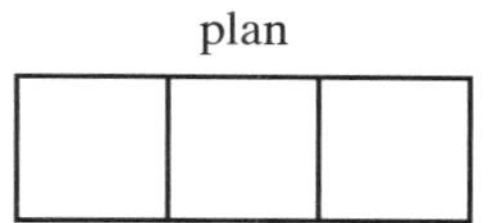

front elevation

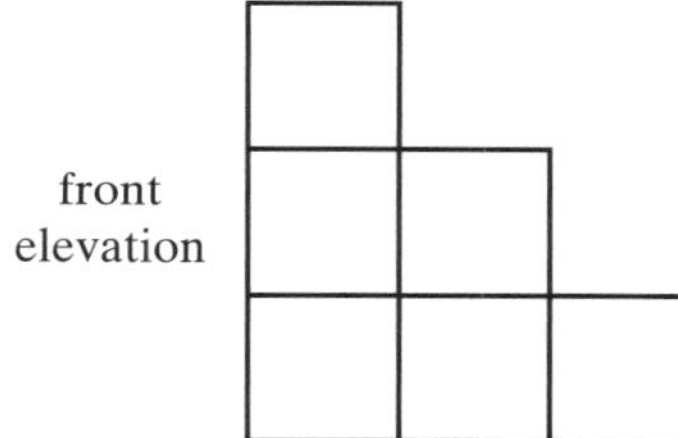

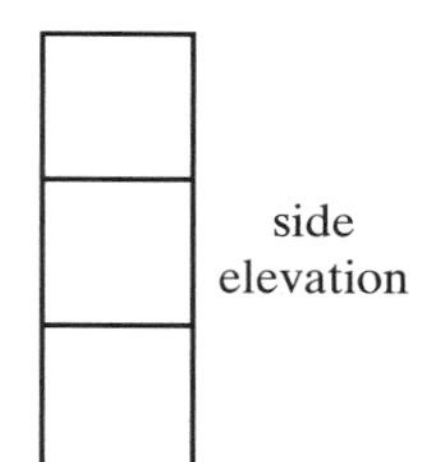

side elevation

4 Here is a net of a cube.
The net is folded to make the cube.
Copy the net.
Two other vertices meet at A.

a) Mark each of them with the letter A.

The length of each edge of the cube is 2 cm

b) Work out the volume of the cube.

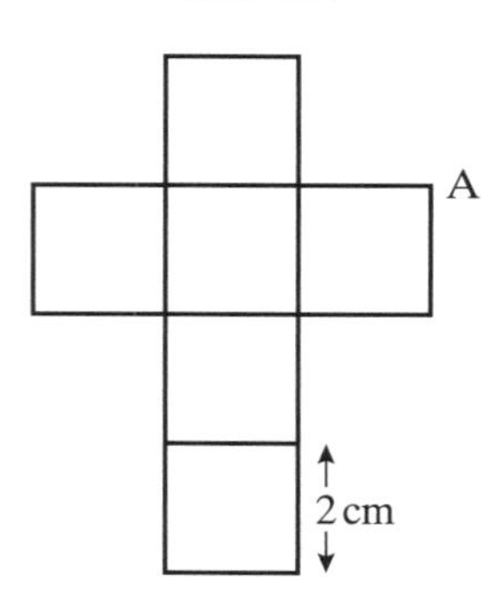

Diagram *not* accurately drawn

[Edexcel]

5 Here are the plan and front elevation of a prism.
The front elevation shows the cross section of the prism.

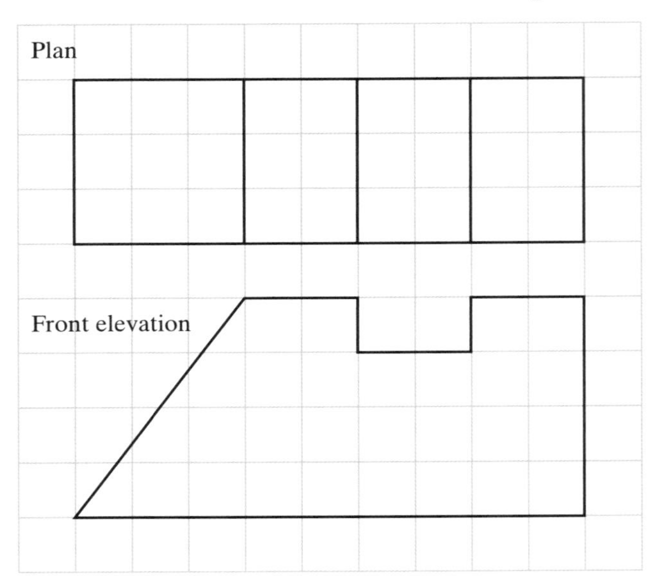

a) On squared paper, draw a side elevation of the prism.
b) Draw a 3-D sketch of the prism. [Edexcel]

6 Use ruler and compasses to construct the perpendicular bisector of the line segment AB.

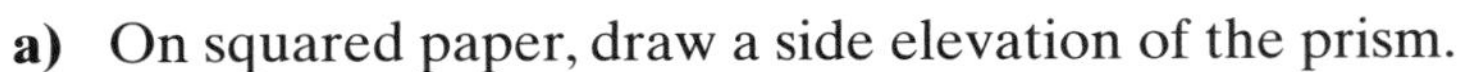

[Edexcel]

7 Use ruler and compasses to construct the bisector of angle ABC.

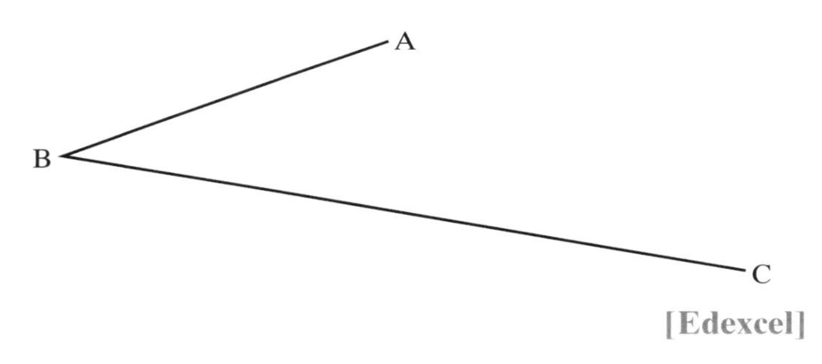

[Edexcel]

8 Draw the locus of all points which are 3 cm away from the line AB.

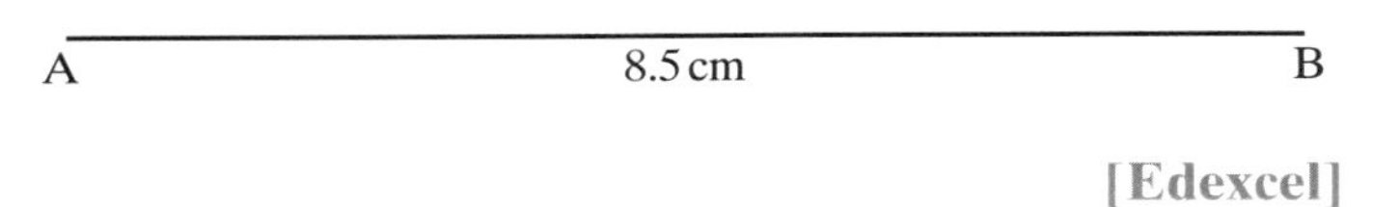

[Edexcel]

You can use your calculator for Questions 9–22.

9 **a)** Measure the length of this line and write it down. ______________________

The line is to be the diameter of a circle.

b) On a copy, mark the centre of the circle with a cross.

c) Draw the circle. **[Edexcel]**

10 The diagrams show some solid shapes and their nets.

An arrow has been drawn from one solid shape to its net.

On a copy, draw an arrow from each of the other solid shapes to their nets.

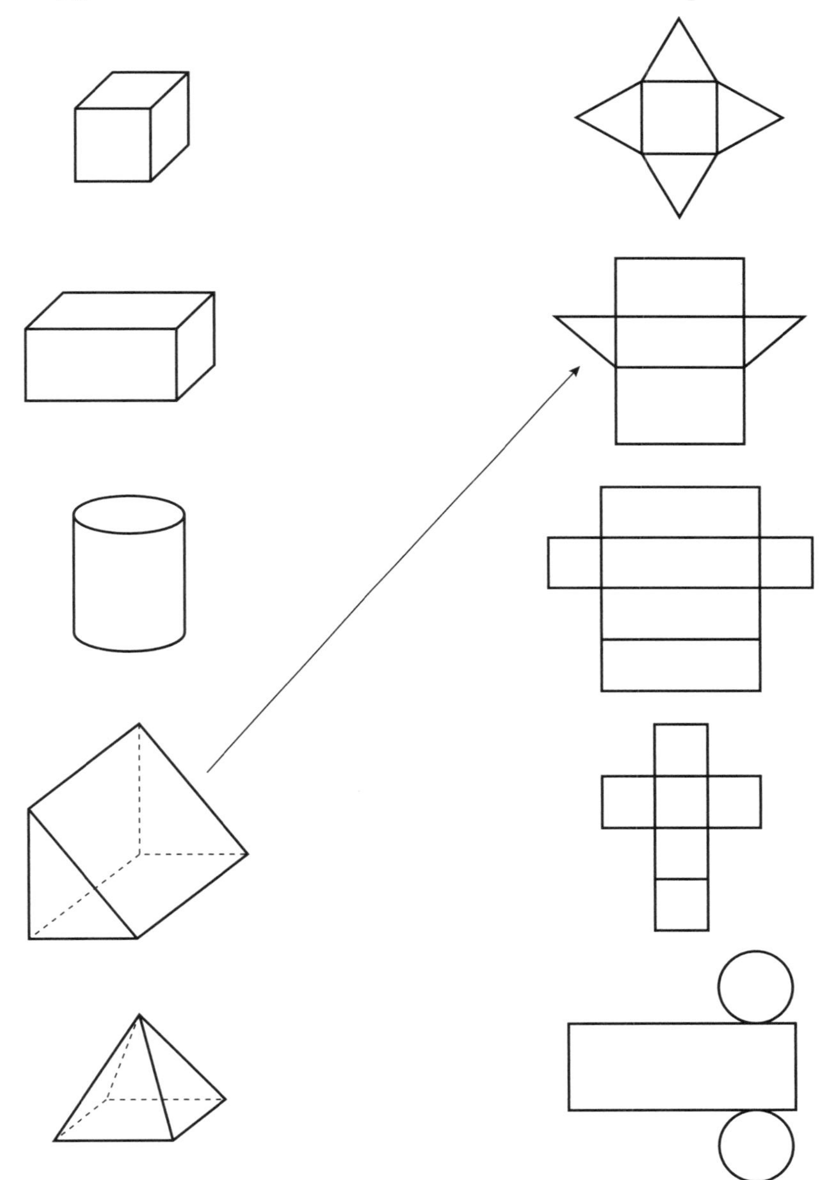

[Edexcel]

11 The diagram shows a triangular prism.
The cross section of the prism is an equilateral triangle.

a) Copy the prism and draw in one plane of symmetry for the prism.
b) Draw a sketch of a net for the triangular prism.
c) Use a ruler and compasses to *construct* an equilateral triangle with sides of 6 centimetres.
You must show all construction lines.

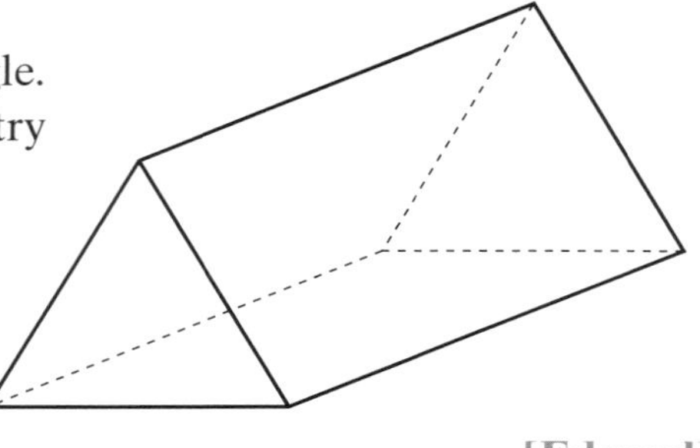

[Edexcel]

12 The diagram shows a sketch of triangle ABC.

BC = 7.3 cm
AC = 8 cm
Angle at C = 38°

a) Make an accurate drawing of triangle ABC.
b) Measure the size of angle at A on your diagram.

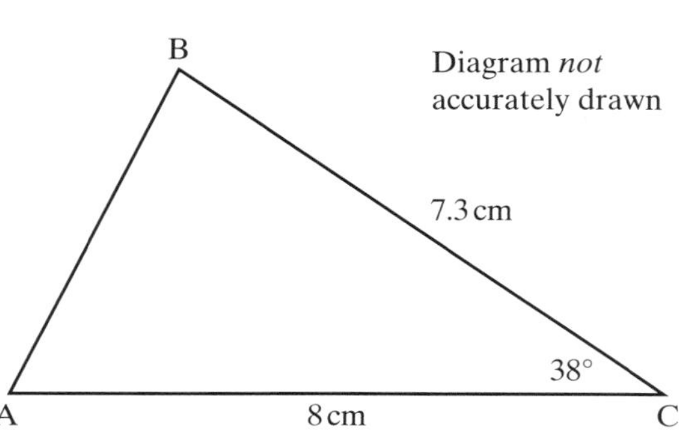

[Edexcel]

13 The diagram shows a solid object.

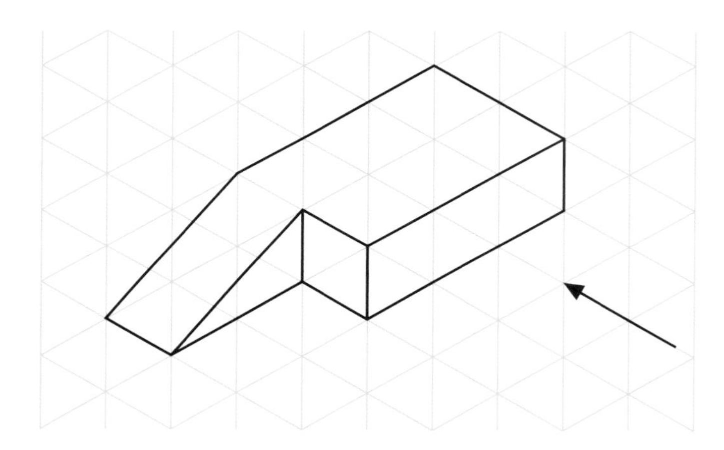

a) Sketch the front elevation from the direction marked with an arrow.
b) Sketch a plan of the solid object.

[Edexcel]

14 On a copy of the diagram, draw the locus of the points, outside the rectangle, that are 3 centimetres from the edges of this rectangle.

[Edexcel]

15 Triangle ABC is shown in the diagram.

a) On a copy of the diagram, draw accurately the locus of the points which are 3 cm from B.

b) On the diagram, draw accurately the locus of the points which are the same distance from BA as they are from BC.

T is a point inside triangle ABC.
T is 3 cm from B.
T is the same distance from BA as it is from BC.

c) On the diagram, mark the point T clearly with a cross. Label it with the letter T.

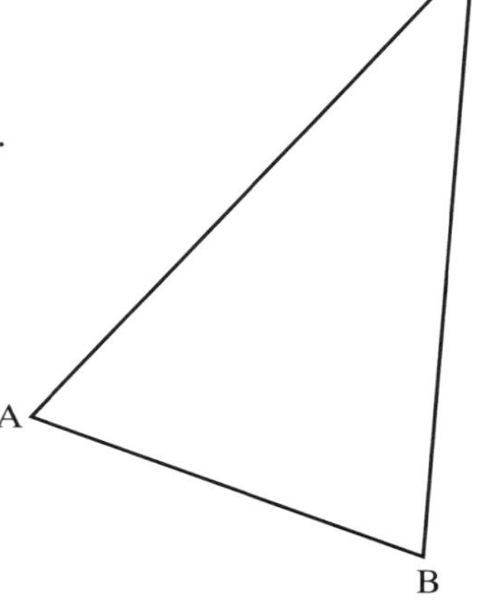

[Edexcel]

16 B is 5 km north of A.
C is 4 km from B.
C is 7 km from A.

Make an accurate scale drawing of triangle ABC.
Use a scale of 1 cm to 1 km

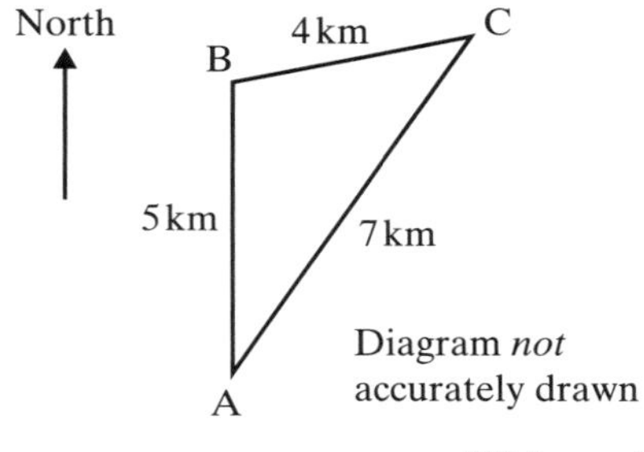

[Edexcel]

17 **a)** Draw accurately, on a copy of the diagram below, the locus of points which are the same distance from the line OA and the line OB.

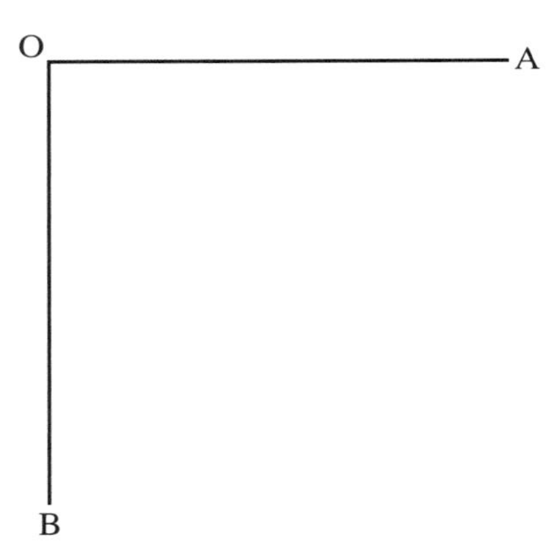

Some points are the same distance from the line OA and the line OB, and are also 4 cm from the point B.

b) Mark the positions of these points with crosses. **[Edexcel]**

18 Here is a sketch of a triangle.

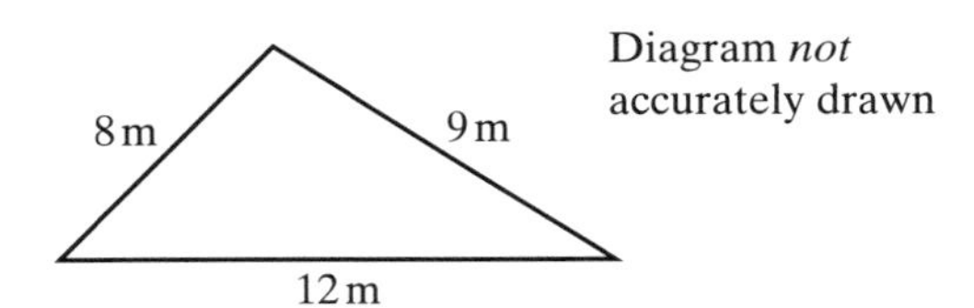

The lengths of the sides of the triangle are 8 m, 9 m and 12 m
Use a scale of 1 cm to 2 m to make an accurate scale drawing of the triangle. **[Edexcel]**

19 Here is a sketch of a triangle.

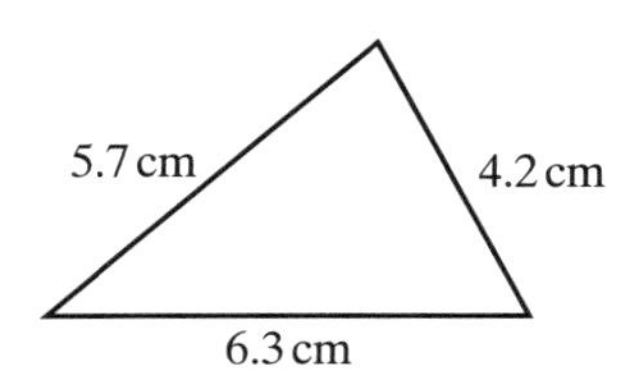

Use a ruler and compasses to *construct* this triangle accurately.
You must show all construction lines. **[Edexcel]**

20 The diagram is a plan of a field drawn to a scale of 1 cm to 20 m

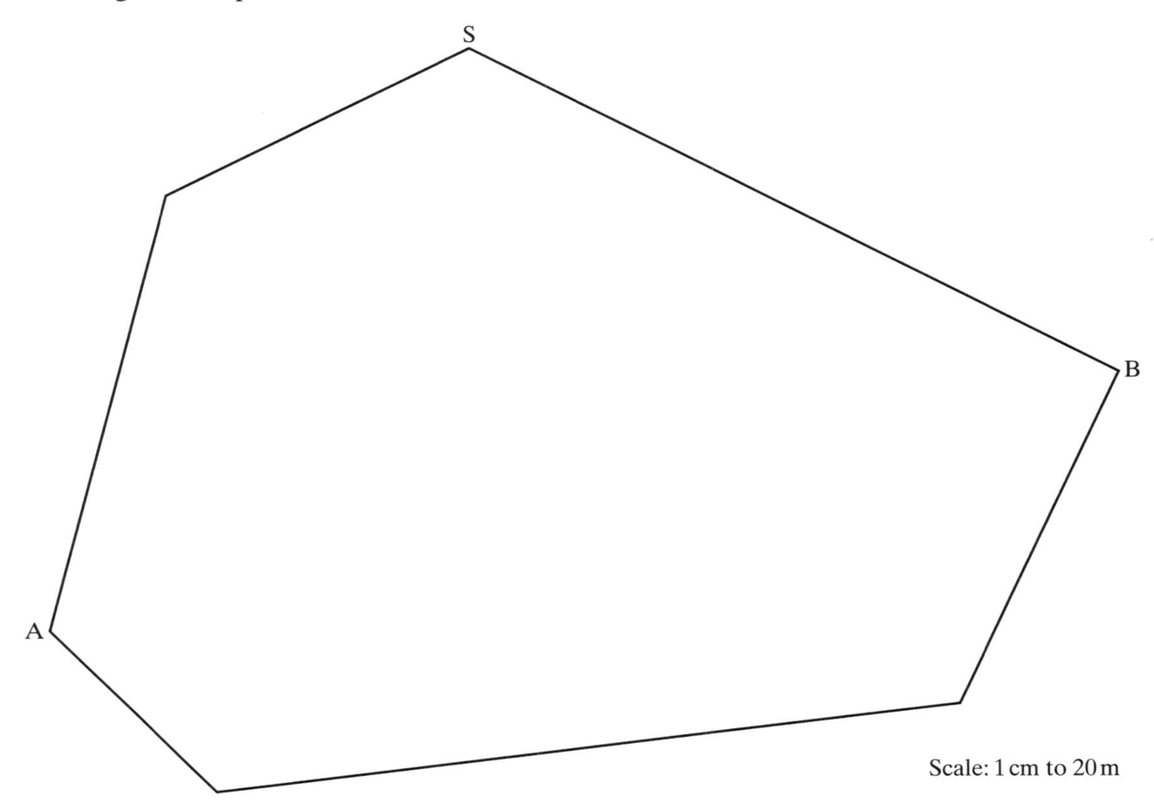

There is a water sprinkler at S.
The sprinkler can water that region of the field which is 60 metres or less from the sprinkler.

a) Shade, on a full size copy of the diagram, the region of the field which is 60 metres or less from the sprinkler.

A farmer is going to lay a pipe to help water the field.
A and B are posts which mark the widest part of the field.
The pipe will cross the field so that it is always the same distance from A as it is from B.

b) On the diagram draw a line accurately to show where the pipe should be laid.

[Edexcel]

21 The diagram represents a triangular garden ABC.
The scale of the diagram is 1 cm to 1 m
A tree is to be planted so that it is:

nearer to AB than to AC
within 5 metres of point A.

On a copy of the diagram, shade the region where the tree may be planted.

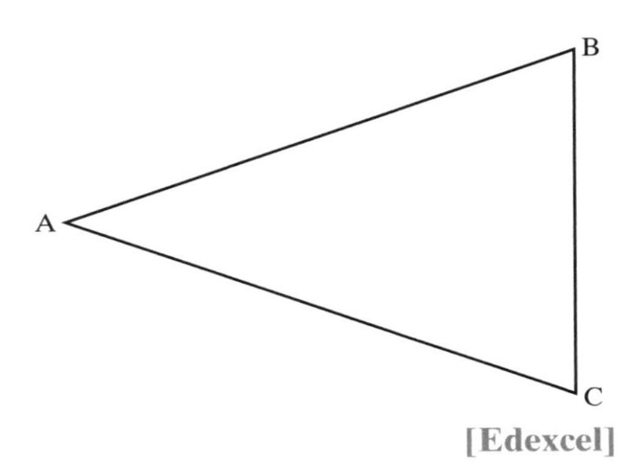

[Edexcel]

22 A map is drawn to a scale of 1 : 25 000
Two schools, A and B, are 12 centimetres apart on the map.
a) Work out the actual distance from A to B.
Give your answer in kilometres.

B is due East of A.
C is another school.
The bearing of C from A is 064°
The bearing of C from B is 312°
b) Copy and complete the scale drawing below.
Mark with a cross the position of the school C.

[Edexcel]

KEY POINTS

1. Bisect means to cut in half.
2. Two lines which are **perpendicular to** each other are at right angles.
3. You need to know how to:
   - construct a triangle given three pieces of information
   - bisect an angle
   - draw a perpendicular bisector
   - drop a perpendicular from a point to a line
   - construct a perpendicular from a point on a line.
4. Always leave your construction marks showing.
5. When a point P moves in the plane, subject to a certain rule, it traces out a path, known as a locus.
6. You should know about the following loci:
   - 1 point → circle
   - 2 points → perpendicular bisector
   - 1 line → 'running track'
   - 2 lines → 'angle bisector'
7. You can represent a 3-D object in 2-D (on paper) in several different ways:
   - sketch
   - isometric drawing
   - net – a net can be folded up to make a 3-D shape
   - plan and elevations.
8. The last of these is where you draw a 2-D view of the object from 3 different directions:
   - above (plan view)
   - from the side (side elevation)
   - from the front (front elevation).

Internet Challenge 18

Polyhedra are 3-D mathematical shapes made up of a number of 2-D plane faces.

Many polyhedra exhibit geometric symmetries of various kinds.

The tetrahedron and the cube are simple examples, but many more exotic polyhedra exist.

Constructing models of them can be quite challenging!

Investigating polyhedra

Use an internet search engine, such as Google, to look for information about Platonic solids.

Then answer these questions:

1 How many Platonic solids are known?
Are we ever likely to find any more?

2 Why are they called Platonic solids?

3 Design nets for each of the Platonic solids, and trace them onto thin card.
Then cut them out, and make some models for your classroom.
Remember to include tabs in suitable places.

Modern footballs are assembled from a net of pentagons and hexagons.

4 Find out how this net is made.

5 Is such a football an example of a Platonic solid?

CHAPTER 19

Collecting data

In this chapter you will **learn how to:**

- distinguish between fair and biased situations
- recognise the difference between primary and secondary data
- avoid common errors when designing questionnaires
- recognise categorical, discrete and continuous data
- use tally charts
- use two-way tables.

You will be **challenged to:**

- investigate how the national census works.

Starter: Tally ho!

Enrico, Myra and Sunita have designed data capture sheets, in the form of tally charts. They are going to ask their school friends a question:

'How many marks did you get in your Key Stage 3 tests?'

Enrico's sheet

0 to 50	
51 to 100	
101 to 150	

Myra's sheet

0
1
2
3
4
5
…
…
150

Sunita s sheet

71 to 80
81 to 90
91 to 100
101 to 110
111 to 120
121 to 130
131 to 140
141 to 150

Here are the responses from their school friends.

127	108	136	99	96	84	80	91	103	112
144	121	149	101	77	89	96	99	108	115

Use a copy of each sheet to tally the data.

Which one seems to work best?

Which is the worst?

19.1 Designing an experiment or survey

My favourite pop group is the most popular in the UK.

Sarah

Girls spend more time on the phone than boys.

Simon

These are both examples of a **hypothesis** (the plural is **hypotheses**).
A hypothesis is a statement which you can **test** to see how true it is.

A **survey** or **experiment** is used to test a hypothesis.
You need to collect data and then analyse it to see if it supports the hypothesis.

Before you collect any data you need to make a plan:

- What are you trying to find out?
- What **population** are you going to investigate?
- Are you going to collect data from every member of the population (a **census**) or just some of them (a **sample survey**)?

So should Simon survey:
- all the girls and boys in the UK?
- all the girls and boys at his school?
- all the girls and boys in his year?

When you use a sample you need to decide how you can collect it efficiently and fairly.

Suppose you wanted to test Sarah's hypothesis.

Here are some possible ways forward.

1 Ask everyone in the UK to tell you who their favourite group is.

2 Ask fifty school friends to tell you who their favourite group is.

3 Ask a local record shop to let you see its sales figures for last week.

4 Telephone a record company and ask them.

Each of these approaches has some advantages and disadvantages.

1 *Ask everyone in the UK to tell you who their favourite band is.*

This is a **fair** way of finding the answer, since it uses responses from everyone in the UK.
In practice, however, it would take too long to be a realistic method.

2 *Ask fifty school friends to tell you who their favourite band is.*

This method generates **primary data**.

You have collected the data.

It is quick and easy, but can be unreliable.
For example, fifty of your school friends are likely to have similar musical tastes to yourself.
So it may not represent the full breadth of national opinion.

This method suffers from **bias**.

3 *Ask a local record shop to let you see its sales figures for last week.*

The data is **secondary data**.

In this case it is likely to be of good quality.

Someone else has collected the data.

The data refers to only one week from only one location in the country, however, so again it could be biased.

4 *Telephone a record company and ask them.*

Again this is secondary data.

However, the data might be **very biased** as the record company may want to promote its own artists and ignore others. This method is unlikely to give statistically useful data.

Primary data is often gathered by doing an experiment or conducting a statistical survey. It is important that this should be done in a fair way.

Surveys often use sets of questions in a leaflet, called a **questionnaire**.

When designing a questionnaire:

- make sure your questions are **relevant** to your hypothesis
- avoid **leading** or **biased** questions
- avoid **open** questions
- use **response boxes** (tick boxes) – at least three in a set
- Make sure that any tick boxes:
  - **1** cover all possible answers
  - **2** don't have overlapping categories.

These are questions that suggest a certain answer, like, 'Don't you agree that smoking is bad for you?'

These are questions which have many different answers, like, 'What do you think of the school canteen?'

Questions which have tick boxes are called **closed** questions.

EXAMPLE

Mary wants to test the hypothesis that students at her school think the library is a good place to study.

She decides to go to the library one lunchtime and ask 50 students the question:

'Don't you agree that the library is a good place to do homework in?'

Write down two reasons why this is not a good way to find out whether students at Mary's school think the library is a good place to study.

SOLUTION

Reason 1 Mary's sample will be biased.

Students who are using the library are more likely to think it is a good place to work.

Reason 2 Mary's question is leading.

It is suggesting that the reply should be 'Yes'.

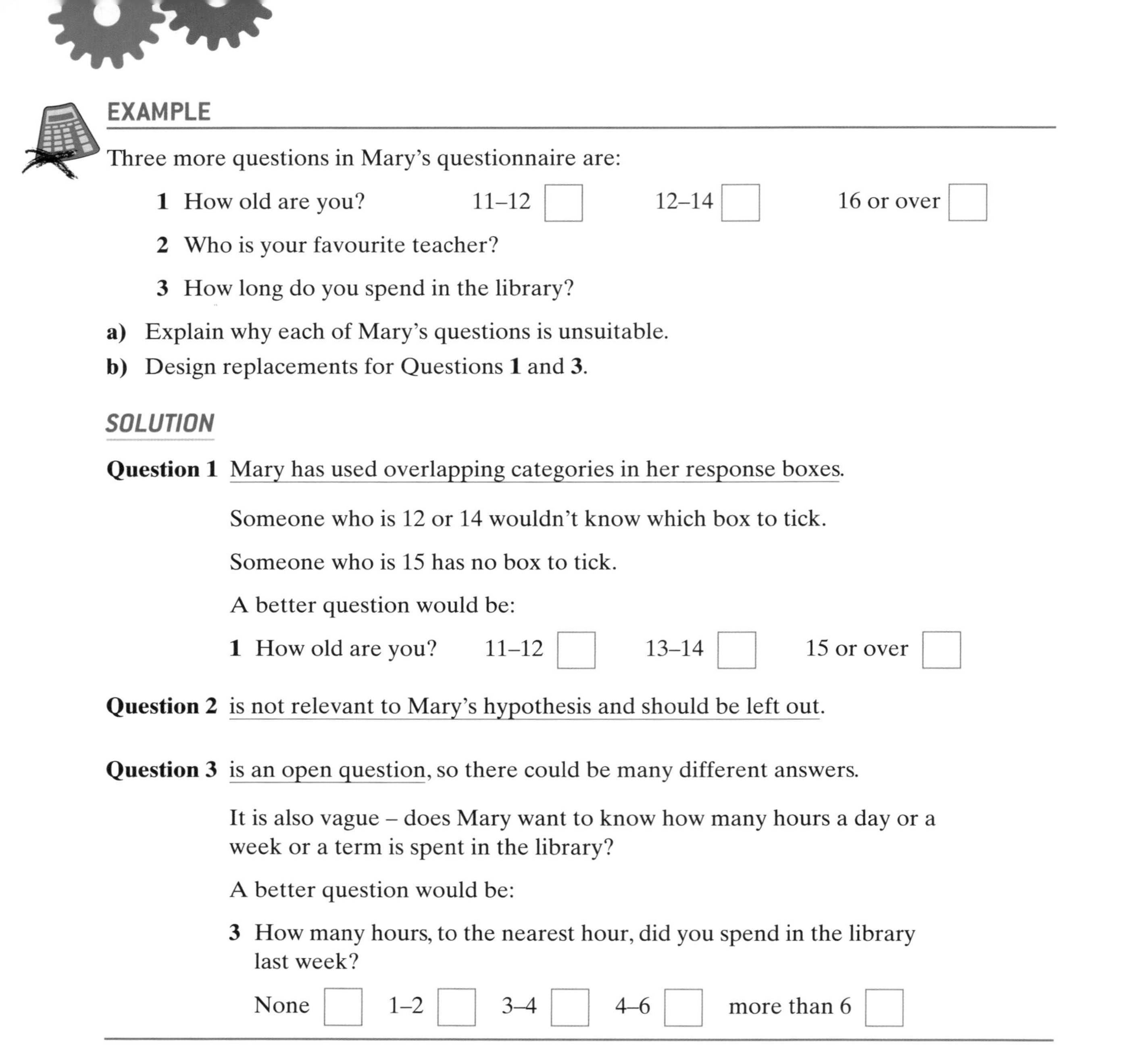

EXAMPLE

Three more questions in Mary's questionnaire are:

1 How old are you? 11–12 ☐ 12–14 ☐ 16 or over ☐

2 Who is your favourite teacher?

3 How long do you spend in the library?

a) Explain why each of Mary's questions is unsuitable.

b) Design replacements for Questions **1** and **3**.

SOLUTION

Question 1 Mary has used overlapping categories in her response boxes.

Someone who is 12 or 14 wouldn't know which box to tick.

Someone who is 15 has no box to tick.

A better question would be:

1 How old are you? 11–12 ☐ 13–14 ☐ 15 or over ☐

Question 2 is not relevant to Mary's hypothesis and should be left out.

Question 3 is an open question, so there could be many different answers.

It is also vague – does Mary want to know how many hours a day or a week or a term is spent in the library?

A better question would be:

3 How many hours, to the nearest hour, did you spend in the library last week?

None ☐ 1–2 ☐ 3–4 ☐ 4–6 ☐ more than 6 ☐

EXERCISE 19.1

1 Fred is planning to carry out a survey of 50 students at his school.
He wants to see whether they think the school is good.
He decides to arrive at 7.30 am and ask the first 50 students who arrive:

'Don't you agree that our school is the best in town?'

a) Write down **two** reasons why this is **not** a good way to find out whether students think that Fred's school is a good school.

b) Describe a better method of selecting 50 students that Fred might use.

c) Write down a replacement question that Fred could use.

2 Kwame is conducting a survey about newspapers.
Here are three questions in his questionnaire.

1: How old are you?

2: What newspaper do you read?

3: Do you agree that newspapers are too expensive?

Yes ☐ No ☐

a) Explain briefly why each of Kwame's questions is unsuitable.
b) Design suitable replacements for each of Kwame's questions.

3 Stephanie is going to find out about the television viewing habits of students at her school.
She has designed three questions.

1: What sex are you?

M ☐ F ☐

2: How old are you?

3: How much television do you watch?

A lot ☐ Not very much ☐

a) Explain briefly why Stephanie's Questions 2 and 3 are unsuitable.
b) Design suitable replacements for Questions 2 and 3 in Stephanie's questionnaire.

4 Bryony wants to find out about the part-time jobs that students have at her school.
She has designed three questions.

1: Where do you work?

2: What is your favourite pop group?

3: How much do you earn per hour?

£5–£6 ☐ £6–£7 ☐ £7 or more ☐

a) Criticise each of Bryony's questions.
b) Design suitable replacements for Questions 1 and 3 in Bryony's questionnaire.

5 Fergus is carrying out a survey.
He wants to use a questionnaire to find out what kind of music students at his school like.
He also wants to see if boys like the same music as girls.
Write down two questions that Fergus might use in his questionnaire.

6 Design three suitable questions that Simon (see page 472) might use to test his hypothesis that girls spend more time using their mobile phones than boys.

7 Jamie and Joe are going to interview 26 pupils from their school. They want to choose the students in as fair a way as possible.

I'm going to take an alphabetical list and pick the first student whose name begins with A, then the first whose name begins with B, and so on. That will give me a representative sample of 26 names.

Jamie

Our school has 26 classes, so I'm just going to pick one randomly from each class.

Joe

Explain carefully whether one of these methods is better than the other one.

19.2 Categorical, discrete and continuous data

The information you collect from a survey or experiment will be one of these types of data – **categorical** or **discrete** or **continuous**.

Categorical data is where the data falls into categories such as colours, pop groups or types of car.

Any set of data which is 'words' rather than 'numbers' will be categorical.
For example:

'The favourite football team of each student in a class'

is categorical data.

Discrete data is numerical (numbers).
However, the numbers can only be of certain values – usually whole numbers.
For example:

'The number of students in each class at school'

is discrete data because you can have 0, 1, 3, 4 … people in a class but not $4\frac{1}{2}$ or 7.3 people.

Continuous data is also numerical.
However this time the numbers can take any value within a certain range.
For example:

'The birth weight of students in a class'

is continuous data because this is numerical data where the values could be of any number (depending on how accurate the scales were!) between, say, 1 kg and 6 kg.

For example, 3.462 kg or 2.9 kg, and so on.

EXERCISE 19.2

State whether the following sets of data would be categorical, discrete or continuous.

1 The number of brothers and sisters that the students in your class have.

2 The time taken to walk to school.

3 The favourite pop group of students at your school.

4 The number of different part-time jobs that students have had.

5 The colours of the cars driving past the school gates in one hour.

6 The makes of the cars driving past the school gates in one hour.

7 The weight of chocolate bars.

8 The height of sunflowers.

9 The number of rooms in each house in a street.

10 The hourly rates for different Saturday jobs.

11 The number of words in the first sentence of each article in a newspaper.

12 The time to the first goal in a netball match.

13 The number of tries scored in a rugby match.

14 The results of a fair coin being thrown 200 times.

15 The results of a fair dice being thrown 200 times.

19.3 Tally charts and frequency tables

The results of a survey can be recorded on a **data collection sheet**.

One way of doing this is to make a **tally chart**, and record the responses as they are received.

You can then work out the totals to make a **frequency table**.

EXAMPLE

Natasha surveys her friends about how many pets they have.
She records her results in a frequency table.

Number of pets	Frequency
0	6
1	14
2	17
3	9
4	5
5	2
6	1

This means that 9 people had 3 pets.

a) How many people did Natasha survey?
b) How many pets do Natasha's friends have altogether?

SOLUTION

a) Find the total frequency:
$6 + 14 + 17 + 9 + 5 + 2 + 1 = 54$
Natasha surveyed 54 people.

This means that 9 people had 3 pets so altogether they had $3 \times 9 = 27$ pets.

b) It will help to add in an extra column on the frequency table.

Number of pets	Frequency	Number of pets × frequency
0	6	$0 \times 6 = 0$
1	14	$1 \times 14 = 14$
2	17	$2 \times 17 = 34$
3	9	$3 \times 9 = 27$
4	5	$4 \times 5 = 20$
5	2	$5 \times 2 = 10$
6	1	$6 \times 1 = 6$
Total	54	111

Add up the numbers in this column.

Altogether, Natasha's friends have 111 pets.

Sometimes it is easier to use a **grouped frequency table**, particularly when there is a lot of data.

EXAMPLE

Martin is doing a survey about homework.

He asks some students at his school to time how long they spent doing their homework one night.

Here are the times, in minutes, that students spent on their homework.

~~17~~	~~23~~	~~32~~	~~12~~	~~9~~	7
45	12	56	30	22	52
20	36	10	0	20	49
51	37	40	59	50	39
55	30	39	26	48	30

Note that these have been crossed off in the data list.

Martin uses this grouped frequency table to record his results.

The first five items of data have been entered in the table for you.

This means the number of minutes is more than or equal to 20 but less than 30…

… so 20, 21, 22, 23, 24, 25, 26, 27, 28 or 29 minutes gets a mark in the tally column.

Time, t (in minutes)	Tally	Frequency
$0 \leq t < 10$	|	
$10 \leq t < 20$	||	
$20 \leq t < 30$	|	
$30 \leq t < 40$	|	
$40 \leq t < 50$		
$50 \leq t < 60$		

a) Copy and complete the grouped frequency table.

b) How many people spent less than half an hour on their homework?

SOLUTION

a)

Time, t (in minutes)	Tally	Frequency
$0 \leq t < 10$	|||	3
$10 \leq t < 20$	||||	4
$20 \leq t < 30$	卌	5
$30 \leq t < 40$	卌 |||	8
$40 \leq t < 50$	||||	4
$50 \leq t < 60$	卌 |	6

Write 卌 and not |||||

b) Add up the frequencies of the first 3 groups: $3 + 4 + 5 = 12$

12 people spent less than half an hour on their homework.

EXERCISE 19.3

Don't use your calculator for Questions 1–4.

1 Kerry carries out a survey about her friends' favourite subjects. She makes a frequency table of her results.

Subject	Tally	Frequency
Maths	~~IIII~~ ~~IIII~~ I	11
Art	~~IIII~~ ~~IIII~~ III	13
P.E.	IIII II	
Drama	~~IIII~~ ~~IIII~~ ~~IIII~~	
Science	IIII	
English	~~IIII~~ ~~IIII~~	

a) Complete Kerry's frequency table.
b) How many people liked English best?
c) Which is the favourite subject of most of Kerry's friends?
d) How many friends did Kerry ask altogether?

2 Alan collects data about the colours of cars passing his house. Here are his results.

Red	Red	Blue	Blue	Red
Blue	Red	White	Blue	Black
Blue	Silver	Silver	Red	Silver
Black	Blue	Red	Silver	Blue
Red	White	Black	Silver	Silver
Blue	Red	Blue	Red	Blue

a) Complete the frequency table to show Alan's results.

Colour of car	Tally	Frequency
Black		
Blue		
Red		
Silver		
White		

b) Which colour car did Alan see most frequently?
c) How many white cars did Alan see?

3 Ivan has carried out a survey to find out the ages of people who use the local swimming pool.
Here are the ages, in years, of the swimmers at the pool.

2	5	30	22	47	55	73	17	26	13
3	8	12	41	56	68	71	84	3	7
9	15	13	15	27	34	47	51	49	14

a) Complete the frequency table to show Ivan's results.

Age	Tally	Frequency
Under 5		
5–10		
11–20		
21–30		
31–40		
41–50		
51–65		
Over 65		

b) How many under-5's were swimming that day?
c) How many people over the age of 50 were swimming?
d) How many swimmers did Ivan survey?

4 The heights of 30 sunflowers were measured.
The results, in centimetres, are shown below

142	196	156	134	187	215	212	103	146	189
132	147	158	173	120	103	204	117	180	198
187	165	143	190	140	136	180	200	126	167

Complete the frequency table to show these results.

Height, h (in cm)	Tally	Frequency
$100 \leqslant h < 120$		
$120 \leqslant h < 140$		
$140 \leqslant h < 160$		
$160 \leqslant h < 180$		
$180 \leqslant h < 200$		
$200 \leqslant h < 220$		

5 Sasha collected some data on the number of phone calls her friends made one evening. Here are the number of phone calls that Sasha's friends made that evening.

1	7	1	3	4	6	2	5	4	7	3	1	2	2	5
4	3	2	6	2	4	3	1	3	2	5	2	1	3	4

a) Copy and complete the frequency table.

b) How many of Sasha's friends made exactly 3 phone calls?

c) How many of Sasha's friends made less than 4 phone calls?

d) How many phone calls did Sasha's friends make in total?

Number of phone calls	Tally	Frequency
1		
2		
3		
4		
5		
6		
7		

19.4 Two-way tables

Data can be displayed using a two-way table.

EXAMPLE

The table shows the results of a survey about the amount of television, to the nearest hour, watched during a typical school week-night by boys and girls from one year group at a school.

	Under 1 hour	1 to 2 hours	3 to 4 hours	Over 4 hours	Total
Boys		10	30	15	60
Girls	10		10	5	
Total		35		20	110

a) Copy the table and fill in the missing values.

b) How many girls watched 3 or more hours of television?

Start with a row or column that has only one missing number – as here – then work around the table.

SOLUTION

a)

	Under 1 hour	1 to 2 hours	3 to 4 hours	Over 4 hours	Total
Boys	5	10	30	15	60
Girls	10		10	5	
Total		35		20	110

	Under 1 hour	1 to 2 hours	3 to 4 hours	Over 4 hours	Total
Boys	5	10	30	15	60
Girls	10	25	10	5	50
Total	15	35	40	20	110

b) $10 + 5 = \underline{15 \text{ girls}}$

EXAMPLE

Beth carries out a survey on the number of brothers and sisters that students in her year group have.

Number of sisters \ Number of brothers	0	1	2	3
0	3	6	5	2
1	7	5	2	4
2	4	3	1	2
3	3	2	1	0

The table shows the number of brothers and sisters that students in one class have.

a) How many students have 2 brothers and 1 sister?

b) How many students have more brothers than sisters?

c) How many students took part in Beth's survey?

SOLUTION

a)

Number of sisters \ Number of brothers	0	1	2	3
0	3	6	5	2
1	7	5	2	4
2	4	3	1	2
3	3	2	1	0

2 students

b)

Number of sisters \ Number of brothers	0	1	2	3
0	3	6	5	2
1	7	5	2	4
2	4	3	1	2
3	3	2	1	0

Add up the highlighted numbers.

$6 + 5 + 2 + 2 + 4 + 2 = 21$ students

c) Add up all the numbers in the table:

$3 + 6 + 5 + 2 + 7 + 5 + 2 + 4 + 4 + 3 + 1 + 2 + 3 + 2 + 1 + 0 = 50$

50 students

EXERCISE 19.4

1 Cassie and Ken have been carrying out a survey.
They asked a sample of householders whether they prefer to shop at the local corner shop or in the out-of-town supermarket.
Here are some of their results, transferred into a two-way table.

a) Complete the two-way table.
b) How many women preferred the supermarket?
c) How many men preferred the local shop?
d) How many people over 30 preferred the supermarket?

	Local shop	Super-market	Total
Men aged 30 or below	16		25
Men aged over 30		6	25
Women aged 30 or below	7	18	
Women aged over 30		15	25
Total	52		100

2 Jamie has collected data on the number of hours that students spent playing computer games in one day.
This two-way table shows some of his results.

	0 hours	1 hour	2 hours	3 hours	4 or more hours	Total
Boys	5	2		10		30
Girls	10		3	5	4	
Total		10	8		12	60

a) Complete the two-way table.
b) How many students spent 4 or more hours playing computer games?
c) How many girls spent 2 or more hours playing computer games?
d) How many boys spent less than 2 hours playing computer games?

3 Sanji has collected data on the way students and teachers travel to school.
This two-way table shows some of his results.

	Walk	Car	Cycle	Train	Bus	Total
Male students	10	2	8	3		
Female students		1	6		4	25
Male teachers		10		8		25
Female teachers	3		1	5	3	25
Total	23		18	22	11	100

a) Complete the two-way table.
b) How many female students walk or cycle to school?
c) How many male teachers use public transport to get to school?
d) How many students walk to school?
e) How many teachers cycle to school?

4 The two-way table shows the number of televisions and the number of DVD players in some households.

a) How many households have 2 televisions and 1 DVD player?

b) How many households have 3 televisions and 2 DVD players?

c) How many households have 3 televisions?

d) How many households have the same number of televisions as DVD players?

e) How many households took part in the survey?

Number of DVD players \ Number of televisions	**0**	**1**	**2**	**3**
0	2	3	5	6
1	0	11	8	12
2	0	0	6	4
3	0	0	0	3

5 The two-way table shows the number of cars and the number of bicycles in some households.

a) How many households have 3 cars and 1 bicycle?

b) How many households have exactly 1 bicycle?

c) How many households have the same number of cars as bicycles?

d) How many households have more cars than bicycles?

Number of bicycles \ Number of cars	**0**	**1**	**2**	**3**
0	0	7	8	3
1	2	8	6	4
2	1	10	15	2
3	0	6	7	1

6 Marco is carrying out a survey to see whether boys and girls like the same sports.
He asks each person in the sample to name their favourite sport.
Marco records the results on a data collection sheet.
Here are his results after asking the first 12 students.

Hockey	Soccer	Tennis	Other	Boys	Girls
𝍸	||	||	|||	𝍸 |||	||||

a) Explain one disadvantage with recording the data in this way.

b) Design an improved data collection sheet that Marco might use.

7 Design a data collection sheet for a survey about the things that teenagers spend their pocket money on.
Trial the sheet with some members of your class, and make any necessary improvements.

REVIEW EXERCISE 19

Don't use your calculator for Questions 1–4.

1 Bob carried out a survey of 100 people who buy tea.
He asked them about the tea they buy the most.
The two-way table gives some information about his results.

	Tea bags	Packet tea	Instant tea	Total
50 g	2	0	5	
100 g	35	20		60
200 g	15			
Total		25		100

Complete a copy of the two-way table. **[Edexcel]**

2 80 students each study one of three languages.
The two-way table shows some information about these students.

	French	German	Spanish	Total
Female	15			39
Male		17		41
Total	31	28		80

Complete a copy of the two-way table. **[Edexcel]**

3 Tony wants to collect information about the amount of homework the students in his class get.
Design a suitable question he could use. **[Edexcel]**

4 Mr Beeton is going to open a restaurant.
He wants to know what type of restaurant people like.
He designs a questionnaire.

a) Design a suitable question that he could use to find out what type of restaurant people like.

He asks his family 'Do you agree that pizza is better than pasta?'
This is **not** a good way to find out what people who might use his restaurant like to eat.

b) Write down **two** reasons why this is **not** a good way to find out what people who might use his restaurant like to eat. **[Edexcel]**

You can use your calculator for Questions 5–12.

5 Daniel carried out a survey of his friends' favourite flavour of crisps.
Here are his results.

Plain	Chicken	Bovril	Salt & Vinegar	Plain
Salt & Vinegar	Plain	Chicken	Plain	Bovril
Plain	Chicken	Bovril	Salt & Vinegar	Bovril
Bovril	Plain	Plain	Salt & Vinegar	Plain

a) Complete the table to show Daniel's results.

b) Write down the number of Daniel's friends whose favourite flavour was Salt & Vinegar.

c) Which was the favourite flavour of most of Daniel's friends?

Flavour of crisps	Tally	Frequency
Plain		
Chicken		
Bovril		
Salt & Vinegar		

[Edexcel]

6 The manager of a school canteen has made some changes.
She wants to find out what students think of these changes.
She uses this question in a questionnaire.

'What do you think of the changes in the canteen?'

☐ Excellent ☐ Very good ☐ Good

a) Write down what is wrong with this question.

This is another question on the questionnaire.

'How much money do you normally spend in the canteen?'

☐ A lot ☐ Not much

b) (i) Write down one thing that is wrong with this question.

(ii) Design a better question for the canteen manager to use.
You should include some response boxes. [Edexcel]

7 Grace and Gemma were carrying out a survey on the food that people eat in the school canteen. Grace suggested the question:

'Which foods do you eat?'

Gemma said that this question was too vague.
Write down two ways in which the question could be improved. [Edexcel]

8 Kim has to carry out a survey into the part-time jobs of all the 16-year-olds in her school. She has to find out:

- what proportion of these 16-year-olds have part-time jobs;
- whether more girls than boys have part-time jobs.

Design two questions which she could include in her questionnaire. [Edexcel]

9 Wayne is going to carry out a survey to record information about the type of vehicles passing the school gate.
Draw a suitable data collection sheet that Wayne could use. [Edexcel]

10 30 pupils were asked about their lunch one day.
The table gives some information about their answers.

	School dinners	Sandwiches	Other	Total
Boys	12	3		16
Girls	8		2	
Total				30

a) Complete the table.
b) How many girls had sandwiches? [Edexcel]

11 The table shows information about the number of certificates awarded to each student in a class last month.

Number of certificates	Number of students
0	14
1	4
2	6
3	3
4	2
5	1

a) How many students were in the class?
b) Work out the total number of certificates awarded to the class last month. [Edexcel]

12 Nazia is going to carry out a survey of the types of videos her friends have watched.
Draw a suitable data collection sheet that Nazia could use. [Edexcel]

KEY POINTS

1. Data may be collected first hand – this is primary data.
2. Alternatively, data may be obtained from a book, over the telephone or from an internet site, where someone else has already collected it for you – this is secondary data.
3. Take care to check the reliability of data from a secondary source.
4. Questionnaires are often used in statistical surveys.
5. A well-designed question may use tick boxes to cover a range of alternative responses, but should avoid suggesting that there is a preferred answer.
6. The word 'biased' is used to mean that something is unfair.
7. A biased sample is one that does not reflect the population fairly.
8. A biased question is one that steers replies towards a favoured response – such questions often begin with a phrase like 'Don't you agree that ...'
9. Categorical data is non-numerical data – for example, eye colour.
10. Discrete data is numerical data which can only be certain values – usually whole numbers – for example, the numbers of hours spent playing computer games.
11. Continuous data is numerical data which could be any value in a given range – for example, heights of students.
12. Data can be collected in a tally chart, a frequency table, a grouped frequency table or a two-way table.

Internet Challenge 19

The National Census 2001

At regular intervals, a national census is held in England and Wales.

Every household is required, by law, to reply to a short questionnaire.

The most recent census took place in 2001.

Use the internet to find the answers to these questions.

1 When was the first ever modern-style national census?

2 When will the next national census take place?

3 What is the maximum penalty for refusing to fill in a census questionnaire?

In the eleventh century, the king's men '… made a survey of all England; of the lands in each of the counties; … there was no single hide nor yard of land … left out.'

4 In what year was this survey completed?

5 By what name is it usually known?

6 Under which king was it commissioned?

The following questions all refer to the 2001 census for England and Wales.
Select the right answer from the alternatives given.

7 How many people in England and Wales are aged 90 or over?
a) 3360 **b)** 33 360 **c)** 336 000

8 What is the approximate total number of households in England and Wales?
a) 15 million **b)** 22 million **c)** 56 million

9 Approximately what proportion of the population of England and Wales reported following no religion?
a) 15% **b)** 25% **c)** 35%

10 About how many people lived in England and Wales in 2001?
a) 52 million **b)** 56 million **c)** 65 million

CHAPTER 20

Working with statistics

In this chapter you will **learn how to:**

- calculate the mean, median and mode
- calculate the range
- construct stem and leaf diagrams
- calculate means from frequency tables
- calculate estimated means from grouped frequency tables
- plot and interpret scatter graphs.

You will also be **challenged to:**

- investigate the populations of countries.

Starter: Sharing the cards

Five number cards are drawn from a pack of cards.

a) Arrange the number cards in order of size. Which number is in the middle?
b) Which number occurs most often?
c) Find the total of the numbers on the cards. Divide the total by 5. What is the answer?

A sixth number card is drawn from the pack. The answers to parts **a)** and **b)** are still the same. When the total of the six cards is divided by 6, the answer is 6

d) What is the number on the sixth card?

A seventh number card is drawn from the pack. The answers to parts **a)** and **b)** are still the same. When the total of the seven cards is divided by 7, the answer is 6

e) What is the number on the seventh card?

20.1 Mean, median and mode

My average GCSE result was a grade C.

On average, it takes me 20 minutes to get the bus to school.

The average employee at our company earns £25 000 a year.

An **average** of a set of data gives you a typical value for that set of data.

Averages can be useful when you want to compare two sets of data – for example, to compare the wages of male and female employees at a company.

There are three different averages that you need to be able to use.

1 Mode This is the most frequent item of data in a set.

2 Median Put all the items of data in order of size and find the middle value.

When there is an even number of data values there will be two values in the middle.

You need to find the value half-way between these two values by:

- adding them together
- dividing the answer by 2

3 Mean To calculate the mean of a set of data:

- add up all the numbers
- divide by how many numbers there are.

This is more formally written as:

$$\text{mean} = \frac{\text{sum of the data values}}{\text{number of data values}}$$

EXAMPLE

Simon records how long it takes him to get the bus to school every day for 10 days.
Here are his results:

15 minutes	47 minutes	18 minutes	16 minutes
19 minutes	13 minutes	17 minutes	13 minutes
22 minutes	20 minutes		

a) Find:
(i) the mean, **(ii)** the median, **(iii)** the mode of Simon's journey times.

b) Which average has Simon on page 492 used when he say it takes an average of 20 minutes to get to school?

SOLUTION

a) (i) $\text{mean} = \dfrac{\text{sum of the data values}}{\text{number of data values}}$

Add up all the numbers … and divide by how many numbers there are.

$$\text{mean} = \frac{15 + 47 + 18 + 16 + 19 + 13 + 17 + 13 + 22 + 20}{10}$$

$$\text{mean} = \frac{200}{10}$$

The mean is 20 minutes.

(ii) Write all the data values in order of size:

13 13 15 16 17 18 19 20 22 47

There is an even number of items of data (10) so there will be two values 'in the middle'.

Find the middle values by counting in from each end – it helps to cross out the items one by one from each end as you work your way in towards the middle:

~~13~~ ~~13~~ ~~15~~ ~~16~~ 17 18 ~~19~~ ~~20~~ ~~22~~ ~~47~~

So we need to find the value halfway between 17 and 18:

$$\frac{17 + 18}{2} = 17.5$$

This is the same as the mean of the two numbers.

The median is 17.5 minutes.

(iii) 13 is the only data value that appears more than once.
The mode is 13 minutes.

b) Simon has used the mean.

Look at the example on page 493.

Simon has used the mean to work out his average journey time. However, Simon took less than 20 minutes on most days to get to school.

On one day he took more than twice as long to get to school (47 minutes).
This value of 47 minutes makes the mean higher than it would otherwise have been.

So the mean does not give a good typical value of the data set. When most data values are close together and only one or two are much higher or lower than the rest we say the data is **skewed**.

When data is skewed the mean does not give a good typical value.

In the above example the mode (13 minutes) doesn't give a good typical value either.
13 minutes was Simon's shortest journey time and only happened twice!

The best average to take in this case is the median.
You need to decide which average would be most appropriate – especially for your handling data coursework.

Each average has advantages and disadvantages.

Average	Advantages	Disadvantages
Mean	Useful when the data values are close together. All the data values are used in the calculation.	May be unrepresentative of the data set when the data is skewed by extreme values.
Median	Useful when the data is **skewed**. It can be useful to compare with a middle value (e.g. 'half the class got 70% or more in their mock exam').	Time consuming to calculate when data set is very large. Can't find an estimated value from grouped data.
Mode	Can be used with categorical data ('red is the most popular colour of car'). Useful for opinion polls (i.e. 'most people think…').	Sometimes there isn't a modal value (e.g. all the data values appear once). There can be more than one mode.

EXERCISE 20.1

Don't use your calculator for Questions 1–3.

1 Find **(i)** the mean, **(ii)** the median, **(iii)** the mode of the following sets of data.

a) 2 2 2 3 4 7 8

b) 7 12 14 16 16

c) 7 12 12 13 13 13 14

2 Igor surveys his friends to find out about their favourite sport.
Here are his results.

football	tennis	hockey	football
netball	football	rugby	football
tennis	netball	football	rugby

Which sport is the mode?

3 Saskia surveys her friends about their favourite television programmes.
Here are her results.

Television programme	Frequency
Drama	9
Comedy	14
Crime	17
Soap	16
Documentary	12

Which type of television programme is the mode?

You can use a calculator for Questions 4–7.

4 Find **(i)** the mean, **(ii)** the median, **(iii)** the mode of the following sets of data.

a) 2 4 3 3 7 5 1 0 5 3 0

b) 41 45 52 45 43 45 45 57 50

c) 19 17 12 14 18 18 12 18 17 18 13

5 Find **(i)** the mean, **(ii)** the median, **(iii)** the mode of the following sets of data.

a) 112 110 110 112 111 110

b) 52 49 51 46 51 47 48 51 55 57

c) 17 20 22 24 20 18 28 32

6 A small company advertises for new employees with this advert.
Here are the wages for the employees and owner of the company.

£12 000	£20 000	£15 000
£12 000	£87 000	£18 000
£22 000	£25 000	£14 000

Great Opportunity!
Fantastic pay!
Our average worker earns
£25 000 a year!

a) Work out the mean wage.
b) Find the median wage.
c) Which average has the advertisement used?
Is this a fair average to take?
Give a reason for your answer.

7 Joanne has recorded the temperature at midnight for ten nights during January.
Here are her results.

$-4\,°C \quad -3\,°C \quad -1\,°C \quad 1\,°C \quad 0\,°C \quad 0\,°C \quad -3\,°C \quad 0\,°C \quad 1\,°C \quad -1\,°C$

Find the **a)** mean temperature, **b)** median temperature, **c)** modal temperature.

20.2 Range

Here are the number of goals scored by two football teams over their last ten games:

Ready Rovers	0	3	2	1	0	6	1	2	1	1
Uptown United	2	2	3	3	1	2	1	2	0	1

The mean for Ready Rovers is: $\frac{0+3+2+1+0+6+1+2+1+1}{10} = \frac{17}{10}$

mean $= 1.7$ goals

The mean for Uptown United is: $\frac{2+2+3+3+1+2+1+2+0+1}{10} = \frac{17}{10}$

mean $= 1.7$ goals

Both teams have the same mean.

To compare their results we can work out the **spread** of the data.
One measure of spread is the **range**:

range = largest value − smallest value

Ready Rovers have a range of $6 - 0 = 6$ goals.
Uptown United have a range of $3 - 0 = 3$ goals.

Uptown United have a smaller range.
They are more consistent in the number of goals they score.

EXERCISE 20.2

1 Find **(i)** the mean **(ii)** the median
(iii) the mode **(iv)** the range of the following sets of data.

a)	5	3	5	8	10	8	9	4	6	5
b)	15	13	15	18	20	18	19	14	16	15
c)	55	53	55	58	60	58	59	54	56	55
d)	105	103	105	108	110	108	109	104	106	105

e) What do you notice about your answers?

You can use a calculator for Questions 2 and 3.

2 The data set below gives the total number of goals scored by each of 15 soccer clubs during a season.

27 8 33 11 27 40 18 24 28 30 31 34 36 14 23

a) Work out the range of the data.
b) Write down the mode.
c) Calculate the value of the mean.
d) Find the value of the median.

3 Jay and Felicity want to compare their mock exam results.
Here are their results.

Subject	Maths	Science	English	French	Art	Drama	German	History
Jay	78%	62%	65%	48%	85%	75%	90%	50%
Felicity	66%	73%	67%	60%	70%	75%	70%	82%

a) Work out **(i)** the mean, **(ii)** the median, **(iii)** the range for Jay's results.
b) Work out **(i)** the mean, **(ii)** the median, **(iii)** the range for Felicity's results.
c) Who did better in their mock exams?
Give a reason for your answer.

20.3 Stem and leaf diagrams

When you have collected data it will probably be in a jumbled up order.
Data that has simply been recorded as you collected it is called **raw data**.

In order to see patterns and draw meaningful conclusions, you need to reorganise your data to obtain **processed data**.

The **stem and leaf** diagram is a way of organising a set of **discrete** raw data into an ordered grid.

Each number is broken into two parts.
For example, you might break '23' into a *stem* of 20 plus a *leaf* of 3

Numbers with the same stem are then collected together.

EXAMPLE

Here are the numbers of runs scored by a cricket batsman during his last 20 matches.

23	0	14	28	27	41	22	6	11	18
45	25	15	0	13	29	12	21	40	20

a) Draw a stem and leaf diagram to show this information.
b) Write down the mode.
c) Find the median.

SOLUTION

a) Begin by listing the stems: 0, 10, 20…
are represented with stems of: 0, 1, 2…
Continue all the way up to 4, as there is a '45' in the data set.
Remember to leave space for a 0's stem too.

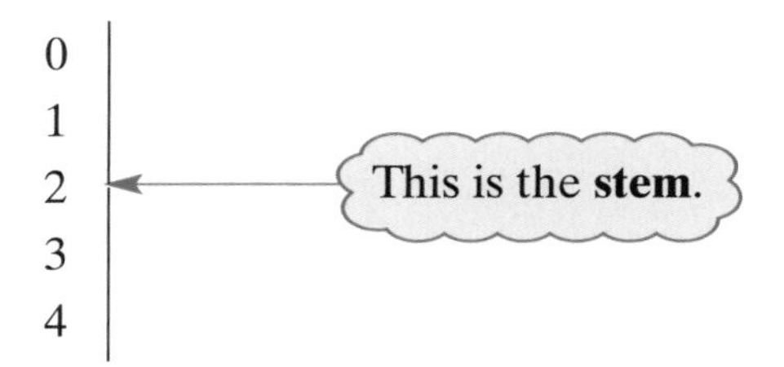

Next, start entering the data.
Think of '23' as 20 plus 3, so the stem is 2 and the leaf is 3
Always add a **key** to show how you have coded this.
This diagram shows the entries for 23 and 0

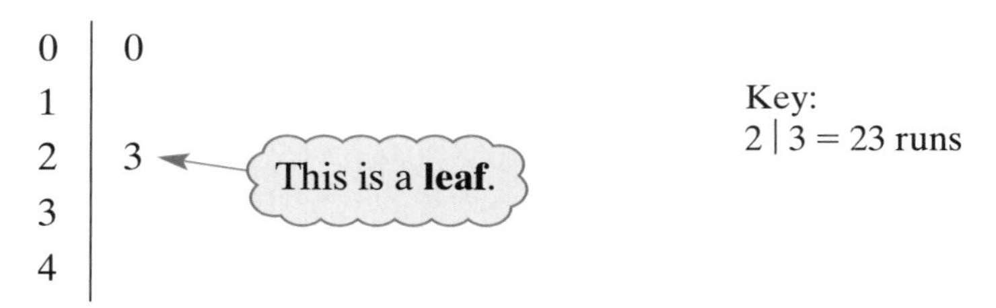

The data is sorted into layers, called **branches**, but the numbers are still jumbled along each branch – this is an **unsorted** stem and leaf diagram.

Stem	Leaves
0	0 6 0
1	4 1 8 5 3 2
2	3 8 7 2 5 9 1 0
3	
4	1 5 0

Key:
2 | 3 = 23 runs

This is a **branch**.

Finally, the branches need to be ordered to make a **sorted** stem and leaf diagram.

In an examination you should always draw a **fully sorted** diagram.

0	0 0 6
1	1 2 3 4 5 8
2	0 1 2 3 5 7 8 9
3	
4	0 1 5

Key:
2 | 3 = 23 runs

b) The mode is 0

c)

0	~~0~~ ~~0~~ ~~6~~
1	~~1~~ ~~2~~ ~~3~~ ~~4~~ ~~5~~ ~~8~~
2	0 1 ~~2~~ ~~3~~ ~~5~~ ~~7~~ ~~8~~ ~~9~~
3	
4	~~0~~ ~~1~~ ~~5~~

Key:
2 | 3 = 23 runs

By counting in from both ends (crossing out as you count inwards) the middle two values are 20 and 21

So the median is 20.5

So far you have found the median by 'counting in' from both ends of a list of data.

When you have a large amount of data it is easier to use a formula.

The median is the $\left(\frac{n+1}{2}\right)$th item of data where n is the number of items of data.

For example, when you have 99 items of data the median will be the $\left(\frac{99+1}{2}\right)$th (i.e. the 50th) item of data, *when the list is in order*.

In this example you have 20 items and the median is the $\left(\frac{20+1}{2}\right) = 10.5$ or the '10.5th' item.

This means that there are two numbers in the middle and that the median is halfway between them – in this case the 10th and 11th items.

Make sure all the 'leaves' are equally spaced.
You can see whether the data is skewed without losing any of the original raw data.

Notice that a 'stem' of 3 is included even though the 'branch' is empty.

The longest branch shows you the **modal class**.
In the last example the modal class is 20–29

This can be a better average to take than the mode.
The mode is not a very useful 'average' for this data set – it would be silly to suggest that the cricketer's average score is 0!
Basically all the scores are different, and it is just a fluke that two of them have coincided to give a misleading mode of 0

If the numbers were 283, 272, 290 and 271, then the stem could be 27, 28, 29, with a key of 28 | 3 being used to show this.

EXERCISE 20.3

1 This stem and leaf diagram shows the ages of the members of a local golf club.

2	1 4 9
3	0
4	1 7 8
5	0 5 6 7 9 2
6	2

Four new members join the club.
Their ages are 35, 37, 43 and 81

a) Draw a new stem and leaf diagram to include the four new members.
b) Add a key to the diagram.
c) Find the median.
d) Find the range.

2 Penny has made an unsorted stem and leaf diagram to show the times taken by members of her class to travel to school each morning.
The times are in minutes.

0	5 1 6
1	4 8 3 1 0
2	4 5 2 1
3	6 9 8

a) Redraw the diagram so that it is fully sorted.
b) Add a key to the new diagram.
c) Find the median.
d) Find the range.

3 Fifteen students have received their results from a French vocabulary test. The scores are out of 50.

36	41	29	32	48	19	36	30
50	35	25	44	31	47	27	

a) Draw a stem and leaf diagram to show this information. Remember to include a key.
b) Work out the range of the scores.
c) Find the median score.

4 Here are the temperatures of a sample of 17 British cities one Saturday. The temperatures are in degrees Fahrenheit.

72	79	68	85	61	88	74	81	73
92	84	49	63	91	86	76	88	

a) Draw a stem and leaf diagram to show this information.
b) Work out the range.
c) Find the median.

5 A box contains electronic components.
They are each supposed to have a resistance of 100 ohms.
Sasha measures each one with a meter.
Here are Sasha's measurements.

97	111	93	105	100	102
100	88	113	96	106	94

a) Draw a stem and leaf diagram to show this information.
b) Find the median of the measurements.

20.4 Calculations with frequency tables

A frequency table is simply a list of possible values with corresponding frequencies that tell you how many times each value occurred.

You can find averages and the range from a frequency table.
Make sure you understand this method.
It is examined regularly in GCSE papers and is a handy skill to apply in your handling data coursework task.

EXAMPLE

Boxes of matches are supposed to contain 50 matches, on average.
Sophie decides to check this figure. She takes a sample of 20 boxes, and counts their contents.

Number of matches	Frequency
48	1
49	5
50	7
51	0
52	5
53	1
54	1

So there are 7 boxes with 50 matches

a) Calculate the value of the mean.
b) Write down the mode.
c) Work out the range of the data.
d) Find the value of the median.

SOLUTION

a) Step 1 Make a new column and call it $x \times f$

So these 7 boxes contain 350 matches altogether.

Step 2 Multiply x and f together:

$48 \times 1 = 48$
$49 \times 5 = 245$
etc.

Number of matches		Frequency, f		$x \times f$
48	$\times$	1	=	48
49	$\times$	5	=	245
50	$\times$	7	=	350
…		…		

Step 3 Add up the numbers in each of the last two columns.

Number of matches	Frequency, f	$x \times f$
48	1	48
49	5	245
50	7	350
51	0	0
52	5	260
53	1	53
54	1	54
	20	1010

There are 20 matchboxes containing 1010 matches altogether.

$$\text{Mean} = \frac{1010}{20}$$

The mean is 50.5 matches.

b) For the mode:

Number of matches	Frequency
48	1
49	5
50	7
51	0
52	5
53	1
54	1

The highest frequency is 7, so the mode is 50

The mode is 50 matches.

c) The lowest value is 48 and the highest is 54
Therefore the range is $54 - 48 = 6$

d) The middle number in a data set is the **median**.
There are 20 numbers in the data set.

$\frac{20 + 1}{2} = 10\frac{1}{2}$ so the median is between the 10th and 11th values.

Number of matches	Frequency
48	1
49	5
50	7
51	0
52	5
53	1
54	1

Think about the list of data: the 1st number is 48, the next 5 numbers are all 49

You have got $1 + 5 = 6$ numbers – not half way yet.

The next 7 numbers are all 50
$1 + 5 + 7 = 13$, which is more than half way through the list of items, so the median must be 50

The median is 50

EXERCISE 20.4

1 Kate is an estate agent.
She gathers some data on the number of bedrooms each house that she is selling has.

Number of bedrooms	Frequency
1	10
2	22
3	43
4	13
5	9
6	3
Total	

a) How many houses did Kate survey?
b) What is the modal number of bedrooms?
c) Find the median number of bedrooms.
d) Work out the mean number of bedrooms.
e) What is the range of Kate's data?

You can use your calculator for Questions 2 and 3.

2 Emily surveys cars driving past her school into town in the morning.
She counts the number of occupants of each car.
The frequency table shows her results.

Number of occcupants	Frequency
1	7
2	14
3	10
4	9
5 or more	0
Total	

a) How many cars did Emily survey?

The average number
of people in each car
is 5

b) Explain why Emily must be wrong.
c) Find the modal number of occupants per car.
d) Find the range of the data.
e) Work out the mean number of occupants per car.
f) Find the median number of occupants per car.

3 John is surveying the wildlife on a local river.
He records the number of eggs in some birds' nests.

Number of eggs	Frequency
1	4
2	7
3	10
4	16
5	25
6	17
7	10
8	5
9	6
Total	

a) How many nests did John survey?
b) What is the modal number of eggs in a nest?
c) Find the median number of eggs in a nest.
d) Work out the mean number of eggs in a nest.
e) What is the range of John's data?

20.5 Calculations with grouped frequency tables

The methods used in the previous section can be adapted to deal with grouped discrete data.

Your calculations might only provide estimates of the average or spread, since the raw values have been lost in the grouping.

EXAMPLE

The marks for Year 10's end-of-term mathematics test are shown in the table.

Mark, m	Frequency, f
0 to 4	0
5 to 9	1
10 to 15	5
16 to 19	23
20 to 24	24
25 to 29	18
30 to 34	15
35 to 39	14
40 or more	0

This is grouped discrete data.

a) Calculate an estimate of the value of the mean.
b) Write down the modal class.
c) Work out an estimate for the range of the data.
d) In which class does the median lie?

SOLUTION

a)

Mark, m	Frequency, f	Midpoint, x	$x \times f$
0 to 4	0 ×	2 =	0
5 to 9	1 ×	7 =	7
10 to 14	5 ×	12 =	60
…	…		

The 5 values between 10 and 14 are treated as if they are all equal to the midpoint, 12

The midpoint is 12 as $\frac{10 + 14}{2} = 12$

Mark, m	Frequency, f	Midpoint, x	$x \times f$
0 to 4	0	2	0
5 to 9	1	7	7
10 to 14	5	12	60
15 to 19	23	17	391
20 to 24	24	22	528
25 to 29	18	27	486
30 to 34	15	32	480
35 to 39	14	37	518
40 or more	0	–	0
	100		2470

The mean is only an estimate because midpoints have been used in the calculations – the actual values are not available.

Estimated mean $= \frac{2470}{100}$

Add up the totals in the f and the $x \times f$ columns.

Note that even though it says 'estimated' we do *not* round the answer.

The estimated mean is 24.7

b) The modal class is found by looking for the highest frequency:

Mark, m	Frequency, f
0 to 4	0
5 to 9	1
10 to 15	5
16 to 19	23
20 to 24	24
25 to 29	18
30 to 34	15
35 to 39	14
40 or more	0

The data is grouped so you can't work out a single value as the mode – so the modal group is used instead.

24 is the highest frequency, so the modal class is 20 to 24.

The modal class is 20 to 24

c) Estimated range = 37 − 2 = 35

So between the 50th and 51st item of data.

d) The median is the $\left(\frac{100 + 1}{2}\right)$th item of data.

By counting through the frequencies in the table, we can see that both the 50th and 51st numbers lie in the 20 to 24 group.

So the median lies in the 20 to 24 group.

Essentially the same method is used when dealing with grouped continuous data.

However, the notation used for writing the class intervals is different.

EXAMPLE

The resistances of a sample of 50 electronic components are measured.

The table shows the results.

Resistance, R	Frequency, f
$80 \leq R < 90$	10
$90 \leq R < 100$	23
$100 \leq R < 110$	11
$110 \leq R < 130$	6

a) Calculate an estimate of the mean resistance.
b) State the class interval that contains the median resistance.
c) Explain whether it is possible for the range to be 48

SOLUTION

a)

Resistance, R	Frequency, f	Midpoint, x	$x \times f$
$80 \leq R < 90$	10	85	850
$90 \leq R < 100$	23	95	2185
$100 \leq R < 110$	11	105	1155
$110 \leq R < 130$	6	120	720
	50		4910

$$\text{Estimated mean} = \frac{4910}{50}$$

$$= 98.2$$

The estimated mean is 98.2

b) The median will be in the 25th/26th position, and from the table the median lies in $90 \leq R < 100$

c) It is possible for the range to be 48 (e.g. $128 - 80$)

EXERCISE 20.5

1 Meera has measured the lengths of a sample of cucumbers from her stall. The frequency table shows her results.

Length of cucumber (L) in cm	Frequency (f)
$20 \leqslant L < 22$	2
$22 \leqslant L < 24$	5
$24 \leqslant L < 26$	8
$26 \leqslant L < 28$	4
$28 \leqslant L < 30$	1
Total	

a) State the modal class.

b) Find the class interval in which the median lies.

c) Estimate the mean length of a cucumber in this sample. Show all your working clearly.

You can use your calculator for Questions 2–4.

2 Maurizio records the number of people at work in his department each day. The frequency table shows his results.

Number of people at work	Frequency	Midpoint
25 to 29	3	
30 to 34	7	
35 to 39	11	
40 to 44	4	
Total		

a) Copy the table, and fill in the midpoint values in the third column.

b) Use the table to help you calculate an estimate of the mean number of people at work each day.

c) State the modal class.

d) Find the class interval which contains the median.

e) Benoit says, 'The range is 20.'
Explain why Benoit must be wrong.

3 A registrar records the ages of men who married in his office one week. The frequency table shows her results.

Age (A) in years	Frequency
$20 \leqslant A < 30$	4
$30 \leqslant A < 40$	7
$40 \leqslant A < 50$	6
$50 \leqslant A < 60$	3
$60 \leqslant A < 70$	1
Total	

a) State the modal class.

b) Find the class interval which contains the median.

c) Work out an estimate of the mean age of the men who married that week.
Give your answer correct to 3 significant figures.

4 Sean sets his friends a puzzle, which they solve under timed conditions.
The frequency table shows the times taken.
a) State the modal class.
b) Find the class interval which contains the median.
c) Calculate an estimate of the mean time taken to solve the puzzle.
d) Explain briefly why your answer can only be an estimate.

Time (T) in minutes	Frequency
$5 \leqslant T < 10$	2
$10 \leqslant T < 15$	5
$15 \leqslant T < 20$	9
$20 \leqslant T < 25$	7
$25 \leqslant T < 30$	4
Total	27

20.6 Scatter diagrams

Scatter graphs are used for data that contains pairs of values, such as:

- heights and weights of children at a school
- daily rainfall and hours of sunshine
- GCSE students' marks in Maths and length of forearm.

To make a scatter graph, you plot one value along the x axis and the other along the y axis.

The pattern that results gives you some idea of whether there is any **correlation** (link) between the data.

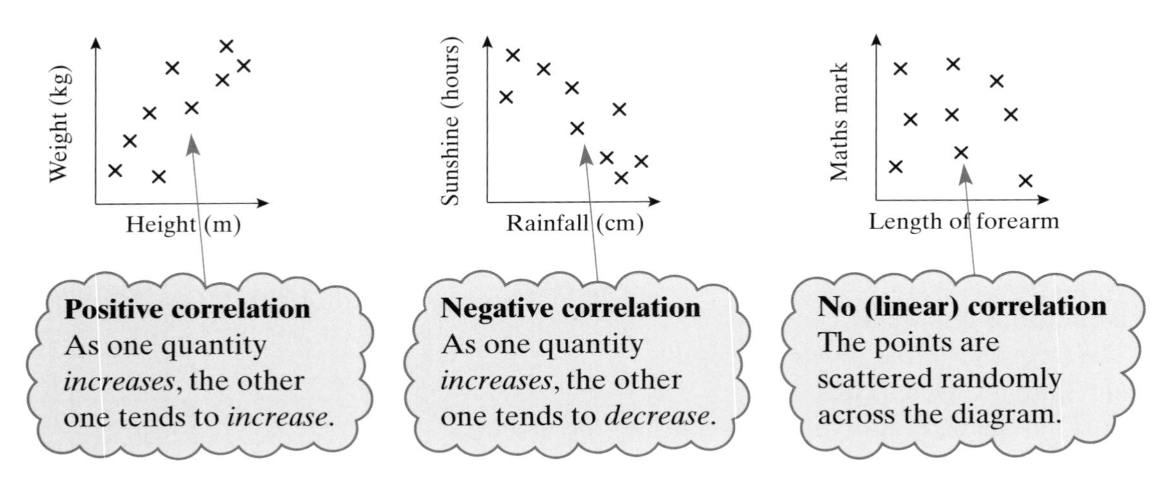

Weight (kg)
Height (m)
Positive correlation
As one quantity increases, the other one tends to increase.
Sunshine (hours)
Rainfall (cm)
Negative correlation
As one quantity increases, the other one tends to decrease.
Maths mark
Length of forearm
No (linear) correlation
The points are scattered randomly across the diagram.

When there is correlation, you can show the overall trend by drawing a **line of best fit**.

This should not be forced to go through all the points.

It should have roughly the same number of points on either side of the line.

EXAMPLE

The table shows the average temperatures in winter and summer for ten cities across Europe.

Winter temperature, °C	2	3	6	7	8	8	12	13	13	14
Summer temperature, °C	15	17	20	17	25	21	23	26	25	24

a) Plot the data on a scatter graph.
b) Describe the relationship between winter and summer temperatures.
c) Add a line of best fit to your graph.

Another city has an average winter temperature of 10 °C
d) Use your line of best fit to estimate its average summer temperature.

SOLUTION

a)

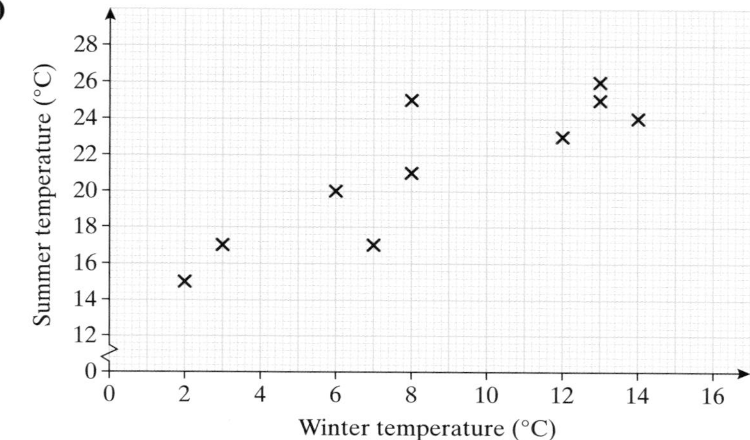

b) The winter and summer temperatures show positive correlation.

c)

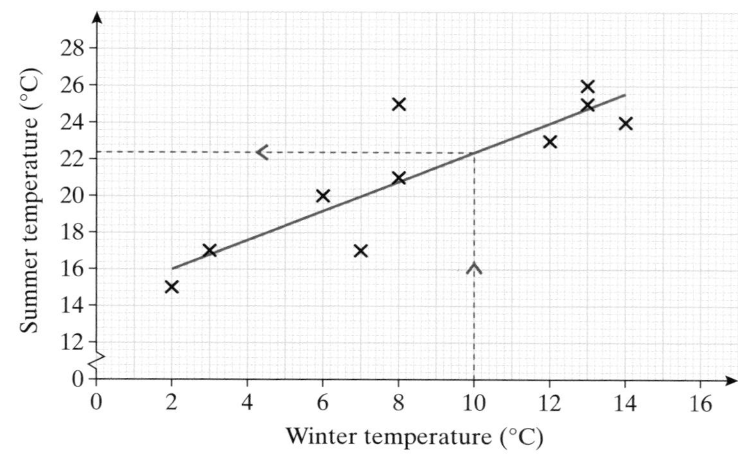

d) As shown on the graph above, the summer temperature is estimated at 22.4 °C

EXERCISE 20.6

1 Match together these graphs and their descriptions:

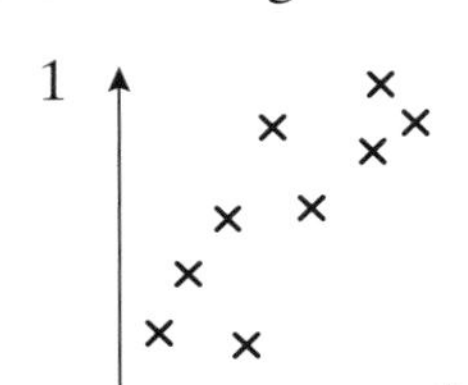

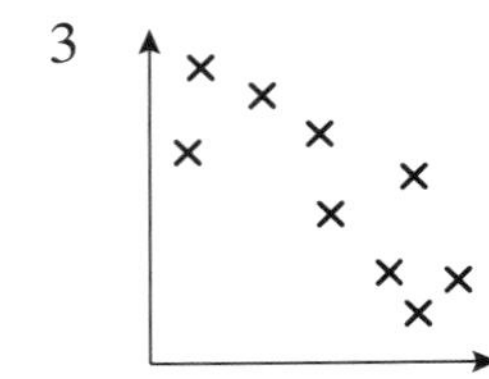

3

A IQ against weight
B Value of car against age of car
C Height of oak tree against age of oak tree

2 The scatter graph shows some information about the marks of six students in two examinations – Paper 1 and Paper 2.

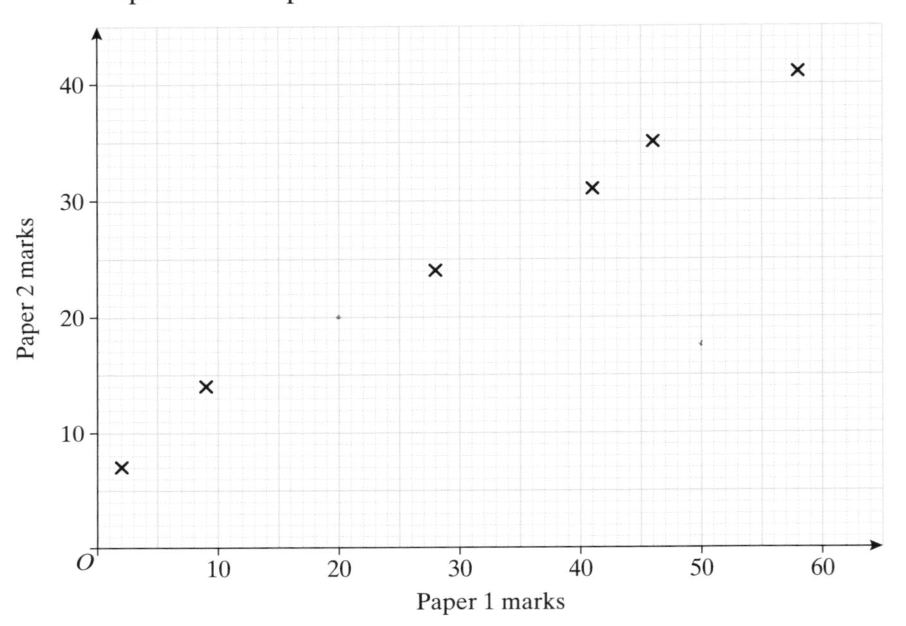

The table shows the marks for Paper 1 and Paper 2 for two more students, A and B.

	Student A	Student B
Paper 1 mark	20	50
Paper 2 mark	20	35

a) On a copy of the scatter graph, plot the information from the table.
b) Describe the correlation between the marks on Paper 1 and Paper 2.
c) Draw a line of best fit on the diagram.

Another student has a Paper 2 mark of 30
d) Use your line of best fit to estimate the Paper 1 mark for this student. [Edexcel]

3 The table shows the number of units of electricity used in heating a house on ten different days and the average temperature for each day.

Average temperature (°C)	6	2	0	6	3	5	10	8	9	12
Units of electricity used	28	38	41	34	31	31	22	25	23	22

a) Complete the scatter graph to show the information in the table.
The first six points have been plotted for you.

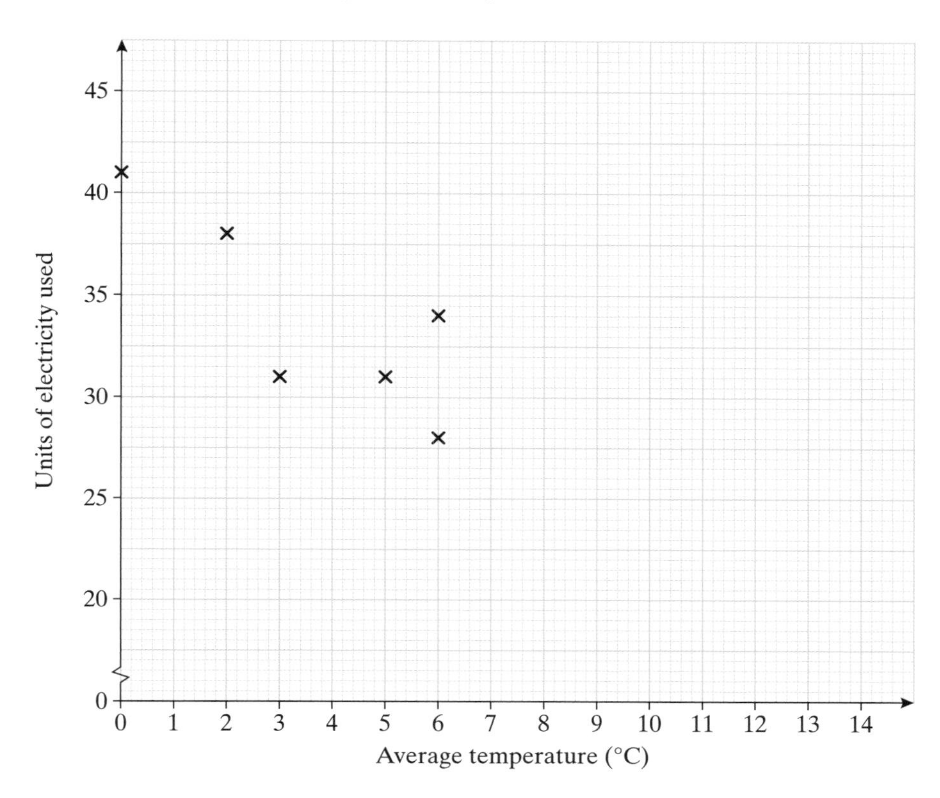

b) Describe the correlation between the number of units of electricity used and the average temperature.

c) Draw a line of best fit on your scatter graph.

d) Use your line of best fit to estimate:

(i) the average temperature if 35 units of electricity are used

(ii) the units of electricity used if the average temperature is 7 °C **[Edexcel]**

4 The scatter graph shows information about fourteen countries.
For each country, it shows the birth rate and the life expectancy, in years.

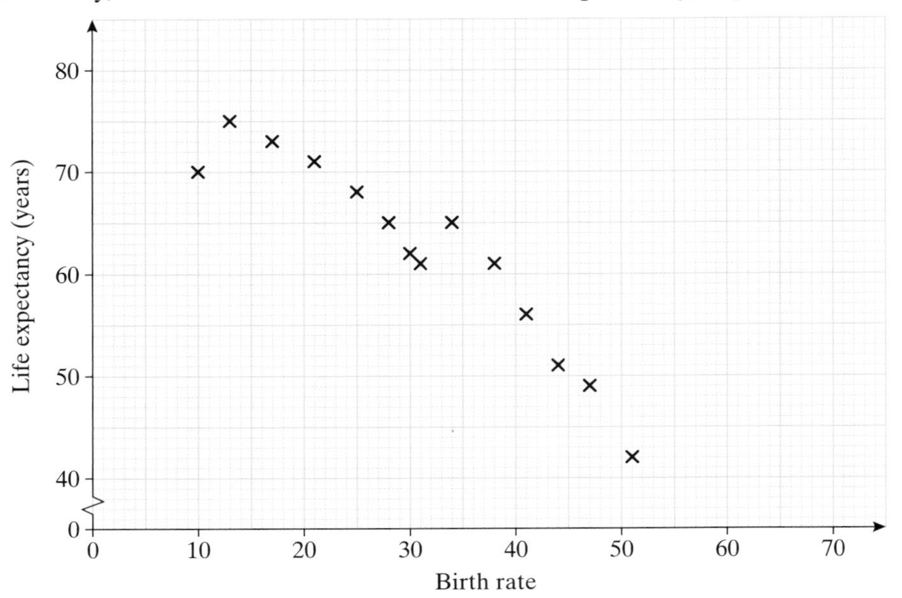

a) Draw a line of best fit on the scatter graph.

The birth rate in another country is 42

b) Use your line of best fit to estimate the life expectancy in that country.

The life expectancy in another country is 66 years.

c) Use your line of best fit to estimate the birth rate in that country. **[Edexcel]**

5 a) Here is a scatter graph.
One axis is labelled 'weight'.

(i) For this graph state the type of correlation.

(ii) From this list choose an appropriate label for the other axis.

shoe size   length of hair
height   hat size   length of arm

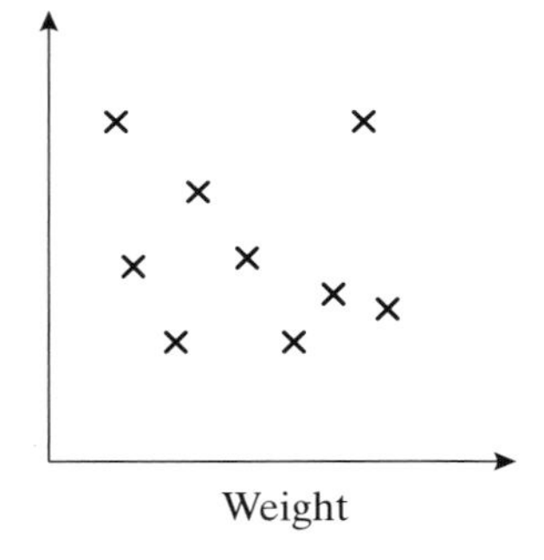

b) Here is another scatter graph with one axis labelled 'weight'.

(i) For this graph state the type of correlation.

(ii) From this list choose an appropriate label for the other axis.

shoe size   distance around neck
waist measurement   GCSE Maths mark

[Edexcel]

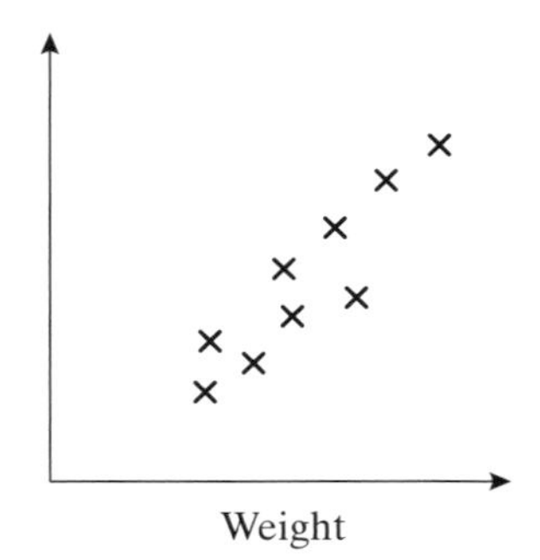

6 On seven days, Helen recorded the time, in minutes, it took a 2 cm ice cube to melt.
She also recorded the temperature, in °C, on that day.
Some of her results are shown in the scatter diagram below.

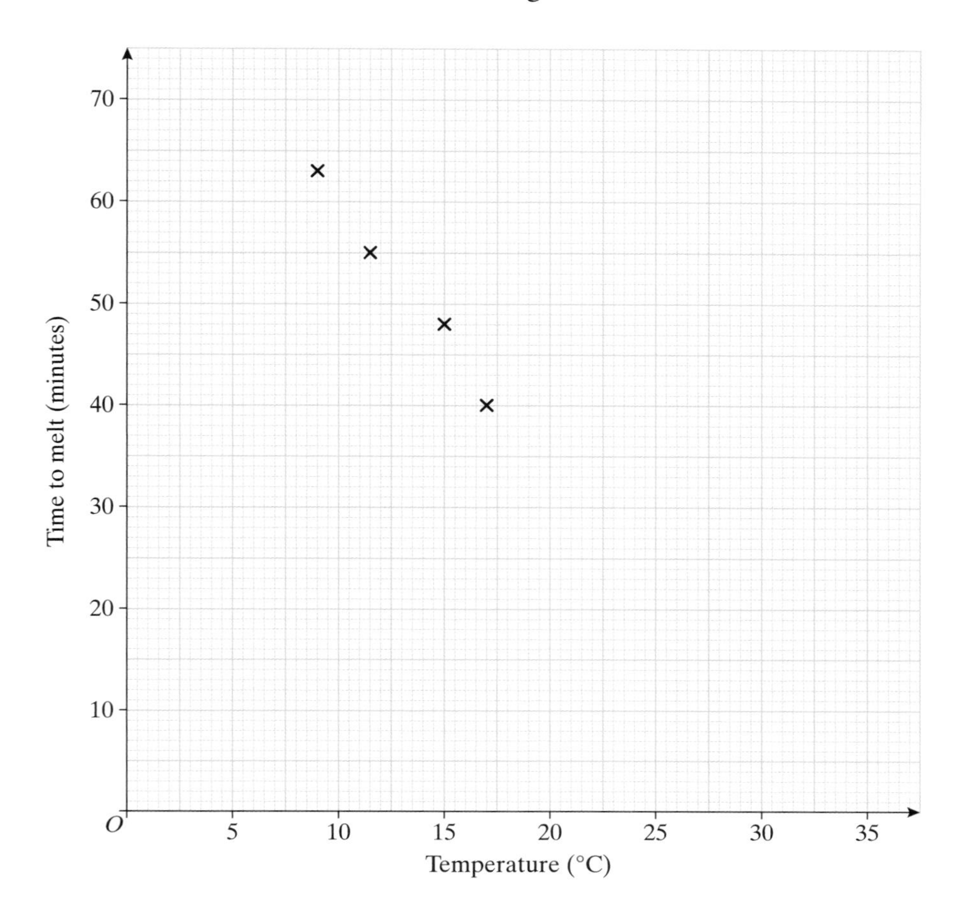

All of her results are shown in the table below.

Temperature (°C)	9	11.5	15	17	20	21	26
Time (minutes)	63	55	48	40	30	25	12.5

a) Complete the scatter diagram.

b) Describe the relationship between the temperature and the time it takes a 2 cm ice cube to melt.

c) Draw a line of best fit on the scatter diagram.

d) Use your line of best fit to estimate the time it took for a 2 cm ice cube to melt when the temperature was 13 °C

e) Use your line of best fit to estimate the temperature when a 2 cm ice cube took 19 minutes to melt.

f) Explain why the line of best fit could not be used to estimate the time it took a 2 cm ice cube to melt when the temperature was 35 °C **[Edexcel]**

7 The table shows the number of pages and the weight, in grams, for each of 10 books.

Number of pages	80	130	100	140	115	90	160	140	105	150
Weight	160	270	180	290	230	180	320	270	210	300

a) Complete the scatter graph to show the information in the table.
The first six points in the table have been plotted for you.

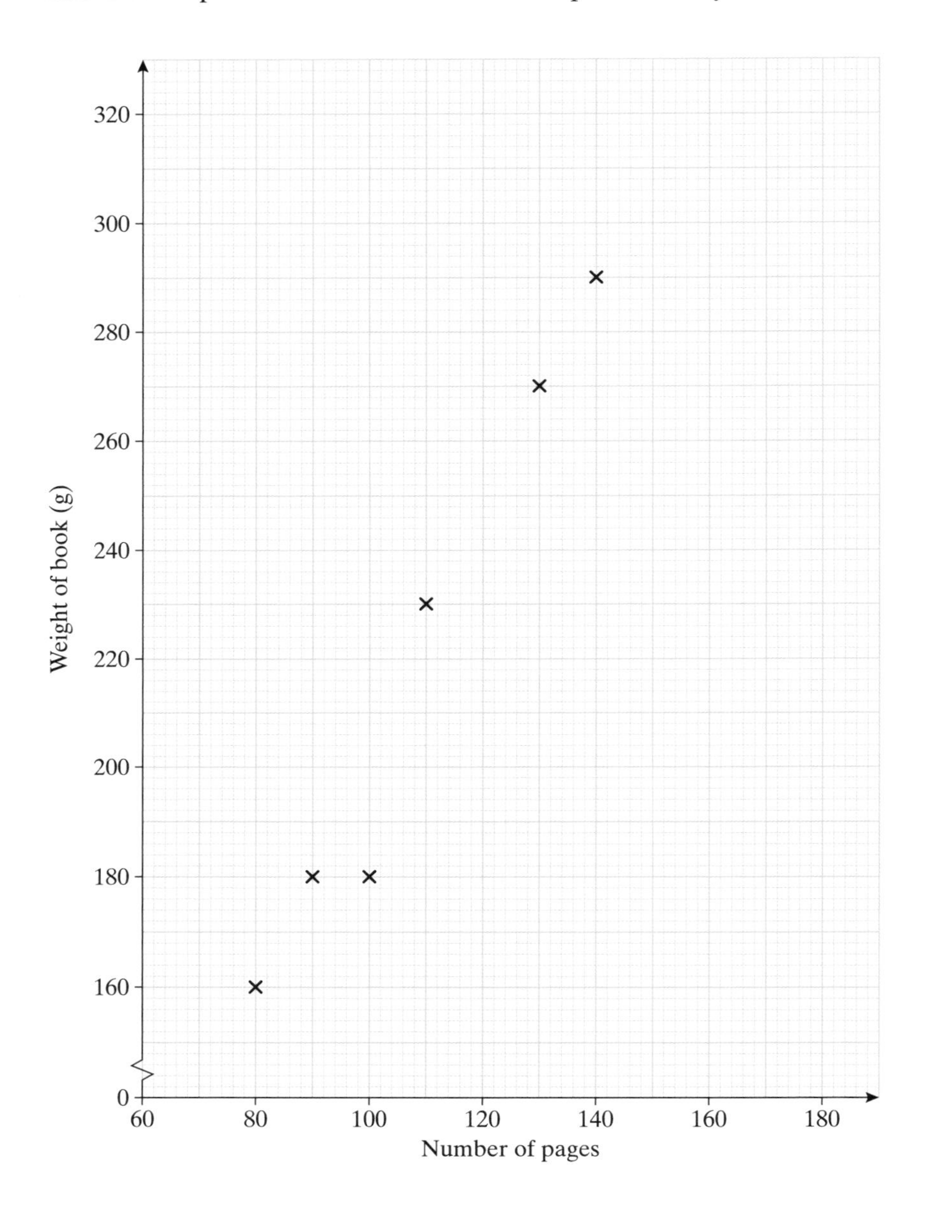

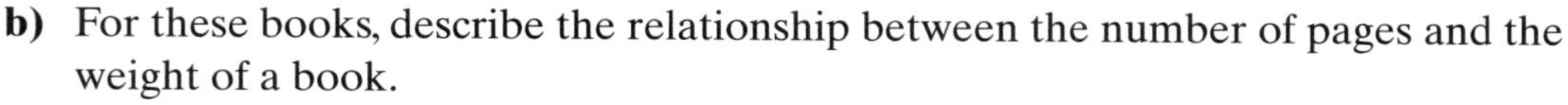

b) For these books, describe the relationship between the number of pages and the weight of a book.
c) Draw a line of best fit on the scatter diagram.
d) Use your line of best fit to estimate:
  (i) the number of pages in a book of weight 280 g
  (ii) the weight of a book with 120 pages. [Edexcel]

REVIEW EXERCISE 20

Don't use your calculator for Questions 1–5.

1 Joanna made a list of the ages of the children in a playgroup.

4 3 1 4 2 4 4 2 1 2

a) Find the median age of the children in the playgroup.
b) Find the range of the ages of the children in the playgroup. [Edexcel]

2 Shirin recorded the number of students late for school each day for 21 days.
The stem and leaf diagram shows this information.

Number of students late

Stem	Leaf
0	4 5 7 8 8 9
1	2 2 5 6 6 7 7 9 9 9
2	0 1 3 4 6

Key:
1 | 4 means 14 students late

a) Find the median number of students late for school.
b) Work out the range of the number of students late for school. [Edexcel]

3 Rosie has 10 boxes of drawing pins.
She counted the number of drawing pins in each box.
The table gives information about her results.

Number of drawing pins	Frequency
29	2
30	5
31	2
32	1

a) Write down the modal number of drawing pins in a box.
b) Work out the range of the number of drawing pins in a box.
c) Work out the mean number of drawing pins in a box. [Edexcel]

4 The scatter graph shows some information about six new-born baby apes.
For each baby ape, it shows the mother's leg length and the baby ape's birth weight.

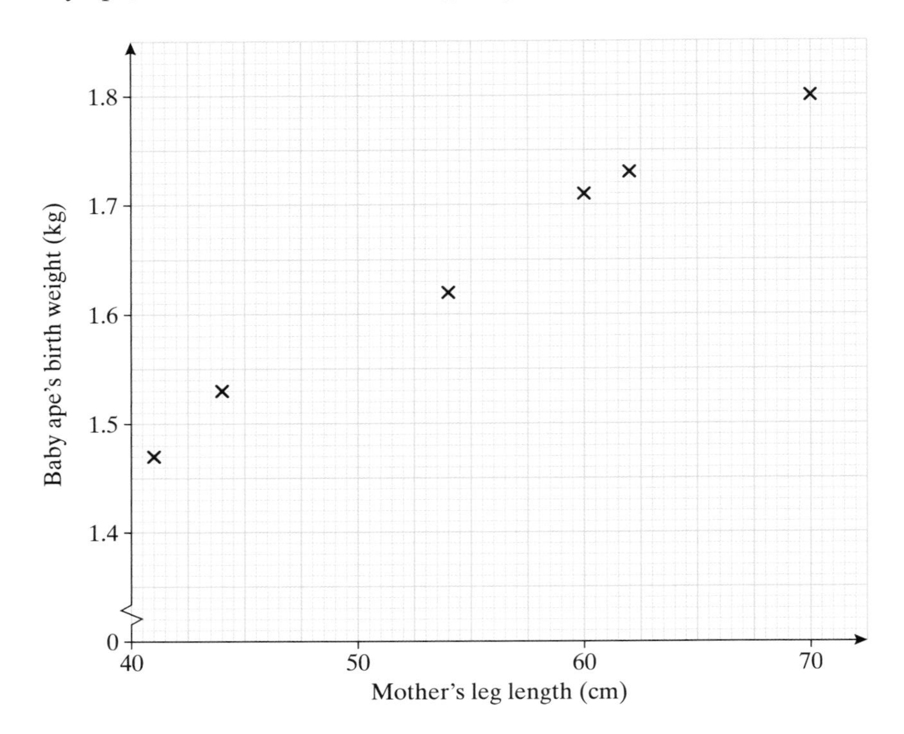

The table shows the mother's leg length and the birth weight of two more baby apes.

Mother's leg length (cm)	50	65
Baby ape's birth weight (kg)	1.6	1.75

a) On the scatter graph, plot the information from the table.

b) Describe the **correlation** between a mother's leg length and her baby ape's birth weight.

c) Draw a line of best fit on the diagram.

A mother's leg length is 55 cm

d) Use your line of best fit to estimate the birth weight of her baby ape. **[Edexcel]**

5 20 students scored goals for the school hockey team last month.
The table gives information about the number of goals they scored.

Goals scored	Number of students
1	9
2	3
3	5
4	3

a) Write down the modal number of goals scored.
b) Work out the range of the number of goals scored.
c) Work out the mean number of goals scored. **[Edexcel]**

You can use your calculator for Questions 6–13.

6 A rugby team plays 10 games.
Here are the number of points they scored.

13 23 15 12 8 19 23 15 37 15

a) Write down the mode.
b) Work out the range.
c) Work out the mean. **[Edexcel]**

7 Andy did a survey of the number of cups of coffee some pupils in his school had drunk yesterday.
The frequency table shows his results.

Number of cups of coffee	Frequency
2	1
3	3
4	5
5	8
6	5

a) Work out the number of pupils that Andy asked.

Andy thinks that the average number of drinks pupils in his survey had drunk is 7
b) Explain why Andy cannot be correct. **[Edexcel]**

8 The scatter graph shows information about the number of donkey rides on Blackpool beach, and the number of hours of sunshine on each day.

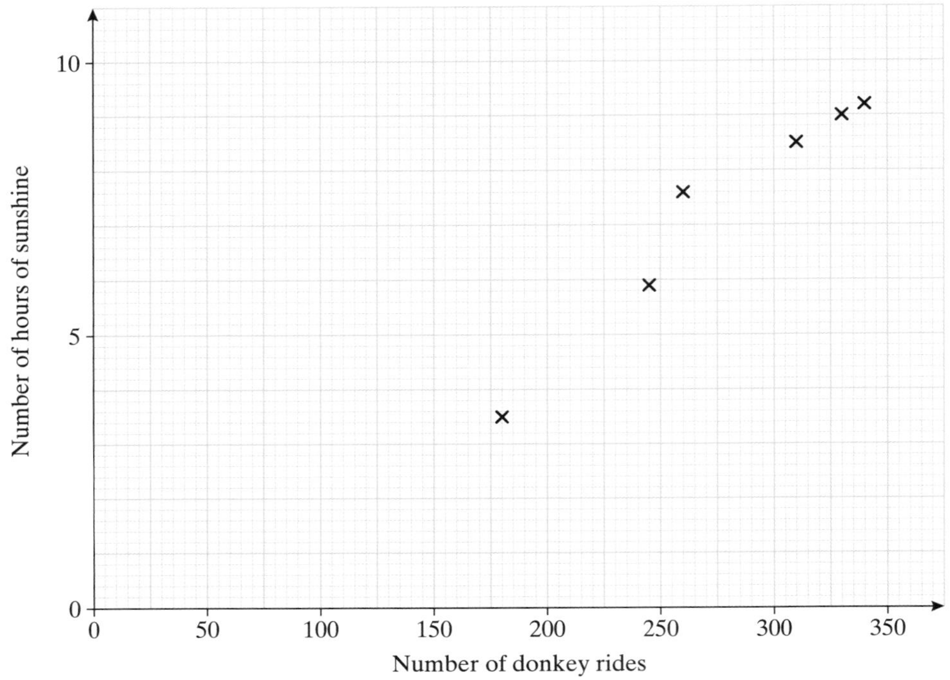

The table shows the number of donkey rides and the number of hours of sunshine on four other days.

Number of donkey rides	130	185	210	220
Number of hours sunshine	3	4	5	5.5

a) On the scatter graph, plot the information from the table.

b) Describe the relationship between the number of donkey rides and the number of hours of sunshine.

c) Draw a line of best fit on the scatter graph.

d) Use your line of best fit to estimate:
(i) the number of donkey rides when there are 7 hours of sunshine
(ii) the number of hours of sunshine on a day when there were 200 donkey rides.

[Edexcel]

9 Jan measures the heights, in millimetres, of 20 plants in her greenhouse.
Here are her results.

178	189	147	147	166
167	153	171	164	158
189	166	165	155	152
147	158	148	151	172

Draw a stem and leaf diagram to show this information. **[Edexcel]**

10 Here are times, in minutes, taken to change some tyres.

5	10	15	12	8
20	33	15	25	10
7	20	35	24	15
8	10	20	16	10

Draw a stem and leaf diagram to show these times. **[Edexcel]**

11 The table shows information about the number of hours that 120 children used a computer last week.

Number of hours (h)	Frequency
$0 < h \leqslant 2$	10
$2 < h \leqslant 4$	15
$4 < h \leqslant 6$	30
$6 < h \leqslant 8$	35
$8 < h \leqslant 10$	25
$10 < h \leqslant 12$	5

Work out an estimate for the mean number of hours that the children used a computer.
Give your answer correct to two decimal places. **[Edexcel]**

12 A garage keeps records of the costs of repairs to its customers' cars.
The table gives information about the costs of repairs which were less than £250 in one week.

Cost (£C)	Frequency
$0 < C \leqslant 50$	4
$50 < C \leqslant 100$	8
$100 < C \leqslant 150$	7
$150 < C \leqslant 200$	10
$200 < C \leqslant 250$	11

a) Find the class interval in which the median lies.

There was only one further repair that week, not included in the table.
That repair cost £1000
Dave says 'The class interval in which the median lies will change'.

b) Is Dave correct?
Explain your answer.

[Edexcel]

13 35 students with Saturday jobs took part in a survey.
They were asked the hourly rate of pay for their jobs.
This information is shown in the grouped frequency table below.
Work out an estimate for the mean hourly rate of pay.
Give your answer to the nearest penny.

Hourly rate of pay (£x)	Frequency
$3.00 < x \leqslant 3.50$	1
$3.50 < x \leqslant 4.00$	2
$4.00 < x \leqslant 4.50$	4
$4.50 < x \leqslant 5.00$	7
$5.00 < x \leqslant 5.50$	19
$5.50 < x \leqslant 6.00$	2

[Edexcel]

KEY POINTS

1 There are three averages you need to know:
   - mode – the most frequent item of data
   - median – put all the items of data in order of size and find the middle value
   - mean – add up all the numbers and divide by how many numbers there are.

2 The range is the difference between the largest data value and the smallest.

3 In order to see patterns in discrete data sets it helps to organise the raw data in some way.
   A good method is the stem and leaf diagram – you should:
   - always draw a fully sorted diagram
   - remember to add a key
   - make sure that the vertical columns of figures line up
   - not put commas in a stem and leaf diagram.

4 Another good way to organise data is by using a frequency table.
   This works for discrete data, grouped discrete data and grouped continuous data.

5 You can calculate the mean (or an estimate of the mean) from a frequency table.

6 Scatter diagrams are useful for data that comes in pairs.

7 Correlation is either positive, negative or there is no (linear) correlation.

8 When there is positive or negative correlation, a line of best fit can be drawn – it passes among the points with roughly an equal number of points on each side of the line.
   The line of best fit should not be forced through any particular point.

9 The line of best fit can be used to estimate missing values.

Internet Challenge 20

Populations

Here is a list of the twelve most populous countries in the world, in decreasing order.

	Name of country	Population in millions	Area in millions of square kilometres
1	China		
2	India		
3	United States of America		
4	Indonesia		
5	Brazil		
6	Pakistan		
7	Bangladesh		
8	Russia		
9	Nigeria		
10	Japan		
11	Mexico		
12	Philippines		

Make a copy of this table on a computer spreadsheet, such as Excel.

Then use an internet search engine to find the missing values for the other two columns, and fill them in.

Use the spreadsheet's Chart Wizard or equivalent to make a scatter graph.

Is there any correlation between the population of a country and its area?

CHAPTER 21

Presenting data

In this chapter you will **learn how to:**

- draw and interpret pictograms, bar charts and pie charts
- draw and interpret frequency polygons and histograms
- use statistics from real-life situations.

You will also be **challenged to:**

- investigate quotations on statistics.

Starter: Lies, damned lies and statistics

Some people think that statistics can be used to deliberately mislead a reader.

Each of these statistical diagrams is misleading.
See if you can spot how.

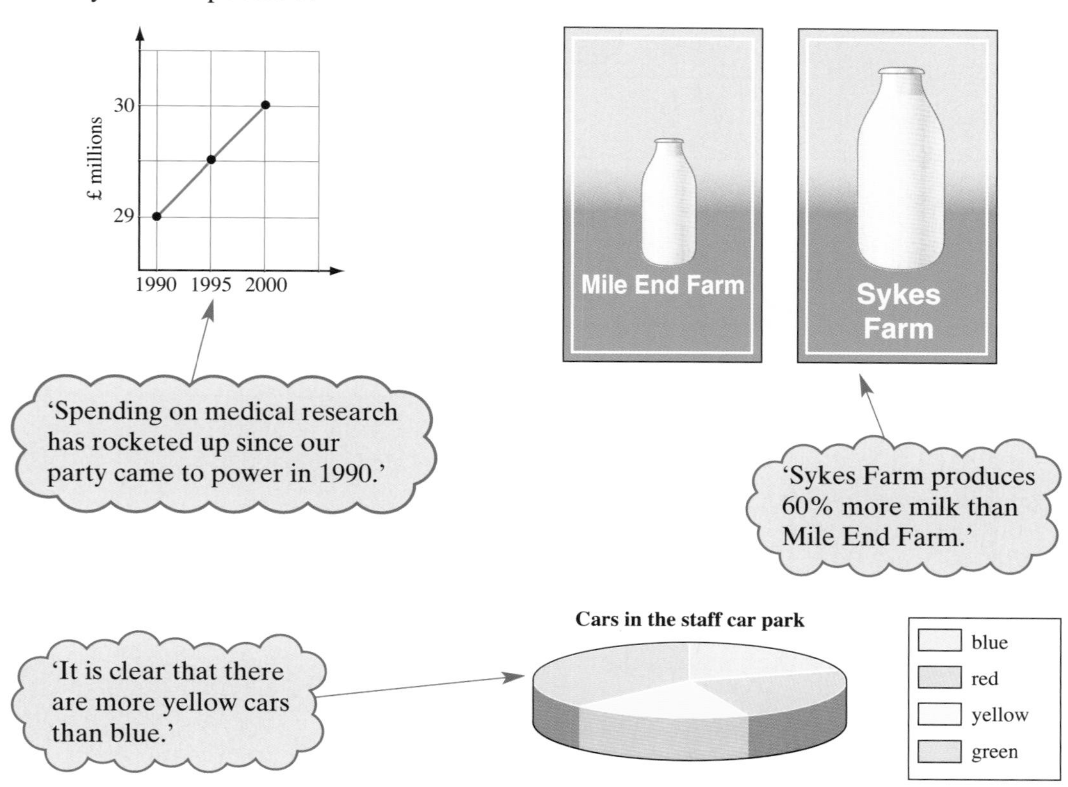

21.1 Pictograms and bar charts

A **pictogram**, or **bar chart**, can be used to represent categorical data.

EXAMPLE

Kelly has carried out a survey on some students' favourite leisure activities.

Here are her results.

Activity	Frequency
Read a book	50
Watch television	140
Cinema	70
Go shopping	40
Visit friends	80
Play sport	150

a) Draw a pictogram of Kelly's results.

Use ☺ to represent 20 people.

b) Which activity is the most popular?

SOLUTION

a)

Read a book	
Watch television	
Cinema	
Go shopping	
Visit friends	
Play sport	

50 = 20 + 20 + 10
So you need two whole faces and one half face.

b) The most popular activity is playing sport.

When drawing a pictogram:

- choose a simple picture which can be easily reproduced
- include a key
- keep all the pictures the same size
- keep all the spaces between the pictures the same.

EXAMPLE

Draw a bar chart of Kelly's data.

SOLUTION

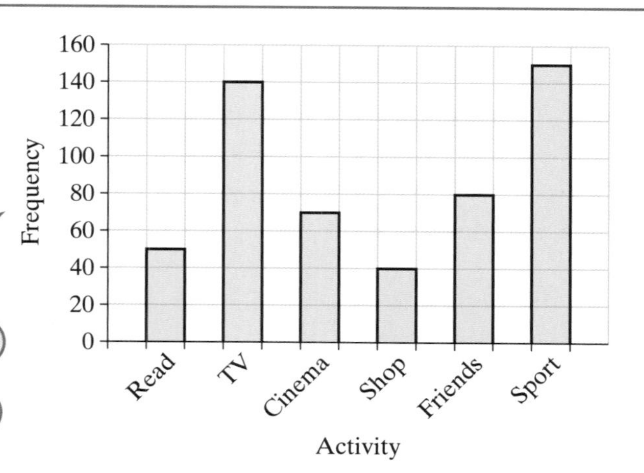

The frequency tells you how many people like each activity.

When drawing a bar chart:

- label the vertical axis
- keep all the bars the same width
- label each bar
- keep the spaces between the bars the same.

A **dual bar chart** is used to compare two sets of data on the same graph.

EXAMPLE

Kelly has carried out a survey on some students' favourite leisure activities.

Here are her results.

Draw a dual bar chart of Kelly's data.

Activity	Number of girls	Number of boys
Read a book	30	20
Watch television	55	85
Cinema	18	52
Go shopping	23	17
Visit friends	42	38
Play sport	70	80

SOLUTION

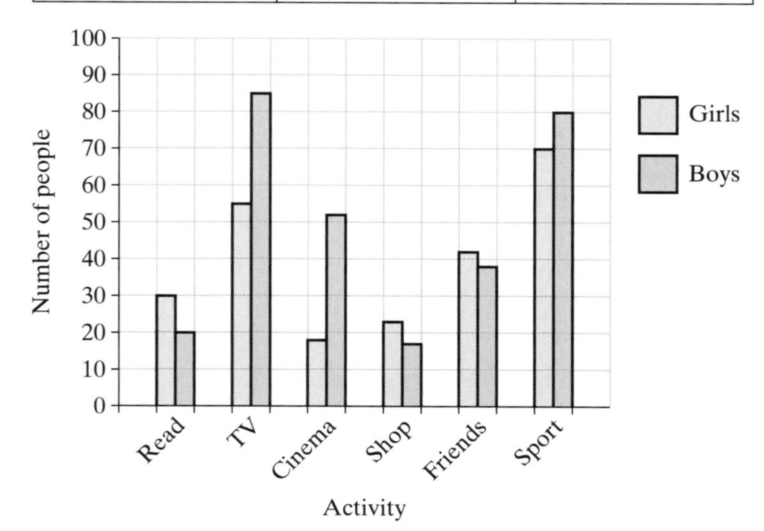

It is easy to compare.

A **compound bar chart** is used to compare two (or more) sets of data on the same graph.

EXAMPLE

Draw a compound bar chart of Kelly's data.

SOLUTION

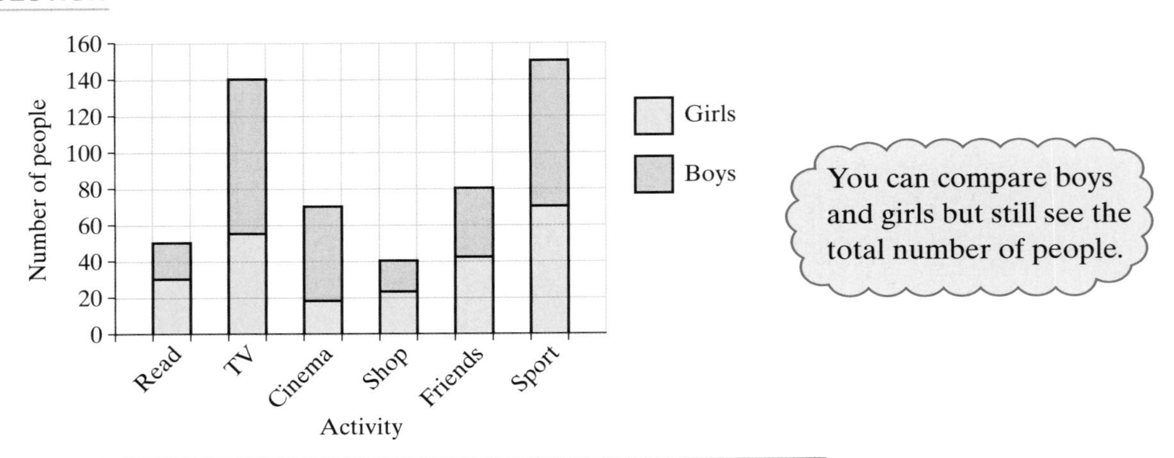

EXERCISE 21.1

1 Simon wrote down the number of hours he spent playing computer games for one week. The bar chart shows his results.

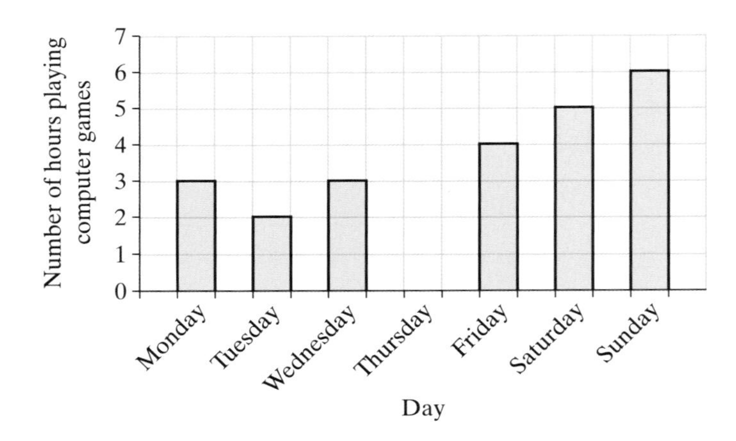

a) On which day did Simon spend the most number of hours playing computer games?

b) How many hours did Simon spend playing computer games on Saturday?

c) There were two days on which Simon spent the same number of hours playing computer games.
Write down the names of these two days.

d) How many hours did Simon spend playing computer games during that week?

2 A music store uses a pictogram to show the number of CDs sold over a six week period.

Key: = 100 CDs

Week 1	
Week 2	
Week 3	
Week 4	
Week 5	
Week 6	

a) How many CDs were sold in week 1?
b) In which week was the smallest number of CDs sold?
c) How many CDs were sold in week 5?
d) How many CDs were sold in week 2?

The music store has data for two more weeks.

Week number	Number of CDs sold
Week 7	300
Week 8	575

e) Add this data to a copy of the pictogram.

3 Fiona is treasurer for the school fete.
She has produced a bar chart to compare how much money has been raised over the last two years.

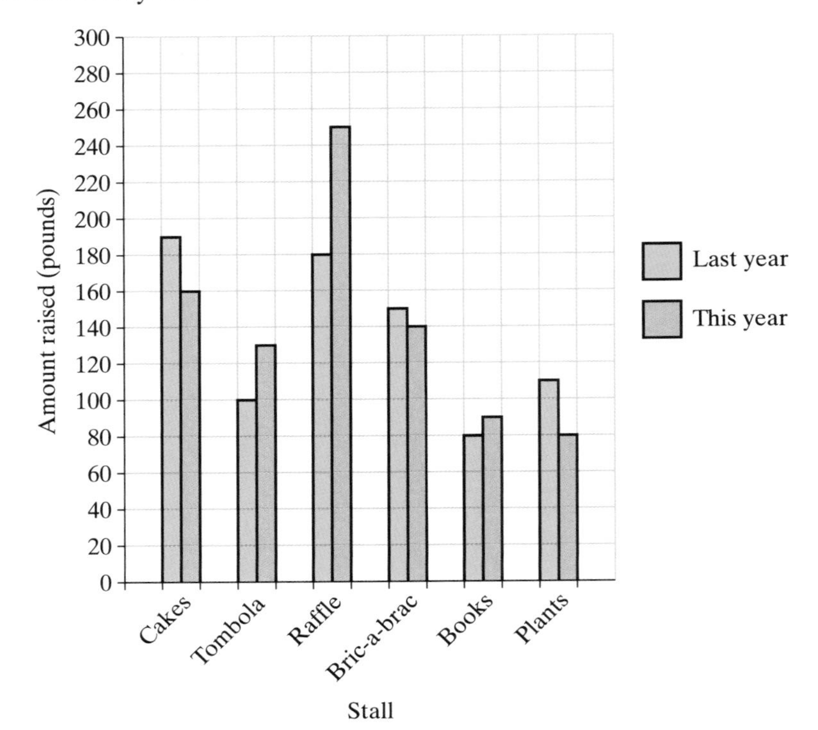

a) Write down the name of the stall which raised the most money:
 (i) this year
 (ii) last year.

b) Write down the name of the stall which raised the least money:
 (i) this year
 (ii) last year.

c) Write down how much money the cake stall raised:
 (i) this year
 (ii) last year.

d) **(i)** In which year was the most money raised?
 (ii) How much more money was raised that year?

4 Sam works in a restaurant.
One week he collects some data on the meals chosen by customers.
Sam makes a bar chart of his results.

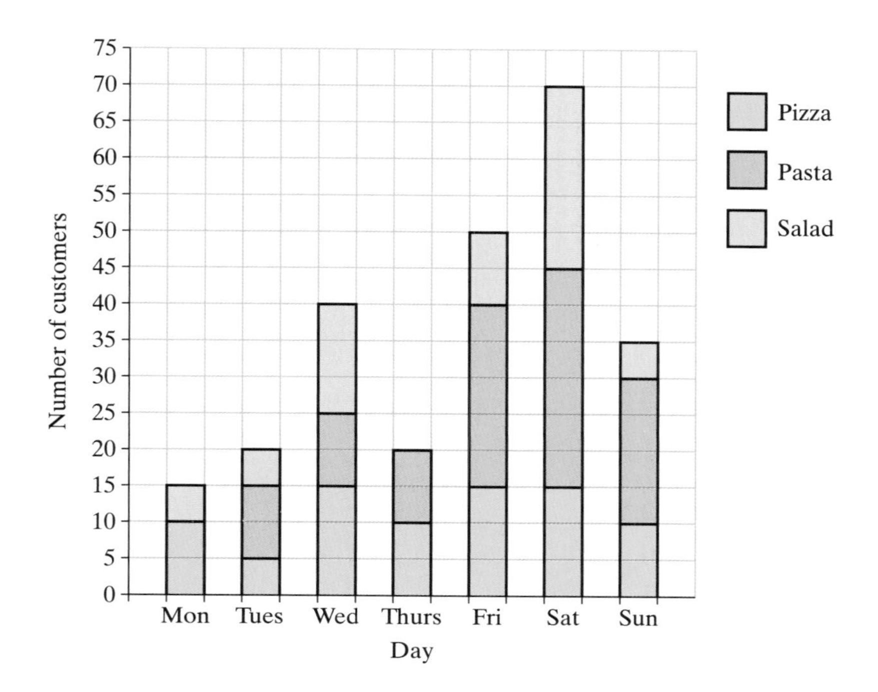

a) On which day was the highest number of meals sold?
b) On which day were no salads sold?
c) On which day were no pasta dishes sold?
d) There are two days when the same number of meals was sold.
Write down the names of these two days.
e) How many meals were sold on Sunday?
f) How many pizzas were sold on Monday?
g) How many pasta dishes were sold on Friday?
h) How many salads were sold on Saturday?

5 The manager of a bookshop collects some data on the number of sales one Monday.

Draw a pictogram of this information.

Use [symbol] to represent 4 sales.

Type of book	Number of sales
Fiction	16
Children's books	20
Biographies	18
Science fiction	5
Crime fiction	10
Non-fiction	13

6 Mary counts the number of late students at her school every morning for a week.

Here are her results.

Draw a bar chart to show this information.

Day	Number of late students
Monday	5
Tuesday	8
Wednesday	3
Thursday	0
Friday	7

21.2 Pie charts

A **pie chart** is used to represent categorical data.
You need to be able to draw and interpret pie charts.

EXAMPLE

Isobel surveys 60 people to find out their favourite sport.
Here are her results.
Draw a pie chart of this information.

Sport	Number of people
Rugby	5
Football	20
Tennis	6
Hockey	15
Netball	5
Swimming	9

SOLUTION

There are 360° in the whole pie chart.

So 60 people are represented by 360°

$\div 60$ $\div 60$

1 person is represented by 6°

Step 1 Work out how many degrees represent 1 person.

Sport	Number of people	Angle
Rugby	5	$5 \times 6° = 30°$
Football	20	$20 \times 6° = 120°$
Tennis	6	$6 \times 6° = 36°$
Hockey	15	$15 \times 6° = 90°$
Netball	5	$5 \times 6° = 30°$
Swimming	9	$9 \times 6° = 54°$
Total	60	360°

Step 2 Work out the angle of each sector – check that your angles add up to 360°

Step 3 Draw pie chart. Make sure you include a key or label each sector.

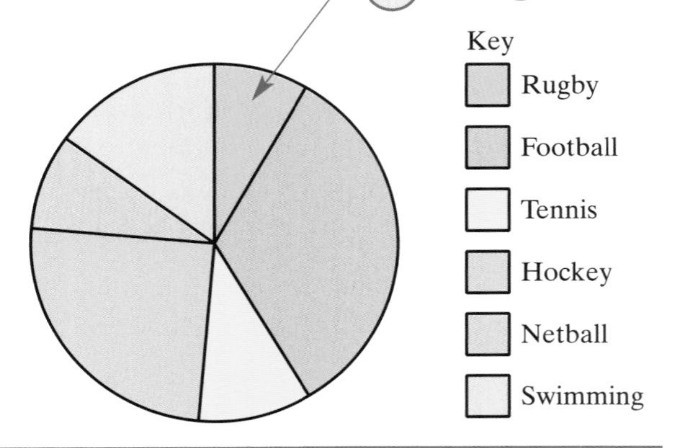

Sometimes the angles don't turn out to be whole numbers.
Round the angles to the nearest degree to draw the pie chart.

EXAMPLE

Scott wrote down the colours of the first 120 cars to pass him one morning.

He drew a pie chart of his results.

a) Which colour is the mode?
b) What fraction are black cars?
c) How many green cars pass Scott?

SOLUTION

The largest sector (or slice) is the mode.

a) Red

Measure the angle of the black sector.

b) The black sector has an angle of 60°

There are 360 degrees in the whole pie chart.

So the fraction which is black is $\frac{60}{360} = \frac{1}{6}$

Divide 'top' and 'bottom' of the fraction by 60 to write it in its lowest terms.

c) The green sector has an angle of 45°

There are 360 degrees in the whole pie chart .

Measure the angle of the green sector.

So the fraction which is green is $\frac{45}{360}$

But there are 120 cars altogether, so there are $\frac{45}{360} \times 120 = 15$ green cars.

There are 15 green cars.

EXERCISE 21.2

1 A pet rescue centre keeps a record of the number of animals at the centre.

Draw a pie chart to illustrate this information.

Animal	Frequency
Dogs	40
Cats	60
Rabbits	20

2 The table shows the results of the last 18 games played by the football team Rovers United.

Draw a pie chart to illustrate this information.

Result	Number of games
Won	12
Lost	4
Drawn	2

3 Paul has carried out a survey on people's favourite television stations.

Draw a pie chart to illustrate this information.

Television station	Frequency
BBC1	17
BBC2	6
ITV	23
Channel 4	14

4 The table shows the UK holiday destinations of the customers of a travel agency.

Draw a pie chart to illustrate this data.

England	35%
Scotland	18%
N. Ireland	17%
Wales	30%

5 Myra earns £36 000 a year.
This table shows how Myra spends her money.

Draw a pie chart to illustrate this data.

Taxes and national insurance	£12 000
Mortgage and bills	£8 000
Holidays	£2 000
Food and drink	£4 000
Other	£6 000
Savings	£4 000

6 Gina draws a pie chart to show how she spends a typical day.

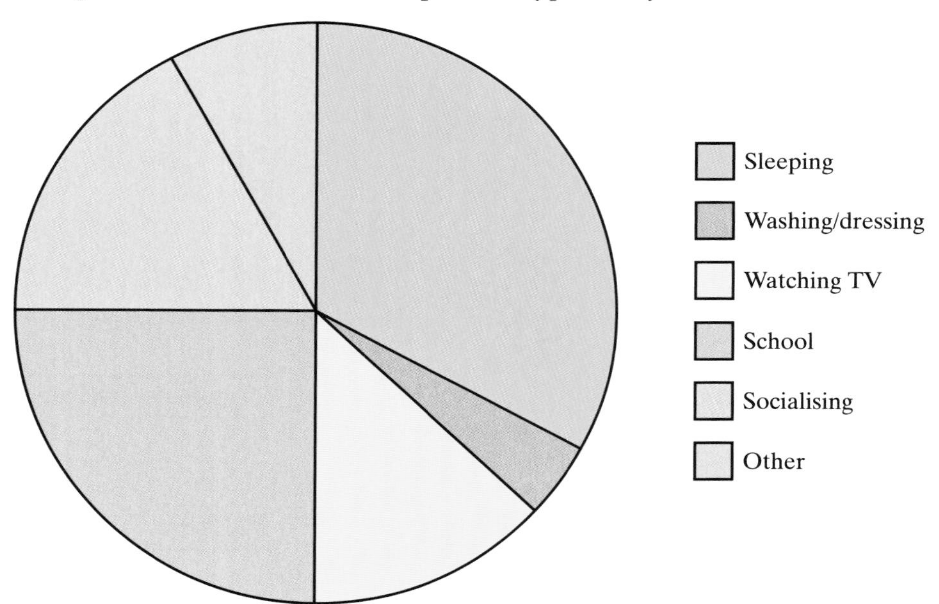

a) Which activity is the mode?
b) Work out how many hours Gina spends:
(i) at school **(ii)** sleeping **(iii)** socialising **(iv)** washing/dressing.

7 Harry is doing some handling data coursework.
He records the number of pages in each section of a broadsheet newspaper and a tabloid newspaper.
The pie charts show his results.

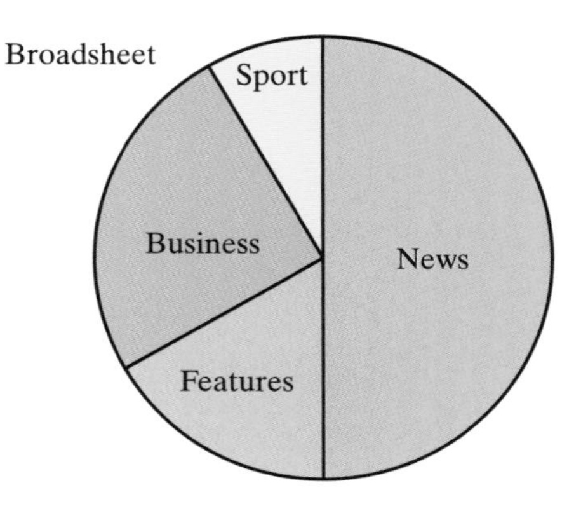

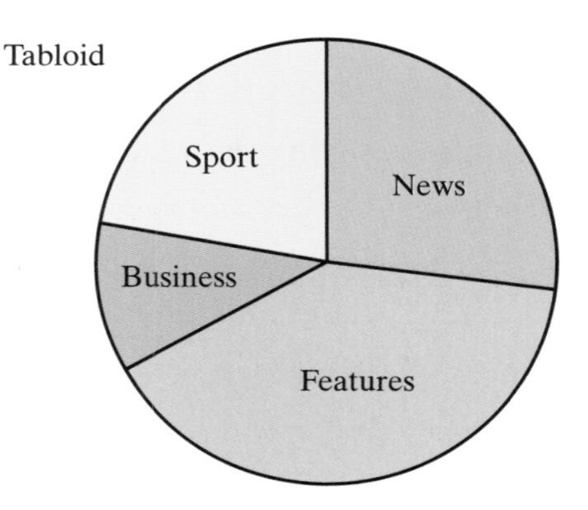

The broadsheet has 60 pages.
The tabloid has 45 pages.

a) Write down the section which is the mode for the **(i)** broadsheet, **(ii)** tabloid.
b) Work out how many pages each paper devotes to **(i)** sport, **(ii)** features.

21.3 Frequency diagrams

You can use a **vertical line chart** to present discrete data.
The frequency of each item is shown by a vertical line.

EXAMPLE

a) Sophie has collected this data about the number of matches in boxes.
Illustrate the data with a vertical line chart.

Number of matches	Frequency
48	1
49	5
50	7
51	0
52	5
53	1
54	1

b) Why shouldn't you join the tops of the lines?

SOLUTION

a)

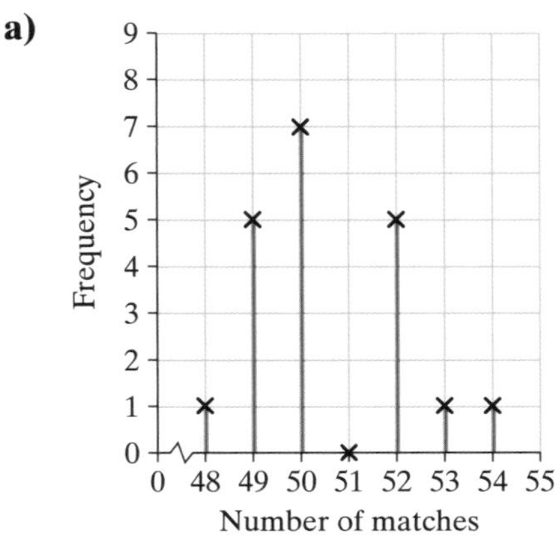

b) You shouldn't join the tops of the lines because the data is discrete.

The intermediate values have no meaning as we are only counting whole matches.

A **histogram** is used to represent grouped continuous data.

Each group is represented by a bar.

A histogram should have:

- a numerical scale along the x axis
- frequency plotted along the y axis
- no gaps between the bars.

EXAMPLE

The resistances of a sample of some electronic components are measured.

The table shows the results.

Draw a histogram for this data.

Resistance, R	Frequency, f
$80 \leqslant R < 90$	10
$90 \leqslant R < 100$	23
$100 \leqslant R < 110$	11
$110 \leqslant R < 120$	6
$120 \leqslant R < 130$	2

SOLUTION

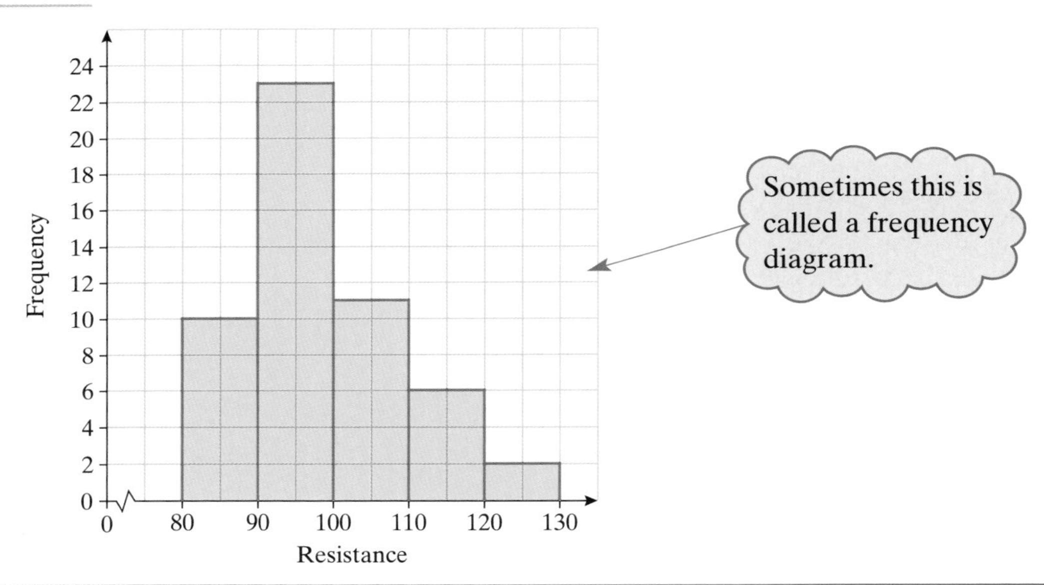

EXERCISE 21.3

Don't use your calulator for Questions 1 and 2.

1 Here are the results of the maths mock exam at Appleton High School.

a) Draw a frequency diagram to show this data.

b) How many students took the mock exam?

c) Write down the modal class interval.

Mark, m	Frequency
$0 < m \leqslant 10$	7
$10 < m \leqslant 20$	15
$20 < m \leqslant 30$	21
$30 < m \leqslant 40$	30
$40 < m \leqslant 50$	18
$50 < m \leqslant 60$	9

2 Liam carries out a survey to find out the average amount of time that some students spend on their mobile phones in a week.

Here are his results.

Draw a histogram of Liam's results.

Average number of hours, h	Frequency
$0 \leqslant h < 4$	5
$4 \leqslant h < 8$	12
$8 \leqslant h < 12$	17
$12 \leqslant h < 16$	15

You can use your calculator for Questions 3 and 4.

3 One morning, Martin carried out a survey of the time, in seconds, between one car and the next car on a road outside his house.

The table shows the results of his survey.

a) Draw a histogram to show this data.

b) Write down the modal class interval.

c) How many cars passed Martin during his survey?

d) Estimate the total time, in seconds, that Martin spent counting cars.

e) Find an estimate for the mean time, in seconds, between cars.

Time, t, in seconds	Frequency
$0 < t \leqslant 20$	15
$20 < t \leqslant 40$	33
$40 < t \leqslant 60$	40
$60 < t \leqslant 80$	9
$80 < t \leqslant 100$	8

4 The midwives in a maternity unit have recorded the weight of 100 babies.

a) Draw a histogram to show this data.

b) Write down the modal class interval.

c) Estimate the mean weight of the babies.

d) Write down the class interval which contains the median.

Weight, w, in kilograms	Frequency
$2 \leqslant w < 2.5$	5
$2.5 \leqslant w < 3$	13
$3 \leqslant w < 3.5$	18
$3.5 \leqslant w < 4$	34
$4 \leqslant w < 4.5$	21
$4.5 \leqslant w < 5$	9

21.4 Frequency polygons

The way in which statistical data is spread out is known as its **distribution**.

Some data sets are distributed uniformly, but others have distinctive peaks and troughs.
The shape of a distribution may be seen by looking at a **frequency polygon**.

A frequency polygon has:

- a numerical scale along the x axis
- frequency along the y axis
- the frequency of each class plotted against the midpoint of the class
- points that are joined up by straight lines.

EXAMPLE

The resistances of a sample of some electronic components are measured.

The table shows the results.

Resistance, R	**Frequency, f**
$80 \leqslant R < 90$	10
$90 \leqslant R < 100$	23
$100 \leqslant R < 110$	11
$110 \leqslant R < 120$	6
$120 \leqslant R < 130$	2

Draw a frequency polygon of this data.

SOLUTION

Resistance, R	**Midpoint**	**Frequency, f**
$80 \leqslant R < 90$	85	10
$90 \leqslant R < 100$	95	23
$100 \leqslant R < 110$	105	11
$110 \leqslant R < 120$	115	6
$120 \leqslant R < 130$	125	2

First you need to work out the midpoint of each class.

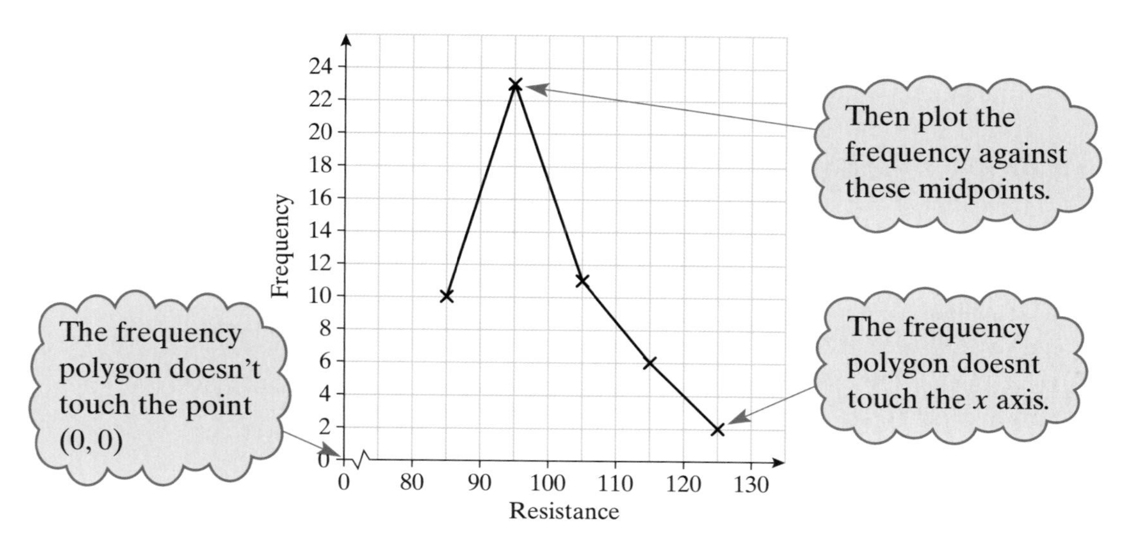

The graph shows how a histogram (frequency diagram) is related to a frequency polygon.

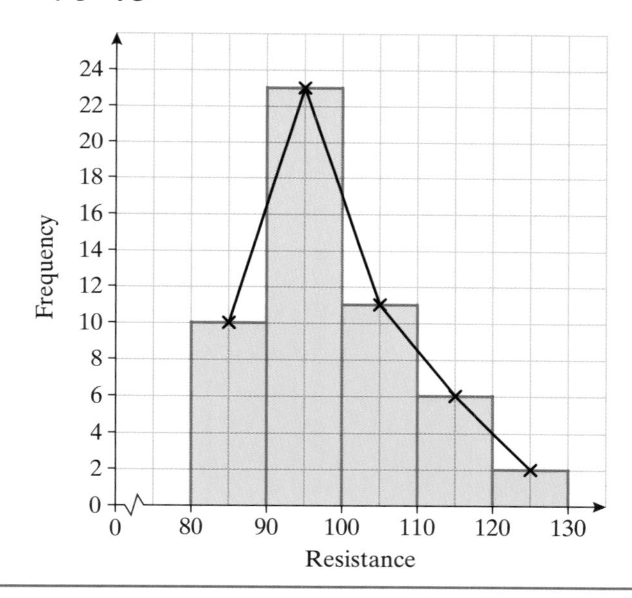

EXERCISE 21.4

1 Draw a frequency polygon for each of the sets of data given in Exercise 21.3

You can use your calculator for Questions 2 and 3.

2 A magazine carried out a survey of the ages of its readers.
The results of the survey are shown in the table.

Age group, y, years	Percentage of readers in this age group
$20 \leqslant y < 30$	17
$30 \leqslant y < 40$	28
$40 \leqslant y < 50$	22
$50 \leqslant y < 60$	18
$60 \leqslant y < 70$	10
$70 \leqslant y < 80$	5

a) Draw a frequency polygon to show this information.
b) Which is the modal class?
c) Find the class interval which contains the median.
d) Find an estimate for the mean age.

3 Jason grows potatoes.
He weighed 100 potatoes and recorded the weights to the nearest gram.
The table shows information about the weights (w) of the 100 potatoes.

Weight, w	Frequency, f
$0 \leqslant w < 20$	0
$20 \leqslant w < 40$	18
$40 \leqslant w < 60$	28
$60 \leqslant w < 80$	25
$80 \leqslant w < 100$	19
$100 \leqslant w < 120$	10

Frequency
30
25
20
15
10
5
O
20 40 60 80 100 120
Weight (w) grams

a) Draw a frequency polygon to show this information on a copy of the grid.
b) Work out an estimate for the mean weight of these potatoes.
c) Find the class interval that contains the median. **[Edexcel]**

21.5 Time-series graphs

EXAMPLE

Marie carries out a survey of the birds in her garden.
Every two hours she counts the number of birds in the garden.

Time	06 00	08 00	10 00	12 00	14 00	16 00	18 00
Number of birds	35	28	18	12	15	21	25

Show Marie's results on a time-series graph.

SOLUTION

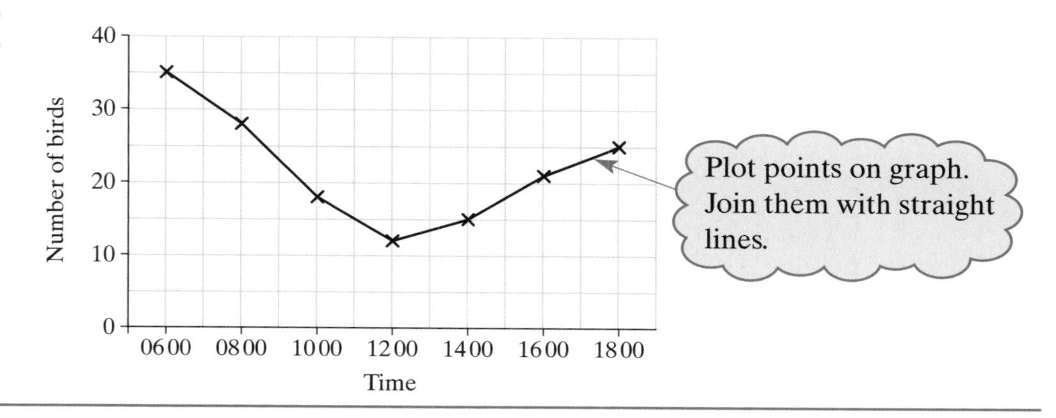

The points have been drawn with straight lines to give a better illustration of Marie's data but the values in between the plotted points have no meaning.

For example, you can't say from the graph how many birds were in the garden at 11 00

EXERCISE 21.5

1 Fern is a nurse.
Every two hours she takes the temperature of the ward to make sure it is warm enough.
Here is a time-series graph of her results.

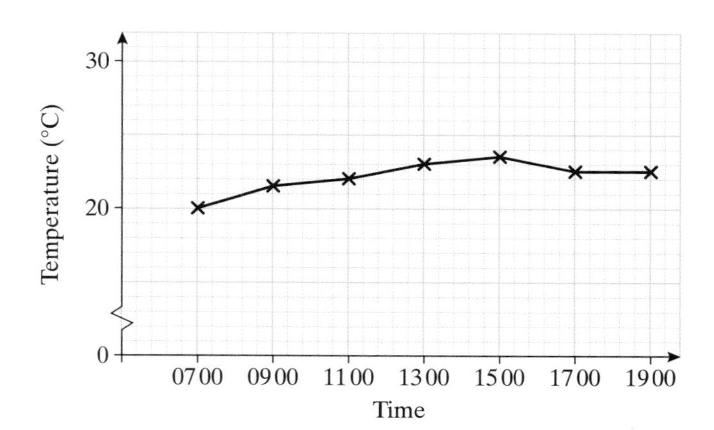

a) What is the range of the temperatures on the ward?
b) Write down the temperature at: **(i)** 11 00, **(ii)** 13 00
c) Fern takes two more readings:
Add these measurements to a copy of Fern's graph.

Time	21 00	23 00
Temperature	23 °C	22 °C

2 The time-series graph shows the average number of hours of sunshine each month for Sunnyside town.

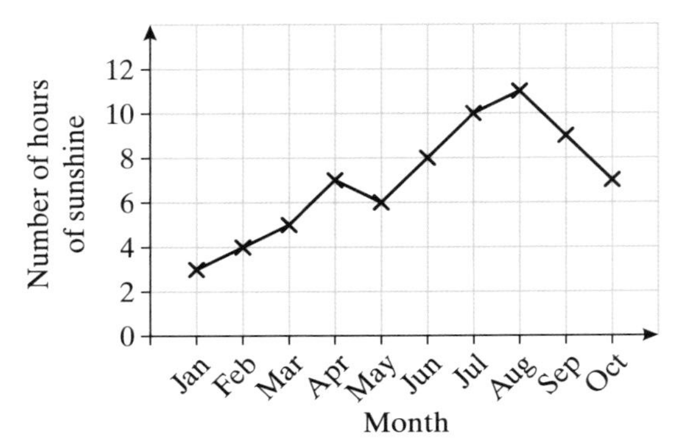

a) What is the range in the average number of hours of sunshine?
b) Which two months had the same average number of hours of sunshine?

Here is the data for November and December.

Month	Nov	Dec
Average number of hours of sunshine	5	4

c) Add this data to a copy of the time-series graph.

3 Appleton High School collect data on the average percentage of students absent for each month throughout the school year.
Here are their results.

Month	Sept	Oct	Nov	Dec	Jan	Feb	Mar	Apr	May	June	July
Percentage of students absent	2%	5%	10%	15%	17%	14%	10%	8%	7%	5%	3%

Draw a time-series graph for this data.

REVIEW EXERCISE 21

Don't use your calculator for Questions 1–5.

1 Here is a pictogram.
It shows the number of boxes of chocolates sold in one week from Monday to Friday.
a) Write down the number of boxes of chocolates sold on: **(i)** Monday, **(ii)** Wednesday.

On Saturday, 100 boxes of chocolates were sold.
b) Show this on a copy of the pictogram.

On Sunday, 55 boxes of chocolates were sold.
c) Show this on the pictogram. **[Edexcel]**

Monday
Tuesday
Wednesday
Thursday
Friday
Saturday
Sunday

Represents 20 boxes of chocolates sold

2 The pictogram shows the number of videos borrowed from a shop on Monday and on Tuesday.

a) Write down the number of videos borrowed on:
(i) Monday **(ii)** Tuesday.

On Wednesday, 40 videos were borrowed.
On Thursday, 15 videos were borrowed.

b) Show this information on the pictogram.

Monday	●● ●● ●●
Tuesday	●● ●● ●
Wednesday	
Thursday	

●● represents 10 videos

[Edexcel]

3 The table gives information about the makes of car in a garage showroom.

Makes of car	Frequency
Ford	2
Toyota	6
Peugeot	10

Draw an accurate pie chart to show this information. **[Edexcel]**

4 The table shows the temperature at midday on each day of a week during winter.

Day	Sun	Mon	Tue	Wed	Thu	Fri	Sat
Temperature (°C)	6	8	6	7	8	8	7

a) Work out the median temperature.

The graph shows the temperature from 07 00 to 11 00 during one day.

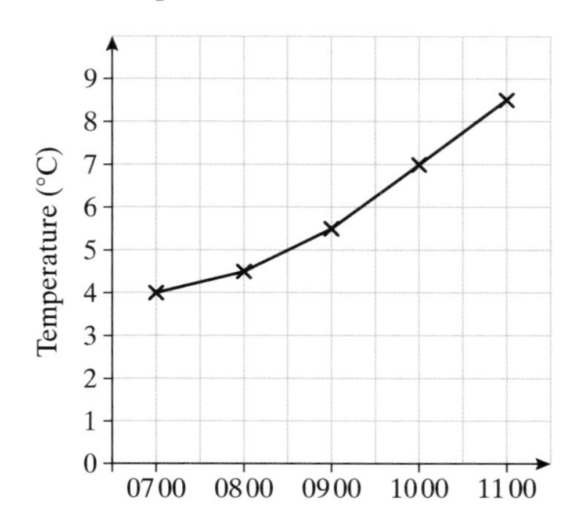

b) What was the temperature at 10 00?

c) What was the temperature at 08 00? **[Edexcel]**

5 32 students took an English test. There were 25 questions in the test. The grouped frequency table gives information about the number of questions the students answered.

Number of test questions answered	Frequency
1–5	1
6–10	3
11–15	9
16–20	8
21–25	11

a) Write down the modal class interval.

b) Write down the class interval which contains the median.

c) Draw a frequency polygon to show the information in the table. Use a copy of the grid below.

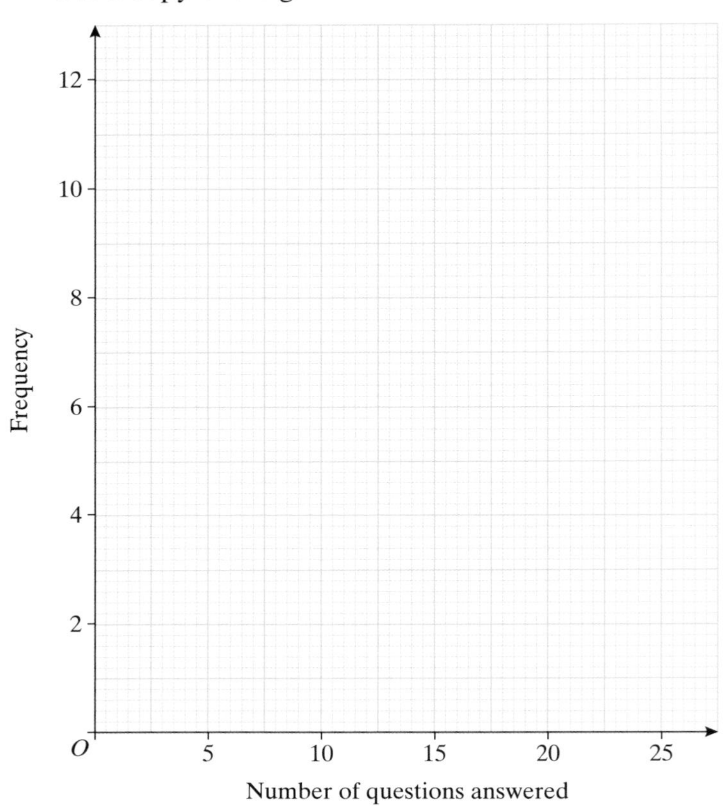

[Edexcel]

You can use your calculator for Questions 6–13.

6 A shop has a sale.

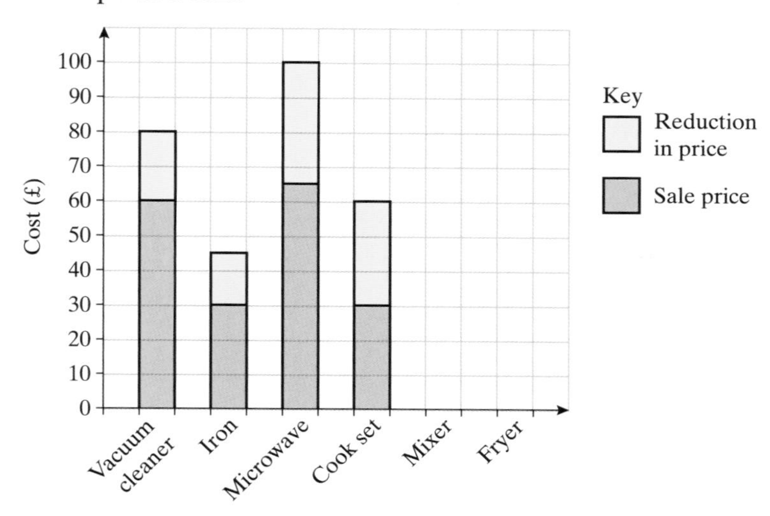

The bar chart shows some information about the sale.
The normal price of a vacuum cleaner is £80
The sale price of a vacuum cleaner is £60
The price of the vacuum cleaner has been reduced from £80 to £60

a) Write the sale price of a vacuum cleaner as a fraction of its normal price. Give your answer in its simplest form.

b) Find the reduction in price of the iron.

c) Which two items have the same sale price?

d) Which item has the greatest reduction in price?

Mixer	
Normal price	£90
Sale Price	£70

Fryer	
Normal price	£85
Sale Price	£70

e) Complete a copy of the bar chart for the mixer and the fryer. [Edexcel]

7 The table gives information about the medals won by Austria in the 2002 Winter Olympic Games.

Medal	Frequency
Gold	3
Silver	4
Bronze	11

Draw an accurate pie chart to show this information. [Edexcel]

8 Ray and Clare are pupils at different schools. They each did an investigation into their teachers' favourite colours.

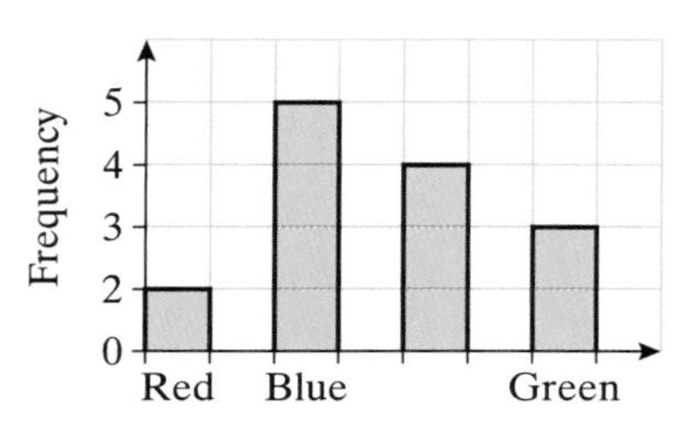

Here is Ray's bar chart of his teachers' favourite colours.

a) Write down two things that are wrong with Ray's bar chart.

Clare drew a bar chart of her teachers' favourite colours. Part of her bar chart is shown.

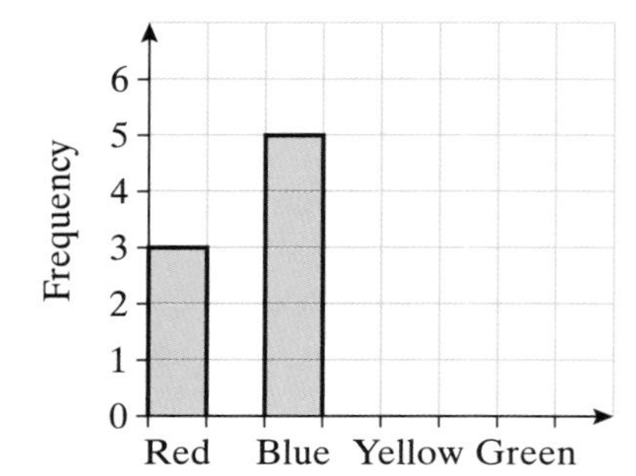

4 teachers said that yellow was their favourite colour.
2 teachers said that green was their favourite colour.

b) Complete Clare's bar chart.

c) Which colour was the mode for the teachers that Clare asked?

d) Work out the number of teachers that Clare asked.

e) Write down the fraction of the number of teachers that Clare asked who said that red was their favourite colour.

[Edexcel]

9 The graph shows the percentage of trains each month that arrived on time from August to March.

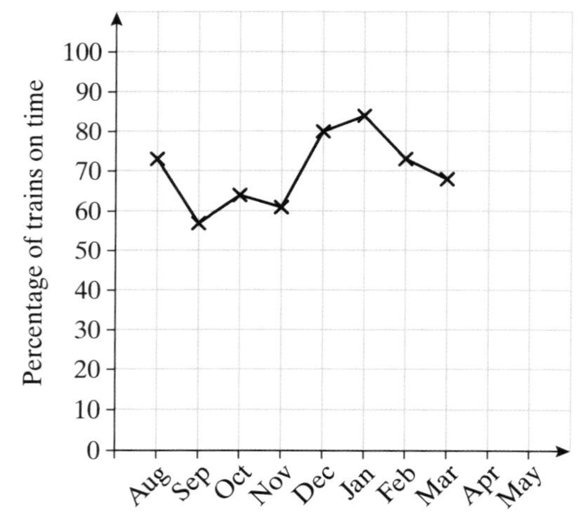

a) Use the graph to write down the:

(i) percentage of trains which arrived on time in December

(ii) lowest percentage of trains which arrived on time.

The percentage for April was 70% and for May was 62%.

b) Copy and complete the graph for April and May.

[Edexcel]

10 Bhavana asked some people which region their favourite football team came from.
The table shows her results.

Region	Frequency
Midlands	22
London	36
Southern England	8
Northern England	24

Complete an accurate pie chart to show these results.
Use a copy of the circle given here.

Midlands

[Edexcel]

11 The bar chart shows the average number of hours of sunshine each day in London and in Corfu each month from April to September.

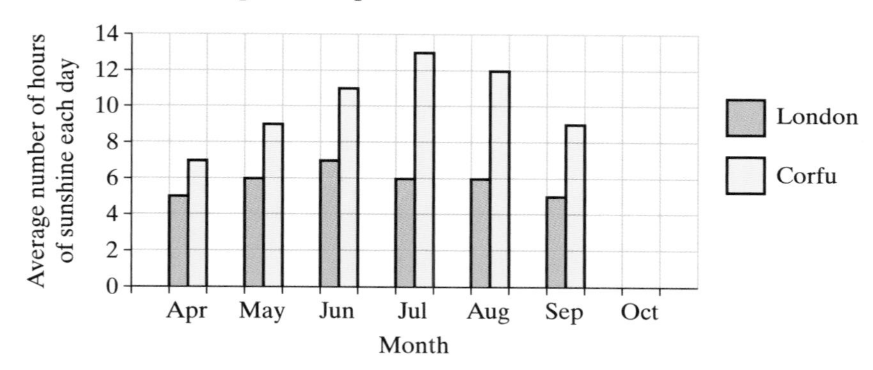

a) Write down the average number of hours of sunshine each day in London in August.

b) Write down the average number of hours of sunshine each day in Corfu in September.

c) Write down the name of the month in which the average number of hours of sunshine each day in London was 7

In October, the average number of hours of sunshine each day in London is 3 hours.
In Corfu, it is 6 hours.

d) Draw two bars to show this information on the bar chart.

There are 30 days in September.
Work out the **total** number of hours of sunshine in Corfu in September. [Edexcel]

12 The table shows the frequency distribution of student absences for a year.
Draw a frequency polygon for this frequency distribution.

Absences, d days	Frequency
$0 \leqslant d < 5$	4
$5 \leqslant d < 10$	6
$10 \leqslant d < 15$	8
$15 \leqslant d < 20$	5
$20 \leqslant d < 25$	4
$25 \leqslant d < 30$	3

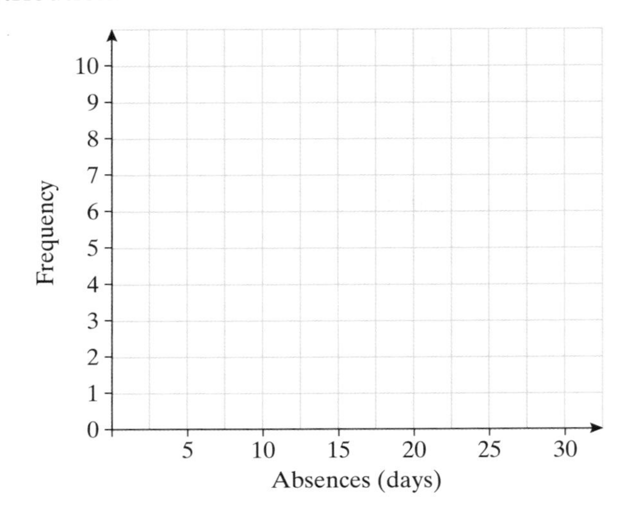

[Edexcel]

13 At a restaurant, waiters recorded the lengths of time that customers had to wait for their meals.

a) Draw a histogram of this data.

b) Write down the modal class interval.

c) Find an estimate for the mean time that customers had to wait.

d) Write down the class interval which contains the median.

Time, t, in minutes	Frequency
$0 < t \leqslant 5$	4
$5 < t \leqslant 10$	7
$10 < t \leqslant 15$	13
$15 < t \leqslant 20$	9
$20 < t \leqslant 25$	5
$25 < t \leqslant 30$	2

[Edexcel]

KEY POINTS

1 A bar chart, pie chart or pictogram can be used to represent categorical data.

2 When drawing a pictogram:
- choose a simple picture which can easily be reproduced
- include a key
- keep all the pictures the same size
- keep all the spaces between the pictures the same.

3 When drawing a bar chart:
- label the vertical axis
- label each bar
- keep all the bars the same width
- keep the spaces between the bars the same.

4 A dual (multiple) or compound barchart may be used to compare two sets of data.

5 The angles in a pie chart add up to 360°

6 Use a vertical line chart to represent ungrouped discrete data.

7 Grouped data can be displayed on a histogram:
- each group is represented by a bar
- there should be a numerical scale along the x axis
- frequency should be plotted along the y axis
- there should be no gaps between the bars.

8 The values from a frequency table can be plotted in a zig-zag graph called a frequency polygon.
Plot the midpoints of each frequency against class.

9 Data which changes over time can be displayed on a time-series graph.

Internet Challenge 21

Statistical quotes

In the Starter for this chapter, you saw ways in which data may be presented in a misleading way.

Some people think that statistics can be used deliberately to mislead.

Here are some famous quotations on this theme.

Use the internet to find out who said or wrote each one.

1. 'There are three kinds of lies: lies, damned lies and statistics.'
2. 'First get your facts; then you can distort them at your leisure.'
3. 'The pure and simple truth is rarely pure and never simple.'
4. 'Then there was the man who drowned crossing a stream with an average depth of six inches.'
5. 'There are two kinds of statistics: the kind you look up and the kind you make up.'
6. 'You know how dumb the average guy is? Well, by definition, half of them are even dumber than that.'
7. 'Statistical thinking will one day be as necessary for efficient citizenship as the ability to read and write.'
8. 'A statistician is a man who comes to the rescue of figures that cannot lie for themselves.'
9. 'A single death is a tragedy, a million deaths is a statistic.'
10. 'Everything should be made as simple as possible, but not simpler.'

CHAPTER 22

Probability

In this chapter you will **revise earlier work on:**

- basic probability.

You will **learn how to:**

- use theoretical and experimental probability
- make estimates for the results of a statistical experiment.

You will **learn that:**

- P(not A) = 1 − P(A)
- results of two or more events can be listed systematically
- probabilities for a full set of mutually exclusive events add up to 1
- P(A or B) = P(A) + P(B) when A and B are mutually exclusive

You will be **challenged to:**

- investigate words and phrases used in probability.

Starter: Fair game

1 Look at these games.
Play them with a partner.
Are they fair?
Give a reason for your answer.

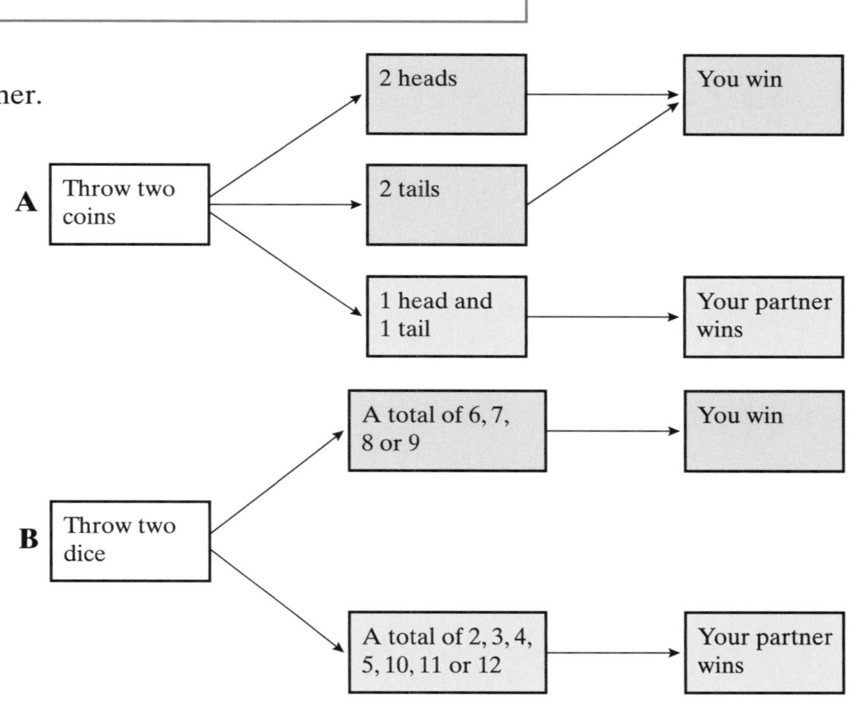

2 You can randomly generate numbers between 1 and 1000 by throwing a 10-sided dice (numbered 0 to 9) three times. The 1st throw gives you the hundreds, the 2nd throw the tens and the 3rd the units.

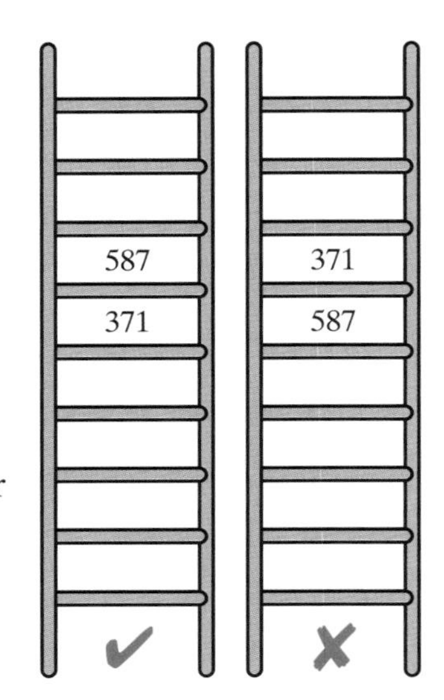

Play this game with a partner.

- Generate a random number between 1 and 1000.
- Each player writes the number on their copy of the ladder:

  You can't place a smaller number above a larger number

  You can't move a number once you have written it on the ladder.
- You lose when you can't enter the random number on your ladder.

What strategies did you use for placing the numbers on the ladder?

22.1 Using equally likely outcomes

If an experiment has a number of different outcomes, you can use **probability** to describe how likely the different outcomes are.

Probability is a number between 0 and 1

It is written as either a fraction (e.g. $\frac{1}{2}$) or a decimal (e.g. 0.5)

Sometimes percentages are used (e.g. 50%)

An outcome with a probability of 0 is impossible.

An outcome with a probability of 1 is certain.

The probability of a particular outcome can be shown on a probability line like this:

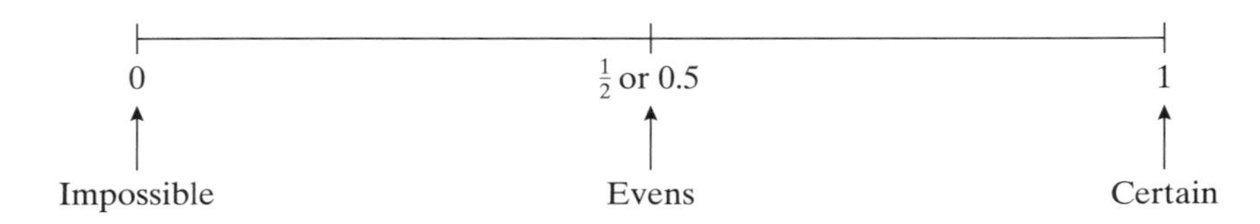

EXAMPLE

A bag contains 20 red counters and 20 blue counters.
One counter is selected at random.

a) What is the probability that the counter is red or blue?
b) What is the probability that the counter is black?
c) What is the probability that the counter is red?

SOLUTION

a) P(red or blue counter) = 1 ← This outcome (or event) is certain.

b) P(black counter) = 0 ← This outcome (or event) is impossible.

c) P(red counter) = 0.5 ← Half the counters are red.

$$\text{Probability is defined as } \frac{\text{number of favourable outcomes}}{\text{total number of outcomes}}$$

EXAMPLE

The whole numbers 1 to 10 are written on ten slips of paper.
The slips are folded and placed in a hat.
One slip is removed at random.
Find the probability that it is:

a) 6 **b)** at least 8 **c)** 12

SOLUTION

a) There is only 1 way of getting the number 6
There are 10 numbers altogether.

$$P(6) = \frac{1}{10}$$

b) There are 3 ways of getting at least 8 – namely 8, 9 or 10
There are 10 numbers altogether.

$$P(8, 9 \text{ or } 10) = \frac{3}{10}$$

c) There are 0 ways of getting 12

$$P(12) = \frac{0}{10} = 0$$

When you conduct a probability experiment, or trial, then a particular result (A) must either happen or not happen. So:

P(A does happen) + P(A does not happen) = 1

This leads to the very useful result that:

P(A does not happen) = 1 − P(A does happen)

EXAMPLE

The probability that it will snow on Christmas Day is 0.15
Find the probability that it will not snow on Christmas Day.

SOLUTION

P(it will snow) + P(it will not snow) = 1
Therefore P(it will not snow) = 1 − P(it will snow)
= 1 − 0.15
= 0.85

EXERCISE 22.1

1 The colours of the rainbow are red, orange, yellow, green, blue, indigo and violet.
Crystal writes each of the colours of the rainbow on a slip of paper, and puts the seven slips in a bag.
She then chooses a slip of paper at random.
Find the probability that the colour written on it is: **a)** red **b)** not blue **c)** brown.

2 A fair coin is thrown once.
On a copy of the probability line below mark with a letter T the probability that the coin will show 'tails'.

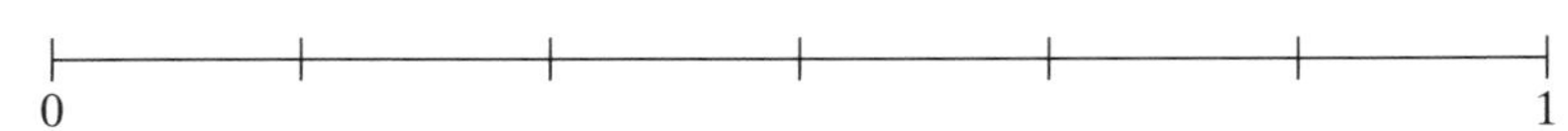

3 A fair dice has sides numbered 1, 2, 3, 4, 5 and 6.
The dice is thrown once.
On a copy of the probability line below:

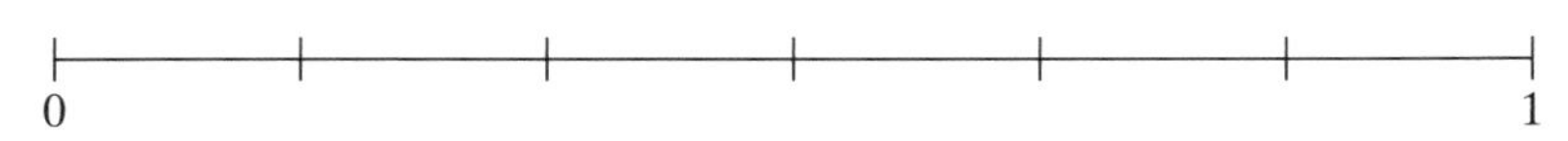

a) mark with a letter A the probability that the dice will land on a 4
b) mark with a letter B the probability that the dice will land on an odd number
c) mark with a letter C the probability that the dice will land on a 7
d) mark with a letter D the probability that the dice will land on a number **less than** 7
e) mark with a letter E the probability that the dice will **not** land on a 6

4 A five-sided spinner has sides 1, 2, 3, 4 and 5
The spinner is spun once.
On a copy of the probability line below:

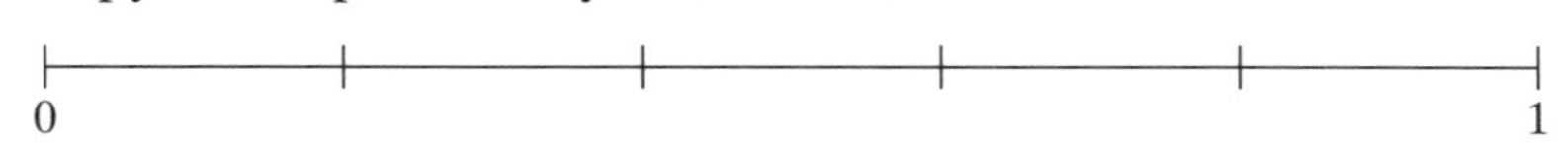

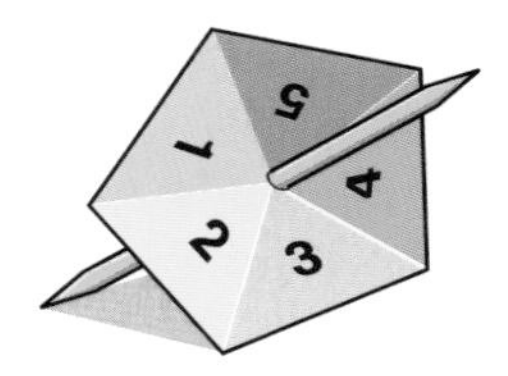

a) mark with a letter *A* the probability that the spinner will land on a 2
b) mark with a letter *B* the probability that the spinner will land on an even number
c) mark with a letter *C* the probability that the spinner will land on a 4 **or less**
d) mark with a letter *D* the probability that the spinner will land on a 6
e) mark with a letter *E* the probability that the spinner will **not** land on an even number.

5 The probability that Mark cycles to school is 0.65
What is the probability that Mark doesn't cycle to school?

6 A bag contains 40 balls.
There are 15 green balls and 9 yellow balls.
The rest of the balls are red.
A ball is chosen at random.
Find the probability that this ball is:
a) green
b) not yellow
c) red.

7 The probability that I win a game of tennis when I play against my friend Tim is 0.2
Work out the probability that I do not win when I play him.

8 A bag contains 30 balls.
10 of them are red, the rest are yellow or blue.
There are three times as many yellow balls as blue balls.
A ball is chosen at random.
a) Work out the probability that it is red.
b) Work out the probability that it is not red.
c) Work out the probability that it is blue.

22.2 Two-way tables

Special care should be taken when finding probabilities from a two-way table.

EXAMPLE

The table shows the number of boys and girls in Years 7, 8 and 9 at a local school.

	Year 7	Year 8	Year 9	Total
Boys	72	66	70	208
Girls	48	44	60	152
Total	120	110	130	360

a) Find the probability that a randomly chosen member of Year 8 is a boy.
b) Find the probability that a randomly chosen boy is a member of Year 8.
c) Find the probability that a randomly chosen member of the school is a Year 8 boy.

SOLUTION

a) There are 110 students in Year 8.
66 of these are boys.

Therefore $P(\text{boy}) = \frac{66}{110}$

$= \frac{3}{5}$

You can simplify this fraction by dividing both 'top' and 'bottom' by 22

b) There are 208 boys.
66 of these are in Year 8.

Therefore $P(\text{Year } 8) = \frac{66}{208}$

$= \frac{33}{104}$

Divide both 'top' and 'bottom' by 2

c) There are 360 members of the school.
66 of these are Year 8 boys.

Therefore $P(\text{Year 8 boy}) = \frac{66}{360}$

$= \frac{11}{60}$

Divide both 'top' and 'bottom' by 6

EXERCISE 22.2

Don't use your calculator for Questions 1 and 2.

1 The diagram shows some shapes.

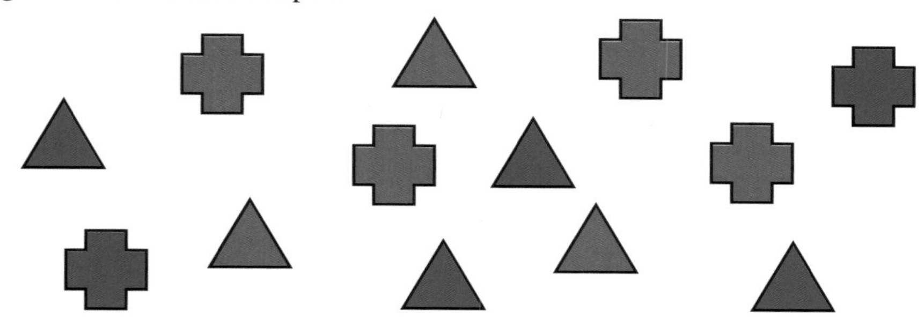

a) Complete a copy of the table to show the number of shapes in each category.

	Red	Blue	Total
Triangle			
Cross			
Total			

One of the shapes is chosen at random.

b) Write down the probability that the shape is a triangle.

c) Write down the probability that the shape is a cross.

d) Write down the probability that the shape is a red triangle.

2 The two-way table shows some information about the numbers of books in a classroom library.

	Science	History	Total
Hardback	10	18	
Paperback	20		
Total	30	50	80

a) Copy and complete the two-way table.

b) A science book is chosen at random.
Work out the probability that it is a hardback.

c) A paperback is chosen at random.
Work out the probability that it is a history book.

You can use your calculator for Questions 3 and 4.

3 A small school made a record of whether its pupils arrived on time or late yesterday.
It also noted what method of transport they used.
The two-way table shows the results.

	Walk	Bus	Other	Total
Late	3	21	5	29
On time	47	19	45	111
Total	50	40	50	140

a) Find the probability that a pupil who walked arrived late.

b) Find the probability that a pupil who arrived late came by bus.

c) The headmaster says these figures prove that pupils don't make enough of an effort to get to school on time.
Do you agree or disagree?
Explain your reasoning.

4 A café keeps records of how many drinks it sells one day.

a) Copy and complete the two-way table.

	Tea	Coffee	Other	Total
Morning	78	32		110
Afternoon	22		20	
Total		80	20	200

b) Find the probability that a randomly chosen drink sold in the morning was coffee.

c) Find the probability that an afternoon drink is tea.

22.3 Possibility space diagrams

Some questions on probability can be solved by first listing all the possible outcomes.

EXAMPLE

Three fair coins are thrown together.

a) Write down all the possible outcomes of throwing three fair coins.

b) Write down the probability of getting:
(i) 3 heads **(ii)** no heads **(iii)** 2 heads and a tail.

SOLUTION

a)

Coin 1	Coin 2	Coin 3
H	H	H
H	H	T
H	T	H
H	T	T
T	T	T
T	T	H
T	H	T
T	H	H

You need to work systematically so that you don't miss any out.

b) (i) There are 8 possible outcomes.
Only 1 of these is '3 heads' (H, H, H)

So P(3 heads) $= \frac{1}{8}$

(ii) There are 8 possible outcomes.
Only 1 of these is 'no heads' (T, T, T)

So P(no heads) $= \frac{1}{8}$

(iii) There are 8 possible outcomes.
3 of these are 'two heads and a tail'

So P(two heads and a tail) $= \frac{3}{8}$

Coin 1	Coin 2	Coin 3
H	H	H
H	H	T
H	T	H
H	T	T
T	T	T
T	T	H
T	H	T
T	H	H

Sometimes drawing up a two-way table helps.

EXAMPLE

Janine has two fair spinners.
Spinner A has the numbers 1, 2, 3
Spinner B has the numbers 3, 4, 5, 6
Janine spins both spinners.
She adds the scores together.

a) Draw a table to show all the possible total scores.
b) Work out the probability that Janine's total is 7

SOLUTION

a)

		Spinner B			
	+	**3**	**4**	**5**	**6**
Spinner A	**1**	4	5	6	7
	2	5	6	7	8
	3	6	7	8	9

This is called a **possibility space diagram.**

b) There are 12 possible outcomes in the table.
3 of them show a total of 7

So $P(\text{total of } 7) = \frac{3}{12}$

$= \frac{1}{4}$

EXERCISE 22.3

1 Two fair dice are thrown together.
The scores are added together.

a) Complete the table to show all the possible outcomes.
b) What is the total number of possible outcomes?
c) Write down the probability of getting a total of:
(i) 12 **(ii)** 7 **(iii)** 4 **(iv)** not 8
(v) an odd number **(vi)** 7 or more

+	1	2	3	4	5	6
1	2					
2						
3					8	
4		6				
5						
6						12

2 A fair coin and a fair dice are thrown together.

a) Write down all the possible outcomes of throwing a fair coin and a fair dice.
b) Write down the probability of getting:
(i) a 'head' and a '6'
(ii) a 'tail' and an 'even number'.

3 Two fair dice are thrown together.
The scores are multiplied together.

a) Complete the table to show all the possible outcomes.
b) What is the total number of possible outcomes?
c) Write down the probability of getting a product of:
(i) 12 **(ii)** 18 **(iii)** 24 **(iv)** not 10
(v) an even number **(vi)** 18 or more

×	**1**	**2**	**3**	**4**	**5**	**6**
1	1					
2						
3					15	
4		8				
5						
6						36

4 Penny is a cook in a café.
Each day she randomly selects a starter, main course and dessert from this list.

Starter	**Main course**	**Dessert**
Soup	Roast beef	Apple pie
–	Pasta	Chocolate cake
–	Fish	–

a) List all the possible outcomes of randomly selecting a starter, main course and desert.
b) Write down the probability that Penny will serve **i)** fish, **ii)** pasta and cake, **iii)** soup.

5 Two fair dice are thrown together.
The difference between the two scores is found.

a) Complete the table to show all the possible outcomes.
b) What is the total number of possible outcomes?
c) Write down the probability of getting a difference of:
(i) 0 **(ii)** 4 **(iii)** 3 **(iv)** not 0
(v) an odd number **(vi)** 2 or more

Difference	**1**	**2**	**3**	**4**	**5**	**6**
1	0					5
2						
3					2	
4						
5			2			
6	5				1	

6 Ben picks one shape from Box A.
He then picks one shape from Box B.

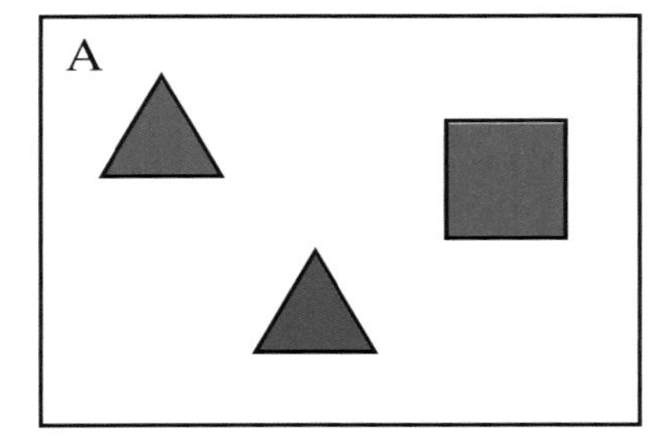

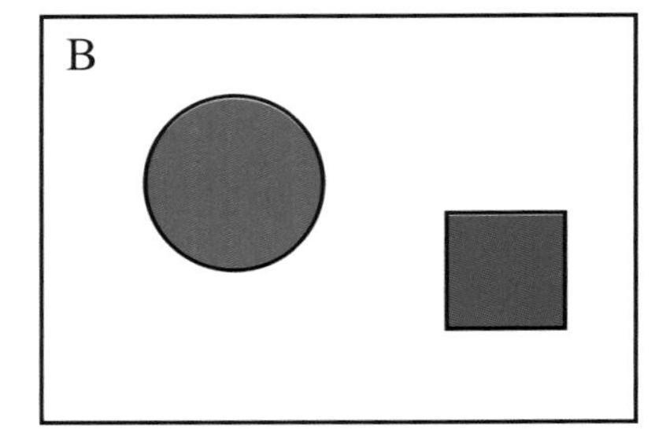

One pair he could pick is (red triangle, blue square)

a) List all the pairs Ben could pick.
b) What is the probability that Ben picks (red triangle, blue square)?
c) What is the probability that Ben picks a blue triangle?
d) What is the probability that Ben picks a blue square?

7 Sarah picks one number from each of Boxes A, B and C.

A	B	C
7, 1	2, 6	4, 3, 9

One trio she could pick is (1, 2, 3)

a) List all the trios Sarah could pick.

b) What is the probability that Sarah picks (1, 2, 3)?

c) What is the probability that Sarah picks a '1'?

d) What is the probability that Sarah picks a '2'?

e) What is the probability that Sarah picks a '1' and a '2' ?

f) What is the probability that Sarah picks a '3'?

22.4 Experimental probability

When an experiment is repeated (or trialled) you can work out the probability of a certain number of successes using the formula.

> Expectation = probability of event × number of trials

EXAMPLE

A fair dice is thrown 60 times.

a) Work out an estimate for the number of times the dice will show a '5'

A coin is thrown 30 times.
It lands 'heads' 13 times.

b) Is it a fair coin?

SOLUTION

a) $\text{P(score of 5)} = \frac{1}{6}$

Number of trials = 60

$$\text{So expectation} = \frac{1}{6} \times 60$$
$$= \underline{10}$$

b) Yes, it is likely to be a fair coin.

Throwing a coin is a random process, so it is possible for a fair coin not to land 'heads' in exactly half the trials.

Not all probability problems lend themselves to a theoretical approach.

For example, if you throw a drawing pin in the air and want to know the probability of it landing point up, there is no obvious theoretical method.

Instead, you would do an experiment, and use the results to calculate an experimental probability.

The more trials you carry out the more accurate your estimate of the probability is likely to be.

EXAMPLE

A drawing pin is thrown in the air 20 times.
It lands 'point up' 13 times and 'point down' 7 times.

a) Calculate the experimental probability that a single throw results in the pin landing 'point up'.

b) The pin is then thrown 300 times.
Work out an estimate for the number of times the pin will land 'point down'.

SOLUTION

a) The pin lands 'point up' on 13 times out of 20 trials.

$$P(\text{point up}) = \frac{13}{20}$$

This is the **relative frequency**.

b) $\frac{7}{20}$ of the throws were 'point down', so work out $\frac{7}{20}$ of 300

$$\frac{1}{20} \text{ of } 300 = 300 \div 20 = 15$$

So $\frac{7}{20}$ of $300 = 15 \times 7 = 105$

So the estimated number of 'point down' results is 105 out of 300 trials.

EXERCISE 22.4

Don't use your calculator for Questions 1–4.

1 A fair coin is thrown 400 times.
How many times would you expect it to show 'heads'?

2 The probability that Sean has toast for breakfast is 0.2
How many times would you expect Sean to have toast in 10 days?

3 Mary has a biased 4-sided spinner.
Its sides are numbered 1, 2, 3 and 4
Mary spins the spinner 50 times.
Here are her results.

Result	Frequency
1	10
2	15
3	5
4	20

Write down the experimental probability of the spinner landing on:

a) a '2' **b)** a '3' **c)** an even number.

4 The probability of a biased coin showing 'heads' is 0.7

a) What is the probability of the coin showing 'tails'?

b) The coin is thrown 30 times.
Work out an estimate for the number of times it will show 'tails'.

You can use your calculator for Questions 5–8.

5 The probability that Cecily will hand in her maths homework is $\frac{19}{20}$

a) What is the probability that Cecily does not hand in her maths homework?

Over a year Cecily has been set 100 homeworks in maths.

b) Work out an estimate for the number of homeworks Cecily owes her maths teacher.

6 Tim throws 50 darts aimed at the bull on a dartboard.
He hits the bull with 12 of his throws.

a) Calculate the probability that Tim will hit the bull.

Tim throws more darts at the bull.
He throws 400 darts in total.

b) Estimate the number of times Tim hits the bull.

7 Margaret is checking a book for spelling mistakes.
She looks at a random sample of 50 pages.
She finds spelling mistakes on 4 of the pages.

a) Work out the probability that a randomly chosen page contains spelling mistakes.

The book contains 350 pages altogether.

b) Estimate the total number of pages that contain spelling mistakes.

8 In class 10G at Mountview School there are 18 boys and 12 girls.

a) A pupil is chosen at random from class 10G.
Find the probability that it is a boy.

b) There are 720 pupils altogether at Mountview School.
Estimate the total number of girls at the school.

c) Explain why your estimate might not be very reliable.

22.5 Mutually exclusive outcomes

Suppose you roll a dice.
The possible outcomes, or **events**, are scores of 1, 2, 3, 4, 5 and 6

These are said to be **mutually exclusive** – when one happens, the others do not.

Outcomes of experiments are not always mutually exclusive.

For example, if you draw a card from a pack, one outcome is that it might be a heart, and another is that it might be a king.
These outcomes are not mutually exclusive, since you can have a card that is both a heart and a king.

When the events A and B are mutually exclusive, then:

$$P(A \text{ or } B) = P(A) + P(B)$$

If you have a set of outcomes that cover all the possible results of your experiment, then they are said to be **exhaustive**.

The scores 1, 2, 3, 4, 5, 6 obtained by rolling a dice are exhaustive because no other score is possible.

When you have a set of outcomes that are **mutually exclusive and exhaustive**, then the corresponding probabilities must all add up to 1

EXAMPLE

The breakfast menu at a works canteen is always one of four options. Some of these options are more likely to be on the menu than others. The table shows the options available on any day, together with three of the four probabilities.

Food	Sausages	Fish fingers	Bacon & eggs	Cereal
Probability	0.3	0.4	0.1	

a) Copy the table, and fill in the value of the missing probability.
b) Find the probability that the breakfast available on a randomly chosen day is:
(i) cereal **(ii)** sausages or bacon & eggs **(iii)** not fish fingers.

SOLUTION

a) The probabilities have to add up to give 1

So $0.3 + 0.4 + 0.1 + \text{P(cereal)} = 1$

$0.8 + \text{P(cereal)} = 1$

So $\underline{\text{P(cereal)} = 0.2}$

Food	Sausages	Fish fingers	Bacon & eggs	Cereal
Probability	0.3	0.4	0.1	**0.2**

b) **(i)** From the table, P(cereal) = 0.2

(ii) P(sausages or bacon & eggs) $= 0.3 + 0.1 = \underline{0.4}$

(iii) P(not fish fingers) $= 1 - 0.4 = \underline{0.6}$

EXERCISE 22.5

1 Laura either walks to school, cycles to school or catches the bus.
The probability that Laura walks to school is 0.4
The probability that Laura cycles to school is 0.5
What is the probability that Laura catches the bus?

2 John has porridge, toast, or cereal for breakfast.
The probability that John has porridge is 0.25
The probability that John has toast is 0.3
What is the probability that John has cereal for breakfast?

3 A 3-sided spinner always shows one of red, green or blue.
The probability that the spinner lands on red is $\frac{1}{7}$
The probability that the spinner lands on green is $\frac{2}{7}$

a) What is the probability that the spinner lands on blue?

b) The spinner is spun 70 times.
Work out an estimate for the number of times it lands on blue.

4 A bag contains red, blue, green and yellow counters.
A counter is drawn at random.
The table shows the probabilities of the counter being each of the four colours.

Colour	Red	Blue	Green	Yellow
Probability	0.4	0.1		0.2

a) What is the probability of the counter being green?

b) What is the probability of the counter being blue or yellow?

c) What is the probability of the counter not being red?

5 A biased dice shows scores of 1, 2, 3, 4, 5 or 6 with these probabilities.

Score	1	2	3	4	5	6
Probability	0.1	0.1	0.1	0.1	0.2	0.4

The dice is rolled once.
Find the probability that the score obtained is:
a) 5 **b)** not 2 **c)** 3 or 4 **d)** an even number.

6 Tomorrow night Ginny is going to go to the cinema, or go out for a pizza, or stay in.
She will do exactly one of those three things.
The incomplete table shows some probabilities.

Activity	Cinema	Pizza	Stay in
Probability	0.25	0.45	

a) What is the probability that Ginny stays in?
b) Which of the three things is Ginny most likely to do?
c) Find the probability that she does not go to the cinema.

7 There are four types of bird in my garden.
The four types of bird are not all equally likely to be seen.
The table shows the probability that a randomly observed bird is of a particular type.

Type of bird	Blackbird	Sparrow	Starling	Robin
Probability	0.35	0.25		0.1

a) Copy and complete the probability table.
b) Which type of bird is the most common in my garden?

A bird is observed at random.
c) Find the probability that it is:
(i) not a blackbird **(ii)** a sparrow or a starling.

8 When Ricky plays computer chess he wins, draws or loses.
The probability that he wins is 0.4 and the probability that he loses is 0.5
Find the probability that, in a randomly chosen game:
a) Ricky draws **b)** Ricky draws or loses.

9 A biased spinner gives scores of 1, 2, 3 or 4
The probability of getting 1 is 0.2
The probability of getting 2 is 0.3
The probability of getting 3 is 0.1
a) Calculate the probability of getting an odd score.
b) Work out the probability of getting a score of 4

10 Beth always eats one bowl of cereal for breakfast.
The probability that she chooses muesli is 0.24
The probability that she chooses cornflakes is 0.18
a) Work out the probability that she chooses muesli or cornflakes.
b) Work out the probability that she does not choose either of these cereals.

11 Alexei has a tin of crayons.
Each crayon is one of red, blue, yellow or green.
Alexei chooses a crayon at random from the tin.
The table shows the probability that a randomly chosen crayon is a particular colour.

Colour of crayon	Red	Blue	Yellow	Green
Probability	0.3	0.4	x	x

a) Find the probability that the crayon is red or blue.
b) Find the probability that the crayon is yellow or green.

The number of yellow crayons is the same as the number of green crayons.
c) Find the probability that the crayon is green.

12 A college contains students who are in their first year, second year or third year.
36% of the students are in their first year, and 33% are in their second year.
This information may be shown in a probability table, as below.

Year group	First year	Second year	Third year
Probability	36%	33%	

a) Copy and complete the probability table.

400 students from the college attend a Saturday night rock concert.
b) Estimate the number of first year students who attend the concert.

13 A bag contains red, green, blue and yellow counters.
In a probability experiment, one counter
is chosen at random, and removed from the bag.
Its colour is noted, and it is returned to the bag.
The table shows some probabilities for this experiment.

Colour	Red	Green	Blue	Yellow
Probability	0.08	0.44	0.28	

a) Find the probability of obtaining a yellow counter.

The experiment is carried out 250 times.
b) Estimate the number of times a blue counter is obtained.

REVIEW EXERCISE 22

Don't use your calculator for Questions 1–6.

1 Some bulbs were planted in October.
The ticks in the table show the months in which each type of bulb grows into flowers.

Bulb	Jan	Feb	Mar	Apr	May	Jun
Allium					✓	✓
Crocus	✓	✓				
Daffodil		✓	✓	✓		
Iris	✓	✓				
Tulip				✓	✓	

a) In which months do tulips flower?
b) What type of bulb flowers in March?
c) In what month do most types of bulb flower?
d) Which type of bulb flowers in the same month as the allium?

Ben puts one of each type of bulb in a bag.
He takes a bulb from the bag without looking.
e) **(i)** Write down the probability that he will choose a crocus bulb.
(ii) On the probability scale, mark with a cross the probability that he will take a bulb which flowers in February.

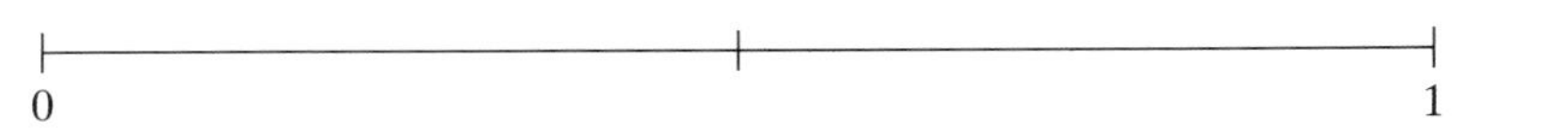

[Edexcel]

2 Lucy has a bag of £1 coins.
5 of the coins are dated 1998
6 of the coins are dated 1999
The other 9 coins are dated 2000
Lucy chooses one of the coins at random from the bag.
What is the probability that she will choose a coin dated 2000? [Edexcel]

3 There are two counters in a bag.
One is black and the other is white.
Abi takes a counter from the bag without looking.

a) On the probability line below mark with B the probability that she will take a black counter.

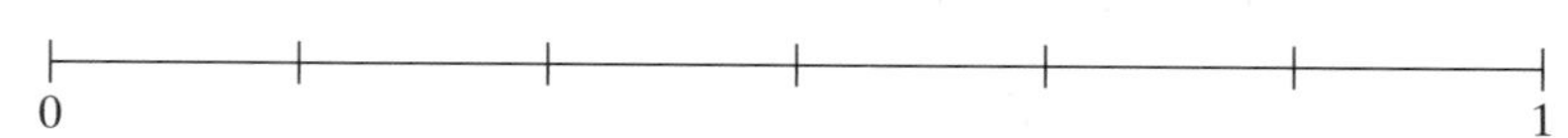

A fair dice has six faces numbered 1, 2, 3, 4, 5, 6
Robert throws the dice and notes its score, and then takes a counter from the bag and notes its colour.
One possible outcome is (1, black)

b) **(i)** List all the possible outcomes.
(ii) Use the list to work out the probability that Robert will get (1, black) **[Edexcel]**

4 The probability that a biased dice will land on a four is 0.2
Pam is going to roll the dice 200 times.
Work out an estimate for the number of times the dice will land on a four. **[Edexcel]**

5 Here is a 4-sided spinner.
The sides are labelled 1, 2, 3 and 4
The spinner is biased.
The probability that the spinner will land on each of the numbers 1 to 3 is given in the table.

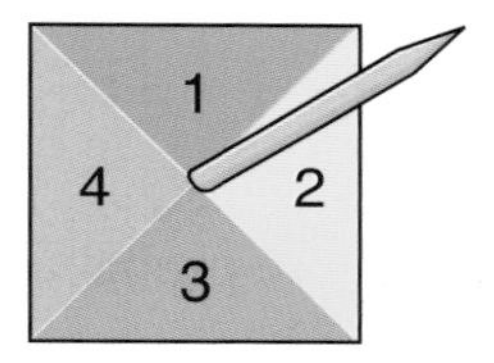

Number	1	2	3	4
Probability	0.3	0.4	0.1	

Sabia spins the spinner once.

a) **(i)** Work out the probability that the spinner will land on the number 4.
(ii) Write down the number that the spinner is most likely to land on.

Nick spins the spinner 100 times.

b) Work out an estimate for the number of times the spinner will land on the number 2. **[Edexcel]**

6 Imran plays a game of chess with his friend.
A game of chess can be won or drawn or lost.
The probability that Imran wins the game of chess is 0.3
The probability that Imran draws the game of chess is 0.25
Work out the probability that Imran loses the game of chess. **[Edexcel]**

You can use your calculator for Questions 7–21.

7 Some students each choose one colour from

black purple red white

The pie chart shows information about their choices.

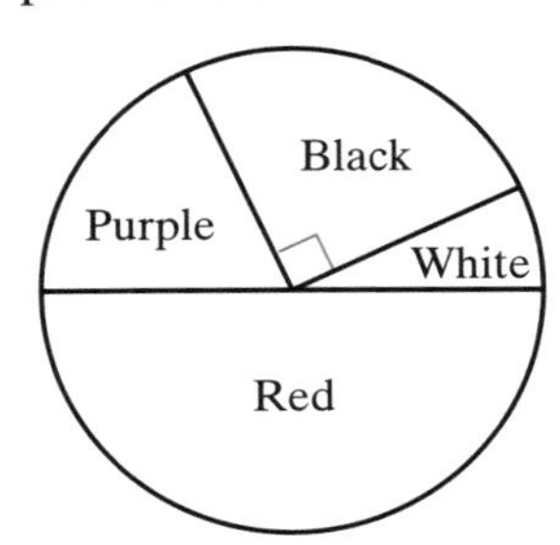

One of these students is to be selected at random.

a) On a copy of the probability line below:

(i) mark with a letter **R** the probability that the student chose red

(ii) mark with a letter **B** the probability that the student chose black.

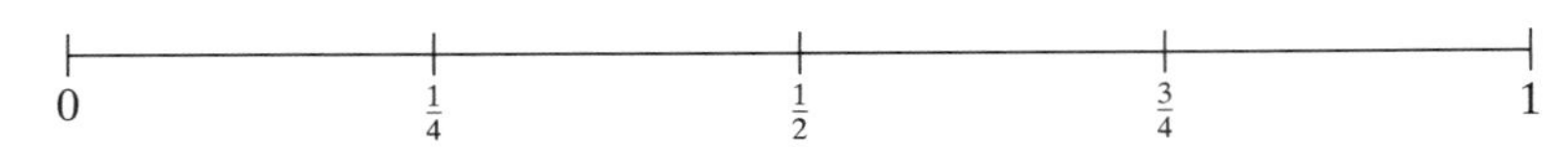

On the copy of the probability line shown below, Emma marked, with the letter **G**, the probability that the student chose green.

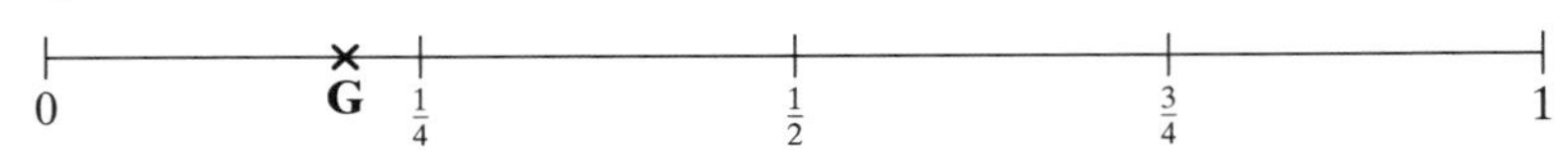

b) Explain why Emma is wrong. **[Edexcel]**

8 The diagram shows some shapes.

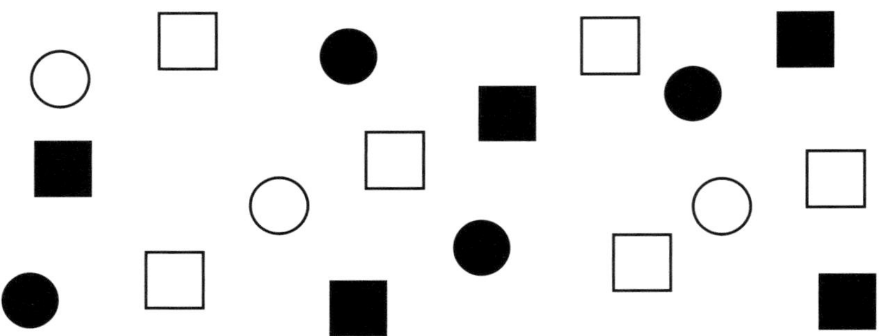

a) Complete a copy of the table to show the number of shapes in each category.

	White	**Black**
Circle		
Square		

One of the shapes is chosen at random.

b) Write down the probability that the shape will be a black square. **[Edexcel]**

9 Shreena has a bag of 20 sweets.
10 of the sweets are red.
3 of the sweets are black.
The rest of the sweets are white.
Shreena chooses one sweet are random.
What is the probability that Shreena will choose a:
a) red sweet
b) white sweet? **[Edexcel]**

10 A bag contains coloured beads.
A bead is selected at random.
The probability of choosing a red bead is $\frac{5}{8}$.
Write down the probability of choosing a bead that is not red from the bag. **[Edexcel]**

11 A game is played with two spinners.
You multiply the two numbers on which the spinners land to get the score.

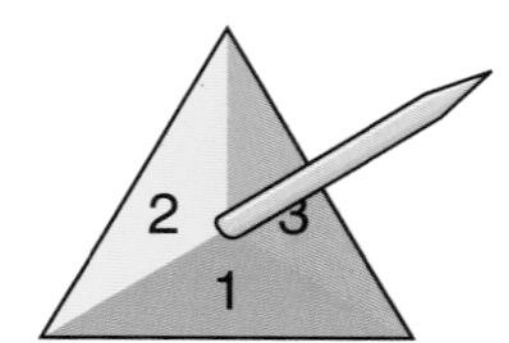

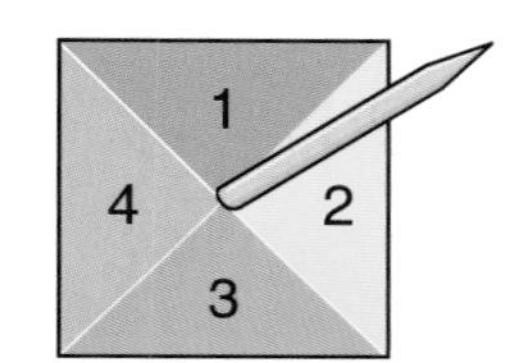

a) Complete the table to show all the possible scores.
One score has been done for you.

	Spinner B				
	×	1	2	3	4
Spinner A	1				
	2				8
	3				

b) Work out the probability of getting a score of 6.
c) Work out the probability of getting a score that is an odd number. **[Edexcel]**

12 Asif has a box of 25 pens.
12 of the pens are blue.
8 of the pens are black.
The rest of the pens are red.
Asif chooses one pen at random from the box.
What is the probability that Asif will choose:
a) a blue pen
b) a red pen. **[Edexcel]**

13 There are 20 bubble gum sweets in a bubble gum machine.
The colour of each bubble gum can be red or green or yellow or white.
This pictogram shows the number of bubble gums of each colour in the bubble gum machine.

Red	● ●
Green	● ● ● ● ●
Yellow	● ● ●
White	● ● ● ● ● ● ● ● ● ●

Bob buys one bubble gum sweet at random from the machine.

a) Write down the probability that he will get:
  (i) a white bubble gum
  (ii) a black bubble gum
  (iii) either a red or a green bubble gum.

b) Write down the probability that he will not get a green bubble gum. **[Edexcel]**

14 Peter has a fair dice and a fair spinner.
The dice has 6 faces numbered from 1 to 6
The spinner has 4 sides numbered from 1 to 4
Peter is going to throw the dice and spin the spinner.
He will add the scores to get a total.

a) Copy and complete the table to show all the possible outcomes.

+	1	2	3	4	5	6
1	2					
2						
3					8	
4		6				

b) Work out the probability that he will get:
  (i) a total of 7
  (ii) a total of less than 6
  (iii) an even total
  (iv) a total of more than 7

15 60 British students each visited one foreign country last week.
The two-way table shows some information about these students.

	France	Germany	Spain	Total
Female			9	34
Male	15			
Total		25	18	60

a) Copy and complete the two-way table.

One of these students is picked at random.
b) Write down the probability that the student visited Germany last week. **[Edexcel]**

16 Mark throws a fair coin.
He gets a head.
Mark's sister then throws the same coin.
a) What is the probability that she will get a head?

Mark throws the coin 30 times.
b) Explain why he may not get exactly 15 heads and 15 tails. **[Edexcel]**

17 Chris is going to roll a biased dice.
The probability he will get a six is 0.09
a) Work out the probability that he will not get a six.

Chris is going to roll the dice 30 times.
b) Work out an estimate for the number of sixes he will get. **[Edexcel]**

18 Mr Brown chooses one book from the library each week.
He chooses a crime novel or a horror story or a non-fiction book.
The probability he chooses a horror story is 0.4
The probability he chooses a non-fiction book is 0.15
Work out the probability Mr Brown chooses a crime novel. **[Edexcel]**

19 A computer game selects letters, from the whole alphabet, for players to use in the game.
The table shows the probability of the letters A, E, I, O and U being chosen.

A	E	I	O	U
0.1	0.1	0.07	0.05	0.07

One letter is chosen.
a) Find the probability of the letter U not being chosen.
b) Work out the probability of the letters A or E or I or O or U being chosen.

300 letters are chosen for the game.
c) Work out an estimate for the numbers of times the letter O will be chosen. **[Edexcel]**

20 Here is a 5-sided spinner.
Its sides are labelled 1, 2, 3, 4, 5
Alan spins the spinner and throws a coin.
One possible outcome is (3, heads)
a) List all the possible outcomes.

The spinner is biased.
The probability that the spinner will land on each of the numbers 1 to 4 is given in the table.

Number	1	2	3	4	5
Probability	0.36	0.1	0.25	0.15	

Alan spins the spinner once.
b) **(i)** Work out the probability that the spinner will land on 5
(ii) Write down the probability that the spinner will land on 6
(iii) Write down the number that the spinner is most likely to land on.
(iv) Work out the probability that the spinner will land on an even number.

[Edexcel]

21 A bag contains counters which are white or green or red or yellow.
The probability of taking a counter of a particular colour at random is:

Colour	White	Green	Red	Yellow
Probability	0.15	0.25		0.4

Laura is going to take a counter at random and then put it back in the bag.
a) **(i)** Work out the probability that Laura will take a red counter.
(ii) Write down the probability that Laura will take a blue counter.

Laura is going to take a counter from the bag at random 100 times.
Each time she will put the counter back in the bag.
b) Work out an estimate for the number of times that Laura will take a yellow counter.

[Edexcel]

KEY POINTS

1 Probability is defined as:

$$\text{Probability} = \frac{\text{number of favourable outcomes}}{\text{total number of outcomes}}$$

2 Probability is a number between 0 and 1

3 Probabilities are written as a fraction (e.g. $\frac{1}{2}$) or a decimal (e.g. 0.5) or a percentage (e.g. 50%)

4 Probabilities can be represented on probability lines:

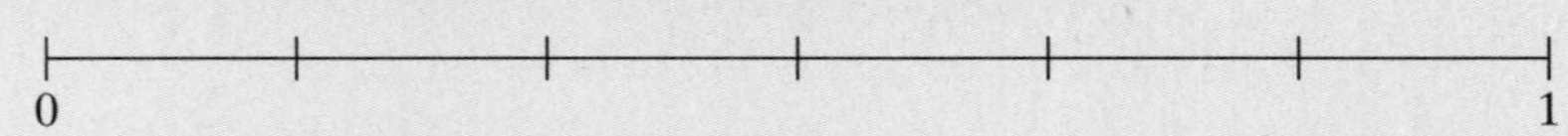

- an event with a probability of 0 is impossible
- an event with a probability of 1 is certain.

5 For a particular event, A:

P(A does not happen) = 1 − P(A does happen).

6 You can estimate how many times an event is likely to occur using the formula:

expectation = probability of event × number of trials.

7 Two events are mutually exclusive if they cannot both happen at the same time.

8 The probabilities of all the possible outcomes of an experiment add up to 1

Probability wordsearch

Here are some words and phrases that are used in probability.

Find them in the wordsearch diagram, then write down the meaning of each one.

You may need to use the internet to check some of the meanings.

D	U	W	A	F	J	Q	Y	U	I	F	F	E	N	E	C	B
B	I	A	S	F	S	F	J	T	R	I	A	L	D	E	K	L
T	S	R	M	M	A	A	L	C	L	I	U	O	R	A	V	A
Y	T	R	S	K	Y	I	S	P	C	V	E	T	E	G	H	Y
F	S	A	E	V	I	R	P	O	C	E	A	J	D	R	E	V
Y	W	I	X	N	W	B	Q	I	S	I	E	E	H	A	K	T
E	C	I	D	J	N	A	O	U	N	D	B	F	R	N	E	N
J	A	N	Y	E	V	I	B	R	S	I	I	E	H	D	B	O
D	X	D	F	R	T	Y	P	O	N	G	O	R	F	O	T	T
G	O	E	C	S	E	E	J	S	C	R	U	M	E	M	W	A
B	H	P	E	S	X	J	I	U	R	E	C	F	U	T	Y	T
I	W	E	V	E	N	C	H	A	N	C	E	J	U	N	G	N
D	U	N	E	N	D	I	V	E	R	T	W	Q	U	I	R	E
E	E	D	V	Q	P	R	E	M	O	C	T	U	O	T	Y	V
A	R	E	Y	H	A	L	F	O	M	O	T	O	G	G	R	E
L	F	N	O	D	E	Y	I	D	X	V	E	N	T	Y	S	R
M	U	T	U	A	L	L	Y	E	X	C	L	U	S	I	V	E

BIAS
CERTAIN
DICE
EVEN CHANCE
EVENT
FAIR
INDEPENDENT
MUTUALLY EXCLUSIVE
OUTCOME
RANDOM
SPINNER
TRIAL

Index